AF353034

Symmetries in Quantum Mechanics

Symmetries in Quantum Mechanics

Editor

G. Jordan Maclay

MDPI • Basel • Beijing • Wuhan • Barcelona • Belgrade • Manchester • Tokyo • Cluj • Tianjin

Editor
G. Jordan Maclay
Quantum Fields LLC
USA

Editorial Office
MDPI
St. Alban-Anlage 66
4052 Basel, Switzerland

This is a reprint of articles from the Special Issue published online in the open access journal *Symmetry* (ISSN 2073-8994) (available at: https://www.mdpi.com/journal/symmetry/special_issues/Symmetries_Quantum_Mechanics).

For citation purposes, cite each article independently as indicated on the article page online and as indicated below:

LastName, A.A.; LastName, B.B.; LastName, C.C. Article Title. *Journal Name* **Year**, *Volume Number*, Page Range.

ISBN 978-3-0365-2694-2 (Hbk)
ISBN 978-3-0365-2695-9 (PDF)

Cover image courtesy of G. Jordan Maclay

Contents

About the Editor

G. Jordan Maclay is the Chief Scientist at Quantum Fields LLC, a contract research organization he founded in 1999 after retiring from the University of Illinois at Chicago in 1998. Prof. Maclay is interested in fundamental quantum phenomena, particularly those related to quantum fluctuations of the electromagnetic field, such as Casimir forces and radiative shifts, and how these phenomena might be used to accomplish engineering objectives. He published the first quantum field theoretical calculation of the Casimir force, introducing the Zeta function method, with his thesis advisor Prof. Lowell S. Brown in 1969. Prof. Maclay and his students did the first work incorporating Casimir forces into microelectromechanical systems (MEMS) in 1995. He has worked extensively in the area of chemical microsensors. He received his PhD in physics from Yale University in 1972. He recently authored the book "Transformations: Poetry and Art," to be published in 2022. Currently he is working on a book on the symmetries of the hydrogen atom.

Editorial

Special Issue: "Symmetries in Quantum Mechanics"

G. Jordan Maclay [1,2]

[1] Quantum Fields LLC, St. Charles, IL 60174, USA; jordanmaclay@quantumfields.com
[2] Department of Electrical Engineering and Computer Science, University of Illinois at Chicago, Chicago, IL 60607, USA

Citation: Maclay, G.J. Special Issue: "Symmetries in Quantum Mechanics". *Symmetry* **2021**, *13*, 1620. https://doi.org/10.3390/sym13091620

Received: 27 August 2021
Accepted: 31 August 2021
Published: 3 September 2021

This Special Issue "Symmetries in Quantum Mechanics" describes research using two of the most fundamental probes we have in nature. The 11 papers discuss phenomena in atoms, galaxies, and people, which is a testimonial to the breadth of the influence of symmetry and quantum mechanics. The papers can be broadly divided into four categories: 1. Fundamentals of quantum systems, 2. Algebraic methods in quantum mechanics, 3. Teleportation and scattering, and 4. Cosmology.

We offer some brief comments on these papers:

Fundamentals of Quantum Systems: In their paper "Symmetry, Transactions, and the Mechanism of Wave Function Collapse", Professors Cramer and Mead address persistent fundamental questions in quantum mechanics using the transactional theory of quantum mechanics [1]. This theory interprets the psi and psi* wavefunctions as retarded and advanced waves moving in opposite time directions that meet to form a quantum "handshake". Applying this formalism, they carefully model the transfer of energy from an excited H atom to a nearby H atom in its ground state as a process that evolves continuously in time, without having to invoke any adhoc assumptions such as a collapse of the wavefunction. This is a welcome milestone in understanding fundamental processes in quantum mechanics.

Professor Jussi Lindgren and collaborator Jukka Liukkonen give a derivation of the Uncertainty Principle that is based on the effect of the stochastic fluctuations of spacetime, described in terms of stochastic optimal control theory [2]. This innovative approach, described in their paper, "The Heisenberg Uncertainty Principle as an Endogenous Property of Stochastic Optimal Control Systems in Quantum Mechanics", frees the description of the uncertainly principle from the properties of the measuring apparatus, and makes it a natural consequence of the approach to equilibrium for this stochastic system. The results are generally consistent with Bohmian mechanics, but provide a simpler mechanism, and suggest an objective, realistic interpretation of quantum mechanics.

In a unique contribution titled "Phishing for (Quantum-Like) Phools-Theory and Experimental Evidence", Prof. Ariane Lambert-Mogiliansky and her doctoral student Adrian Calmettes have applied quantum-like decision theory to explore the well-documented influence of distraction from irrelevant information on decision making, and conducted an experiment with 1253 respondents to verify their model [3]. They develop their theory using a quantum cognition model in which the quantum state represents the decision makers representation of the world, classically the beliefs. Uncertainty is seen as a quantum effect, and different influences, informative or distractions, are akin to complementary variables is quantum mechanics. This paper highlights the breadth of the influence of quantum theory in diverse fields.

Prof. Garrett Moddel and his students have built an entirely new type of optical micro-device with the striking property of having a measurable current present with no applied voltage that is described in their paper "Optical-Cavity-Induced Current" [4]. The metal-insulator-metal tunneling device has an optical cavity on only one side so the MIM device is exposed to quantum vacuum fluctuations with a highly asymmetric mode distribution on opposite sides, which may induce asymmetric hot electron currents, resulting in a net current. Very extensive tests were conducted over a period of several years to eliminate a broad variety of artifacts and verify the extraordinary experimental results. Over the years, I have seen many attempts to obtain useful energy by manipulations of vacuum fluctuations, but this is the first device I have seen that I think may represent a real breakthrough.

Algebraic Methods in Quantum Mechanics: Post-doctoral researcher Marisol Bermudez-Montana and collaborators develop in their paper "Algebraic DVR Approaches Applied to Describe the Stark Effect" two algebraic approaches using a discrete variable representations (DVR) to compute matrix representations of three dimensional Hamiltonians of the hydrogen atom and the harmonic oscillator in a simple form [5]. The elegant methods are applied to compute the Stark effect.

In the paper "Dynamical Symmetries of the H Atom, One of the Most Important Tools of Modern Physics: SO(4) to SO(4,2), Background, Theory, and Use in Calculating Radiative Shifts", Prof. Jordan Maclay has given a clear and comprehensive presentation of the group theory for the H atom in which all generators are manifestly Hermitean and expressed in terms of the canonical momenta and positions and the normal inner product is used [6]. Emphasis is on the physics of the hydrogen atom. He has used a complete set of basis states of the inverse of the coupling constant, which have the same quantum numbers as the usual bound states, allowing a uniform treatment of the bound and scattering states. With a new approach to calculating radiative shifts using SO(4,2) symmetries, he has derived a novel generating function for the non-relativistic Lamb shifts for different states.

Teleportation and Scattering: Professors Carlos Cardoso-Isidoro and Francisco Delgado describe in "Symmetries in Teleportation Assisted by N-Channels under Indefinite Causal Order and Post-Measurement" a method to enhance the information transfer in quantum teleportation using a superposition of channels of indefinite causal order [7]. A weak measurement can provide further improvement.

In their paper "Permutation Symmetry in Coherent Electron Scattering by Disordered Media," Professors Elena Orlenke and Fedor Orlenko compute the scattering cross section for inelastic scattering from atoms in disordered matter, taking into account the indistinguishability of electrons in the atomic core and in the incident beam [8]. As a consequence, different results are predicted for scattering from He triplet and singlet states.

Cosmology: The paper "An Invariant Characterization of the Levi-Civita Spacetimes" by Cooper Watson, a doctoral student of Prof. Gerald Cleaver at Baylor University, and his collaborators, reviews and analyzes different space–time solutions of Einstein's equations of general relativity published by Tullio Levi-Civita from 1917–19 [9]. These little known solutions have, until now, been generally unavailable to English speaking researchers. These results have been cast in modern language and related to other GR results. Today these solutions may be of use in understanding aspects of gravitational waves.

Prof. Abhik Sanyal and collaborators discuss inflation in terms of a Scalar-Tensor theory of gravity in "Inflation with the Scalar-Tensor Theory of Gravity" [10]. The parameters of scalar-tensor theories, which are generalizations of the Brans–Dicke theory, are determined for the particular case of standard non-minimal coupling by exploiting a symmetry associated with a conserved current. All the parameters are determined and are shown to be consistent with the Planck survey data from 2018.

Prof. Pinto gives a careful review in "Gravitational Dispersion Forces and Gravity Quantization" [11] of the history of electrodynamic dispersion forces, detailing the classical and semi-classical explanations and the later quantum developments, and contrasts this development to the purely quantum discussions of gravitational dispersion forces, which, however, can also be considered within a semi-classical framework. He sheds light on the common assumption that the gravitational field is necessarily a quantized field, and suggests that data do not support the conclusion that the gravitational field must be quantized.

We thank all the authors for their important and welcome contributions to "Symmetries in Quantum Mechanics," we thank all the reviewers and editors for their invaluable and timely help insuring the quality of the papers, and we thank the excellent and responsive staff at MDPI, especially Mia Guo, for making this issue a success.

Funding: This research received no external funding.

Conflicts of Interest: The author declares no conflict of interest.

References

1. Cramer, J.G.; Mead, C.A. Symmetry, Transactions, and the Mechanism of Wave Function Collapse. *Symmetry* **2020**, *12*, 1373. [CrossRef]
2. Lindgren, J.; Liukkonen, J. The Heisenberg Uncertainty Principle as an Endogenous Property of Stochastic Optimal Control Systems in Quantum Mechanics. *Symmetry* **2020**, *12*, 1533. [CrossRef]
3. Lambert-Mogiliansky, A.; Calmettes, A. Phishing for (Quantum-like) Phools-Theory and Experimental Evidence. *Symmetry* **2021**, *13*, 162. [CrossRef]
4. Moddel, G.; Weerakkody, A.; Doroski, D.; Bartusiak, D. Optical-Cavity-Induced Current. *Symmetry* **2021**, *13*, 517. [CrossRef]
5. Bermudez-Montana, M.; Rodriguez-Arcos, M.; Lemus, R.; Arias, J.M.; Gomez-Camacho, J.; Orgaz, E. Algebraic DVR Approaches Applied to Describe the Stark Effect. *Symmetry* **2020**, *12*, 1719. [CrossRef]
6. Maclay, G.J. Dynamical Symmetries of the H Atom, One of the Most Important Tools of Modern Physics: SO(4) to SO(4,2), Background, Theory, and Use in Calculating Radiative Shifts. *Symmetry* **2020**, *12*, 1323. [CrossRef]
7. Cardoso-Isidoro, C.; Delgado, F. Symmetries in Teleportation Assisted by N-Channels under Indefinite Causal Order and Post-Measurement. *Symmetry* **2020**, *12*, 1904. [CrossRef]
8. Orlenko, E.V.; Orienko, F.E. Permutation Symmetry in Coherent Electron Scattering by Disordered Media. *Symmetry* **2020**, *12*, 1971. [CrossRef]
9. Watson, C.K.; Julius, W.; Gorban, M.; McNutt, D.D.; Davis, E.W.; Cleaver, G.B. An Invariant Characterization of the Levi-Civita Spacetimes. *Symmetry* **2021**, *13*, 1469. [CrossRef]
10. Saha, D.; Sanyal, S.; Sanyal, A.K. Inflation with the Scalar-Tensor Theory of Gravity. *Symmetry* **2020**, *12*, 1267. [CrossRef]
11. Pinto, F. Gravitational Dispersion Forces and Gravity Quantization. *Symmetry* **2021**, *13*, 40. [CrossRef]

Article

Symmetry, Transactions, and the Mechanism of Wave Function Collapse

John Gleason Cramer [1],* and Carver Andress Mead [2]

[1]   Department of Physics, University of Washington, Seattle, WA 98195, USA
[2]   California Institute of Technology, Pasadena, CA 91125, USA; carver@caltech.edu
*   Correspondence: jcramer@uw.edu

Received: 17 June 2020; Accepted: 14 August 2020; Published: 18 August 2020

Abstract: The Transactional Interpretation of quantum mechanics exploits the intrinsic time-symmetry of wave mechanics to interpret the ψ and ψ^* wave functions present in all wave mechanics calculations as representing retarded and advanced waves moving in opposite time directions that form a quantum "handshake" or transaction. This handshake is a 4D standing-wave that builds up across space-time to transfer the conserved quantities of energy, momentum, and angular momentum in an interaction. Here, we derive a two-atom quantum formalism describing a transaction. We show that the bi-directional electromagnetic coupling between atoms can be factored into a matched pair of vector potential Green's functions: one retarded and one advanced, and that this combination uniquely enforces the conservation of energy in a transaction. Thus factored, the single-electron wave functions of electromagnetically-coupled atoms can be analyzed using Schrödinger's original wave mechanics. The technique generalizes to any number of electromagnetically coupled single-electron states—no higher-dimensional space is needed. Using this technique, we show a worked example of the transfer of energy from a hydrogen atom in an excited state to a nearby hydrogen atom in its ground state. It is seen that the initial exchange creates a dynamically unstable situation that avalanches to the completed transaction, demonstrating that wave function collapse, considered mysterious in the literature, can be implemented with solutions of Schrödinger's original wave mechanics, coupled by this unique combination of retarded/advanced vector potentials, without the introduction of any additional mechanism or formalism. We also analyze a simplified version of the photon-splitting and Freedman–Clauser three-electron experiments and show that their results can be predicted by this formalism.

Keywords: quantum mechanics; transaction; Wheeler–Feynman; transactional interpretation; handshake; advanced; retarded; wave function collapse; collapse mechanism; EPR; HBT; Jaynes; NCT; split photon; Freedman–Clauser; nonlocality; entanglement

1. Introduction

Quantum mechanics (QM) was never properly finished. Instead, it was left in an exceedingly unsatisfactory state by its founders. Many attempts by highly qualified individuals to improve the situation have failed to produce any consensus about either (a) the precise nature of the problem, or (b) what a better form of QM might look like.

At the most basic level, a simple observation illustrates the central conceptual problem:

An excited atom somewhere in the universe transfers *all* of its excitation energy to another single atom, independent of the presence of the vast number of alternative atoms that could have received all or part of the energy. The obvious "photon-as-particle" interpretation of this situation has a one-way symmetry: The excited source atom is depicted as emitting a particle, a *photon* of electromagnetic energy that is somehow oscillating with angular frequency ω while moving in a particular direction.

The photon is depicted as carrying a quantum of energy $\hbar\omega$, a momentum $\hbar\omega/c$, and an angular momentum $\hbar$ through space, until it is later absorbed by some unexcited atom. The emission and absorption are treated as independent isolated events without internal structure. It is insisted that the only real and meaningful quantities describing this process are *probabilities*, since these are measurable. The necessarily abrupt change in the quantum wave function of the system when the photon arrives (and an observer potentially gains information) is called "wave function collapse" and is considered to be a mysterious process that the founders of QM found it necessary to "put in by hand" without providing any mechanism. [The missing mechanism behind wave function collapse is sometimes called "the measurement problem", particularly by acolytes of Heisenberg's knowledge interpretation. In our view, measurement requires wave function collapse *but does not cause it*.] [Side comments will be put in square brackets]

Referring to statistical quantum theory, which is reputed to apply only to *ensembles* of similar systems, Albert Einstein [1] had this to say:

> *"I do not believe that this fundamental concept will provide a useful basis for the whole of physics."*

> *"I am, in fact, firmly convinced that the essentially statistical character of contemporary quantum theory is solely to be ascribed to the fact that this [theory] operates with an incomplete description of physical systems."*

> *"One arrives at very implausible theoretical conceptions, if one attempts to maintain the thesis that the statistical quantum theory is in principle capable of producing a complete description of an individual physical system ..."*

> *"Roughly stated, the conclusion is this: Within the framework of statistical quantum theory, there is no such thing as a complete description of the individual system. More cautiously, it might be put as follows: The attempt to conceive the quantum-theoretical description as the complete description of the individual systems leads to unnatural theoretical interpretations, which become immediately unnecessary if one accepts the interpretation that the description refers to ensembles of systems and not to individual systems. In that case, the whole 'egg-walking' performed in order to avoid the 'physically real' becomes superfluous. There exists, however, a simple psychological reason for the fact that this most nearly obvious interpretation is being shunned—for, if the statistical quantum theory does not pretend to describe the individual system (and its development in time) completely, it appears unavoidable to look elsewhere for a complete description of the individual system. In doing so, it would be clear from the very beginning that the elements of such a description are not contained within the conceptual scheme of the statistical quantum theory. With this. one would admit that, in principle, this scheme could not serve as the basis of theoretical physics. Assuming the success of efforts to accomplish a complete physical description, the statistical quantum theory would, within the framework of future physics, take an approximately analogous position to the statistical mechanics within the framework of classical mechanics. I am rather firmly convinced that the development of theoretical physics will be of this type, but the path will be lengthy and difficult."*

> *"If it should be possible to move forward to a complete description, it is likely that the laws would represent relations among all the conceptual elements of this description which, per se, have nothing to do with statistics."*

In what follows we put forth a simple approach to *describing the individual system (and its development in time)*, which Einstein believed was missing from statistical quantum theory and which must be present before any theory of physics could be considered to be complete.

The way forward was suggested by the phenomenon of *entanglement*. Over the past few decades, many increasingly exquisite Einstein–Podolsky–Rosen [2] (EPR) experiments [3–11] have demonstrated that multi-body quantum systems with separated components that are subject to conservation laws exhibit a property called "quantum entanglement" [12]: Their component wave functions are inextricably locked together, and they display a nonlocal correlated behavior enforced over an arbitrary interval of space-time without any hint of an underlying mechanism or any show of respect for our cherished classical "arrow of time." Entanglement is the most mysterious of the many so-called "quantum mysteries."

It has thus become clear that the quantum transfer of energy must have quite a different symmetry from that implied by this simple "photon-as-particle" interpretation. Within the framework of statistical QM, the intrinsic symmetry of the energy transfer and the mechanisms behind wave function collapse and entanglement have been greatly clarified by the **Transactional Interpretation of quantum mechanics** (TI), developed over several decades by one of us and recently described in some detail in the book *The Quantum Handshake* [12]. [We note that Ruth Kastner has extended her "probabilist" variant of the TI, which embraces the Heisenberg/probability view and characterizes transactions as events in many-dimensional Hilbert space, into the quantum-relativistic domain [13,14] and has used it to extend and enhance the "decoherence" approach to quantum interpretation [15]].

This paper begins with a tutorial review of the TI approach to a credible photon mechanism developed in the book *Collective Electrodynamics* [16], followed by a deeper dive into the electrodynamics of the quantum handshake, and finally includes descriptions of several historic experiments that have excluded entire classes of theories. We conclude that the approach described here has *not* been excluded by any experiment to date.

1.1. Wheeler–Feynman Electrodynamics

The Transactional Interpretation was inspired by classical time-symmetric Wheeler–Feynman electrodynamics [17,18] (WFE), sometimes called "absorber theory." Basically, WFE assumes that electrodynamics must be time-symmetric, with equally valid retarded waves (that arrive *after* they are emitted) and advanced waves (that arrive *before* they are emitted). WFE describes a "handshake" process accounting for emission recoil in which the emission of a retarded wave stimulates a future absorber to produce an advanced wave that arrives back at the emitter at the instant of emission. WFE is based on electrodynamic time symmetry and has been shown to be completely interchangeable with conventional classical electrodynamics in its predictions.

WFE asserts that the breaking of the intrinsic time-symmetry to produce the electromagnetic arrow of time, i.e., the observed dominance of retarded radiation and absence of advanced radiation in the universe, arises from the presence of more absorption in the future than in the past. In an expanding universe, that assertion is questionable. One of us has suggested an alternative cosmological explanation [19], which employs advanced-wave termination and reflection from the singularity of the Big Bang.

1.2. The Transactional Interpretation of Quantum Mechanics

The Transactional Interpretation of quantum mechanics [12] takes the concept of a WFE handshake from the classical regime into the quantum realm of photons and massive particles. The retarded and advanced waves of WFE become the quantum wave functions ψ and ψ^*. Note that the complex conjugation of ψ^* is in effect the application of the Wigner time-reversal operator, thus representing an advanced wave function that carries negative energy from the present to the past.

Let us here clarify what an *interpretation* of quantum mechanics actually is. An interpretation serves the function of explaining and clarifying the formalism and procedures of its theory. In our view, the mathematics is (and should be) exclusively contained in the formalism itself. The interpretation should not introduce additional variant formalism. [We note, however, that this principle is violated by the Bohm–de Broglie "interpretation" with its "quantum potentials" and uncertainty-principle-violating trajectories, by the Ghirardi–Rimini–Weber "interpretation" with its nonlinear stochastic term, and by many other so-called interpretations that take the questionable liberty of modifying the standard QM formalism. In that sense, these are alternative variant quantum theories, *not* interpretations at all.]

The present work is a calculation describing the formation of a transaction that was inspired by the Transactional Interpretation but has not previously been a part of it. In Section 12 below, we discuss how the TI is impacted by this work. We use Schrödinger's original wave mechanics formalism *with the inclusion of retarded and advanced electromagnetic four-potentials* to describe and illuminate the processes of transaction formation and the collapse of the wave function. We show that this approach can provide a detailed mathematical description of a "quantum-jump" in which what seems to be a photon is emitted by one hydrogen atom in an excited state and excites another hydrogen atom initially in its ground state. Thus, the mysterious process of wave function collapse becomes just a phenomenon involving an exchange of advanced/retarded electromagnetic waves between atomic wave functions described by the Schrödinger formalism.

As illustrated schematically in Figure 1, the process described involves the initial existence in each atom of a very small admixture of the wave function for the opposite state, thereby forming two-component states in both atoms. This causes them to become weak dipole radiators oscillating at the same difference-frequency ω_0. The interaction that follows, characterized by a retarded-advanced exchange of 4-vector potentials, leads to an exponential build-up of a transaction, resulting in the complete transfer of one photon worth of energy $\hbar\omega_0$ from one atom to the other. This process is described in more detail below.

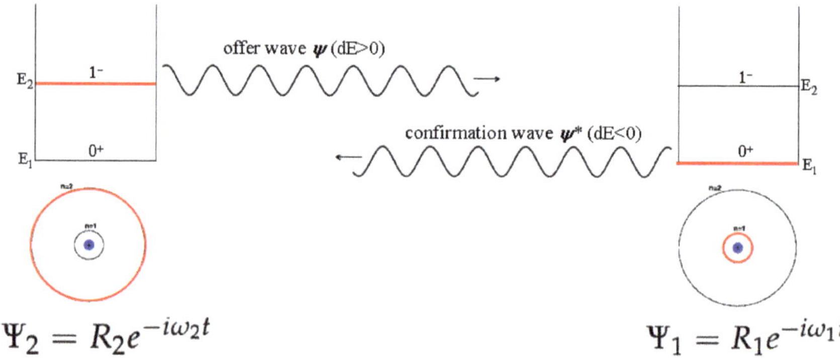

Figure 1. Model of transaction formation: An emitter atom 2 in a space-antisymmetric excited state of energy E_2 and an absorber atom 1 in a space-symmetric ground state of energy E_1 both have slight admixtures of the other state, giving both atoms dipole moments that oscillate with the same difference-frequency $\omega_0 = \omega_2 - \omega_1$. If the relative phase of the initial small offer wave ψ and confirmation wave $\psi*$ is optimal, this condition initiates energy transfer, which avalanches to complete transfer of one photon-worth of energy $\hbar\omega_0$.

2. Physical Mechanism of the Transfer

The standard formalism of QM consists of a set of arbitrary rules, conventionally viewed as dealing only with probabilities. When illuminated by the TI, that formalism hints at an underlying physical mechanism that might be understood, in the usual sense of the concept *understood*. The first glimpse of such an understanding, and of the physical nature of the transactional symmetry, was suggested by Gilbert N. Lewis in 1926 [20,21], the same year he gave electromagnetic energy transfer the name "photon":

"It is generally assumed that a radiating body emits light in every direction, quite regardless of whether there are near or distant objects which may ultimately absorb that light; in other words that it radiates 'into space'..."

"I am going to make the contrary assumption that an atom never emits light except to another atom..."

> *"I propose to eliminate the idea of mere emission of light and substitute the idea of transmission, or a process of exchange of energy between two definite atoms... Both atoms must play coordinate and symmetrical parts in the process of exchange..."*

> *"In a pure geometry it would surprise us to find that a true theorem becomes false when the page upon which the figure is drawn is turned upside down. A dissymmetry alien to the pure geometry of relativity has been introduced by our notion of causality."*

In what follows, we demonstrate that the pair of coupled Schrödinger equations describing the two atoms, as coupled by a relativistically correct description of the electromagnetic field, exhibit a unique solution. This solution has exactly the symmetry of the TI and thus provides a *physically understandable* mechanism for the experimentally observed behavior: Both atoms, in the words of Lewis, *"play coordinate and symmetrical parts in the process of exchange"*.

The solution gives a smooth transition in each of the atomic wave functions, brought to abrupt closure by the highly nonlinear increase in coupling as the transition proceeds. The origin of statistical behavior and "quantum randomness" can be understood in terms of the random distribution of wave-function amplitudes and phases provided by the perturbations of the many other potential recipient atoms; no "hidden variables" are required. Although much remains to be done, these findings might be viewed as a next step towards a physical understanding of the process of quantum energy transfer.

We will close by indicating the deep, fundamental questions that we have not addressed, and that must be understood before anything like a complete physical understanding of QM is in hand.

3. Quantum States

In 1926, Schrödinger, seeking a wave-equation description of a quantum system with mass, adopted Planck's notion that energy was somehow proportional to frequency together with deBroglie's idea that momentum was the propagation vector of a wave and crafted his wave equation for the time evolution of the wave function Ψ [22]:

$$-\frac{\hbar}{2m\,i}\nabla^2\Psi + \frac{q\,V}{i\,\hbar}\Psi = \frac{\partial\Psi}{\partial t}. \tag{1}$$

Here, V is the electrical potential, m is the electron mass, and q is the (negative) charge on the electron. Thus, what is the meaning of the wave function Ψ that is being characterized? In modern treatments, Ψ is called a "probability amplitude", which has only a probabilistic interpretation. In what follows, however, we return to Schrödinger's original vision, which provides a detailed physical picture of the wave function and how it interacts with other charges:

> *"The hypothesis that we have to admit is very simple, namely that the square of the absolute value of Ψ is proportional to an electric density, which causes emission of light according to the laws of ordinary electrodynamics."*

That vision has inspired generations of talented conceptual thinkers to invent solutions to technical problems using Schrödinger's approach. Foremost among these was Ed Jaynes who, with a number of students and collaborators, attacked a host of quantum problems in this manner [23–30]. A great deal of physical understanding was obtained, in particular concerning lasers and the coherent optics made possible by them. The theory evolved rapidly and had an enabling role in the explosive progress of that field. Indeed, the continued rapid technical progress into the present is due, in no small part, to the understanding gained through application of the Jaynes way of thinking. A detailed review of the progress up to 1972 was reported in a conference that year [30]. By then this class of theory was called **neoclassical** (NCT) because of its use of Maxwell's equations. While there was no question about the utility of NCT in the conceptualization and technical realization of amazing quantum-optics devices and their application, there was a deep concern about whether it could possibly be correct *at* *the fundamental level*—maybe it was just a clever bunch of hacks. The tension over this concern was a major focus of the 1972 *Third Rochester Conference on Coherence and Quantum Optics* [31], and several experiments testing NCT predictions were discussed there by Jaynes [30].

He ended his presentation this way:

"We have not commented on the beautiful experiment reported here by Clauser [32], which opens up an entirely new area of fundamental importance to the issues facing us..."

"What it seems to boil down to is this: a perfectly harmless looking experimental fact (nonoccurence of coincidences at 90°), which amounts to determining a single experimental point—and with a statistical measurement of unimpressive statistical accuracy—can, at a single stroke, throw out a whole infinite class of alternative theories of electrodynamics, namely all local causal theories."

"...if the experimental result is confirmed by others, then this will surely go down as one of the most incredible intellectual achievements in the history of science, and my own work will lie in ruins."

The experiment he was alluding to was that of Freedman and Clauser [6], and in particular to their observaton of an essentially zero coincidence rate with crossed polarizers. The Freedman–Clauser experiment (see Section 13.3 below), with its use of entangled photon pairs, was the vanguard of an entire new direction in quantum physics that now goes under the rubric of **Tests of Bell's Inequality** [3,4] and/or **EPR experiments** [6–11]. Both the historic EPR experiment and its analysis have been repeated many times with ever-increasing precision, and always with the same outcome: a difinitive violaton of Bell's inequalities. Local causal theories were dead! [Much of the literature on violations of Bell's inequalities in EPR experiments has unfortunately emphasized the refutation of *local hidden-variable theories*. In our view, this is a regrettable historical accident attributable to Bell. *Nonlocal* hidden-variable theories have been shown to be compatible with EPR results. It is *locality* that has been refuted. Entangled systems exhibit correlations that can only be accomodated by quantum nonlocality. The TI supplies the mechanism for that nonlocality.]

In fact, it was the manifest quantum nonlocality evident in the early EPR experiments of the 1970s that led to the synthesis of the transactional interpretation in the 1980s [19,33,34], designed to compactly explain entanglement and nonlocality. This in turn led to the search for an underlying transaction mechanism, as reported in 2000 in *Collective Electrodynamics* [16]. As we detail below, the quantum handshake, as mediated by advanced/retarded electromagnetic four-potentials, provides the effective non-locality so evident in modern versions of these EPR experiments. In Section 13, we analyze the Freedman–Clauser experiment in detail and show that their result is a natural outcome of our approach. Jaynes' work does not lie in ruins—all that it needed for survival was the non-local quantum handshake! What follows is an extension and modification of NCT using a different non-Maxwellian form of E&M [16] and including our non-local Transactional approach. We illustrate the approach with the simplest possible physical arrangements, described with the major goal of conceptual understanding rather than exhaustion. Obviously, much more work needs to be done, which we point out where appropriate.

Atoms

We will begin by visualizing the electron as Schrödinger and Jaynes did: as having a smooth charge distribution in three-dimensional space, whose density is given by $\Psi^*\Psi$. There is no need for statistics and probabilities at any point in these calculations, and none of the quantities have statistical meaning. The probabilistic outcome of quantum experiments has the same origin as it does in all other experiments—random perturbations beyond the control of the experimenter. We return to the topic of probability after we have established the nature of the transaction.

For a local region of positive potential V, for example near a positive proton, the negative electron's wave function has a local potential energy (qV) minimum in which the electron's wave function can form local **bound states**. The spatial shape of the wave function amplitude is a trade-off between getting close to the proton, which lowers its potential energy, and bunching together too much, which increases its ∇^2 "kinetic energy." Equation (1) is simply a mathematical expression of this trade-off, a statement of the physical relation between mass, energy, and momentum in the form of a wave equation.

A discrete set of states called **eigenstates** are standing-wave solutions of Equation (1) and have the form $\Psi = Re^{-i\omega t}$, where R and V are functions of only the spatial coordinates, and the angular frequency ω is itself independent of time. For the hydrogen atom, the potential $V = \epsilon_0 q_p / r$, where q_p is the positive charge on the nucleus, equal in magnitude to the electron charge q. Two of the lowest-energy solutions to Equation (1) with this potential are:

$$\Psi_{100} = \frac{e^{-r}}{\sqrt{\pi}} e^{-i\omega_1 t} \qquad \Psi_{210} = \frac{r e^{-r/2} \cos(\theta)}{4\sqrt{6\pi}} e^{-i\omega_2 t}, \qquad (2)$$

where the dimensionless radial coordinate r is the radial distance divided by the **Bohr radius** a_0:

$$a_0 \equiv \frac{4\pi\epsilon_0 \hbar^2}{mq^2} = 0.0529 \text{ nm}, \qquad (3)$$

and θ is the azimuthal angle from the North Pole ($+z$ axis) of the spherical coordinate system.

The spatial "shape" of the two lowest energy eigenstates for the hydrogen atom is shown in Figure 2. Here, we focus on the excited-state wave function Ψ_{210} that has no angular momentum projection on the z-axis. For the moment, we set aside the wave functions $\Psi_{21\pm1}$, which have $+1$ and -1 angular momentum z-axis projections. Because, for any single eigenstate, the electron density is $\Psi^*\Psi = Re^{i\omega t} Re^{-i\omega t} = R^2$, the charge density is not a function of time, so none of the other properties of the wave function change with time. The individual eigenstates are thus **stationary states**. The lowest energy bound eigenstate for a given form of potential minimum is called its **ground state**, shown on the left in Figure 3. The corresponding charge densities are shown in Figure 4.

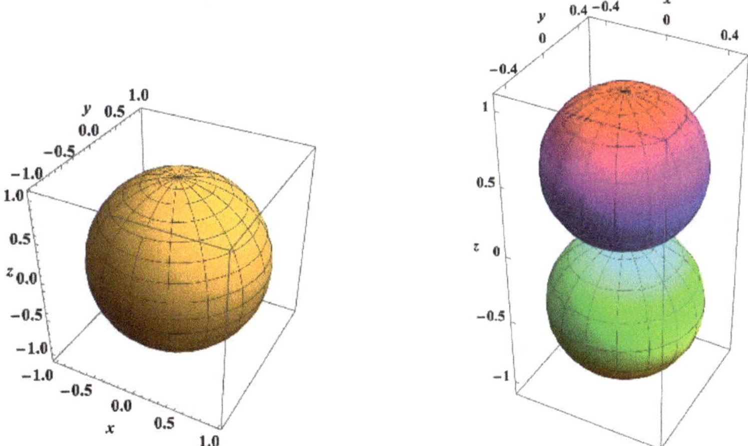

Figure 2. Angular dependence of the spatial wave function amplitudes for the lowest (100, **left**) and next higher (210, **right**) states of the hydrogen atom, plotted as unit radius in spherical coordinates from Equation (2). The 100 wave function has spherical symmetry: positive in all directions. The 210 wave function is antisymmetric along the z-axis, as indicated by the color change. In practice, the direction of the z-axis will be established by an external electromagnetic field, as we shall analyze shortly.

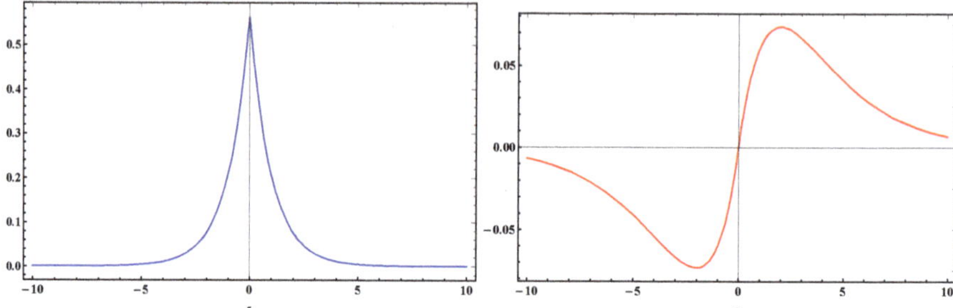

Figure 3. Wave function amplitudes Ψ for the 100 and 210 states, along the z-axis of the hydrogen atom. The horizontal axis in all plots is the position along the z-axis in units of the Bohr radius.

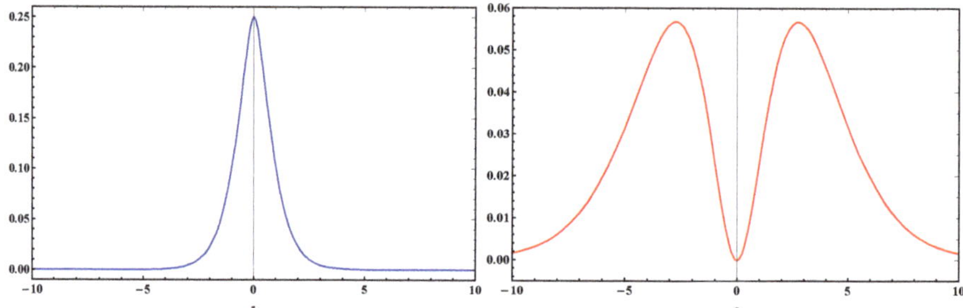

Figure 4. Contribution of $x - y$ "slices" at position z of wave function density $\Psi^*\Psi$ to the total charge or mass of the 100 and 210 states of the hydrogen atom. Both curves integrate to 1.

In 1926, Schrödinger had already derived the energies and wave functions for the stationary solutions of his equation for the hydrogen atom. His physical insight that the absolute square $\Psi^*\Psi$ of the wave function was the *electron density* had enabled him to work out the energy shifts of these levels caused by external applied electric and magnetic fields, the expected strengths of the transitions between pairs of energy levels, and the polarization of light from certain transitions.

These predictions could be compared directly with experimental data, which had been previously observed but not understood. He reported that these calculations were:

"...not at all difficult, but very tedious. In spite of their tediousness, it is rather fascinating to see all the well-known but not understood "rules" come out one after the other as the result of familiar elementary and absolutely cogent analysis, like e.g., the fact that $\int_0^{2\pi} \cos m\phi \, \cos n\phi \, d\phi$ vanishes unless $n = m$. Once the hypothesis about $\Psi^*\Psi$ has been made, no accessory hypothesis is needed or is possible; none could help us if the "rules" did not come out correctly. However, fortunately they do [22,35]."

The Schrödinger/Jaynes approach enables us to describe continuous quantum transitions in an intuitively appealing way: We extend the electromagnetic coupling described in Collective Electrodynamics [16] to the wave function of a single electron, and require only the most rudimentary techniques of Schrödinger's original quantum theory.

4. The Two-State System

The first two eigenstates of the Hydrogen atom, from Equation (2), form an ideal two-state system. We refer to the 100 ground state as State 1, with wave function Ψ_1 and energy E_1, and to the 210 excited state as State 2, with wave function Ψ_2 and energy $E_2 > E_1$:

$$\Psi_1 = R_1 e^{-i\omega_1 t} \qquad\qquad \Psi_2 = R_2 e^{-i\omega_2 t}, \tag{4}$$

where $\omega_1 = E_1/\hbar$, $\omega_2 = E_2/\hbar$, and R_1 and R_2 are real, time-independent functions of the space coordinates. The wave functions represent totally continuous matter waves, and all of the usual operations involving the wave function are methods for computing properties of this continuous distribution. The only particularly quantal assumption involved is that the wave function obeys a **normalization condition**:

$$\int \Psi^* \Psi \, dvol = 1, \tag{5}$$

where the integral is taken over the entire three-dimensional volume where Ψ is non-vanishing. [Envelope functions like R_1 and R_2 generally die out exponentially with distance sufficiently far from the "region of interest", such as an atom. Integrals like this one and those that follow in principle extend to infinity but in practice are taken out far enough that the part being neglected is within the tolerance of the calculation.].

Equation (5) ensures that the total charge will be a single electronic charge, and the total mass will be a single electronic mass.

By construction, the eigenstates of the Schrödinger equation are real and orthogonal:

$$\int R_1 R_2 \, dvol = 0. \tag{6}$$

The first moment $\langle z \rangle$ of the electron distribution along the atom's z-axis is:

$$\langle z \rangle \equiv \int \Psi^* z \Psi \, dvol, \tag{7}$$

In statistical treatments, $\langle z \rangle$ would be called the "expectation value of z", whereas for our continuous distribution it is called the "average value of z" or the "first moment of z." The electron wave function is a *wave packet* and is subject to all the Fourier properties of one, as treated at some length in Ref. [12]. Statistical QM insisted that electrons were "point particles", so one was no longer able to visualize how they could exhibit interference or other wave properties, so a set of rules was constructed to make the outcomes of statistical experiments come out right. Among these was the **uncertainty principle**, which simply restated the Fourier properties of an object described by waves in a statistical context. No statistical attributes are attached to any properties of the wave function in this treatment.

Equation (7) gives the position of the center of negative charge of the electron wave function relative to the positive charge on the nucleus. When multiplied by the electronic charge q, it is called *the electric dipole moment* $q \langle z \rangle$ of the charge distribution of the atom:

$$q \langle z \rangle = q \int \Psi^* z \Psi \, dvol. \tag{8}$$

From Equations (7) and (4), the dipole moment for the ith eigenstate is:

$$q \langle z_i \rangle = q \int \Psi_i^* z \Psi_i \, dvol = q \int R_i^* z R_i \, dvol = q \int R_i^2 z \, dvol. \tag{9}$$

Pure eigenstates of the system will have a definite parity, i.e., they will have wave functions with either even symmetry $[\Psi(z) = \Psi(-z)]$, or odd symmetry $[\Psi(z) = -\Psi(-z)]$. For either symmetry, the integral over $R^2 z$ vanishes, and the dipole moment is zero. We note that, even if the wave function did not have even or odd symmetry, the dipole moment, and all higher moments as well, would be independent of time. By their very nature, eigenstates are stationary states and can be visualized as standing-waves—none of their physical spatial attributes can be functions of time. In order to radiate electromagnetic energy, the charge distribution *must change with time*.

The notion of *stationarity* is the quantum answer to the original question about atoms depicted as electrons orbiting a central nucleus like a tiny Solar System:

Why doesn't the electron orbiting the nucleus radiate its energy away?
In his 1917 book, **The Electron**, R.A. Millikan [36] anticipates the solution in his comment about the

> *"…apparent contradiction involved in the non-radiating electronic orbit—a contradiction which would disappear, however, if the negative electron itself, when inside the atom, were a ring of some sort, capable of expanding to various radii, and capable, only when freed from the atom, of assuming essentially the properties of a point charge."*

Millikan was the first researcher to directly observe and measure the quantized charge on the electron with his famous oil-drop experiment, for which he later received the Nobel prize. Ten years before the statistical quantum theory was put in place, he had clearly seen that a continuous, symmetric electronic charge distribution would not radiate, and that the real problem was the assumption of a point charge.

5. Transitions

The eigenstates of the system form a complete basis set, so any behavior of the system can be expressed by forming a linear combination (superposition) of its eigenstates.

The general form of such a superposition of our two chosen eigenstates is:

$$\Psi = ae^{i\phi_a}R_1 e^{-i\omega_1 t} + be^{i\phi_b}R_2 e^{-i\omega_2 t}, \tag{10}$$

where a and b are real amplitudes, and ϕ_a and ϕ_b are real constants that determine the phases of the oscillations ω_1 and ω_2.

Using $\omega_0 = \omega_2 - \omega_1$ and $\phi = \phi_a - \phi_b$, the charge density ρ of the two-component-state wave function is:

$$\begin{aligned}
\rho &= q\Psi^*\Psi \\
\frac{\rho}{q} &= \left(ae^{-i\phi_a}R_1 e^{i\omega_1 t} + be^{-i\phi_b}R_2 e^{i\omega_2 t}\right)\left(ae^{i\phi_a}R_1 e^{-i\omega_1 t} + be^{i\phi_b}R_2 e^{-i\omega_2 t}\right) \\
&= a^2 R_1^2 + b^2 R_2^2 + \left(ae^{-i\phi_a}be^{i\phi_b}e^{-i\omega_0 t} + be^{-i\phi_b}ae^{i\phi_a}e^{i\omega_0 t}\right)R_1 R_2 \\
&= a^2 R_1^2 + b^2 R_2^2 + ab\left(e^{i(\phi_b-\phi_a)}e^{-i\omega_0 t} + e^{i(\phi_a-\phi_b)}e^{i\omega_0 t}\right)R_1 R_2 \\
&= a^2 R_1^2 + b^2 R_2^2 + ab\left(e^{-i(\omega_0 t-\phi)} + e^{i(\omega_0 t+\phi)}\right)R_1 R_2 \\
&= a^2 R_1^2 + b^2 R_2^2 + 2ab R_1 R_2 \cos(\omega_0 t + \phi).
\end{aligned} \tag{11}$$

Thus, the charge density of the two-component wave function is made up of the charge densities of the two separate wave functions, shown in Figure 4, plus a term proportional to the product of the two wave function amplitudes. It reduces to the individual charge density of the ground state when $a = 1, b = 0$ and to that of the excited state when $a = 0, b = 1$. The product term, shown in green in Figure 5, is the only non-stationary term; it oscillates at the transition frequency ω_0. The integral of the total density shown in the right-hand plot is equal to 1 for any phase of the cosine term, since there is only one electron in this two-component state.

All the $\Psi^*\Psi$ plots represent the density of negative charge of the electron. The atom as a whole is neutral because of the equal positive charge on the nucleus. The dipole is formed when the center of charge of the electron wave function is displaced from the central positive charge of the nucleus.

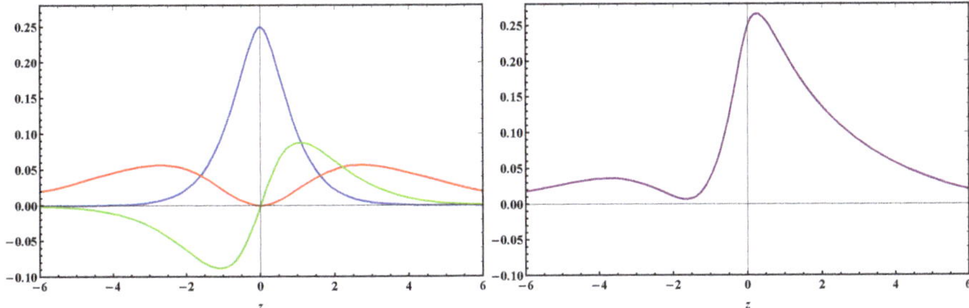

Figure 5. Left: Plot of the three terms in the wave-function density in Equation (11) for an equal $\left(a = b = 1/\sqrt{2}\right)$ superposition of the ground state (R_1^2, blue) and first excited state (R_2^2, red) of the hydrogen atom. The green curve is a snapshot of the time-dependent $R_1 R_2$ product term, which oscillates at difference frequency ω_0. **Right**: Snapshot of the total charge density, which is the sum of the three curves in the left plot. The magnitudes plotted are the contribution to the total charge in an $x - y$ "slice" of $\Psi^* \Psi$ at the indicated z coordinate. All plots are shown for the time such that $\cos(\omega_0 t + \phi) = 1$. The horizontal axis in each plot is the spatial coordinate along the z-axis of the atom, given in units of the Bohr radius a_0. Animation here [37] (see Supplementary Materials).

The two-component wave function must be normalized, since it is the state of one electron:

$$\int \Psi^* \Psi \, dvol = 1$$

$$= \int \left(ae^{-i\phi_a} R_1 e^{i\omega_1 t} + be^{-i\phi_b} R_2 e^{i\omega_2 t}\right)$$

$$\left(ae^{i\phi_a} R_1 e^{-i\omega_1 t} + be^{i\phi_b} R_2 e^{-i\omega_2 t}\right) dvol \tag{12}$$

$$= a^2 \int R_1^2 \, dvol + b^2 \int R_2^2 dvol$$

$$+ \left(ae^{-i\phi_a} be^{i\phi_b} e^{-i\omega_0 t} + be^{-i\phi_b} ae^{i\phi_a} e^{i\omega_0 t}\right) \int R_1 R_2 dvol.$$

Recognizing from Equations (5) and (6) that the individual eigenfunctions are normalized and orthogonal:

$$\int R_1^2 \, dvol = 1 \qquad \int R_2^2 \, dvol = 1 \qquad \int R_1 R_2 \, dvol = 0. \tag{13}$$

Equation (12) becomes

$$\int \Psi^* \Psi \, dvol = 1 = a^2 + b^2. \tag{14}$$

Thus, a^2 represents the fraction of the two-component wave function made up of the lower state Ψ_1, and b^2 represents the fraction made up of the upper state Ψ_2. The total energy E of a system whose wave function is a superposition of two eigenstates is:

$$E = a^2 E_1 + b^2 E_2. \tag{15}$$

Using the normalization condition $a^2 + b^2 = 1$ and solving Equation (15) for b^2, we obtain:

$$b^2 = \frac{E - E_1}{E_2 - E_1}. \tag{16}$$

In other words, b^2 is just the energy of the wave function, normalized to the transition energy, and using E_1 as its reference energy. Taking E_1 as our zero of energy and $E_0 = E_2 - E_1$, Equation (16) becomes:

$$E = E_0 b^2 \quad \Rightarrow \quad \frac{\partial E}{\partial t} = E_0 \frac{\partial \left(b^2\right)}{\partial t}. \tag{17}$$

Defining:

$$d_{12} \equiv 2q \int R_1 R_2 \, z \, d\text{vol} = 2q \, \langle z \rangle_{\text{max}} , \tag{18}$$

the dipole moment $q \langle z \rangle$ of such a superposition can, from Equation (11), be written:

$$q \langle z \rangle = d_{12} ab \, \cos(\omega_0 t + \phi). \tag{19}$$

The factor d_{12} we call the **dipole strength** for the transition. When one R is an even function of z and the other is an odd function of z, as in the case of the 100 and 210 states of the hydrogen atom, then d_{12} is nonzero, and the transition is said to be **electric dipole allowed**. When both R_1 and R_2 are either even or odd functions of z, $d_{12} = 0$, and the transition is said to be **electric dipole forbidden**.

Even in this case, some other moment of the distribution generally will be nonzero, and the transition can proceed by magnetic dipole, magnetic quadrupole, or other higher-order moments.

For now, we will concentrate on transitions that are electric dipole allowed.

We have the time dependence of the electron dipole moment $q \langle z \rangle$ from Equation (19), from which we can derive the velocity and acceleration of the charge:

$$
\begin{aligned}
q \langle z \rangle &= d_{12} ab \, \cos(\omega_0 t + \phi) \\
q \frac{\partial \langle z \rangle}{\partial t} &= -\omega_0 d_{12} ab \, \sin(\omega_0 t + \phi) + d_{12} \cos(\omega_0 t + \phi) \frac{\partial (ab)}{\partial t} \\
&\approx -\omega_0 d_{12} ab \, \sin(\omega_0 t + \phi) \\
q \frac{\partial^2 \langle z \rangle}{\partial t^2} &\approx -\omega_0^2 d_{12} ab \, \cos(\omega_0 t + \phi),
\end{aligned}
\tag{20}
$$

where the approximation arises because we will only consider situations where the coefficients a and b change slowly with time over a large number of cycles of the transition frequency: $\left(\frac{\partial (ab)}{\partial t} \ll ab \, \omega_0 \right)$.

The motion of the electron mass density endows the electron with a momentum $\vec{p}$:

$$\vec{p} = m\vec{v} \quad \Rightarrow \quad p_z = m \frac{\partial \langle z \rangle}{\partial t} \approx -\frac{m}{q} \omega_0 d_{12} ab \, \sin(\omega_0 t + \phi). \tag{21}$$

6. Atom in an Applied Field

Schrödinger had a detailed physical picture of the wave function, and he gave an elegant derivation of the process underlying the change of atomic state mediated by electromagnetic coupling. [The original derivation in Ref. [38], p. 137, is not nearly as readable as that in Schrödinger's second and third 1928 lectures [39], where the state transition is described in Section 9 starting at the bottom of page 31, for which the second lecture is preparatory. There he solved the problem more generally, including the effect of a slight detuning of the field frequency from the atom's transition frequency.] Instead of directly tackling the transfer of energy between two atoms, he considered the response of a single atom to a small externally applied vector potential field $\vec{A}$. He found that the immediate effect of an applied vector potential is to change the momentum p of the electron wave function:

$$
\begin{aligned}
p_z &= m \frac{\partial \langle z \rangle}{\partial t} - q A_z \\
\frac{\partial p_z}{\partial t} &= m \frac{\partial^2 \langle z \rangle}{\partial t^2} - q \frac{\partial A_z}{\partial t}.
\end{aligned}
\tag{22}
$$

Thus, the quantity $-q \frac{\partial A_z}{\partial t}$ acts as a *force*, causing an *acceleration* of the electron wave function.

This is the physical reason that $-\frac{\partial A_z}{\partial t}$ can be treated as an *electric field* $\mathcal{E}_z$. [At a large distance from an overall charge-neutral charge distribution like an atom, the longitudinal gradient of the scalar potential just cancels the longitudinal component of $\partial \vec{A}/\partial t$, so what is left is $\vec{\mathcal{E}} = -\partial A_{\perp}/\partial t$, which is purely transverse.].

$$\mathcal{E}_z = -\frac{\partial A_z}{\partial t}. \tag{23}$$

In the simplest case, the $q A_z$ term makes only a tiny perturbation to the momentum over a single cycle of the ω_0 oscillation, so its effects will only be appreciable over many cycles.

We consider an additional simplification, where the frequency of the applied field is exactly equal to the transition frequency ω_0 of the atom:

$$A_z = A \cos\left(\omega_0 t\right) \quad \Rightarrow \quad -\frac{\partial A_z}{\partial t} = \mathcal{E}_z = \omega_0 A \sin\left(\omega_0 t\right). \tag{24}$$

In such evaluations, we need to be very careful to identify *exactly which energy* we are calculating:

The electric field is merely a bookkeeping device to keep track of the energy that an electron in one atom exchanges with another electron in another atom, in such a way that the total energy is conserved. We will evaluate how much energy a given electron gains from or loses to the field, recognizing that the negative of that energy represents work done by the electron on the source electron responsible for the field. The force on the electron is just $q\mathcal{E}_z$. Because $\mathcal{E}_z = \omega_0 A \sin\left(\omega_0 t\right)$, for a stationary charge, the force is in the $+z$ direction as much as it is in the $-z$ direction, and, averaged over one cycle of the electric field, the work averages to zero. However, if the charge itself oscillates with the electric field, it will gain energy ΔE from the work done by the field on the electron over one cycle:

$$\frac{\Delta E}{\text{cycle}} = \int q \mathcal{E}_z \, dz = \int_0^{2\pi/\omega_0} q \mathcal{E}_z \frac{\partial \langle z \rangle}{\partial t} \, dt, \tag{25}$$

where $\langle z \rangle$ is the z position of the electron center of charge from Equation (20).

When the electron is "coasting downhill" *with* the electric field, it gains energy and ΔE is positive. When the electron is moving "uphill" *against* the electric field, the electron loses energy and ΔE is negative.

As long as the energy gained or lost in each cycle is small compared with E_0, we can define a continuous **power** (rate of change of electron energy), which is understood to be an average over many cycles. The time required for one cycle is $2\pi/\omega_0$, so Equation (25) becomes:

$$\frac{\partial E}{\partial t} = \frac{\omega_0 \Delta E}{2\pi} = \frac{\omega_0}{2\pi} \int_0^{2\pi/\omega_0} q \mathcal{E}_z \frac{\partial \langle z \rangle}{\partial t} \, dt = \frac{1}{2\pi} \int_0^{2\pi} q \mathcal{E}_z \frac{\partial \langle z \rangle}{\partial t} \, d(\omega_0 t). \tag{26}$$

7. Electromagnetic Coupling

Because our use of electromagnetism is conceptually quite different from that in traditional Maxwell treatments (including Jaynes' NCT), we review here the foundations of that discipline from the present perspective. [A more detailed discussion from the present viewpoint is given in Mead, **Collective Electrodynamics** [16]. The standard treatment is given in Jackson, **Classical Electrodynamics, 3rd Edition**, Chapter 8 [40].] It is shown in Ref. [16] that electromagnetism is of totally quantum origin. We saw in Equation (22) that it is the vector potential $\vec{A}$ that appears as part of the momentum of the wave function, signifying the coupling of one wave function to one or more other wave functions. Thus, to stay in a totally quantum context, we must work with electromagnetic relations based on the vector potential and related quantities. The entire content of electromagnetism is contained in the relativistically-correct Riemann–Sommerfeld second-order differential equation:

$$\left(\nabla^2 - \frac{\partial^2}{\partial t^2} \right) \mathbf{A} = -\mu_0 \mathbf{J}, \tag{27}$$

where $\mathbf{A} = [\vec{A}, V/c]$ is the four-potential and $\mathbf{J} = [\vec{J}, c\rho]$ is the four-current, $\vec{A}$ is the vector potential, V is the scalar potential, $\vec{J}$ is the physical current density (no displacement current), and ρ is the physical charge density, all expressed in the same inertial frame.

Connection with the usual electric and magnetic field quantities $\vec{\mathcal{E}}$ and $\vec{B}$ is as follows:

$$\vec{\mathcal{E}} = -\nabla V - \frac{\partial \vec{A}}{\partial t} \qquad \vec{B} = \nabla \times \vec{A}. \tag{28}$$

Thus, once we have the four-potential $\mathbf{A}$, we can derive any electromagnetic relations we wish.

Equation (27) has a completely general Green's Function solution for the four-potential $\mathbf{A}(t)$ at a point in space, from four-current density $\mathbf{J}(r, t)$ in volume elements $d$vol at at distance r from that point:

$$\mathbf{A}(t) = \frac{\mu_0}{4\pi} \int \left. \frac{\mathbf{J}(r, t')}{r} \right|^{t' = t \pm \frac{r}{c}} d\text{vol}, \tag{29}$$

where r is the distance from element $d$vol to the point where $\mathbf{A}$ is evaluated, assumed large compared to the size of the atomic wave functions, and c is the speed of light.

Equation (29) is the first fundamental equation of electromagnetic coupling: The vector potential, which will appear as part of an electron's momentum, is simply the sum of all current elements on that electron's light cone, each weighted inversely with its distance from that electron. The second-order nature of derivatives in Equation (27) does not favor any particular sign of space or time. Thus, the four-potential from a current element on the *past* light cone of the electron $(t - r/c)$ will be "felt" by the electron at later time t, and is termed a **retarded** field. Conversely, the four-potential from a current element on the *future* light cone of the electron $(t + r/c)$ will be "felt" by the electron at earlier time t, and is termed an **advanced** field. Historically, with rare exception, advanced fields have been discarded as non-physical because evidence for them has been explained in other ways. We shall see that modern quantum experiments provide overwhelming evidence for their active role in **quantum entanglement.**

Equation (29) can be expressed in terms of more familiar E&M quantities:

$$\vec{A}(t) = \frac{\mu_0}{4\pi} \int \left. \frac{\vec{J}(r, t')}{r} \right|^{t' = t \pm \frac{r}{c}} d\text{vol} \qquad\qquad V(t) = \frac{\mu_0 c^2}{4\pi} \int \left. \frac{\rho(r, t')}{r} \right|^{t' = t \pm \frac{r}{c}} d\text{vol} \tag{30}$$

If the current density $\vec{J}$ is due to the movement of a small, unified "cloud" of charge, as is the case for the wave function of an atomic electron, and the motion of the wave function is non-relativistic, the $\vec{J}$ integral just becomes the movement of the center of charge relative to its average position at the nucleus:

$$\vec{A}(t) \approx \frac{\mu_0}{4\pi} \int \left. \frac{\rho \vec{v}(r, t')}{r} \right|^{t' = t \pm \frac{r}{c}} d\text{vol} \approx \frac{\mu_0}{4\pi} \left. \frac{q \vec{v}(r, t')}{r} \right|^{t' = t \pm \frac{r}{c}} \tag{31}$$

If, as we have chosen previously, the motion is in the z direction,

$$A_z(t) \approx \frac{q\mu_0}{4\pi r} \left. \frac{\partial \langle z(r, t') \rangle}{\partial t'} \right|^{t' = t \pm \frac{r}{c}} \tag{32}$$

If we use the current element as our origin of time, the signs are reversed:

$$A_z(t') \Big|^{t' = t \mp \frac{r}{c}} \approx \frac{q\mu_0}{4\pi r} \frac{\partial \langle z(r, t) \rangle}{\partial t} \tag{33}$$

In this case, $t + r/c$ represents the retarded field and $t - r/c$ represents the advanced field. We shall use these two forms for the simple examples presented below.

An important difference between standard Maxwell E&M practice and our use of the four-potential to couple atomic wave functions is highlighted by Wheeler and Feynman [17]:

"There is no such concept as 'the' field, an independent entity with degrees of freedom of its own."

The field has no degrees of freedom of its own. It is simply a convenient bookkeeping device for keeping track of the total effect of an arbitrary number of charges on a particular charge distribution in some region of space. The general form of interaction energy E is given by:

$$E = \int (\vec{A} \cdot \vec{J} + \rho V)\, d\text{vol} \tag{34}$$

Equation (34) is the second fundamental equation of electromagnetic coupling. The interaction described in Section 6 is a simplified case of this relation for a single atom. The energy minimum created by the positive nucleus is the V, and the ρ is the charge distribution of the electron wave function. The $\vec{A}$ from a second atom is very small, and is assumed to not change the eigenfunctions.

From Equations (34) and (29), we see that, when applied to two atoms, what is being described is a way to factor a bi-directional connection between them so that each can be analyzed separately by Schrödinger's equation as a one-electron problem, using the vector potential from the other as part of its energy.

The fact that the four-potential field from a charge is defined everywhere on its light cone does not imply that it is "radiating into space", carrying energy with it. Energy is only transferred at the position of another charge. Since all charges are the finite charge densities of wave functions, **there are no self-energy infinities in this formulation**.

One widely-held viewpoint treats the "quantum vacuum" as being made up of an infinite number of quantum harmonic oscillators. The problem with this view is that each such oscillator would have a zero-point energy that would contribute to the energy density of space in any gravitational treatment of cosmology. Even when the energies of the oscillators are cut off at some high value, the contribution of this "dark energy" is 120 orders of magnitude larger than that needed to agree with astrophysical observations. Such a disagreement between theory and observation (called the "cosmological constant problem"), even after numerous attempts to reduce it, is *many orders of magnitude worse than any other theory-vs-observation discrepancy in the history of science!* However, somehow this viewpoint remains a central part of the standard model of particle physics and standard practice in QM.

Our approach does not suffer from this serious defect, since its vacuum has no degrees of freedom of its own. Where, then, is radiated energy going if an atom's excitation decays and does not interact locally? The obvious candidate is the enormous continuum of states of matter in the early universe, source of the **cosmic microwave background**, to which atoms here and now are coupled by the quantum handshake. For independent discussions from the two of us, see Ref. [16], p. 94 and Ref. [19].

8. Two Coupled Atoms

The central point of this paper is to understand the *photon* mechanism by which energy is transferred from a *single* excited atom (atom α) to another *single* atom (atom β) initially in its ground state.

We proceed with the simplest and most idealized case of two identical atoms, where:

(1) Excited atom α will start in a state where $b \approx 1$ and a is very small, but never zero because of its ever-present random statistical interactions with a vast number of other atoms in the universe, and

(2) Likewise, atom β will start in a state where $a \approx 1$ and b is very small, but never zero for the same reason.

Thus, each atom starts in a two-component state that has an oscillating electrical current described by Equation (20):

$$q\frac{\partial \langle z_\alpha \rangle}{\partial t} \approx -\omega_0 d_{12} a_\alpha b_\alpha \sin(\omega_0 t)$$
$$q\frac{\partial \langle z_\beta \rangle}{\partial t} \approx -\omega_0 d_{12} a_\beta b_\beta \sin(\omega_0 t + \phi),$$
(35)

where we have taken excited atom α as our reference for the phase of the oscillations ($\phi_\alpha = 0$), and the approximation assumes that a and b are changing slowly on the scale of ω_0.

Although that random starting point will contain small excitations of a wide range of phases, we simplify the problem by assuming the following:

All of the vector potential A_β at atom α is supplied by atom β,
All of the vector potential A_α at atom β is supplied by atom α,
The dipole moments of both atoms are in the z direction,
The atoms are separated by a distance r in a direction orthogonal to z,

The vector potential at distance r from a small charge distribution oscillating in the z-direction is from Equation (32):

$$A_z(t) = \frac{q\mu_0}{4\pi r}\frac{\partial \langle z(r,t') \rangle}{\partial t'}\bigg|^{t'=t\pm\frac{r}{c}}.$$
(36)

Since all electron motions and fields are in the z direction, we can henceforth omit the z subscript.

When the distance r is small compared with the wavelength, i.e., $r \ll 2\pi c/\omega_0$, the delay r/c can be neglected. Since atomic dimensions are of the order of 10^{-10} m and the wavelength is of the order of 10^{-7} m, this case can be accommodated. We shall find that the results we arrive at here are directly adaptable to the centrally important case in which the atoms are separated by an arbitrarily distance, which will be analyzed in Section 9. Using Equation (23) and (32), the vector potentials, and hence the electric fields, from the two atoms become:

$$A_\alpha \approx \frac{q\mu_0}{4\pi r}\frac{\partial \langle z_\alpha \rangle}{\partial t} \quad \Rightarrow \quad \mathcal{E}_\alpha = -\frac{\partial A_\alpha}{\partial t} \approx -\frac{q\mu_0}{4\pi r}\frac{\partial^2 \langle z_\alpha \rangle}{\partial t^2}$$
$$A_\beta \approx \frac{q\mu_0}{4\pi r}\frac{\partial \langle z_\beta \rangle}{\partial t} \quad \Rightarrow \quad \mathcal{E}_\beta = -\frac{\partial A_\beta}{\partial t} \approx -\frac{q\mu_0}{4\pi r}\frac{\partial^2 \langle z_\beta \rangle}{\partial t^2}.$$
(37)

When atom α is subject to electric field $\mathcal{E}_\beta$ and atom β is subject to electric field $\mathcal{E}_\alpha$, the energy of both atoms will change with time in such a way that the total energy is conserved. Thus, the superposition amplitudes a and b of both atoms change with time, as described by Equation (17) and (26), from which:

$$\frac{\partial E_\alpha}{\partial t} \approx \frac{1}{2\pi}\int_0^{2\pi} q\mathcal{E}_\beta \frac{\partial \langle z_\alpha \rangle}{\partial t}\, d(\omega_0 t) = -\frac{q^2\mu_0}{8\pi^2 r}\int_0^{2\pi}\frac{\partial^2 \langle z_\beta \rangle}{\partial t^2}\frac{\partial \langle z_\alpha \rangle}{\partial t}\, d(\omega_0 t)$$
$$\frac{\partial E_\beta}{\partial t} \approx \frac{1}{2\pi}\int_0^{2\pi} q\mathcal{E}_\alpha \frac{\partial \langle z_\beta \rangle}{\partial t}\, d(\omega_0 t) = -\frac{q^2\mu_0}{8\pi^2 r}\int_0^{2\pi}\frac{\partial^2 \langle z_\alpha \rangle}{\partial t^2}\frac{\partial \langle z_\beta \rangle}{\partial t}\, d(\omega_0 t).$$
(38)

From Equation (38), using the $\langle z \rangle$ derivatives from Equation (20):

$$\frac{\partial E_\alpha}{\partial t} \approx -\frac{\mu_0}{8\pi^2 r}\int_0^{2\pi}\left(-\omega_0^2 d_{12} a_\beta b_\beta \cos(\omega_0 t + \phi)\right)\left(-\omega_0 d_{12} a_\alpha b_\alpha \sin(\omega_0 t)\, d(\omega_0 t)\right)$$
$$\approx -\frac{\mu_0\omega_0^3 d_{12}^2 a_\beta b_\beta a_\alpha b_\alpha}{8\pi^2 r}\int_0^{2\pi}\cos(\omega_0 t + \phi)\sin(\omega_0 t)\, d(\omega_0 t) = \frac{\mu_0\omega_0^3 d_{12}^2 a_\beta b_\beta a_\alpha b_\alpha}{8\pi r}\sin(\phi)$$
$$\frac{\partial E_\beta}{\partial t} \approx -\frac{\mu_0}{8\pi^2 r}\int_0^{2\pi}\left(-\omega_0^2 d_{12} a_\alpha b_\alpha \cos(\omega_0 t)\right)\left(-\omega_0 d_{12} a_\beta b_\beta \sin(\omega_0 t + \phi)\, d(\omega_0 t)\right)$$
$$\approx -\frac{\mu_0\omega_0^3 d_{12}^2 a_\alpha b_\alpha a_\beta b_\beta}{8\pi^2 r}\int_0^{2\pi}\cos(\omega_0 t)\sin(\omega_0 t + \phi)\, d(\omega_0 t) = -\frac{\mu_0\omega_0^3 d_{12}^2 a_\alpha b_\alpha a_\beta b_\beta}{8\pi r}\sin(\phi).$$
(39)

These equations describe energy transfer between the two atoms in either direction, depending on the sign of $\sin(\phi)$. For transfer from atom α to atom β, $\partial E_\alpha / \partial t$ is negative. Since this transaction dominated all competing potential transfers, its amplitude will be maximum, so $\sin(\phi) = -1$.

If the starting state had been atom β in the excited ($b \approx 1$) state, the $\sin(\phi) = +1$ choice would have been appropriate:

$$\text{Using}: \quad \sin(\phi) = -1 \quad \text{and} \quad P_{\alpha\beta} \equiv \frac{\mu_0 \omega_0^3 d_{12}^2}{8\pi r}, \tag{40}$$

This rate of transferred energy is calculated for two isolated atoms suspended in space. We have no experimental data whatsoever for such a situation. All optical experiments are done with some **optical system** between the two atoms. Even the simplest such arrangement couples the two atoms *orders of magnitude* better than the simple $1/r$ dependence in Equation (40) would indicate. We take up the enhancement due to an intervening optical system in Section 10. Any such enhancement merely provides a constant multiplier in $P_{\alpha\beta}$. In any case, Equation (39) becomes:

$$\frac{\partial E_\alpha}{\partial t} = E_0 \frac{\partial b_\alpha^2}{\partial t} = -P_{\alpha\beta}\, a_\beta b_\beta a_\alpha b_\alpha \qquad\qquad \frac{\partial E_\beta}{\partial t} = E_0 \frac{\partial b_\beta^2}{\partial t} = P_{\alpha\beta}\, a_\alpha b_\alpha a_\beta b_\beta. \tag{41}$$

Remembering that our total starting energy was E_0 that b^2 is the energy of the electron in units of E_0 referred to the ground state, that energy is conserved by the two atoms during the transfer, and that the wave functions are normalized:

$$E_0\left(b_\alpha^2 + b_\beta^2\right) = E_0 \quad \Rightarrow \quad b_\alpha^2 + b_\beta^2 = 1 \quad \Rightarrow \quad b_\beta = \sqrt{1 - b_\alpha^2}$$
$$a_\alpha^2 + b_\alpha^2 = 1 \quad \Rightarrow \quad a_\alpha = \sqrt{1 - b_\alpha^2} = b_\beta \tag{42}$$
$$a_\beta^2 + b_\beta^2 = 1 \quad \Rightarrow \quad a_\beta = \sqrt{1 - b_\beta^2} = b_\alpha,$$

after which substitutions Equation (41) becomes:

$$\frac{\partial\left(b_\alpha^2\right)}{\partial t} = b_\alpha^2\left(1 - b_\alpha^2\right)/\tau, \quad \text{where the transition time scale is } \tau \equiv \frac{E_0}{P_{\alpha\beta}}. \tag{43}$$

This has the solution plotted in Figure 6:

$$b_\alpha^2 = a_\beta^2 = \frac{1}{e^{t/\tau} + 1} \qquad\qquad a_\alpha^2 = b_\beta^2 = \frac{1}{e^{-t/\tau} + 1}. \tag{44}$$

Note that this waveform is that of an *individual interaction* and has no probabilistic meaning. It was the subject of many intense discussions about NCT in general, including a quite detailed one in [30]. We refrain from such discussion here because the dependence of the dipole moment with superposition makeup is not the only nonlinearity in the problem. The self-focusing nature of the matched advanced/retarded electromagnetic solutions, described in Section 11, may be an even larger nonlinearity in many cases. Although its time dependence is much more difficult to estimate, it will almost certainly make the individual quantum transition much more abrupt.

The direction and magnitude of the entire energy-transfer process is governed by the relative phase ϕ of the electric field and the electron motion in *both atoms*: When the electron motion of *either atom is in phase* with the field, the field transfers energy to the electron, and the field is said to **excite** the atom. When the the electron motion has *opposite phase* from the field, the electron transfers energy "to the field", and the process is called **stimulated emission**.

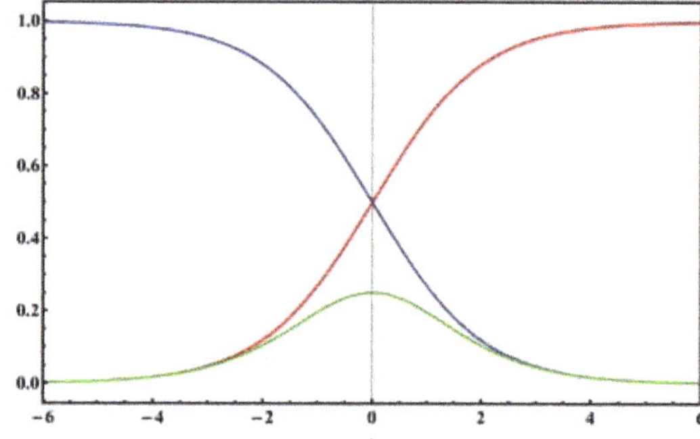

Figure 6. Squared state amplitudes for atom α: b_α^2 (blue) and $a_\alpha^2 = b_\beta^2$ (red) for the Photon transfer of energy $E_0 = \hbar\omega_0$ from atom α to atom β, from Equation (44). Using the lower state energy as the zero reference, $E_0 b^2$ is the energy of the state. The green curve shows the normalized power radiated by the atom α and absorbed by atom β, from Equation (43). The optical oscillations at ω_0 are not shown, as they are normally many orders of magnitude faster than the transition time scale τ. The time t is in units of τ. In the next section, we will find that atoms spaced by an arbitrary distance exhibit transactions of exactly the same form.

Therefore, for the photon transaction to proceed the field from atom α must have a phase such that it "excites" atom β, while the field from atom β must have a phase such that it absorbs energy and "de-excites" atom α. In the above example, that unique combination occurs when $\sin(\phi) = -1$.

This dependence on phase makes a transaction exquisitely sensitive to the frequency match between atoms α and β. The frequency $\omega_0 \approx 10^{16}/$ s, so for transitions in the nanosecond range, a mismatch of one part in 10^7 can cause a transaction to fail.

8.1. Competition between Recipient Atoms

As discussed at the end of Section 7, the bi-directional vector potential coupling has allowed us to analyze the problem of two coupled atoms, which looks like a two-electron problem, as two coupled one-electron problems, and therefore is treatable using Schrödinger's wave mechanics. The approach generalizes, so we can now examine the important three-atom case of a single excited atom α that is equally coupled to two ground-state atoms $\beta 1$ and $\beta 2$. Recipient atoms $\beta 1$ and $\beta 2$ have the same level spacing as source atom α. For atom $\beta 1$, this corresponds to transition frequency ω_0, and we assume a relative phase of $\sin(\phi) = -1$. However, atom $\beta 2$ is moving and has a slightly Doppler-shifted transition frequency $\omega_0 + \Delta\omega$. We assume that $\beta 2$ has the same structure and intial phase as $\beta 1$ at time $t = 0$. Thence, Equation (41), with the transition time scale $\tau \equiv E_0/P_{0\beta 1} = E_0/P_{0\beta 2}$, becomes:

$$\tau\frac{\partial b_{\beta 1}^2}{\partial t} = a_\alpha b_\alpha \left(a_{\beta 1} b_{\beta 1}\right) \qquad \tau\frac{\partial b_{\beta 2}^2}{\partial t} = a_\alpha b_\alpha \left(a_{\beta 2} b_{\beta 2} \cos(\Delta\omega t)\right) \qquad \tau\frac{\partial b_\alpha^2}{\partial t} = -a_\alpha b_\alpha \left(a_{\beta 1} b_{\beta 1} + a_{\beta 2} b_{\beta 2} \cos(\Delta\omega t)\right) \tag{45}$$

As with Equation (42), wave functions are normalized and energy is conserved:

$$a_\alpha^2 + b_\alpha^2 = 1 \qquad a_{\beta 1}^2 + b_{\beta 1}^2 = 1 \qquad a_{\beta 2}^2 + b_{\beta 2}^2 = 1 \qquad b_\alpha^2 + b_{\beta 1}^2 + b_{\beta 2}^2 = 1. \tag{46}$$

After these substitutions, we obtain two simultaneous differential equations in $b_\beta^2 1$ and $b_\beta^2 2$:

$$\tau\frac{\partial b_{\beta 1}^2}{\partial t} = \sqrt{b_{\beta 1}^2 \left(1 - b_{\beta 1}^2\right)\left(1 - b_{\beta 1}^2 - b_{\beta 2}^2\right)\left(b_{\beta 1}^2 + b_{\beta 2}^2\right)}$$

$$\tau\frac{\partial b_{\beta 2}^2}{\partial t} = \sqrt{b_{\beta 2}^2 \left(1 - b_{\beta 2}^2\right)\left(1 - b_{\beta 1}^2 - b_{\beta 2}^2\right)\left(b_{\beta 1}^2 + b_{\beta 2}^2\right)} \cos(\Delta\omega t)$$

$$\tag{47}$$

The solutions of Equation (47) are shown in Figure 7 for two very small values of $\Delta\omega$:

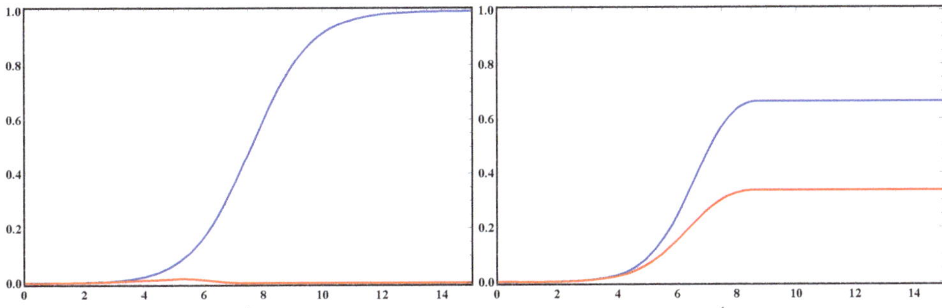

Figure 7. Squared excited state amplitudes for recipient atoms $\beta1$ and $\beta2$: $b_{\beta1}^2$ (blue) and $b_{\beta2}^2$ (red), for the photon transfer of energy $E_0 = \hbar\omega_0$ from atom α. The time t is in units of τ. The left plot, for $\Delta\omega = 0.3/\tau$, shows both recipient atoms being equally excited at the beginning, but the slip in phase of the red $\beta2$ atom causes it to rapidly lose out, so the blue $\beta1$ atom hogs all the energy and proceeds to become fully excited, much as if it were the "only atom in town". Its curve is nearly identical with that for the isolated recipient atom in Figure 6. The right plot, for $\Delta\omega = 0.15/\tau$, shows a totally different story. Its red $\beta2$ atom transition frequency is just close enough to that of source atom 0 that it "hangs in there" during the transition period and ends up partially excited, leaving blue $\beta1$ atom partially excited as well. The smaller the "slip" frequency $\Delta\omega$, the closer are the post-transition excitations of the two recipient atoms. "Split-photons" of this kind are observed in the Hanbury–Brown–Twiss Effect described in Section 13. As noted above, the frequency window for such events is *extraordinarily narrow*, typically of order $10^{-7}\omega_0$. Doppler shift of this magnitude requires velocity ≈ 30 m/s. Room temperature thermal velocities of gasses are typically tens to hundreds of times this value, which would eliminate such competition. Thus, complete transactions are the most common, with "split" transactions relatively rare and are likely to end as HBT-type four-atom events.

Here, again the transaction time scale τ is not to be confused with the time scale for initiating a transaction after atom α is excited, which is a question of probability. We make no pretense here of deriving probability results for a random population of atoms, but, from these results, we can imagine what one might look like: The random starting point for the transaction involving one excited atom will contain small excitations of a wide range of phases. Equation (43) is a highly nonlinear equation—the amplitude of each of those excitations will initially grow exponentially at a rate proportional to its own phase match. Thus, the excitation of a random recipient atom that happens to have $\sin(\phi) \approx -1$ will win in the race and become the dominant partner in the coordinated oscillation of both atoms. *Thus, we have conceptualized the source of the intrinsic randomness within quantum mechanics, an aspect of statistical QM that has been considered mysterious since its inception in the 1920s.*

Each wave function will thus evolve its motion to follow the applied field to its maximum resonant coupling and we can take $\sin(\phi) = -1$ in these expressions, which we have done in Equation (41), Figure 6 and Equation (45). [What we have not done is to derive the full second-order nature of phase locking in this arrangement. That analysis is rendered much more difficult by the potentially *huge* amplification due to the self-focusing nature of the bi-directional electromagnetic coupling described in Section 10. Thus, a full derivation remains open to future generations.]

From the TI point of view, all three atoms start in stable states, with each having extremely small admixtures of the other state, so that they have very small dipole moments oscillating with angular frequency $\omega_0 \approx (E_2 - E_1)/\hbar$. We assume that in source atom α this admixture creates an initial positive-energy offer wave that interacts with the small dipole moments of absorber atoms $\beta1$ and $\beta2$ to transfer positive energy, and that in atoms $\beta1$ and $\beta2$ this admixture creates initial negative-energy confirmation waves to the excited emitter atom α that interact with the small dipole moment of emitter atom α to transfer negative energy, as shown schematically in Figure 1. As a result of the mixed-energy superposition of states as shown in Figure 5, all three atoms oscillate with very nearly the same frequency ω_0 and act as coupled dipole resonators.

Energy transferred from source atom α to both recipient atoms $\beta 1$ and $\beta 2$ causes an increase in both minority states of the superposition, thus increasing the dipole moment of all three states, thereby increasing the coupling and, hence, the initial rate of energy transfer. This behavior is self-reinforcing for any atom that can stay in phase, giving the transition its initial exponential character. In the usual case, only one atom is sufficiently well frequency matched to stay in phase for the entire duration of the transition, the unfortunate runner-up is rudely driven out of the competition, and the winner drives the transaction to its conclusion, as shown in the left panel of Figure 7.

In the presence of near-equal competition, one competitor either loses out to a competing transaction, or, in case of a tie, results in a "split-photon", as shown in the right panel of Figure 7. That situation represents an *intermediate* state with actively oscillating dipole moments, as discussed above, in which the two confirmation waves, in the phase at the source, can precipitate the de-excitation of an additional excited atom to create a final HBT four-atom event.

The universe is full of similar atoms, all with slightly different transition frequencies due to random velocities. There are also random perturbation by waves from other systems that can randomly drive the exponential instability in either direction. This random environment is the source of the intrinsic randomness in quantum processes. Ruth Kastner [41] attributes intrinsic randomness to "spontaneous symmetry breaking", which could split a "tie" in the absence of environmental factors.

We note here that the probability of the transition must depend on two things: the strength of the electromagnetic coupling between the two states, and the degree to which the wave functions of the initial states are superposed. The magnitude of the latter must depend on the environment, in which many other atoms are "chattering" and inducing state-mixing perturbations. The more potential partner atoms there are per unit energy, the greater the probability of a perfect match. Thus, we see the emergence of **Fermi's "Golden Rule"** [42], the assertion that a transition probability in a coupled quantum system depends on the strength of the coupling and the density of states present to which the transition can proceed. The emergence of Fermi's Golden Rule is an unexpected gift delivered to us by the logic of the present formalism.

It is certainly not obvious a priori that the Schrödinger recipe for the vector potential in the momentum (Equation (22)), together with the radiation law from a charge in motion (Equation (33)), would conspire to enable the composition of the superposed states of two electromagnetically coupled wave functions to reinforce in such a way that, from the asymmetrical starting state, the energy of one excited atom could *explosively and completely* transfer to the unexcited atom, as shown in Figure 6 and Figure 7.

If nature had worked a slightly different way, an interaction between those atoms might have resulted in a different phase, and no full transaction would have been possible. The fact that transfer of energy between two atoms has this nonlinear, self-reinforcing character makes possible arrangements like a *laser*, where many atoms in various states of excitation participate in a glorious dance, all participating at exactly the same frequency and locked in phase.

Why do the signs come out that way? No one has the slightest idea, but the behavior is so remarkable that it has been given a name: Photons are classified as *bosons*, meaning that they behave that way!

That remarkable behavior is not due to any "particle-like" quantization of the electromagnetic field. Quantization of the photon energy is a result of the discrete nature of electron states in atoms.

The movement of an electron in a superposed state couples to another such electron electromagnetically. It is essential that this electromagnetic coupling is bi-directional in space-time to conserve energy in the transaction. The statistical QM formulation needed some mechanism to finalize a transaction and did not recognize the inherent nonlinear positive-feedback that nature built into a pair of coupled wave functions. Therefore, the founders had to put in wave-function collapse "by hand", and it has mystified the field ever since. The NCT formulations *did* understand the inherent nonlinear positive-feedback that nature built into a pair of coupled wave functions, but postulated a unidirectional "Maxwell" treatment of the electromagnetic field that *did not conserve energy*, as we now discuss.

9. Two Atoms at a Distance

We saw that for two atoms to exchange energy, the vector potential A field at atom β must come from atom α, the A field at atom α must come from atom β, and the oscillations must stay in coherent phase, with a particular phase relation during the entire transition. *This phase relation must be maintained even when the two atoms are an arbitrary distance apart.* This is the problem we now address.

To be definite, we consider the case where the two atoms are separated along the x-axis, atom α at $x = 0$ in the excited state and atom β at $x = r$ in its ground state, so their separation r is orthogonal to the z-directed current in the atoms. The "light travel time" from atom α to atom β is thus $\Delta t = r/c$. What is observed is that the energy radiated by atom α at time t is absorbed by atom β at time $t' = t + \Delta t$:

$$\left. \frac{\partial E_\beta(r, t')}{\partial t'} \right|^{t' = t + \Delta t} = -\frac{\partial E_\alpha(0, t)}{\partial t} \tag{48}$$

This behavior is familiar from the behavior of a "particle", which carries its own degrees of freedom with it: It leaves $x = 0$ at time t and arrives at $x = r$ at time $t' = t + \Delta t$ after traveling at velocity c. Thus, Lewis's "photon" became widely accepted as just another particle, with degrees of freedom of its own. We shall see that this assumption violates a wide range of experimental findings.

For atom β, Equation (38) becomes:

$$\frac{\partial E_\beta(r, t')}{\partial t'} = \frac{1}{2\pi} \int_0^{2\pi} -q\mathcal{E}_\alpha(r, t') \frac{\partial \langle z_\beta(r, t') \rangle}{\partial t'} d(t') \qquad \text{where} \qquad t' = t + \Delta t \tag{49}$$

The retarded field from atom α interacts with the motion of the electron in atom β. The only difference from our zero-delay solution is that the energy transfer has its time origin shifted by $\Delta t = r/c$.

This result has required that we choose a *positive* sign for the $\mp r/c$ in Equation (33). By doing that, we are building in an "arrow of time", a preferred time direction, in the otherwise even-handed formulation. In particular, we are assuming that the retarded solution transfers positive energy. So far, everything is familiar and consistent with commonly held Maxwell notions: A retarded solution carrying energy with it. However, we saw that the source atom required a matched vector potential to lose energy.

The standard picture leaves no way for atom α to lose energy to atom β. It does not conserve energy!

When energy is transferred between two atoms spaced apart on the x-axis, the field amplitude must be "*coordinate and symmetrical*" as Lewis described. The field $\mathcal{E}_\alpha(x = r)$ at the second atom due to the current in the first must be exactly equal in magnitude to the the field $\mathcal{E}_\beta(x = 0)$ at the first atom due to the current in the second, but separated in time by Δt: For atom α, Equation (38) becomes

$$\frac{\partial E_\alpha(0, t)}{\partial t} = \int_0^{2\pi} -q\mathcal{E}_\beta(0, t) \frac{\partial \langle z_\alpha \rangle}{\partial t} dt \tag{50}$$

Thus, the field $\mathcal{E}_\beta$ from atom β, which arises from the motion of its electron at time $t' = t + \Delta t$, must arrive at atom α at time t, *earlier* than its motion by Δt. The only field that fulfills this condition is the *advanced* field from atom β, signified by choosing a negative sign for the $\mp r/c$ in Equation (33). That choice uniquely satisfies the requirement for conservation of energy. It also builds complementary "arrows of time" into the formulation—we assume that the advanced solution transfers negative energy to the past and the retarded solution transfers positive energy to the future. These two assumptions create a new non-local "handshake" symmetry that is not expressed in conventional Maxwell E&M.

Once these choices for the $\mp r/c$ in Equation (33) are made, the resulting equations for each of the energy derivatives in Equation (39) are unchanged when $t' = t + \Delta t$ is substituted for t in the expression for $\partial E_\beta / \partial t$. Thus, each transition proceeds in the local time frame of its atom—for all the world (except for amplitude) as if the other atom were local to it. This "locality on the light cone" is the meaning of Lewis' comment:

> "In a pure geometry it would surprise us to find that a true theorem becomes false when the page upon which the figure is drawn is turned upside down. A dissymmetry alien to the pure geometry of relativity has been introduced by our notion of causality."

The dissymmetry that concerned Lewis has been eliminated.

This conclusion is completely consistent with the 1909 formulation of Einstein [43], who was critical of the common practice of simply ignoring the advanced solutions for electromagnetic propagation:

> "I regard the equations containing retarded functions, in contrast to Mr. Ritz, as merely auxiliary mathematical forms. The reason I see myself compelled to take this view is first of all that those forms do not subsume the energy principle, while I believe that we should adhere to the strict validity of the energy principle until we have found important reasons for renouncing this guiding star."

After defining the retarded solution as f_1, and the advanced solution as f_2, he elaborates:

> "Setting $f(x,y,z,t) = f_1$ amounts to calculating the electromagnetic effect at the point x,y,z from those motions and configurations of the electric quantities that took place prior to the time point t.' "Setting $f(x,y,z,t) = f_2$, one determines the electromagnetic effects from the motions and configurations that take place after the time point t."

> "In the first case the electric field is calculated from the totality of the processes producing it, and in the second case from the totality of the processes absorbing it...
> Both kinds of representation can always be used, regardless of how distant the absorbing bodies are imagined to be."

The choice of advanced or retarded solution cannot be made a priori: It depends upon the *boundary conditions* of the particular problem at hand. The quantum exchange of energy between two atoms just happens to require one advanced solution carrying negative energy and one retarded solution carrying positive energy to satisfy its boundary conditions at the two atoms, which then guarantees the conservation of energy.

Thus, the even-handed time symmetry of Wheeler–Feynman electrodynamics [17,18] and of the Transactional Interpretation of quantum mechanics [12], as most simply personified in the two-atom photon transaction considered here, arises from the symmetry of the electromagnetic propagation equations, with boundary conditions imposed by the solution of the Schrödinger equation for the electron in each of the two atoms, as foreseen by Schrödinger. We see that the missing ingredients in previous failed attempts, by Schrödinger and others, to derive wave function collapse from the wave mechanics formalism were that advanced waves were not explicitly used as a part of the process.

To keep in touch with experimental reality, we return to our two H atoms spaced a distance r apart. We can estimate the "transition time" τ from Equations (41) and (40):

$$\frac{\partial E_\beta}{\partial t} = P_{\alpha\beta}\, a_\alpha b_\alpha a_\beta b_\beta = \frac{\mu_0 \omega_0^3 d_{12}^2}{8\pi r}\, a_\alpha b_\alpha a_\beta b_\beta. \tag{51}$$

From the green curve in Figure 5, we can estimate the dipole strength, which is q times the "length" between the positive and negative "charge lumps", say $d_{12} \approx 3qa_0$. At the steepest part of the transition, all the a and b terms will be $1/\sqrt{2}$, so

$$\left.\frac{\partial E_\beta}{\partial t}\right|_{\max} \approx \frac{\mu_0 \omega_0^3 (3qa_0)^2}{32\pi r}. \tag{52}$$

From any treatment of the hydrogen spectrum, we obtain, for the 210→100 transition:

$$E_0 = \hbar\omega_0 = \frac{9q^2}{128\pi\epsilon_0 a_0} \tag{53}$$

so the transition time will be:

$$\tau_{12} \approx \frac{E_0}{\left.\frac{\partial E_\beta}{\partial t}\right|_{max}} \approx \frac{r}{4\,a_0^3\omega_0^3\epsilon_0\mu_0} = \frac{r\,c^2}{4\,a_0^3\omega_0^3} \approx r \times 0.04 \,\frac{\text{sec}}{\text{m}}, \tag{54}$$

Thus, if the assumption of the $1/r$ dependence of the vector potential (r dependence of the transition time) were the whole picture, it would take $1/25$ of a second for a transaction to complete if the atoms were suspended one meter apart. Such a long transition time would allow the excited atom's energy to be frittered away by the many possible competitive paths, thus making any modern optical experiment virtually impossible, so *no experiments are done that way!* In real experiments, atoms are coupled by some *optical system*, composed of lenses, mirrors, and the like. That optical system will have some **solid angle** containing paths from one atom to the other. The effect of the optical system is to replace the $1/r$ dependence with Solid Angle$/\lambda$, as described in Section 10. Thus, the 0.04 s transition given by Equation (54) for two isolated atoms 1 m apart becomes 2×10^{-9} s when a 1 steradian optical system is used.

Once again, we caution that the time estimated here is *the time course of the single transaction after a handshake is formed*, which *must not* be confused with the probabilistic time for a transaction to be initiated after excitation of the source atom.

10. Paths of Interaction

We have developed a simple conceptual understanding of how a single quantum $\hbar\omega_0$ of energy is transferred from one isolated atom to another by way of a "photon" transaction. Real experiments with such transactions measure the statistics of many such events as functions of intensity, polarization, time delay, and other variables. Much has been discovered in the process, some results quite surprising, as described for the Freeman–Clauser experiment in Section 13. Thus, the time has come for us to discuss, at a conceptual level, where the probabilities come from. In the wonderful little book *QED* [44], our Caltech colleague the late Richard Feynman gives a synopsis of the method by which light propagating along multiple paths initiates a transaction, which he calls an event:

> *"Grand Principle: The probability of an event is equal to the square of the length of an arrow called the 'probability amplitude.'…"*

> *"General Rule for drawing arrows if an event can happen in alternative ways: Draw an arrow for each way, and then combine the arrows ('add' them) by hooking the head of one to the tail of the next."*

> *"A 'final arrow' is then drawn from the tail of the first arrow to the head of the last one."*

> *"The final arrow is the one whose square gives the probability of the entire event."*

Feynman's "arrow" is familiar to every electrical engineer as a **phasor**, introduced in 1894 by Steinmetz [45,46] as an easy way to visualize and quantify phase relations in alternating-current systems. In physics, the technique is known as the **sum over histories** and led to **Feynman path integrals**. His "probability amplitude" is the amplitude of our vector potential, whose square is the **probability** of a photon.

Feynman then illustrates his Grand Principle with simple examples how a source of light S at one point in space and time influences a receptor of that light P at another point in space and time, as shown in Figure 8. It is somewhat unnerving to many people to learn from these examples that the resultant intensity is dependent on *every possible path* from S to P. We strongly recommend that little book to everyone. That discussion, as well as what follows, details the behavior of highly coherent electromagnetic radiation with a well-defined, highly stable frequency ω and wavelength λ.

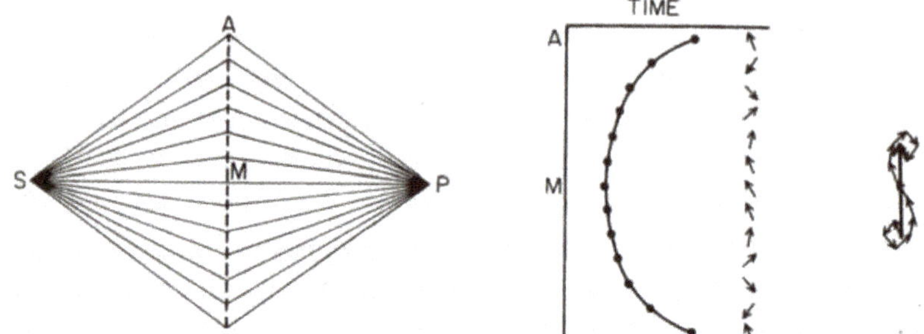

Figure 8. All the paths from coherent light source S to detector P are involved in the transfer of energy. The solid curve on the "TIME" plot shows the propagation time, and hence the accumulated phase, of the corresponding path. Each small arrow on the "TIME" plot is a **phasor** that shows the magnitude (length) and phase (angle) of the contribution of that path to the resultant total vector potential at P. The "sea horse" on the far right shows how these contributions are added to form the total amplitude and phase of the resultant potential. (From Fig. 35 in Feynman's **QED**).

Of course, we have all been taught that light travels in straight lines which spread out as they radiate from the source, and so the resultant intensity decreases as $1/r^2$, where r is the distance from the source S. However, if the light intensity at P depends on *all of the paths*, how can this $1/r^2$ dependence come about? Well, let's follow Feynman's QED logic: [If the reader does not have a copy of QED handy, there is a condensed version in Chapter I-26 of the *Feynman Lectures on Physics* at https://www.feynmanlectures.caltech.edu/I_26.html].

We can see from the "seahorse" phasor diagram at the right of Figure 8 that the vast majority of the length of the resultant arrow is contributed by paths very close to the straight line S-M-P. Thus, let's make a rough estimate of how many paths there are near enough to "count." We can see from the diagram that, once the little arrows are plus or minus 90° from the phase of the straight line, additional paths just cause the resultant to wind around in a tighter and tighter circle, making no net progress. Thus, the uppermost and lowermost paths that "count" are about a quarter wavelength longer than the straight line. Let's use r for the straight-line distance S-P, λ for the wavelength, and y for the vertical distance where the path intersects the midline above M. Then, Pythagoras tells us that the length $l/2$ of either segment of the path is

$$\frac{l}{2} = \sqrt{\left(\frac{r}{2}\right)^2 + y^2} \tag{55}$$

Therefore, the entire path length l is

$$l = \sqrt{r^2 + 4y^2} = r\sqrt{1 + 4\frac{y^2}{r^2}} \tag{56}$$

We are particularly interested in atoms at a large distance from each other, and will guess that this means that y is very small compared to r, so all the paths involved are very close to the straight line. We can check that assumption later. Since $y^2/r^2 \ll 1$, we can expand the square root:

$$l \approx r\left(1 + 2\frac{y^2}{r^2}\right) = r + 2\left(\frac{y^2}{r}\right) \tag{57}$$

Thus, the outermost path that contributes is $2y^2/r$ longer than the straight line path. We already decided that maximum extra length of a contributing path would be about a quarter wavelength:

$$2\left(\frac{y^2}{r}\right) \approx \frac{\lambda}{4} \quad \Rightarrow \quad \left(\frac{y}{r}\right)^2 \approx \frac{\lambda}{8r} \tag{58}$$

We can now check our assumption that $y^2/r^2 \ll 1$. If $r = 1$ m and our $210 \rightarrow 100$ transition has $\lambda \approx 10^{-7}$ m, then $y^2/r^2 \approx 10^{-8}$, so our assumption is already very good, and gets better rapidly as r gets larger.

How do we estimate the number of paths from S to P? Well, no matter how we choose the path spacing radiating out equally in all directions from S, the number of paths that "count" will be proportional to the solid angle subtended by the outermost such paths. Paths outside that "bundle" will have phases that cancel out as Feynman describes. The angle of the uppermost path is y/r. Paths also radiate out perpendicular to the page to the same extent, so the total number that "count" goes as the solid angle $= \pi(y/r)^2$. Feynman tells us that the resultant amplitude A is proportional to the total number of paths that "count," so we conclude from Equation (58):

$$A \propto \text{Solid Angle} = \pi\left(\frac{y}{r}\right)^2 \approx \frac{\pi\lambda}{8r} \tag{59}$$

Thus, this is the fundamental origin of the $1/r$ law for amplitudes.

The intensity is proportional to the square of the amplitude, and therefore goes like $1/r^2$, as we all learned in school. Thus, instead of lines of energy radiating out into space in all directions, Feynman's view of the world encourages us to visualize the source of electromagnetic waves as "connected" to each potential receiver by all the paths that arrive at that receiver in phase. Just to convince us that all this "path" stuff is real, Feynman gives numerous fascinating examples where the $1/r^2$ law doesn't work at all. Our favorite is shown in Figure 9:

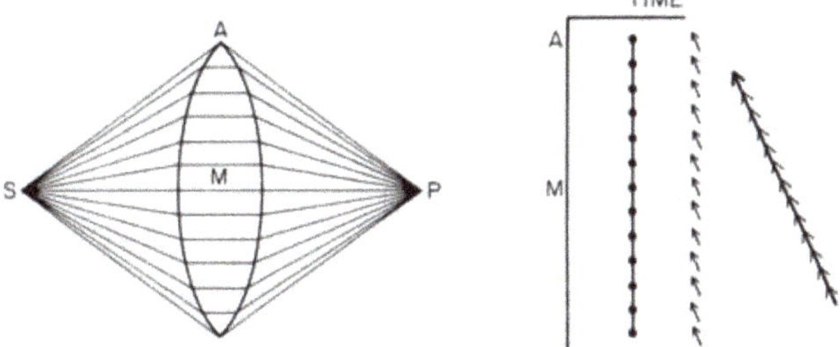

Figure 9. Situation identical to Figure 8 but with a piece of glass added. The shape of the glass is such that all paths from the source S reach P in phase. The result is an *enormous* increase in the amplitude reaching P. (From Fig. 36 in Feynman's **QED**).

The piece of glass does not alter the amplitude of any individual path very much—it might lose a few percent due to reflection at the surfaces. However, it *slows down* the speed of propagation of the light. In addition, the thickness of the glass has been tailored to slow the shorter paths more than the longer paths, so *all paths take exactly the same time*. The net result is that *the oscillating potential propagating along every path reaches P in phase with all the others!* Now, we are adding up all the little phasor arrows *and they all point in the same direction!* The amplitude is enhanced by this little chunk of glass by the factor:

$$\text{EnhancementFactor} = \frac{\text{Lens Solid Angle}}{\text{Solid Angle from Equation (59)}} \approx \frac{8r}{\pi\lambda}\text{Lens Solid Angle} \tag{60}$$

For the arrangement shown in Figure 9, at the size it is printed on a normal-sized page, $r \approx 5$ cm, so from From Equation (59), the solid angle of the "bundle" of paths without the glass was of the order of 10^{-7}. The solid angle enclosing the paths through the glass lens is ≈ 1 steradian (sr), so the little piece of glass has increased the potential at P by *seven orders of magnitude*, and the intensity of the light by *fourteen orders of magnitude!* In this and the more general case, we get an important principle:

$$\textbf{Insertion of an optical system replaces the } \frac{1}{\textbf{r}} \textbf{ factor by } \frac{8}{\pi\lambda} \times \textbf{Lens Solid Angle} \qquad (61)$$

Returning to our two H atoms spaced 1 m apart, we found in Equation (54) that, using the standard $1/r$ potential, the "transition time" for the quantum transaction was ≈ 0.04 s. For the ω_0 wavelength the Rate Enhancement Factor with a 1st optical system is $\approx 2 \times 10^7$, thereby shortening the transition time τ by a factor of $\approx 5 \times 10^{-8}$, making $\tau \approx 2$ ns.

We have learned an important lesson from Feynman's characterization of propagation phenomena: Changing the configuration of components of the arrangement in what appear to be innocent ways can make *drastic* differences to the resultant potential at certain locations. The reader will find many other eye-opening examples in QED and FLP I-26. We will find in Section 11 that two atoms in a "quantum handshake" form a pattern of paths that greatly increases the potential by which the atoms are coupled, and hence can shorten the transition beyond what is possible with just the optical system.

All the results in statistical QM are probabilities because Heisenberg denied that there was any physics in the transactions. That denial has left the field in a conceptual mess. There is no doubt that statistical QM makes it easy to calculate probabilities of a wide variety of experimental outcomes, and that these predicted outcomes overwhelmingly agree with reality. However that discipline is, by design, powerless to provide reasoning for *how* those outcomes come about. The object of this paper is to *understand the individual transaction, not to calculate probabilities.* Thus, the times quoted above are the times required for the individual event, once initiated, *not* the time constant of some statistical distribution. **We deal with a realm of which statistical QM denies the existence.**

11. Global Field Configuration

We are now in a position to visualize the field configuration for the quantum exchange of energy between two atoms, as analyzed in Section 9, using the locations and coordinated defined there. From Equations (37) and (35), and using $\sin(\phi) = -1$, the total field is composed of the sum of the retarded solution A_α, at distance r_α from atom α and the advanced solution A_β, at distance r_α from atom β:

$$A_\alpha \propto \frac{1}{r_\alpha}\frac{\partial\langle z_\alpha\rangle}{\partial t} = -\frac{1}{r_\alpha}\sin\left(\omega_0(t - r_\alpha/c)\right)$$
$$A_\beta \propto \frac{1}{r_\beta}\frac{\partial\langle z_\beta\rangle}{\partial t} \propto \frac{1}{r_\beta}\cos\left(\omega_0(t + \Delta t + r_\beta/c)\right) \qquad (62)$$

Including both x and y coordinates in the distances r_α and r_β from the two atoms, the vector potentials from the two atoms anywhere in the $x - y$ plane are

$$A_\alpha(x,y,t) \propto -\frac{1/\tau}{\sqrt{x^2 + y^2}}\sin\left(\omega_0\left(t - \frac{\sqrt{x^2 + y^2)}}{c}\right)\right)$$
$$A_\beta(x,y,t) \propto \frac{1/\tau}{\sqrt{(x - \Delta x)^2 + y^2}}\cos\left(\omega_0\left((t + \Delta t) + \frac{\sqrt{(x - \Delta x)^2 + y^2)}}{c}\right)\right) \qquad (63)$$

An example of the total vector potential $A_{\text{tot}} = A_\alpha + A_\beta$ along the x-axis is shown in Figure 10.

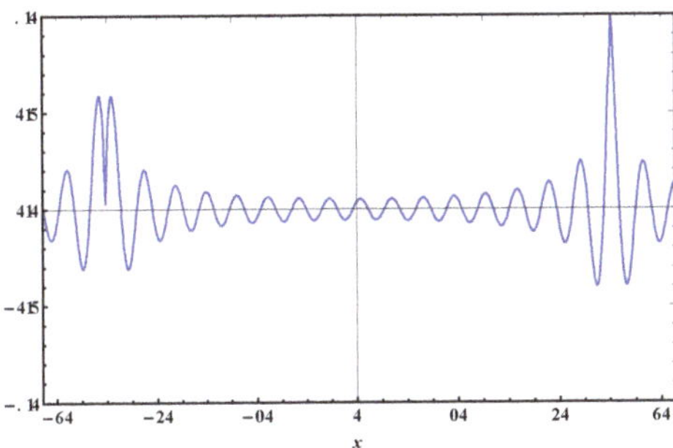

Figure 10. Normalized vector potential along the x-axis (in wavelength$/2\pi$) between two atoms in the "quantum handshake" of Equation (63). The wave propagates smoothly from atom α (left) to atom β (right). Animation here [37]. (see Supplementary Materials).

A "snapshot" of the potential of Equation (63) at one particular time for the full $x - y$ plane is shown in Figure 11.

Figure 11. Two atoms in the "quantum handshake" of Equation (63). Animation here [37]. (see Supplementary Materials).

The still image in this figure looks like a typical interference pattern from two sources—a "standing wave." There are high-amplitude regions of constructive interference which appear light blue and yellow on this plot. These are separated from each other by low-amplitude regions of destructive interference, which appear green. In a standing wave, these maxima would oscillate at the transition frequency, with no net motion. The animation, however, shows a totally different story: Instead of oscillating in place as they would in a standing wave, *the maxima of the pattern are moving steadily from the source atom (left) to the receiving atom (right)*. This movement is true, not only of the maxima between the two atoms, but of maxima well above and below the line between the two atoms. These maxima can be thought of as Feynman's paths, each carrying energy along its trajectory from atom α to atom β. For those readers that do not have access to the animations, the same story is illustrated by a stream-plot of the Poynting vector in the x-y plane, shown in Figure 12:

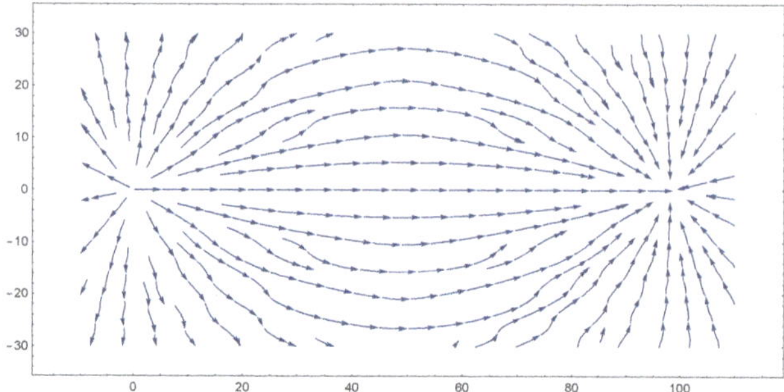

Figure 12. Poynting vector stream lines of the "quantum handshake" of Equation (63).

We can get a more precise idea of the phase relations by looking at the zero crossings of the potential at one particular time, as shown in Figure 13. Paths from atom 1 to atom 2 can be traced through either the high-amplitude regions or the low-amplitude regions. The paths shown in Figure 13 are traced through high-amplitude regions.

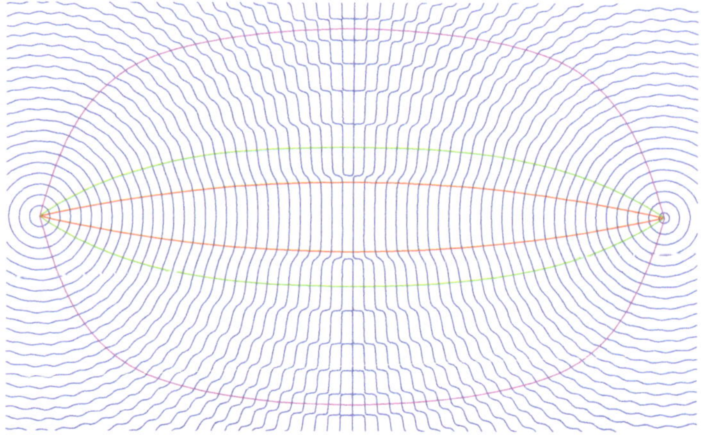

Figure 13. The zero crossings of the handshake vector potential at $t = 0$. Paths near the axis (between the two red lines) are responsible for the conventional $1/r$ dependence of the potential. Paths shown through high-amplitude regions have an even number of zero crossings, and thus the potentials traversing these paths all arrive in phase, thus adding to the central potential.

The central set of paths, delimited by the red lines, are responsible for the conventional $1/r$ dependence of the potential, as described with respect to Equation (59). Working outward from there, each high-amplitude path region is separated from the next by a slim low-amplitude region. It is a remarkable property of this interference pattern that each low-amplitude path has π more phase accumulation along it than the prior high-amplitude path and π less than the next high-amplitude path. The low-amplitude paths are the ones that contribute to the "de-phasing" in this arrangement, but they are very slim and of low amplitude, so they don't de-phase the total signal appreciably. In addition, the phases of the paths through the high-amplitude regions are separated by $2\pi n$, where n is an integer. **All waves propagating from atom α to atom β along high-amplitude paths arrive in phase!**

In Feynman's example shown in Figure 8, there are an equal number of paths of any phase, so every one has an opposite to cancel it out. In Figure 9, the lens makes all paths have *equal time delay*, which then enables them to all arrive with the same phase.

The phase coherence of the advanced-retarded handshake creates a pattern of potentials that has a unique property: It is not like either of Feynman's examples in Figure 8 or Figure 9. Its high-amplitude paths do arrive *in phase*, but by a completely different mechanism. It all starts with the bundle of paths between the red lines, which has the $1/r$ amplitude, just as if there were no quantum mechanism. Then, as the handshake begins to form, additional paths are drawn into the process. The process is self-reinforcing on two levels—the increase in dipole moment and the increase in number of paths that arrive in phase. Paths that formerly would have cancelled the in-phase ones are "squeezed" into extremely narrow regions, all of low amplitude, as can be seen in Figure 13. Thus, a large fraction of the solid angle around the atoms is available for the in-phase high-amplitude paths.

For optical systems with large solid angle, the self-focusing enhancement may still be noticeable in the shape of the transition waveform, but for one-sided systems, like an astronomical telescope, we expect it to be be dominant.

Following Feynman's program has led us to conclude that: The vector potential from all paths sum to make a highly-amplified connection between distant atoms. The advanced-retarded potentials form nature's very own phase-locked loop, which forms nature's own *Giant Lens in the Sky!*

The consequences of this fact are staggering: Once an initial handshake interference pattern is formed between two atoms that have their wave functions synchronized, the strength grows explosively: Not only because the dipole moment of each atom grows exponentially, but, in addition, a substantial fraction of the possible interaction paths between the two atoms propagate through high-amplitude regions, *independent of the distance between them!* Although we have not worked out the difficult second-order dynamics of phase-locking between coupled atoms, we believe that here is the solution to the long-standing mystery of the "collapse of the wave function" of the "photon".

The interaction depends critically on the advanced-retarded potential handshake to keep all paths in phase. Ordinary propagation over very long paths becomes "de-phased" due to the slightest variations of the propagation properties of the medium. By contrast, the advanced and retarted fields are precise negative images of each other on exactly the same light cone, so the phase of high-amplitude handshake paths are always related by an even number of π to the phase of other such paths. Paths having odd numbers of π phase are always of low amplitude, and do not cancel the even-π phase paths as they would in a one-way propagating wave like Feynman used in his illustrations. [A detailed analysis of these properties has not been done. It is a wonderful project for the future.]

The interaction proceeds in the local time frame of each atom because they are linked with the advanced-retarded potential. The waves carrying positive energy from emitter to absorber are retarded waves with positive transit time; they reach the absorber after a single transit time $\Delta t = r/c$. Once they have established a phase-coherent "handshake" connection, Lewis' *"coordinate and symmetrical"* advanced waves with negative transit time are launched toward the emitter, arriving at the precise time and in the precise phase to withdraw energy from the emitter. During the transaction, as long as the "handshake" connection is active, any change in the state of one atom will be directly reflected in the state of the other. Aside from the time-of-flight propagation time to establish the "handshake", there is no additional "round trip" time delay in the quantum-jump process, which proceeds as if the two atoms were local to each other.

Thus, the Transactional Interpretation allows us to conceptualize Niels Bohr's "instantaneous" quantum jump [47] concept that Schrödinger, who expected time-extended classical transitions, found impossible to accept [48].

12. Relevance to the Transactional Interpretation

The calculation that we have presented here, with its even-handed treatment of advanced and retarded four-potentials, was inspired by WFE and the Transactional Interpretation, but it also provides interesting insights that clarify and modify the mechanism by which a transaction forms. Wheeler–Feynman electrodynamics suggests that a retarded wave, arriving at a potential absorber, stimulates that entity to generate a canceling retarded wave accompanied by an advanced wave. The TI suggests that this advanced wave arrives back at the emitter with amplitude $\psi\psi*$, a relation suggesting the Born rule.

However, from our calculation, we see that a slightly different process is described. Both emitter and absorber initially have small admixtures of the opposite eigenstate, giving them dipole moments that oscillate with the *same* difference-frequency ω_0. The two oscillations each have an environment-induced random phase. If the phases have the correct relation [$\sin(\phi) \approx -1$], the dipole moments of both atoms initially increase exponentially, the system becomes a phase-locked loop, and it avalanches to a final state of multi-path energy transfer that satisfies all boundary conditions. In that scenario, the initial confirmation wave is likely to have an amplitude much weaker than WFE would suggest, and the quantity $\psi\psi*$ becomes that used by Schrödinger to provide the electron density function.

How can a linear system generate such nonlinear behavior? While the Schrödinger equation is indeed linear, Equation (43) governing the evolution of transaction formation and wave function collapse is highly nonlinear. Thus, the TI's assertion that the offer wave "stimulates the generation of the confirmation wave" must be modified. Rather, advanced and retarded potentials, boundary conditions at both ends, and a fortuitous matching of phase trigger the nonlinear avalanche in both atoms and brings about the transaction.

We also note that the advanced and retarded waves do not carry "information" in the usual sense in either time direction, but only deliver a pair of oscillating four-potentials to the sites of a pair of oscillating charges, leading dynamically to an initially exponential rise in coupling, a focusing of alternative paths, the formation of a transaction, a transfer of energy, and the enforcement of conservation laws. For the Transactional Interpretation, this phase selection process clarifies the randomizing mechanism by which, in the first stage of transaction formation, the emitter makes a random choice between competing offer waves arriving from many potential absorbers. The offer wave with the best phase is likely to win, even if it comes from far away. [It is sometimes asserted that this handshake situation is only possible in a frozen deterministic four-dimensional "block universe" because of the two-way connections between present and future. We reject this assertion, which is dissected in some detail in Section 9.2 of Ref. [12]. While it is true that the assumption of a block universe would dispel or bypass many of the quantum paradoxes, it would only do so at the terrible price of imposing complete determinism on the universe.].

We saw in the derivation of the coupling of two separated atoms that it was necessary to use the advanced 4-potential *in order to satisfy the law of conservation of energy*. This, not "information transfer", is the role of the quantum handshake in the Transactional Interpretation. The quantum handshakes act to enforce conservation laws and do not form unless all conserved quantities are properly transferred and conserved. This is what is going on in *quantum entanglement*: the separated parts of a quantum system are linked by conservation laws that are enforced by V-shaped three-vertex advanced-retarded quantum handshakes [12] and cannot emerge as a completed transaction unless those conservation laws are satisfied. In this context, we note that the Transactional Interpretation, using such linked advanced-retarded handshakes, is able to explain in detail the behavior of over 26 otherwise paradoxical and mysterious quantum optics experiments and *gedankenexperiments* from the literature. See Chapter 6 of Ref. [12]. If we cannot dismiss the plethora of competing QM interpretations based on their failure in experimental tests, *we should eliminate them when they fail to explain paradoxical quantum optics experiments* (as almost all of them do.)

13. Historic Tests

13.1. The Hanbury–Brown–Twiss Effect and Waves vs. Particles

It is often said that particles in quantum mechanics "travel as waves but arrive as particles". The Hanbury–Brown–Twiss effect [49] (HBT) is an example of this principle. It demonstrates that, in the second-order interference of incoherent wave sources, photons are divisible and are not electromagnetic "billiard balls" that maintain individual identities. The HBT technique was first applied to the measurement of the diameters of nearby stars, e.g., Betelguese, using intensity interferometry with radio waves. The original experiments involved large parabolic radio dishes mounted on rail cars.

A simplified version of an HBT interference measurement is illustrated in Figure 14. Sources 1 and 2 are separated by a distance d_{12}. Both sources emit waves of the same wavelength $\lambda = 2\pi c/\omega$, but are not causally connected. Radiation from the two sources is received by detectors A and B, which are separated by a distance d_{AB}. The line between the sources is parallel to the line between the detectors, and the two lines are separated by a distance L. Hanbury-Brown and Twiss showed that a significant rate of coincident detections—up to a factor of 2 larger than could be ascribed to chance—was observed in this arrangement. For this simple configuration, the probability of a coincident detection in the two detectors, for small d_{AB} and very large L, is proportional to $1 + \cos[2\pi\, d_{AB}\, d_{12}/(\lambda L)]$, which is maximum when $d_{AB} = 0$ and falls off as the detector separation increases, the rate of falloff indicating the value of d_{12}.

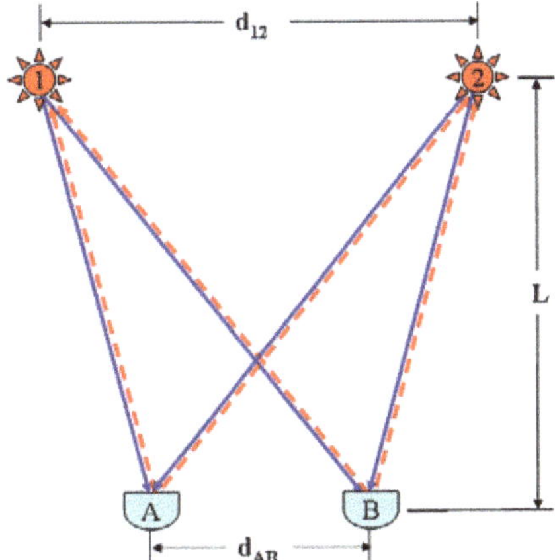

Figure 14. Schematic diagram of the Hanbury–Brown–Twiss effect, with excited atoms 1 and 2 in distant separated sources simultaneously exciting ground state atoms A and B in two separated detectors.

The very fact that coincidences are observed in this experiment reveals a deep truth about electromagnetic coupling: Photons cannot be consistently described as little blobs of mass-energy that travel uniquely from a single source point to a single detector point. In the HBT effect, a whole photon's worth of energy $\hbar\omega$ is assembled at each detector out of fractional energy contributions from each of the two sources.

In the TI description of the HBT event described above, a retarded offer wave is emitted by the source 1 and travels to both detectors A and B. Similarly, a retarded offer wave is emitted by the source 2 and travels to both detectors. Detector A receives a linear superposition of the two offer waves and seeks to absorb the "offered" energy by producing an advanced confirmation wave. If the phases match, as they will in a coincident event, the energy transfer begins with an exponential increase in the dipole moment of each source atom. Atoms in detectors A and B respond similarly, their oscillating dipole moments producing advanced confirmation waves that travel back to the two sources, each of which responds with an increasing dipole moment that enhances the offer waves. A four-atom transaction of the form shown in Figure 14 is formed that removes one photon's worth of energy $\hbar\omega$ from each of the two sources 1 and 2 and delivers one photon's worth of energy $\hbar\omega$ to each of the two detectors A and B.

Neither of the detected "photons" *can be said to have originated uniquely in one of the two sources.* The energy arriving at each detector originated partly in one source and partly in the other. It might be said that each source produced two "fractional photons" and that these fractions from two sources combined at the detector to make a full size "photon". The "particles" transferred have no separate identity that is independent from the satisfaction of the quantum mechanical boundary conditions. The boundary conditions here are those imposed by the HBT geometry, conservation laws, and the detection criteria.

Finally, we note that in many experiments published in the physics literature the HBT effect has been observed and demonstrated not only for photons, but also in the detection of charged π mesons emitted in the ultra-relativistic collision of heavy nuclei. It is observed that the detection probability doubles when the detected particles are close in momentum and position. Furthermore, in nuclear physics experiments with pairs of half-integer-spin neutrons or protons, a Pauli-Exclusion version of the HBT effect has been observed, in which the detection probability drops to zero when the detected particles are close in momentum and position [50]. All of these particle-like entities "travel as waves but arrive as particles".

13.2. Splitting Photons

At the 5th Solvay Conference in 1927, Albert Einstein posed a riddle, sometimes called "Einstein's Bubble Paradox", to the assembled founders of quantum mechanics [12,51]. Einstein's original language was rather convoluted and technical (and in German), but his question can be simply stated as follows:

> *A source emits a single photon isotropically, so that there is no preferred emission direction. According to the quantum formalism, this should produce a spherical wave function Ψ that expands like an inflating bubble centered on the source. At some later time, the photon is detected. Since the photon does not propagate further, its wave function bubble should "pop", disappearing instantaneously from all locations except the position of the detector. In this situation, how do the parts of the wave function away from the detector "know" that they should disappear, and how is it enforced that only a single photon is always detected when only one photon is emitted?*

The implication of his question is that, if a photon is an indivisible particle, it should not be possible to divide one. However, we have already seen in the Hanbury–Brown–Twiss effect that photons are *not* little indivisible billiard balls and that they *can* divide their energy between two receiving atoms. How, then, is it possible that for one photon emitted there is always only one received?

A search for this hypothetical divided-photon behavior is implemented in the setup shown in the left panel in Figure 15, which was enabled once it became possible to build sources of single photons. The idea is that, if a photon is just a short pulse of light, half of it should go through each of the dotted paths, and both halves should be counted at the same time, registering as a coincidence. Of course, the original pulse must have twice the energy required to trigger a detector, so either half by itself would have just enough. However, if the photon was indivisible, as mandated by certain versions of QM, it would make a random decision on which path to take, and no coincidences would be observed. In practice, the number of coincidences is counted for a certain counting period, with the time delay τ between the two detector output pulses as a parameter. Since the photons are generated randomly, the time between successive photons can accidentally range from zero to large, and a plot of the number of correlator outputs vs. time delay τ gives information about the statistics of the source.

Modern versions of the experiment [52] give plots like that shown in the right panel in Figure 15. [Early versions of this kind of experiment suffered from certain defects that made them inconclusive. A definitive version using "heralded photons" was finally accomplished by Clauser in 1973 [53].] Let us look at the experiment from a TI perspective: The source excites one atom, which, due to random coupling, develops a superposition with a tiny presence of ground state and nearly unity excited state. As described for the two-atom photon, the tiny presence in the superposition enables the dipole moment to oscillate, thereby generating a radiating vector potential that propagates along both dotted paths to both detectors. To find a perfectly matched partner atom is rare, but, when one matches up, a quantum handshake grows up connecting them.

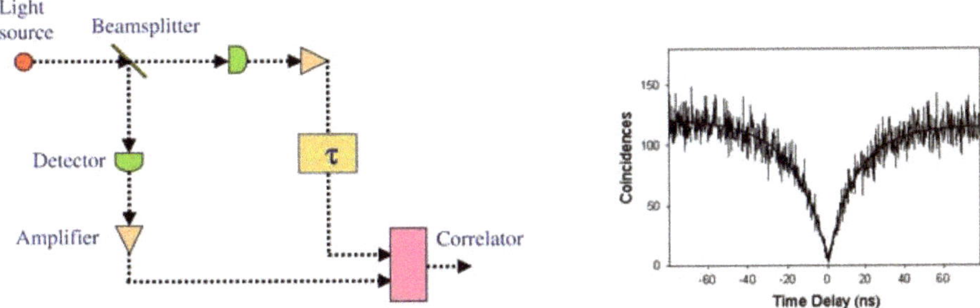

Figure 15. **Left**: Schematic of the photon-splitting experiment. **Right**: Plot of the number of coincidences vs. time delay between the arriving pulses. The source on average generates a photon per ≈ 12 nsec. The finite number counted at zero delay is consistent with the accidental presence of two excited atoms.

There are then three possible outcomes:

1. The handshake goes to completion and the partner atom is in one of the detectors, in which case only the chosen detector registers an output.
2. The handshake goes to completion, but the partner atom is not in one of the detectors. In this case, no output is registered from either detector.
3. The initial stages of a handshake begins in *two* partner atoms, one in each detector. When the source atom has de-excited, both of the detector atoms are left in mixed states with roughly equal components of ground-state and excited state wave functions, as was illustrated in Figure 7. This is not a stable configuration because both of the detector atoms have oscillating dipole moments sending strong "unrequited" advanced confirmation waves. These waves are in phase at and focused on the source, and they are likely to find another well-phased excited-state atom there or nearby that will complete the four-way transaction. Thus, a four-atom HBT event should be created, in which there are two emissions and two detections. Such an unlikely event would register as an "accidental" case of two simultaneous emissions in the same time window. There will never be an event with a single emission and two detections.

Thus, we see that the outcomes of the experiment predicted by TI and QM are essentially identical. Certainly, no solid conclusion can be drawn from this experiment as to whether quantization occurs in the field or in the transaction.

13.3. Freedman–Clauser Experiment

In our introductory discussion of Schrödinger's visualization of his newly-invented Wave Mechanics in Section 3, we described how Clauser and his colleagues, through experiments that were heroic at the time, were able to show that no "local, realistic theory" was compatible with their results [6–11]. We now describe the earliest conclusive version of these EPR experiments and show how our TI approach gives a simple and natural explanation of the otherwise mysterious outcome. A sketch of the arrangement is shown in Figure 16.

The atomic configuration used was the three-level system of the Ca atom, shown at the right of the figure. The atomic wave functions were an upper $4p^2\,S_0$ state Ψ_3 of frequency ω_3, a middle $4p4s\,P_1$ state Ψ_2 of frequency ω_2, and a $4s^2\,S_0$ ground state Ψ_1 of frequency ω_1, so that the cascade starts and ends in an S_0 state of zero angular momentum.

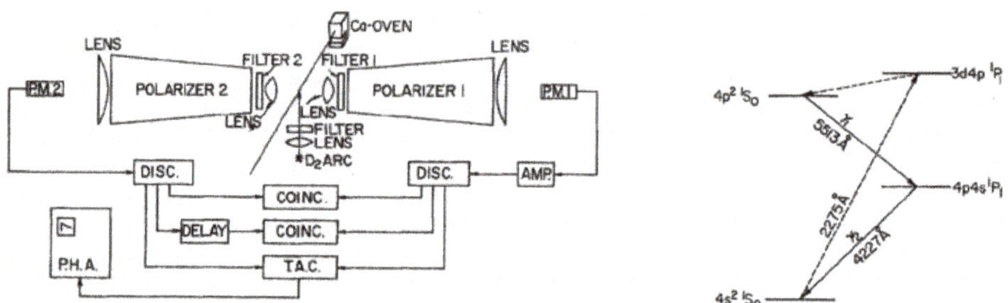

Figure 16. Left: Schematic of the polarization-correlation experiment. Right: Energy levels of the Ca atoms used in this experiment. From Freedman and Clauser 1972 [6].

The state wave functions are of the same form as those given in Equation (2):

$$\Psi_3 = R_3(r)\,e^{-iw_3 t} \qquad \Psi_2 = f_2(r)\cos(\theta)\,e^{-iw_2 t} = R_2(r,\theta)\,e^{-iw_2 t} \qquad \Psi_1 = R_1(r)\,e^{-iw_1 t} \tag{64}$$

The transition state is a superposition of these three states: The analysis is a simple extension of that for the two-level system, starting with Equation (10):

$$\Psi = a e^{i\phi_a} R_1 e^{-iw_1 t} + b e^{i\phi_b} R_2 e^{-iw_2 t} + c e^{i\phi_c} R_3 e^{-iw_3 t}, \tag{65}$$

The charge density of the mixed state is then:

$$\rho = q\Psi^* \Psi$$

$$\frac{\rho}{q} = \left(a e^{-i\phi_a} R_1 e^{iw_1 t} + b e^{-i\phi_b} R_2 e^{iw_2 t} + c e^{-i\phi_c} R_3 e^{iw_3 t} \right) \left(a e^{i\phi_a} R_1 e^{-iw_1 t} + b e^{i\phi_b} R_2 e^{-iw_2 t} + c e^{i\phi_c} R_3 e^{-iw_3 t} \right)$$

$$= a^2 R_1^2 + b^2 R_2^2 + c^2 R_3^2$$

$$+ 2ab R_1 R_2 \cos\left((w_2 - w_1)t + (\phi_b - \phi_a)\right) \tag{66}$$

$$+ 2ac R_1 R_3 \cos\left((w_3 - w_1)t + (\phi_a - \phi_c)\right)$$

$$+ 2bc R_2 R_3 \cos\left((w_3 - w_2)t + (\phi_b - \phi_c)\right)$$

The dipole srength d_{ij} for the three terms is given by straightforward extension of Equation (18)

$$d_{12} = 2q \int R_1 R_2 z \qquad d_{23} = 2q \int R_3 R_2 z \qquad d_{13} = 2q \int R_1 R_3 z = 0 \tag{67}$$

By symmetry around the ($+z$ axis) of the spherical coordinate system, both d_{12} and d_{23} are in the $\bar{z}$ direction determined by the direction of the Ψ_2 wave function. In general, that direction will shift around in space depending on the coupling of the atom to others. However, in any given situation, there is only one z-axis that defines the "North pole" direction, and both dipole moments are oscillating in that direction. Thus, it is the Ψ_2 state, shared by both transactions of the 3-state "cascade" that aligns the linear polarizations of the two interlocking transactions.

13.3.1. Dynamics of the Transaction

The cascade atom is emitted from the oven shown at the top of the figure. It is is initially "pumped" by the D_2 arc when it is centered between the two lenses. The excitation is along the 2275Å dashed path on the energy diagram, from which it relaxes into the $4p^2 \, S_0$ excited state where $a \approx 0$, $b \approx 0$, $c \approx 1$, leaving a small residual admixture of the middle and ground states. By Equation (66) and the obvious generalization of Equation (19), the superposition begins to oscillate with dipole moment:

$$\text{dipole moment} = q \langle z \rangle = d_{12}ab \, \cos(\omega_{12}t + \phi_{ab}) + d_{23}bc \, \cos(\omega_{23}t + \phi_{bc}) \tag{68}$$

where we have included only non-zero time-varying terms terms and have used:

$$\omega_{12} = (\omega_2 - \omega_1) \qquad \omega_{23} = (\omega_3 - \omega_2) \qquad \phi_{ab} = (\phi_b - \phi_a) \qquad \phi_{bc} = (\phi_c - \phi_b) \tag{69}$$

Because $\omega_{12} \neq \omega_{23}$, the two terms in the dipole moment do not interact over periods of many cycles, and each term can couple to a separate atom to form its own quantum handshake. The analysis has already been done starting with Equation (35) and ending with Equation (41), and applies directly to the upper (ω_{23}) transaction with atom we will call α, which has level spacing exactly equal to E_{23}.

Since c and a_α both start almost equal to 1 and have the same derivative, we can set $c = a_\alpha$.

$$\frac{\partial E_\alpha}{\partial t} = -E_{23}\frac{\partial c^2}{\partial t} = -E_{23}\frac{\partial a_\alpha^2}{\partial t} = -P_\alpha \, c \, b \, a_\alpha b_\alpha \sin(\phi_\alpha) = -P_\alpha \, c^2 \, b \, \sqrt{1 - c^2} \, \sin(\phi_\alpha) \tag{70}$$

Similarly, we describe the fraction a^2 of the superposition in the ground state due to the lower 5227Å (ω_{12}) transaction with atom we will call β, which has level spacing exactly equal to E_{12}.

Since a and b_β both start almost equal to 0 and have the same derivative, we can set $a = b_\beta$.

$$\frac{\partial E_\beta}{\partial t} = E_{12}\frac{\partial a^2}{\partial t} = E_{12}\frac{\partial b_\beta^2}{\partial t} = P_\beta \, a \, b \, a_\beta b_\beta \sin(\phi_\beta) = P_\beta \, a^2 \, b \, \sqrt{1 - a^2} \, \sin(\phi_\beta) \tag{71}$$

These relations may then be expressed more compactly as:

$$\frac{\partial c^2}{\partial t} = -\frac{1}{\tau_\alpha} c^2 \, b \, \sqrt{1 - c^2} \qquad \frac{\partial a^2}{\partial t} = \frac{1}{\tau_\beta} a^2 \, b \, \sqrt{1 - a^2} \qquad a^2 + b^2 + c^2 = 1 \tag{72}$$

where $1/\tau_\beta = P_\beta \sin(\phi_\beta)/E_{12}$ and $1/\tau_\alpha = P_\alpha \sin(\phi_\alpha)/E_{23}$ express the respective strength of coupling to each atom, and the last relation constrains the superposition to contain exactly one electron. Equation (72) is identical in form to Equation (43) with the exception of the shared state amplitude b occurring in both derivatives.

The behavior of this arrangement, shown in Figure 17, is very instructive. To review in brief: The cascade atom is prepared by providing the energy to promote the electron to the upper state: $c \approx 1$, $a \approx 0$, $b \approx 0$. The preparation is never perfect, so there is always a small residual of b and a components in the initial superposition. The small admixture of b and c, by Equation (68), creates an oscillating dipole moment at frequency ω_{23}, the amplitude of which is shown as the red curve in the right panel of Figure 17. That oscillating dipole moment initiates a vector potential "offer wave" which propagates outward according to Equation (37). When that vector potential couples to the wave function of another atom of the same frequency, its wave function oscillates with the vector potential in the correct phase to withdraw energy according to Equation (50). The vector potential from that oscillation propagates *backwards in time*, so it is "felt" by the cascading atom as if it had been there all along.

The Transactional Interpretation is applied to the Freedman–Clauser experiment as shown in Figure 18. Two-way handshakes between the source and the two polarimeters are joined at the source and must satisfy the boundary condition, based on conservation of angular momentum in a system that begins and ends in a state of zero angular momentum that the polarizations must match for the two offer waves emitted back-to-back. This V-shaped transaction holds the key to understanding the mechanism behind quantum nonlocality in EPR expertiments.

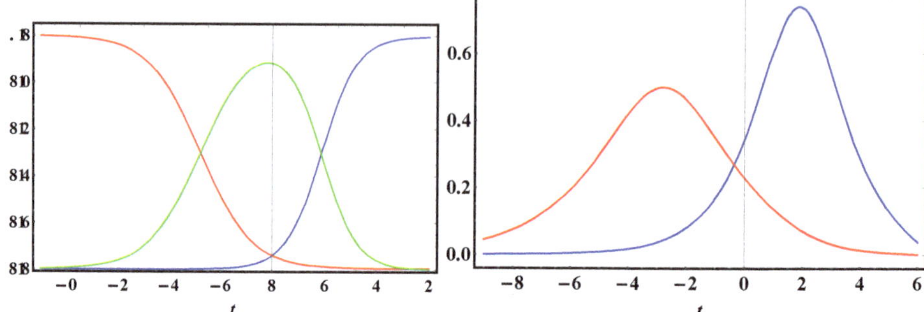

Figure 17. **Left**: Superposition contributions of the upper state c^2 (red), shared middle level b^2 (green) and ground state a^2 (blue). **Right**: Amplitude of dipole oscillations due to upper transition at ω_{23} (red) and lower transition at ω_{12} (blue), from. The horizontal axis in both plots is time in units of $\tau_\alpha = 1.5\,\tau_\beta$.

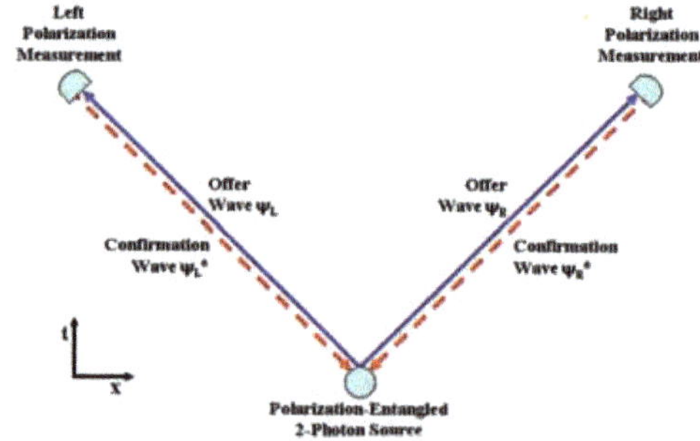

Figure 18. Three-vertex transaction formed by a detection event in the Freedman–Clauser experiment. Linked transactions between the source and the two polarimeters cannot form unless the source boundary condition of matching polarization states is met.

It was pointed out by one of our reviewers that the kind of description just given might appeal to readers who are not accustomed to standard relativistic space-time diagrams, on which events on the "light cone" are local in the sense that $r^2 - c^2 t^2 = 0$, and that we had, in some measure, ignored G.N. Lewis' chiding quoted earlier "A dissymmetry alien to the pure geometry of relativity has been introduced by our notion of causality." Thus, let's try it again:

A quantum handshake is an antisymmetric bidirectional electromagnetic connection between two atoms on a light cone, whose direction of time is the direction of positive energy transfer.

Either way we look at it, the cascade atom and atom α are locked in phase and amplitude at frequency ω_{23}, and the locked amplitude of oscillation of the wave functions of the two atoms is growing with time. In the process, the z-axes of both atoms becomes better and better aligned. Meanwhile, the small superposition amplitudes b and a are growing, thus developing a growing oscillation at ω_{12}, shown by the blue curve in the right panel of Figure 17. Since both the upper and lower levels are S states, they have no effect on the orientation of the oscillation, which is determined by the shared middle level, which is a P state that has a definite direction in space. That direction determines the direction of oscillation of any superposition involving that P state. The vector potential from the nascent ω_{12} oscillation recruits a willing atom β whose level spacing is precisely match to ω_{12}, and forms an embryonic quantum handshake of its own whose z-axis is already determined by the fully developed ω_{23} oscillation. From there, both transactions were completed, with z axes aligned, in very much the same way we have described for a single photon.

13.3.2. The NCT-Killer Result

The unique aspect of the Freedman–Clauser experiment was separating the back-to-back paths of the two wave propagations with filters that selected the ω_{23} path to the left and the ω_{12} path to the right. Each path was equipped with its own glass-plate-stack linear polarizer, whose angle could be adjusted. The number of coincidences of signals from the left and right photomultipliers P.M.1 and P.M.2 were plotted as a function of the angle ϕ between the polarization axes of the polarizers, and is shown in Figure 19.

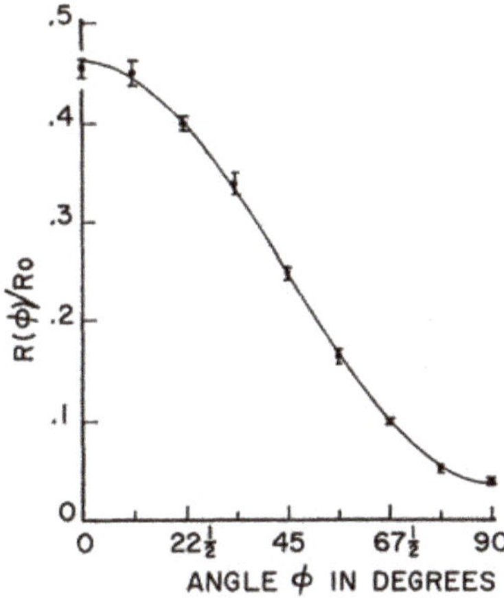

Figure 19. Coincidence rate vs. angle ϕ between the polarizers, divided by the rate with both polarizers removed. The solid line is the prediction by quantum mechanics, calculated using the measured efficiencies of the polarizers and solid angles of the experiment.

The key aspect of Figure 19 is that at $\phi = 90°$ the coincidence rate drops to essentially zero. The graph does show a small non-zero coincidence rate at 90°, but this is because of the imperfect linear polarization discriminations of the glass-plate-stack polarizing filters and the finite solid angles of the detection paths. Jaynes' NCT approach failed to reproduce [32] this result. That failure, in the view of most of the field, "falsified" the NCT approach to quantum phenomena and caused it to be subsequently ignored.

We can understand this result very simply from our TI perspective, reasoning directly from Figure 17: By the time the blue ω_{12} transition is just getting started, the red ω_{23} transition is well along, and has connected to a partner in P.M.2 through polarizer 2. Since the polarizer only transmits a vector potential aligned with its axis of polarization, the z axes of both the cascade atom and atom α will be well aligned with polarizer 2. Now, as the blue oscillation just begins to build, its axis, as part of the same P wave function, will also be aligned with polarizer 2, at rotation angle θ_2.

The way that these glass-plate-stack polarizers work is that they only pass the component of propagating vector potential along their axis of polarization, the orthogonal component being reflected out of the direction of propagation. Thus, the fraction of θ_2 polarized vector potential that can pass through a perfect θ_1 oriented polarizer 1 is just $\cos(\theta_2 - \theta_1)$. Thus, the amplitude of the nascent ω_{12}, θ_2 polarized "offer wave" propagating outward from the cascade atom through polarizer 1 will be proportional to $\cos(\theta_2 - \theta_1)$. By Feynman's *Grand Principle*, the probability amplitude of an "offer wave" actually forming a transaction is proportional to the amplitude of the vector potential. Since, in this example, the ω_{23} transaction in P.M.2 has already formed, the slightly later formation of an ω_{12} transaction in P.M.1 will count as a coincidence. The probability of coincidence counts will therefore be proportional to the square of the probably amplitude which, for this case, is $\cos^2(\theta_2 - \theta_1)$ which, when corrected as noted, gives the solid line in Figure 19 and is zero when $\phi = 90°$. Once again, the bi-directional non-local nature of the quantum handshake predicts the observed outcome of this historic experiment.

14. Conclusions

The development of our understanding of quantum systems began with a physical insight of deBroglie: Momentum was the wave vector of a propagating wave of some sort. Schrödinger is well known for developing a sophisticated mathematical structure around that central idea, only a shadow of which remains in current practice. What is less well known is that Schrödinger also developed a deep understanding of the *physical meaning* of the mathematical quantities in his formalism. That physical understanding enabled him to *see* the mechanism responsible for the otherwise mysterious quantum behavior. Meanwhile, Heisenberg, dismissing visual pictures of quantum processes, had developed a matrix formulation that dealt with only the probabilities of transitions—what he called "measurables." It looked for a while as if we had two competing quantum theories, until Schrödinger and Dirac showed that they gave the same answers. However, the stark contrast between the two approaches was highlighted by the ongoing disagreement in which Bohr and Heisenberg maintained that the transitions were events with no internal structure, and therefore there was nothing left to be understood, while Einstein and Schrödinger believed that the statistical formulation was only a stopgap and that a deeper understanding was possible and was urgently needed. This argument still rages on.

Popular accounts of these ongoing arguments, unfortunately, usually focus on the 1930 Solvay Conference confrontation between Bohr and Einstein that was centered around Einstein's clock paradox, a clever attempted refutation of the uncertainty principle [51]. Einstein is generally considered to have lost to Bohr because he was "stuck in classical thinking." However, as detailed in **The Quantum Handshake** [12], Einstein's effort was doomed from the start and was also beside the point. The uncertainty principle is simply a Fourier-algebra property of *any system described by waves.* Both parties to the Solvay argument lacked any real clarity as to how to handle the intrinsic wave nature of matter. In the introduction, we quoted Einstein's deepest concern with statistical QM:

There must be a deeper structure to the quantum transition.

Back in 1926, the field was faced with a choice: Schrödinger's wave function in three-dimensional space, or Heisenberg and Born's matrices, in as many dimensions as you like. The choice was put forth clearly by Hendrik Lorentz [54] in a letter to Schrödinger in May, 1926:

"If I had to choose now between your wave mechanics and the matrix mechanics, I would give the preference to the former because of its greater intuitive clarity, so long as one only has to deal with the three coordinates x,y,z. If, however, there are more degrees of freedom, then I cannot interpret the waves and vibrations physically, and I must therefore decide in favor of matrix mechanics. However, your way of thinking has the advantage for this case too that it brings us closer to the real solution of the equations; the eigenvalue problem is the same in principle for a higher dimensional q-space as it is for a three-dimensional space.

"There is another point in addition where your methods seem to me to be superior. Experiment acquaints us with situations in which an atom persists in one of its stationary states for a certain time, and we often have to deal with quite definite transitions from one such state to another. Therefore, we need to be able to represent these stationary states, every individual one of them, and to investigate them theoretically. Now a matrix is the summary of all possible transitions and it cannot at all be analyzed into pieces. In your theory, on the other hand, in each of the states corresponding to the various eigenvalues, E plays its own role."

Thus, the real choice was between the intuitive clarity of Schrödinger's wave function and the ability of Heisenberg–Born matrix mechanics to handle more degrees of freedom. That ability was immediately put to the test when Heisenberg [55] worked out the energy levels of the helium atom, in which two electrons shared the same orbital state and their correlations could not be captured by wave functions with only three spatial degrees of freedom. That amazing success set the field on the path of eschewing Schrödinger's views and moving into multi-dimensional Hilbert space, which was further ossified by Dirac and von Neumann. Schrödinger's equation had been demoted to a bare matrix equation, engendering none of the *intuitive clarity*, the ability to *interpret the waves and vibrations physically* so treasured by Lorentz.

The matrix formulation of statistical QM, as now universally taught in physics classes, saves us the "tedious" process of analyzing the details of the transaction process. That's the good news. The bad news is that *it actively prevents us from learning anything about the transaction process, even if we want to!*

What has been left out is, as Einstein said, any *"description of the individual system"*.

Thus, it was left to the more practical-minded electrical engineers and applied scientists to resurrect, each in their own way, Schrödinger's way of thinking because they needed a *"description of the individual system"* to make progress. Electrons in conductors were paired into standing waves, which could carry current when the propagation vector of one partner was increased and that of the other partner decreased. Energy gaps resulted from the interaction of the electron wave functions with the periodic crystal lattice. Those same electron wave functions can "tunnel" through an energy gap in which they decay exponentially with distance. The electromagnetic interaction of the collective wave functions in superconducting wires leads to a new formulation of the laws of electromagnetism without the need for Maxwell's equations [16]. The field of Quantum Optics was born. Conservation of momentum became the matching of wavelengths of waves such that interaction can proceed. When one such wave is the wave function of an electron in the conduction band and the other is the wave function of a hole in the valence band of a semiconductor, matching of the wavelengths of electron, hole, and photon leads to light emission near the band-gap energy.

When that emission intensity is sufficient, the radiation becomes coherent—a semiconductor laser. These insights, and many more like them, have made possible our modern electronic technology, which has transformed the entire world around us.

Each of them requires that, as Lorentz put it: *we... represent these stationary states, every individual one of them, and to investigate them theoretically.*

Each of them also requires that we analyze the transaction involved very much the way we have done in this paper.

What we have presented is a detailed analysis of the most elementary form of quantum transition, indicating that the simplest properties of solutions of Schrödinger's equation for single-electron atomic states, the conservation of energy, and a symmetry property of relativistic laws of electromagnetic propagation, together with Feynman's insight that all paths should be counted, give a unique form to the photon transaction between two atoms.

We have extended this approach to experiments involving three atoms. The reason we can treat situations with more than one electron using a wave-mechanics Schrödinger equation that only works for one electron is that the non-local bi-directional electromagnetic coupling between wave functions can be factored into a retarded wave propagating forward in time and an advanced wave propagating backward in time, the vector potential of each partner in a photon transaction being incorporated in the opposite partner's one-electron Schrödinger equation.

These calculations are, of course, not general proofs that in every system the offer/confirmation exchange always triggers the formation of a transaction. They do, however, represent demonstrations of that behavior in tractable cases and constitute prototypes of more general transactional behavior. They further demonstrate that the transaction model is implicit in and consistent with the Schrödinger wave mechanics formalism, and they demonstrate how transactions, as a space-time standing waves connecting emitter to absorber, can form.

We see that the missing ingredients in previous failed attempts by others to derive wave function collapse from the standard quantum formalism were:

1. Advanced waves were not explicitly used as part of the process.
2. The "focusing" property of the advanced-retarded radiation pattern had not been identified.

Although many complications are avoided by the simplicity of these two-atom and three-atom systems, they clearly illustrate that *there is internal structure to quantum transitions* and that *this structure is amenable to physical understanding*. Each of them is an example of Einstein's *"description of the individual system"*. Through the Transactional Interpretation, the standard quantum formalism is seen as an ingenious shorthand for arriving at probabilities without wading through the underlying details that Schrödinger described as "tedious".

Although the internal mechanism detailed above is of the simplest form, it describes the most mysterious behavior of quantum systems coupled at a distance, as detailed in [12]. All of these behaviors can be exhibited by single-electron quantum systems coupled electromagnetically. The only thing "mysterious" about our development is our unorthodox use of the advanced-retarded electromagnetic solution to conserve energy and speed up the transition. Therefore, we have learned some interesting things by analyzing these simple transactions!

This experience brings us face to face with the obvious question: **What if Einstein was right?**

If there is internal structure to these simple quantum transitions, there must also be internal structure to the more complex questions involving more than one electron, which cannot be so simply factored!

In this case, we should find a way to look for it. That would require that we effectively time-travel back to 1926 and *grok* the questions those incredibly talented scientists were grappling with at the time.

To face into the *conceptual* details of questions involving an overlapping multi-electron system is a daunting task that has defeated every attempt thus far. We strongly suspect that the success achieved by the matrix approach—adding three more space dimensions and one spin dimension for each additional electron—came at the cost of being "lost in multi-dimensional Hilbert space." Heisenberg's triumph with the helium atom led into a rather short tunnel that narrowed rapidly in the second row of the periodic table.

Quantum chemists work with complex quantum systems that share many electrons in close proximity, and thus must represent many overlapping degrees of freedom. Their primary goal is to find the ground state of such systems. Lorentz's hope—that the intuitive insights of Schrödinger's wave function in three dimensions would *bring us closer to the real solution* in systems with more than one electron—actually helped in the early days of quantum chemistry: Linus Pauling visualized chemical bonds that way, and made a lot of progress with that approach. It is quite clear that the covalent bond has a wave function in three dimensions, even if we don't yet have a fully "quantum" way of handling it in three dimensions. The Hohenberg–Kohn theorems [56] demonstrate that the ground-state of a many-electron system is uniquely determined by its electron density, which depends on only three spatial coordinates. Thus, the chemists have a three-dimensional wave function for many electrons! They use various approaches to minimize the total energy, which then gives the best estimate of the true ground state.

These approaches have evolved into **Density Functional Theory** (DFT), and are responsible for amazingly successful analyses of an enormous range of complex chemical problems. The original Thomas–Fermi–Hohenberg–Kohn idea was to make the Schrödinger equation just about the 3d density. The practical implementations do not come close to the original motivation because half-integer spin, Pauli exclusion, and 3N dimensions are still hiding there. DFT, as it stands today, is a practical tool for generating numbers rather than a fundamental way of thinking. Although it seems unlikely at present that a more intuitive view of the multi-electron wave function will emerge from DFT, the right discovery of how to adapt 3D thinking to the properties of electron pairs could be a major first step in that direction.

When we look at even the simplest two-electron problems, we see that our present understanding uses totally ad hoc rules to eliminate configurations that are otherwise sensible: The most outrageous of these is the **Pauli Exclusion Principle**, most commonly stated as: *Two electrons can only occupy the same orbital state if their spins are anti-parallel.*

It is the reason we have the periodic table, chemical compounds, solid materials, and electrical conductors. It is just a rule, with no underlying *physical* understanding. We have only mathematics to cover our ignorance of *why it is true physically.* [The matrix formulation of QM has a much fancier mathematical way of enforcing this rule. The associated quantum field theory axiom is *not* a physical understanding.]

There is no shame in this—John Archibald Wheeler said it well:

"We live on an island surrounded by a sea of ignorance.

As our island of knowledge grows, so does the shore of our ignorance."

The founders made amazing steps forward in 1926.

Any forward step in science *always* opens new questions that we could not express previously. However, we need to make it absolutely clear *what it is that we do not yet understand:*

- We do not yet understand the mechanism that gives the 3D wave function its "identity", which causes it to be normalized.
- We do not yet have a *physical* picture of how the electron's wave function can be endowed with half-integer "spin", which why it requires a full 720° (twice around) rotation to bring the electron's wave function back to the same state, why both matter and electron antimatter states exist, and why the two have opposite parity.
- We do not yet have a *physical* understanding of how two electron wave functions interact to enforce Pauli's Exclusion Principle.

However, our analysis *has* allowed us to understand *conceptually* several things that have been hidden under the statistics: We saw that the Bose–Einstein property of photons can be understood as arising from the symmetry of electromagnetic coupling together with the movement of electron charge density of a superposed state. There was nothing "particle-like" about the electromagnetic coupling. Indeed, the two-way space-time symmetry of the photon transaction cannot really be viewed as the one-way symmetry of the flight of a "photon particle." Thus, looking at the mechanism of the "wave-function collapse" gives us a deeper view of the "boson" behavior of the photon: The rate of growth of oscillation of the superposed state is, by Equation (26), proportional to the oscillating electromagnetic field. When the oscillating currents of all the atoms are in phase, the amplitudes of their source contributions add, and any new atom is correspondingly more likely to synchronize its contribution at that same phase.

Thus, the "magic" bosonic properties of photons, including the quantization of energy $\hbar\omega$ and tendency to fall into the same state, are simply properties of single-electron systems coupled electromagnetically: Their two-way space-time symmetry is in no way "particle-like." It seems as though there is, after all, a fundamental conceptual difference between "matter" and "coupling".

Perhaps, it is the stubborn determination of theoretical physicists to make everything into particles residing in a multi-dimensional Hilbert space that has delayed for so long Lorentz's *greater intuitive clarity*—a deeper *conceptual, physical understanding* of simple quantum systems.

Thanks to modern quantum optics, we are experimentally standing on the shoulders of giants: We can now routinely realize radio techniques, such as phase-locked loops, at optical frequencies. The old argument that "everything is just counter-clicks" just doesn't cut it in the modern world!

Given the amazing repertoire of these increasingly sophisticated experiments with coherent optical-frequency quantum systems, many of the "mysterious" quantum behaviors seem more and more physically transparent when viewed as arising from the transactional symmetry of the interaction, rather than from the historic "photon-as-particle" view. The bottom line is that Schrödinger wave mechanics can easily deal with issues of quantum entanglement and nonlocality in atomic systems coupled by matched advanced/retarded 4-potentials. It remains to be seen whether this wave-based approach can be extended to systems involving the emission/detection of quark-composites or leptons.

Our Caltech colleague Richard Feynman left a legacy of many priceless quotations; a great one is:

> "However, the *Real Glory* of *Science*
> is that we can *Find* a *Way* of *Thinking*
> such that the *Law* is *Evident!*"

What he was describing is **Conceptual Physics**.

From our new technological vantage point, it is possible to develop Quantum Science in this direction, and make it accessible to beginning students.

We urge new generations of talented researchers to take this one on.

Be Fearless—as they were in 1926!

Supplementary Materials: The following are available online at http://www.mdpi.com/2073-8994/12/8/1373/s1.

Author Contributions: Both authors contributed to this work in roughly equal amounts. It started as a short internal report written by J.G.C. discussing Mathematica calculations based on C.A.M.'s book. Following this, C.A.M. made major contributions in expansion to the present length and produced most of the discussion of equations and formalism. All authors have read and agreed to the published version of the manuscript.

Funding: This research received no external funding.

Acknowledgments: The authors are grateful to Jamil Tahir-Kheli for sharing his unmatched knowledge and understanding of physics history, and his deeply insightful critique of the approach developed here, during many thoughtful discussions down through the years. We thank Ruth Kastner and Gerald Miller for helpful comments and for rubbing our noses in the probabilistic approach to QM. We are particularly grateful to Nick Herbert for asking about transition time in our calculations, which led us to important new insights. David Feinstein, Glenn Keller, Ed Kelm, and Lloyd Watts caught many bugs and offered helpful suggestions. We thank Jordan Maclay for asking us to write this paper for a special issue of *Symmetry* and for making useful comments and suggestons. Finally, we thank our four anonymous reviewers, whose penetrating comments and questions led to major improvements in this paper.

Conflicts of Interest: The authors declare that there are no conflict of interest.

Abbreviations

The following abbreviations are used in this manuscript:

DFT	density functional theory
EPR experiment	Einstein, Podolsky, and Rosen experiment demonstrating nonlocality
NCT	neoclassical theory, i.e., Schrödinger's wave mechanics plus Maxwell's equations
QM	quantum mechanics
TI	the Transactional Interpretation of quantum mechanics [12]
WFE	Wheeler–Feynman electrodynamics

References

1. Einstein, A. Reply to Bohr. In *The Tests of Time: Readings in the Development of Physical Theory*; Lisa M.D., Arthur, F.G., Glenn, N.S., Eds.; Princeton University Press, Princeton, NJ, USA, 2003; ISBN 978-0691090849.
2. Einstein, A.; Podolsky, B.; Rosen, N. Can Quantum-Mechanical Description of Physical Reality Be Considered Complete? *Phys. Rev.* **1935**, *47*, 777–785. [CrossRef]
3. Bell, J.S. On the Einstein Podolsky Rosen paradox. *Physics* **1964**, *1*, 195. [CrossRef]
4. Bell, J.S. On the Problem of Hidden Variables in Quantum Mechanics. *Rev. Mod. Phys.* **1966**, *38*, 447. [CrossRef]
5. Camhy-Val, C.; Dumont, A.M. Absolute Measurement of Mean Lifetimes of Excited States Using the Method of Correlated Photons in a Cascade. *Astron. Astrophys.* **1970**, *6*, 27.
6. Freedman, S.J.; Clauser, J.F. Experimental Test of Local Hidden-Variable Theories. *Phys. Rev. Lett.* **1972**, *28*, 938. [CrossRef]
7. Fry, E.S.; Thompson, R.C. Experimental Test of Local Hidden-Variable Theories. *Phys. Rev. Lett.* **1976**, *37*, 465. [CrossRef]
8. Aspect, A.; Grangier, P.; Roger, G. Experimental Investigation of a Polarization Correlation Anomaly. *Phys. Rev. Lett.* **1981**, *47*, 460. [CrossRef]

9. Clauser, J.F. Experimental Investigation of a Polarization Correlation Anomaly. *Phys. Rev. Lett.* **1976**, *36*, 1223–1226. [CrossRef]

10. Aspect, A.; Grangier, P.; Roger, G. Experimental Realization of Einstein-Podolsky-Rosen-Bohm Gedankenexperiment: A New Violation of Bell's Inequalities. *Phys. Rev. Lett.* **1982**, *49*, 91. [CrossRef]

11. Aspect, A.; Dalibard, J.; Roger, G. Experimental Test of Bell's Inequalities Using Time-Varying Analyzers. *Phys. Rev. Lett.* **1982**, *49*, 1804. [CrossRef]

12. Cramer, J.G., *The Quantum Handshake—Entanglement, Nonlocality and Transactions*; Springer: Berlin/Heidelberg, Germany, 2016; ISBN 978-3-319-24640-6.

13. Kastner, R.E. *The Transactional Interpretation of Quantum Mechanics: The Reality Of Possibility*; Cambridge University Press: London, UK, 2012; ISBN 978-1108407212.

14. Kastner, R.E.; Cramer, J.G. Quantifying Absorption in the Transactional Interpretation. *Int. J. Quantum Found.* **2018**, *4*, 210–222.

15. Kastner, R.E. Decoherence and the Transactional Interpretation. *Int. J. Quantum Found.* **2020**, *6*, 24–39.

16. Mead, C. *Collective Electrodynamics: Quantum Foundations of Electromagnetism*; The MIT Press: Cambridge, MA, USA, 2000; ISBN 0-262-13378-4.

17. Wheeler, J.A.; Feynman, R.P. Interaction with the Absorber as the Mechanism of Radiation. *Rev. Mod. Phy.* **1945**, *17*, 157. [CrossRef]

18. Wheeler, J.A.; Feynman, R.P. Classical Electrodynamics in Terms of Direct Interparticle Action. *Rev. Mod. Phys.* **1949**, *21*, 425. [CrossRef]

19. Cramer, J.G. The Arrow of Electromagnetic Time and Generalized Absorber Theory. *Found. Phys.* **1983**, *13*, 887. [CrossRef]

20. Lewis, G.N. The Nature of Light. *Proc. Natl. Acad. Sci.* **1926**, *12*, 22–29. [CrossRef]

21. Lewis, G.N. The conservation of photons. *Nature* **1926**, *118*, 874–875. [CrossRef]

22. Erwin Schrödinger. An Undulatory Theory of the Mechanics of Atoms and Molecules. *Phys. Rev.* **1926**, *28*, 1049–1070. [CrossRef]

23. Jaynes, E.T. *Some Aspects of Maser Theory*; C Microwave Laboratory Report No. 502; Stanford University: Stanford, CA, USA, 1958; Available online: http://faculty.washington.edu/jcramer/TI/Jaynes_1958.pdf (accessed on 17 August 2020).

24. Jaynes, E.T.; Cummings, F.W. Comparison of Quantum and Semiclassical Radiation Theory with Application to the Beam Maser. *Proc. IEEE* **1963**, *51*, 89–109. [CrossRef]

25. Lamb, W.E., Jr.; Scully, M.O. The photoelectric effect without photons. In *Polarization: Matiere et Rayonnement*; Bederson, H., Ed.; Presses universitaires de France: Paris, France, 1969.

26. Crisp, M.D.; Jaynes, E.T. Radiative Effects in Semiclassical Theory. *Phys. Rev.* **1969**, *179*, 1253.[CrossRef]

27. Stroud, C.R., Jr; Jaynes, E.T. Long-Term Solutions in Semiclassical Radiation Theory. *Phys. Rev. A* **1970**, *1*, 106. [CrossRef]

28. Leiter, D. Comment on the Characteristic Time of Spontaneous Decay in Jaynes's Semiclassical Radiation Theory. *Phys. Rev A* **1970**, *2*, 259. [CrossRef]

29. Jaynes, E.T. Reply to Leiter's Comment. *Phys. Rev A* **1970**, *2*, 260. [CrossRef]

30. Jaynes, E.T. Survey of the Present Status of Semiclassical Radiation Theories. In *Coherence and Quantum Optics: Proceedings of the Third Rochester Conference on Coherence and Quantum Optics held at the University of Rochester, June 21–23, 1972*, Mandel, L., Wolf, E., Eds.; Springer: Berlin/Heidelberg, Germany, 1973.

31. Mandel, L.; Wolf, E. (Eds.) *Coherence and Quantum Optics: Proceedings of the Third Rochester g on Coherence and Quantum Optics held at the University of Rochester, June 21–23, 1972*; Springer: Berlin/Heidelberg, Germany, 1973.

32. Clauser, J.F. Experimental Limitations to the Validity of Semiclassical Radiation Theories. In *Coherence and Quantum Optics: Proceedings of the Third Rochester Conference on Coherence and Quantum Optics held at the University of Rochester, June 21–23, 1972*, Mandel, L.; Wolf, E. Eds.; Springer: Berlin/Heidelberg, Germany, 1973.

33. Cramer, J.G. Generalized Absorber Theory and the Einstein Podolsky Rosen Paradox. *Phys. Rev. D* **1980**, *22*, 362–376. [CrossRef]

34. Cramer, J.G. The Transactional Interpretation of Quantum Mechanics. *Rev. Mod. Phys.* **1986**, *58*, 647–687. [CrossRef]

35. Erwin Schrödinger. *Collected Papers on Wave Mechanics*, 3rd ed.; AMS Chelsea Publishing: Providence, RI, USA, 1982; p. 220.

36. Millikan, R.A. *The Electron*; University of Chicago Press: Chicago, IL, USA, 1917.

37. The Three .mov Animation Files Need to be Downloaded and Placed in the Same Folder as This .pdf File. Set them to "Repeat". Available online: https://1drv.ms/u/s!Ap3rYYlMocgZgZxju0KRgd5mTjXK9A?e=qR7mQy (accessed on 17 August 2020).

38. Erwin Schrödinger. *Collected Papers on Wave Mechanics*; Blackie & Son Ltd., London, UK, 1928; p. 58.

39. Erwin Schrödinger. *Four Lectures on Wave Mechanics Delivered at the Royal Institution, London, on 5th, 7th, 12th, and 14th March, 1928*; Blackie & Son Limited: London/Scotland, UK, 1928.

40. Jackson, J.D. *Classical Electrodynamics*, 3rd ed.; John Wiley & Sons: New York, NY, USA, 1999; ISBN 978-0471309321.

41. Ruth E. Kastner, *Understanding Our Unseen Reality: Solving Quantum Riddles*; Imperial College Press: London, UK, 2015; ISBN 978-1-787326-695-1.

42. Dirac, P.A.M. The Quantum Theory of Emission and Absorption of Radiation. *Proc. R. Soc.* **1927**, *114*, 243–265.

43. Ritz, W.; Einstein, A. Zur Aufklärung. *Phys. Z.* **1909**, *10*, 323–324.

44. Feynman, R.P. *QED: The Strange Theory of Light and Matter*; Princeton University Press: Princeton, NJ, USA, 1985; ISBN 978-0691024172.

45. Steinmetz, C.P.; Berg, E.G. Complex Quantities and Their Use in Electrical Engineering. In Proceedings of the International Electrical Congress, Chicago, IL, USA, 21–25 August 1893; American Institute of Electrical Engineers: New York, NY, USA, 1894; pp. 33–74.

46. Steinmetz, C.P. *Theory and Calculation of Alternating Current Phenomena*, 3rd. ed.; rev. and enl.; Electrical World and Engineer, Inc.: New York, NY, USA, 1900.

47. Heisenberg, W. Reminiscences from 1926 and 1927. In *Niels Bohr, A Centenary Volume*; French, A.P., Kennedy, P.J., Eds.; Harvard University Press: Cambridge, MA, USA, 1985; ISBN 978-0674624153.

48. Schrödinger, E. Are there quantum jumps? *Br. J. Phil. Sci.* **1952**, *3*, 109–123; 233–242. [CrossRef]

49. Brown, R.H.; Twiss, R.Q. A new type of interferometer for use in radio astronomy. *Philos. Mag.* **1954**, *45*, 663–682. [CrossRef]

50. Padula, S.S. HBT Interferometry: Historical Perspective. *Braz. J. Phys.* **2005**, doi:10.1590/S0103-97332005000100005.

51. Jammer, M. *The Philosophy of Quantum Mechanics: The Interpretations of Quantum Mechanics in Historical Perspective*; John Wiley & Sons.: New York, NY, USA, 1974; ISBN 978-0471439585.

52. Lounis, B.; Bechtel, H.A.; Gerion, D.; Alivisatos, P.; Moerner, W.E. Photon antibunching in single CdSe/ZnS quantum dot fluorescence. *Chem. Phys. Lett.* **2000**, *329*, 399–404.[CrossRef]

53. Clauser, J.F. Experimental distinction between the quantum and classical field-theoretic predictions for the photoelectric effect. *Phys. Rev. D* **1974**, *9*, 853–860. [CrossRef]

54. Lorentz, H.A. *Letter to Schrödinger, 27 May 1926*; Letters on Wave Mechanics: 76; Philosophical Library/Open Road: New York, NY, USA, 2011.

55. Heisenberg, W. Mehrkörperproblem und Resonanz in der Quantenmechanik. *Z. für Physik* **1926**, *38*, 411. [CrossRef]

56. Hohenberg, P.; Kohn, W. Inhomogeneous electron gas. *Phys. Rev.* **1964**, *136*, B864–B871.[CrossRef]

 *symmetry*

Article

The Heisenberg Uncertainty Principle as an Endogenous Equilibrium Property of Stochastic Optimal Control Systems in Quantum Mechanics

Jussi Lindgren [1,*] **and Jukka Liukkonen** [2]

[1] Department of Mathematics and Systems Analysis, Aalto University, 02150 Espoo, Finland
[2] Nuclear and Radiation Safety Authority, STUK, 00880 Helsinki, Finland; jukka.liukkonen@stuk.fi
* Correspondence: jussi.lindgren@aalto.fi

Received: 26 August 2020; Accepted: 15 September 2020; Published: 17 September 2020

Abstract: We provide a natural derivation and interpretation for the uncertainty principle in quantum mechanics from the stochastic optimal control approach. We show that, in particular, the stochastic approach to quantum mechanics allows one to understand the uncertainty principle through the "thermodynamic equilibrium". A stochastic process with a gradient structure is key in terms of understanding the uncertainty principle and such a framework comes naturally from the stochastic optimal control approach to quantum mechanics. The symmetry of the system is manifested in certain non-vanishing and invariant covariances between the four-position and the four-momentum. In terms of interpretation, the results allow one to understand the uncertainty principle through the lens of scientific realism, in accordance with empirical evidence, contesting the original interpretation given by Heisenberg.

Keywords: stochastic optimal control; stochastic mechanics; thermodynamic equilibrium; uncertainty principle

---

1. Introduction

This research article aims to provide an intuitive explanation for the quantum phenomenon of measurement uncertainty and it also explains why the Born rule should hold, i.e., why the probability of finding a test particle should be proportional to the square of the wave function. The article builds on the basic framework presented in the authors' recent article [1] and it improves the framework there. The basic framework is built on the assumption that quantum mechanics should be seen through the framework of stochastic optimal control theory; stochastic dynamic optimization in a coordinate-invariant manner on the Minkowski spacetime. In particular, the algebraic structure including the imaginary units can be understood through this framework. However, even though we have shown how relativistic and non-relativistic wave equations of quantum mechanics are the special cases of the corresponding Hamilton–Jacobi–Bellman transport equation for the value function, little consideration was given to the Born rule and the Heisenberg uncertainty principle. In this article, we consider these issues in more detail as an extension and follow-up.

The Heisenberg uncertainty principle is a concept in quantum mechanics that is as elusive as it is important, as there seems to be no unanimous agreement in the literature on its real meaning, content and implications. Originally, Werner Heisenberg linked the principle with some external interference stemming from the measurement experiment [2]. More recently, Ozawa has linked it more abstractly with measurement errors and disturbances, see [3]. There is also a line of research, which relates the uncertainty principle more in line with the approach of the present authors, see, e.g., [4]. On the other hand, Popper [5] and Ballentine [6] already promoted over 50 years ago an interpretation where the uncertainty principle should be seen merely as a statistical scatter relationship. The approach by Bohm [7] is also in line with a statistical ensemble interpretation.

We argue in this article that indeed a more straightforward statistical interpretation for the uncertainty principle is possible when one considers the conjugate variables as random vectors with some possible linear covariance structure. In this respect, in particular, what seems to be especially fruitful is to consider the optimal spacetime diffusions and the respective Fokker–Planck equations or Kolmogorov forward equations for the stochastic optimal control model. The Fokker–Planck equations determine the transport behavior of the transition probability density, as the test particle undergoes the optimal spacetime diffusion, which minimizes the expected action.

In [1] it was shown that the optimal four-velocity is essentially a four-gradient of the value function J, which is typical for such linear–quadratic (stochastic) optimal control problems. This optimal behavior of the test particle is a random motion, which has a drift into the optimal direction and is disturbed by a Brownian noise. We have argued that the diffusion model might be just a phenomenological model in the sense that it is the spacetime itself that is fluctuating [8] and this can be modeled by vector diffusions on the Minkowski spacetime.

We use the Einstein summation convention throughout the article so that upper indices and lower indices represent contravariant and covariant objects, which are generally different. We use slightly different index notation compared to [1] in order to avoid misunderstandings. We use the normalization that the reduced Planck's constant is unity, $\hbar = 1$.

2. The Evolution of the Transition Probability Density and the Stationary Distribution

As was shown in [1] by the authors, the key equations of quantum mechanics can be obtained as optimality equations of certain coordinate invariant stochastic optimization programs on Minkowski spacetimes. Here, as in that model, the test particle undergoes a diffusion process on the Minkowski spacetime; we can analyze the transition probabilities and possible stationary distributions as well.

We argue that the stationary distribution is indeed the relevant distribution in terms of falsifiable experiments, as in order to test any predictions about observables or expectations, the distribution must be sampled many times, by which time the ensemble distribution is the relevant probability distribution. This is due to the property that a stochastic system with a *confining potential* (and other mild technical requirements) yields a stochastic process, which is *ergodic* and the transition probability converges to the Gibbs distribution *exponentially fast*, see [9]. A confining potential means a potential which loosely guarantees that the test particle is "confined" due to the gradient drift of the stochastic process. In this sense, the stochastic optimal control approach to quantum mechanics is actually quite close conceptually to non-equilibrium and equilibrium statistical mechanics.

The basic framework of the stochastic optimal control problem is similar to the set-up in [1], we have a spacetime diffusion for the test particle:

$$dX^\mu = u^\mu ds + \sigma^\mu dW^\mu \tag{1}$$

where X^μ is the random four-position, the contravariant index $\mu = 0, 1, 2, 3$ represents the spacetime coordinates in the (flat) Minkowski spacetime and s is the proper time. There are four independent standard Brownian motions dW^μ with a scale $\sigma^{\mu\mu} = -\frac{1}{2im}$. We use the notation $\sigma^{\mu\nu} = \sigma^\mu \sigma^\nu$, $\mu = 0, 1, 2, 3$, $\nu = 0, 1, 2, 3$ for the diffusion matrix, so it is a tensor product of two vectors. Note that in the denominator, we have the imaginary unit i. The mass of the test particle is m. A capital letter X refers to the random variable itself, whereas a small x refers to the realization coordinate of the random variable. For convenience, we have used a normalization in which the speed of light is unity, $c = 1$. The four-velocity is represented by u^μ. We use slightly different scales (same for all $\mu = 0, 1, 2, 3$) for the Brownian motions compared to [1], as we believe it makes the set-up even more clear and simple. The diffusion matrix should be chosen in such a way that the Hamilton–Jacobi–Bellman equation can be linearized in the simplest possible manner using the Hopf–Cole transformation, as in [1]. Utilizing Occam's razor, the simplest such structure comes from the option:

$$\sigma^{\mu\mu} = -\frac{1}{2im}, \text{ otherwise zero (independent Brownian motions)} \tag{2}$$

The functional to be optimized is given by:

$$S = \int \int_{\tau}^{T} (\frac{1}{2}mg_{\mu\nu}u^{\mu}u^{\nu} - V(x))dsp(x,s)dV. \tag{3}$$

With the invariant volume form $dV = \sqrt{g}dcx_0dx_1dx_2dx_3$, with the metric determinant g. The metric tensor on the Minkowski spacetime is given by $g_{\mu\nu}$. The transition probability density to the four-position x at proper time s is given by $p(x,s)$. The Lagrangian is the four-dimensional generalization of the usual Lagrangian in classical mechanics in a stochastic environment.

We note that the model, as in [1], has a gradient structure, as the optimal drift (necessary condition for the optimal control problem) is given by:

$$u^{\mu} = -\frac{1}{im}\nabla^{\mu}J, \tag{4}$$

where J is the value function.

The resulting nonlinear Hamilton–Jacobi–Bellman optimality equation for this system, as in [1], is given by:

$$\frac{\partial J}{\partial \tau} - iV(x) - \frac{1}{2im}\nabla_{\mu}J\nabla^{\mu}J + \frac{1}{2}\sigma^2\nabla_{\mu}\nabla^{\mu}J = 0, \tag{5}$$

where $\sigma^2 = \sigma_{\mu}^{\mu} = -\frac{1}{im}$, which is the sum of the diagonal elements of the diffusion matrix. Note that the Minkowski metric implies that the trace includes the minus sign when summing the elements together. The Hamilton-Jacobi-Bellman equation can then be linearized to obtain the Stueckelberg and Schrödinger wave equations, as in [1], when we use the Hopf–Cole transformation $J = \log \varphi$, where φ is the wave function. This implies that the wave function is connected to the optimal expected action.

We then expect the respective Fokker–Planck equation for the optimal stochastic process to yield a stationary solution, which is the well-known Gibbs distribution from equilibrium statistical mechanics and thermodynamics. As such, we can analyze the properties of this diffusion process by indeed considering the respective Fokker–Planck equation, which is the partial differential equation governing the transition probability $p(x,s)$ density, where x is the four-position. The Fokker–Planck equation for the spacetime diffusion is given by:

$$\frac{\partial p}{\partial s} = \nabla_{\mu}\left(-u^{\mu}p + \frac{1}{2}\nabla_{\nu}(\sigma^{\mu\nu}p)\right), \tag{6}$$

where we have used the Einstein summation convention. Although in general we cannot solve explicitly the transition probability density for all proper times s, we can, however, in this case solve the stationary distribution, if the value function J has some mild properties, see [9]. The stationary distribution is obtained by setting the expression in the brackets to be zero. The more general family of stationary distributions can be obtained by setting only the partial derivative of the transition density with respect to proper time to zero. For our purposes, setting the expression in the brackets to zero suffices. As the symmetric diffusion matrix $\left(\sigma^{\mu\mu} = -\frac{1}{2im}, \text{ otherwise zero}\right)$ is constant (no correlated noise), we must have: $\frac{1}{2}\nabla^{\mu}p = \nabla^{\mu}Jp$, where on the right side the gradient operates only on J, from which we can directly deduce that the stationary distribution is indeed the Gibbs distribution:

$$p = \frac{1}{Z}e^{2J}, \tag{7}$$

where Z is a normalization constant (the partition function in statistical mechanics). Note that the value function could be complex-valued in general and therefore we need to require additional properties from the system as we want the stationary probability density to be real. We allow the four-position of the test particle to be complex-valued, but we require that the probability density is real. A complex four-position might also have interesting links with the algebra of octonions.

2.1. The Requirement that the Stationary Density is Real Leads to Born's Rule

Remember that, as in [1], we assume a simple relationship (Hopf–Cole transformation) between the value function J and the wave function φ: $J = \log \varphi$. As it holds that the complex conjugates also have the relationship: $J^* = \log \varphi^*$, it is straightforward to see that if the value function is of the form $J = a + ib$, then:

$$2a = \log \varphi \varphi^*. \tag{8}$$

Note now that as we need to require that the stationary distribution is purely real, then the Gibbs distribution implies that $= \frac{1}{Z}\, \varphi \varphi^*$, i.e., we can deduce that the Born rule must hold. We cannot infer from this that the Born rule holds at all times, but we can guarantee that when sampling from the stationary ensemble, the Born rule holds. This constructive demonstration has an interesting parable with the quantum equilibrium hypothesis in Bohmian mechanics [7].

2.2. The Heisenberg Uncertainty Principle

In this subsection, we show how the stochastic optimal control approach to quantum mechanics can help to understand the meaning of Heisenberg's uncertainty principle. In terms of abstract notions, this principle is usually derived by the assumption that the momentum and position operators do not commute. There is no generally agreed common understanding of the principle beyond of the assumption that conjugate variables such as position and momentum cannot be measured simultaneously to an arbitrary degree of certainty.

If one considers the stochastic approach toquantum mechanics, the uncertainty principle can be understood in an intuitive manner by treating the four-position and four-momentum as random vectors. The uncertainty principle is then a statement about the covariance of these random vectors. As such, it does not necessarily then say anything about how experiments are limited or affected by the experimenter, but instead the principle only observes that the covariance does not vanish between these two random vectors. In its usual form, the uncertainty principle is given as (we have used the normalization $\hbar = 1$):

$$\sigma_X \sigma_P \geq \frac{1}{2}, \tag{9}$$

where the sigmas are the standard deviations of the four-position X and four-momentum P, when using the Born rule in the stationary ensemble. The key is to observe that we can use the Cauchy–Schwartz inequality:

$$\sigma_X \sigma_P \geq \left| Cov(X, P) \right|. \tag{10}$$

So that the uncertainty principle holds, we need to have a non-vanishing absolute covariance of the four-position and the four-velocity (up to a scale). As covariance is defined by

$$Cov(X, P) = E[XP] - E[X]E[P], \tag{11}$$

where $E[[]$ is the mathematical expectation operator, we should be able to show that $E[XP] \neq E[X]E[P]$ where X is the four-position and P is the four-momentum. As we can center the random variables without any loss of generality, it suffices to show that

$$E[XP] \neq 0. \tag{12}$$

Consider now again the stationary distribution. As testing the prediction for an expected value of a random variable makes sense only if we can sample the distribution many times and measure the average (law of large numbers), we argue that in terms of any proper experiment, the uncertainty principle concerns only the stationary ensemble averages. Therefore, we calculate the covariance based on the thermodynamical ensemble distribution, which we showed is the Gibbs distribution. This means that all expectations from now on are calculated with respect to the stationary Gibbs distribution.

Consider first the expectation $E[XP]$. Remembering that $P^\mu = mu^\mu = i\nabla^\mu J$, we can consider the integral:

$$E[XP] = i \int X^\mu \nabla^\mu Jp \sqrt{g}d^4x, \tag{13}$$

where we integrate over the invariant volume form $\sqrt{g}d^4x$. Using the implicit definition of the stationary distribution (stationary solution for the Fokker–Planck equation) $\frac{1}{2}\nabla^\mu p = \nabla^\mu Jp$, we have:

$$E[XP] = -\int X^\mu \nabla^\mu Jpd^4x = -\frac{1}{2}\int X^\mu \nabla^\mu pd^4x. \tag{14}$$

By integration by parts, we have (assuming that the probability flux vanishes at the boundary):

$$-\tfrac{1}{2}\int X^\mu \nabla^\mu pd^4x = \tfrac{1}{2}\int \nabla^\mu X^\mu pd^4x = \\ \tfrac{1}{2}\nabla^\mu X^\mu \int pd^4x = \tfrac{1}{2}\nabla^\mu X^\mu \neq 0.$$

where in the last part we have used the fact that the total probability mass is unity and that the $\mu : th$ component derivative of the $\mu : th$ position vector component spits out constants. This shows that the covariance does not vanish and that Heisenberg's uncertainty principle holds (up to a scale). Note that the expectation is invariant as it is a constant.

3. Discussion and Interpretation of the Result

The uncertainty principle obtained is presented in an objective manner through the properties of the stationary distribution and the gradient structure of the underlying stochastic process. The epistemological limit for measuring conjugate variables, i.e., four-position and four-momentum, can be understood intuitively through the stochastic optimal control framework. Intuitively, the stochastic optimal control approach describes an ensemble of test particles with the optimal gradient drift and undergoing a diffusion, which seeks to find a stationary point for the value function. If one prepares such a system and measures the position and momentum repeatedly in order to test the predictions about observables, one will find that the statistical spreads of the position and momentum of the test particle have an inverse relationship. The existence of the lower limit for the product of standard deviations for the position and momentum does not in this framework, however, preclude the possibility of measuring the position and momentum simultaneously. Based on the present result, it only says that when measurements are repeated, the statistical spread obeys the uncertainty principle. Importantly, the constructive proof in this study does not require any external interference from the measurement apparatus, merely the structure of the gradient drift and the underlying diffusion imply the uncertainty principle.

Indeed, the linear–quadratic structure of the stochastic optimal control problem implies that the optimal four-velocity of the test particle has a gradient structure. The gradient structure is somewhat similar to the guidance equation of the pilot wave in the de Broglie–Bohm approach [7]. We therefore argue that Bohmian mechanics is essentially very close to the present stochastic optimal control approach, and that the *guidance equation* is close to the optimal drift velocity of the present model. Unfortunately, when David Bohm developed his model, stochastic optimal control theory had not been developed yet. Moreover, the *quantum equilibrium* in Bohmian mechanics seems to be quite close to the stationary distribution derived in this study. Indeed, it is shown here that the stochastic process under the optimal gradient drift converges exponentially fast to a stationary distribution (thermodynamical equilibrium), where one has a non-vanishing covariance between the four-position and the four-momentum.

Whereas in Bohmian mechanics the ontology of the theory is rather complicated with *pilot waves*, the present approach is more minimalistic: as the *spacetime has fluctuations* at small scales, the test particle obeys a random walk with a drift in such a way that the drift is a "gradient search", which tries to find a dynamically optimal route for the test particle in order to make the expected action stationary. This system results in an equilibrium distribution for the random walk, and this "thermodynamic ensemble" is what we believe corresponds to the quantum equilibrium in Bohmian mechanics. The model is realistic and objective in line with the inclinations of Karl Popper and the statistical/ensemble interpretation of quantum mechanics. The reason why nature seems to obey variational principles is nevertheless a deep question. Why the drift is a gradient search for the value function is no more or less deep a question than why nature prefers various optimal programs, be it in classical mechanics, electromagnetism or general relativity. In the present model, the test particle has a drift into the direction, in which the value function decreases rapidly. These teleological questions are equally deep and equally valid in the Hamilton–Jacobi formulation of classical mechanics. In this sense, the particle just "knows" the direction in which the expected action decreases most rapidly. Therefore, we do not need any "hidden variables" in quantum mechanics any more than we need such "hidden variables" in classical or statistical mechanics. In the Hamilton–Jacobi approach of classical mechanics, the velocity of the test particle is equivalently linked to the gradient of the value function.

Intuitively, the content of the uncertainty principle is, in the present model, that the product of the standard deviations has a lower bound in terms of the stationary distribution due to the covariance. If one would like to have a small standard deviation for the position, one would need to have a rather steep value function in order to constrain and confine the particle within a small location in terms of the stationary distribution (due to the exponential structure of the Gibbs distribution). A steep and a localized value function J would then imply a large standard deviation for the gradient of the value function. There is then a natural tradeoff between the localizations of four-position and four-momentum (Figure 1). The key is the gradient structure of the stochastic process, where the test particle tries to seek a minimum for the value function as in a "gradient search".

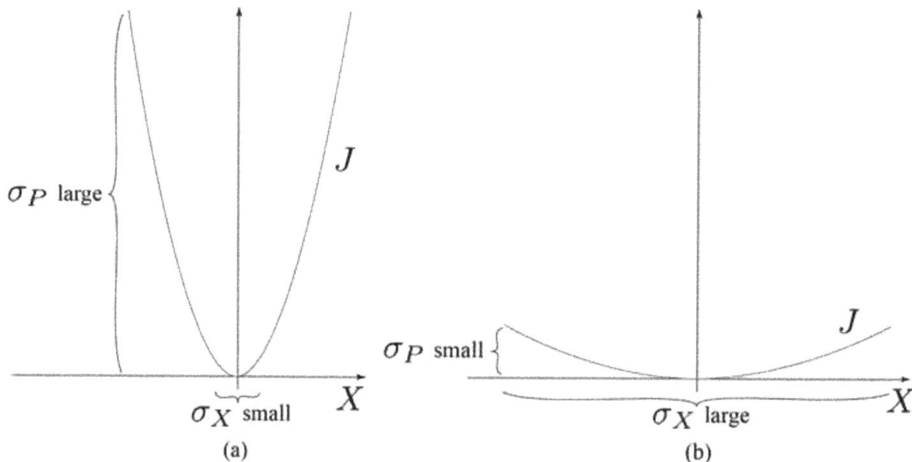

Figure 1. The tradeoff between the localizations of four-position and four-momentum: (**a**) position localized; (**b**) momentum localized.

Therefore, if one seeks an equilibrium system which has a very deep potential well and thus good localization and confinement properties, one needs to accept the large variability in the gradient of the potential (= value) function. This understanding of the uncertainty principle then implies that the lower bound for the product of standard deviations has nothing to do with experiments interfering with the set-up, but it is merely a natural trade-off implied by the linear–quadratic structure of the stochastic optimal control program. Indeed, one can see the framework through the lens of thermodynamics, where the system relaxes towards a thermodynamic equilibrium, which is just the stationary state for the respective Fokker–Planck equation. The results in this paper seem to therefore strongly support the statistical or the ensemble interpretation of quantum mechanics, put forward by, for example, Ballentine [6].

4. Conclusions

We believe that the Heisenberg uncertainty principle can be understood in a fruitful manner by considering it as a stationarity property of stochastic or statistical mechanics. The ontological and epistemological problems related to the uncertainty principle can be mitigated with such an interpretation as in the stochastic optimal control framework; the tradeoff of standard deviations between the four-position and the four-momentum can be understood intuitively through the confinement properties of the value function. The sharp localization, and thus small standard deviation of the test particle four-position, requires a rather steep value function locally, which naturally implies large variability for the gradient of the value function and thus for the four-momentum. Within this interpretation, it still holds that we cannot measure momentum and position simultaneously to an arbitrary degree of precision repetitively, but the uncertainty principle has a new, more intuitive meaning based on scientific realism and objectivism. The interpretation of the uncertainty principle, in which the measurement process itself causes the interference and the uncertainty limit, should be therefore finally abandoned.

The present model allows scientific realism and objectivism in the sense that the test particle obeys an optimal diffusion irrespective of the observer and its position and momentum are uniquely defined and measurable in principle at each point in spacetime, but as we repetitively measure something such as electrons from a beam, the fluctuations of the spacetime induce such randomness to the system that on average we can only "see" the ensemble in the "thermodynamic" equilibrium. In this sense, we argue that this study, together with [1], is very much a constructive model of an objective interpretation of quantum mechanics put forward by Karl Popper [5], among others. Moreover, the interpretation given in this article implies that the uncertainty principle does not logically rule out determinism as such. There is, however, a conceptual connection to Bohmian mechanics, due to the gradient drift structure and the quantum equilibrium, but as we have argued, the diffusion is a phenomenological model for the fluctuation of the spacetime itself. In a sense, the "hidden variables" are not needed, because the medium (spacetime) is stochastic. It is a modeling paradigm, from which one cannot, however, deduce whether the universe is deterministic or not. The optimization procedure, which nature seems to obey, just implies that certain random variables (due to the fluctuations of spacetime) must have a non-vanishing correlation. These conclusions are also in line with rather recent experimental evidence; see [10,11]. The present result seems to also support the paradigm that such uncertainty principles are general features of specific stochastic systems, as has been indicated, for example, in [4]. The symmetries of the present framework are manifested clearly in the property that the covariance of the four-position and the four-momentum is non-vanishing and invariant.

Author Contributions: J.L. (Jussi Lindgren) invented the link between the stationary distribution and the uncertainty principle through the linear–quadratic optimization program; J.L. (Jukka Liukkonen) edited the article and wrote the discussion and conclusions. Both authors reviewed and approved the article. All authors have read and agreed to the published version of the manuscript

Funding: This research received no external funding.

Conflicts of Interest: The authors declare no conflict of interest.

References

1. Lindgren, J.; Liukkonen, J. Quantum mechanics can be understood through stochastic optimization on spacetimes. *Sci. Rep.* **2019**, *9*, 1–8. [CrossRef]
2. Heisenberg, W. Über den anschaulichen inhalt der quantentheoretischen kinematik und mechanik. In *Original Scientific Papers Wissenschaftliche Originalarbeiten*; Springer: Berlin/Heidelberg, Germany, 1985; pp. 478–504.
3. Ozawa, M. Physical content of Heisenberg's uncertainty relation: Limitation and reformulation. *Phys. Lett. A* **2003**, *318*, 21–29. [CrossRef]
4. Koide, T.; Kodama, T. Generalization of uncertainty relation for quantum and stochastic systems. *Phys. Lett. A* **2018**, *382*, 1472–1480. [CrossRef]
5. Popper, K. *Quantum Theory and the Schism in Physics*; Psychology Press: Hove, UK, 1992.
6. Ballentine, L.E. The statistical interpretation of quantum mechanics. *Rev. Mod. Phys.* **1970**, *42*, 358–381. [CrossRef]
7. Bohm, D. A suggested interpretation of the quantum theory in terms of "hidden" variables. I. *Phys. Rev.* **1952**, *85*, 166. [CrossRef]
8. Frederick, C. Stochastic space-time and quantum theory. *Phys. Rev. D* **1976**, *13*, 3183–3191. [CrossRef]
9. Pavliotis, G.A. *Stochastic Processes and Applications, Diffusion Processes, the Fokker-Planck and Langevin Equations*; Springer: New York, NY, USA, 2014.
10. Rozema, L.A.; Darabi, A.; Mahler, D.H.; Hayat, A.; Soudagar, Y.; Steinberg, A.M. Violation of Heisenberg's measurement-disturbance relationship by weak measurements. *Phys. Rev. Lett.* **2012**, *109*, 100404. [CrossRef]
11. Erhart, J.; Sponar, S.; Sulyok, G.; Badurek, G.; Ozawa, M.; Hasegawa, Y. Experimental demonstration of a universally valid error–disturbance uncertainty relation in spin measurements. *Nat. Phys.* **2012**, *8*, 185–189. [CrossRef]

symmetry

Article

Phishing for (Quantum-like) Phools—Theory and Experimental Evidence

Ariane Lambert-Mogiliansky [1,*] and Adrian Calmettes [2]

[1] Department of Economic Theory, Paris School of Economics, 75014 Paris, France
[2] Department of Political Science, The Ohio State University, Columbus, OH 43210, USA; calmettes.1@osu.edu
* Correspondence: alambert@pse.ens.fr
† Phishing for Phools—the Economics of manipulation and deception. 2015.

Abstract: Quantum-like decision theory is by now a theoretically well-developed field (see e.g., Danilov, Lambert-Mogiliansky & Vergopoulos, 2018). We provide a first test of the predictions of an application of this approach to persuasion. One remarkable result entails that, in contrast to Bayesian persuasion, distraction rather than relevant information has a powerful potential to influence decision-making. We first develop a quantum decision model of choice between two uncertain alternatives. We derive the impact of persuasion by means of distractive questions and contrast them with the predictions of the Bayesian model. Next, we provide the results from a first test of the theory. We conducted an experiment where respondents choose between supporting either one of two projects to save endangered species. We tested the impact of persuasion in the form of questions related to different aspects of the uncertain value of the two projects. The experiment involved 1253 respondents divided into three groups: a control group, a first treatment group and the distraction treatment group. Our main result is that, in accordance with the predictions of quantum persuasion but in violation with the Bayesian model, distraction significantly affects decision-making. Population variables play no role. Some significant variations between subgroups are exhibited and discussed. The results of the experiment provide support for the hypothesis that the manipulability of people's decision-making can to some extent be explained by the quantum indeterminacy of their subjective representation of reality.

Keywords: decision-making; uncertainty; persuasion; quantum-like; distraction

Citation: Lambert-Mogiliansky, A.; Calmettes, A. Phishing for (Quantum-Like) Phools—Theory and Experimental Evidence. *Symmetry* **2021**, *13*, 162. https://doi.org/10.3390/sym13020162

Academic Editor: G. Jordan Maclay
Received: 14 December 2020
Accepted: 11 January 2021
Published: 21 January 2021

1. Introduction

Why is the famous P&G (Procter and Gamble) 2010 "Thank you, Mom" advertisement [1] showing devoted mothers supporting young athletes, among the most successful ads of all time? The pervasiveness of informationally irrelevant messaging in advertising is stunning. In this paper, we present and provide experimental evidence for an application of quantum-like decision-making theory that explains why distraction—i.e., addressing informationally irrelevant issues—can be a powerful manipulation technics.

The idea that people are being influenced and manipulated by a systematic exploitation of non rational psychological factors rather that by providing information that is rationally processed, was first forcefully put forward in the seminal book of Vance Packard (1957) "The Hidden Persuaders" [2]. His thesis is that persuador relies on psychiatric and psychological technics to address their message to our "wild and unruly subconscious". Later Cialdini developed a "science of persuasion" based on a general behavioral principles (e.g., bias for reciprocity) that can be exploited to influence people's choice (see e.g., [3,4]). Closer to our approach which focuses on information processing, is an early work by Festinger and Maccoby [5]. They published the first experiment showing that distraction can induce attitude change: a message has larger persuasion power among respondents subjected to distraction. Their idea is that distraction in the course of information processing makes attempts to provide counter arguments less successful. This in turn makes

57

people more vulnerable to the persuador's message. Later, we saw the development of a broad literature in psychology showing that distraction may decrease attention, impair learning and remembering opening up for manipulation [4,6–11]. Failures in information processing are also a the heart of Nobel-prize winner Kahneman's best selling book "Thinking Fast and Slow" [12]. In the last section, we discuss how they relate to our approach to distraction.

More recently, Akerlof and Shiller (2015) provided loads of evidence showing that people are systematically "phished" in economic transactions. The authors suggest that this is due to the significance of the story people tell themselves when making decision i.e., the "narratives" or as they also write "the focus of the mind". They conclude "just change people's focus and you can change the decisions they make" (p. 173 [13]). Emphasizing the significance of the "narratives" is closely related to a rich literature in psychology on framing effects (see among others [14–21]).

Quantum cognition offers an approach to the concept of narratives in terms of perspectives on reality [22]. Formally, a perspective is a coordinate system of a state space and there exists a number of equally valid alternative coordinate systems. Different perspectives can be simultaneously true but not compatible with each other. In this paper, we rely on a formalisation of the concept of narratives in line with the general theory of quantum decision-making. As shown in [23], that theory delivers a power of distraction in the terms of Akerlof and Shiller. The power of distraction arises from non-Bayesian information processing reflecting the mathematical structure of the quantum model. Experimental evidence (see e.g., [24,25]) shows that people oftentimes systematically depart from Bayes' rule when confronted with new information. Cognitive sciences propose a number of alternatives to Bayesianism (see e.g., [26–29]). The attractiveness of the quantum approach is partly due to the fact that quantum mechanics has properties that reminds of the paradoxical phenomena exhibited in human cognition. In addition, quantum cognition has been successful in explaining a wide variety of behavioral phenomena such as disjunction effect, cognitive dissonance or preference reversal (see among others [30–35]). Importantly, there exists by now a fully developed decision theory in the context of non-classical (quantum) uncertainty. Different formulations of that theory exist, including that by Aerts et al. [36]. In this paper, we rely on the formulation developed by Danilov et al. [37,38]. Clearly, the mind is likely to be even more complex than a quantum system, but our view is that the quantum cognitive approach already delivers interesting new insights in particular with respect to persuasion.

In quantum cognition, the object of interest is the decision-maker's mental representation of the world. It is modelled as a quantum-like system represented by its state—a cognitive state which is the equivalent of beliefs in the classical context. In quantum cognition, the decision relevant uncertainty is consequently of non-classical (quantum) nature. As argued in [22] this modelling approach allows capturing widespread cognitive limitations in information processing. The key quantum property that we use is the "Bohr complementarity" of characteristics (properties) of the mental object (representation of the world). The decision-maker cannot consider all properties simultaneously i.e., they cannot have a definite value in his mind. Instead, the decision-maker processes information sequentially moving from one perspective to the other and order matters.

As in the classical context our rational decision-maker uses new information to update her beliefs. A rational quantum-like decision-maker is a decision-maker who has preferences over mental objects representing items (or actions). These mental objects are modelled as quantum-like systems. Her preferences satisfy a number of axioms that secure that they can be represented by an expected utility function. In [38] we learned that a dynamically consistent quantum-like decision-maker updates her beliefs according to Lüder's postulate which in Quantum Mechanics governs state transition following the measurement of a system. In two recent papers, important theoretical results were established. First, it is shown in [39] that in the absence of constraints (on the number of operations that trigger updating), full persuasion applies: Sender can always persuade Receiver to believe anything that he wants. Next, in [23] the same authors investigate a short sequence of operations but in the frame of a simpler task that they call "targeting". The object of "targeting" is the transition of a belief state into another specified target state. The main result of relevance to our issue is that distraction i.e., a test or question that generates irrelevant but "Bohr complementary" information has significant persuasion power. In contrast, a Bayesian decision-maker does not update her beliefs when the information is not relevant to her concern and thus cannot be persuaded in this manner to change her decision.

In the present paper we first formulate a model of quantum-like decision-making in the context of a choice between two uncertain alternatives. The model is used to derive the impact of relevant respectively distractive information on choice behavior. The results are contrasted with those of Bayesian persuasion. A contribution of the paper is to provide a first (illustrative) experimental test of the model's predictions. We opted for a more basic treatment of the data because the quantum persuasion model is so rich that a rigorous estimation of the relevant parameters is beyond the scope of the present paper. The experimental situation that we consider is the following. People are invited to choose between two projects aimed at saving endangered species (elephants and tigers). The selected project will receive a donation of 50 euros (one randomly selected respondent will determine the choice). We consider two perspectives of relevance for the choice: the urgency of the cause and the honesty of the organization that manages the donations. As a first step and in a separate experiment we establish that the two perspectives are incompatible by exhibiting a significant order effect which is the signature of incompatible measurements (see [40]). In the main experiment 1253 respondents are divided into three groups: a control and two treatment groups. They all go through a presentation of the projects and some questions about their preferences. The difference between the groups is that the first treatment group is invited to answer a question about their beliefs of direct relevance to their choice while the second must answer a question that distracts them from what is relevant to their choice. We find that, at the population level, the results are in accordance with the predictions of the quantum model: the distractive question has a significant impact on the respondents choices as compared with both the control group and the other treatment group. The pattern of reactions is disconnected from the thematic content of the distractive information screen which is to be expected when the two perspectives are incompatible. In contrast the question on decision relevant beliefs had no significant impact compared to the control group. The data reveal some significant variation between subgroups with respect to their responsiveness to distraction. In particular, we find that people who care about the urgency of the cause are more responsive to distraction. We argue that this is consistent with the quantum model under the reasonable assumption that those people are more passionate about the issue. The quantum-like working of the mind is expected to be more pronounced for passionate people. This is because standard rational thinking which denies the contextuality of mental representations tends to constrain that spontaneous drive. We conclude with a discussion on rationality in information processing and relate our approach to other prominent behavioral theories.

This paper contributes to the economic literature on persuasion initiated by Kamenica and Gentskow's seminal article "Bayesian Persuasion [41]. More precisely, it contributes to its recent development which introduces various kinds of imperfections in information processing one example is Bloedel and Segal's "Persuasion with rational inattention" [42], which show how Sender optimally exploits Receiver's inattention. Another example is Lipnowsky and Mathevet [43]. Their focus is on how Sender responds to Receiver's problem with temptation and self-control by adapting the signal structure. More closely related to our work is Galperti "Persuasion—the art of changing worldviews" [44]. The author is interested on how a better informed Sender can modify Receiver's incorrect worldview with "surprises" that trigger a change in the support of Receiver's beliefs. Our approach is different because we do not assume that there is a single correct worldview. As shown in [45], a large number of non-bayesian rules systematically distort updated beliefs. However, Kamenica and Gentzkow's concavification argument for optimal persuasion extends to such rules which boil down to introducing some form of bias. Kamenica and Gentzkow's result entails that Senders payoff is concave in Receivers's belief so that Sender's problem can be formulated as the choice of Receiver's posteriors. Our contribution departs more fundamentally from Kamenica and Gentzkow because quantum cognition relies on non-classical (quantum) uncertainty (contextuality) so in particular that result does not apply. This approach reveals a powerful role for distraction in persuasion and we provide some experimental evidence for it.

Our results also contribute to the literature in psychology by offering a novel explanation for the well documented distraction-persuasion nexus. Our study provides support to the thesis that people's propensity to be persuaded is due to the contextuality (intrinsic indeterminacy) of their representation of the world rather than to limited cognitive capacity or to some bias. In so doing our paper contributes to the growing literature in quantum cognition (see recent contributions in [46–49] for other examples on how the (quantum) contextuality approach offers a new paradigm for explaining a variety of behavioral phenomena.

The paper is organized as follows. We first briefly remind of the classical Bayesian persuasion approach. Next we provide a quantum-like model of choice between two uncertain alternatives. We formulate the predictions related to the impact of information on choice behavior. In the second part of the paper, we first describe the experimental set-up used to test the predictions. We thereafter report and inteprete the results from the analysis of the data. We conclude with a discussion of our results in view of some of the existing literature.

2. Quantum Persuasion

2.1. Bayesian Persuasion

Let us first briefly describe Bayesian persuasion an approach developed by Kamenica and Gentzkow [41] in a classical uncertainty setting. The subject matter of the theory of Bayesian persuasion is the use of an "information structure", we shall refer to it a "measurement" (in practical terms, it corresponds to an investigation, a test or a question), that generates new information in order to modify a person's state of beliefs with the intent of making her act in a specific way.

More precisely the setting involves two players Sender and Receiver. Receiver chooses an action among a set of alternatives with uncertain consequences. An action yields consequences for both players. Sender may try to influence Receiver so she chooses an action that is most valuable to him. A crucial element of the Bayesian persuasion approach is that Sender does not choose the information Receiver obtains. If he did that would raise issues of strategic concealment and revelation. Instead Sender chooses an "information structure" (IS) or a measurement that is a test, an investigation or a question. Sender is committed to truthfully reveal the outcome of the IS (e.g., he does not control the entity that performs the study). One example is in lobbying. A pharmaceutical company commissions to a scientific laboratory a specific study of a drug impact, the result of which is delivered to the regulator. Another example, closer to our application here, is a question to Receiver: do you believe (Yes or No) that politicians' climate inaction will lead to global catastrophe under this century? The outcome of any IS is information. In our examples above, it is information about the impact of a drug or about the opinion (beliefs) of Receiver on the responsibility of politicians. This information generally affects Receiver's beliefs which in turn may affect her evaluation of the uncertain choice alternatives and therefore the choices she makes. Sender chooses an IS to move Receiver's (expected) choice closest to his own preferred choice. In the classical context Receiver updates her beliefs using Bayes rule and therefore the power of Sender is constrained by Bayesian plausibility: the fact the expected posteriors must equal the priors.

2.2. The Quantum Persuasion Approach

The quantum persuasion approach has been developed in the same vein as Bayesian persuasion: we are interested in how Sender can use an IS to influence Receiver's choices. A central motivation is that persuasion seems much more influential than what comes out of the Bayesian approach. So instead of assuming that agents are classical Bayesian, it has been proposed that their beliefs are quantum-like. A first line of justification is that people do not make decision based on reality but based on a representation of that reality, a mental object. In quantum cognition, the decision-maker's mental representation is modeled as a quantum-like system and characterized by a cognitive state. The decision relevant uncertainty is therefore of a non-classical (quantum) nature. A second line of motivation is that, as argued in e.g., [30,38], this modeling approach allows capturing widespread cognitive limitations. In particular, the fact that people face difficulties in combining different types of information into a stable picture. Instead, the picture (mental object) that emerges depends on the order in which information is processed. The key quantum property that we appealed to is 'Bohr complementarity' of attributes i.e., that some attributes (or properties) of a mental object may be incompatible in the decision-maker's mind: they cannot have definite value simultaneously. A central implication is that measurements (new information) modifies the cognitive state in a non-Bayesian well-defined manner: the mental representation evolves in response to new information in accordance Lüder's rule which, as shown in [38], secures the dynamic consistency of preferences. As shown in [23,39] a rational quantum-like decision-maker can be manipulated well beyond the limits imposed by Bayesian plausibility. In particular, Sender can exploit the incompatibility properties of certain attributes in Receiver's mind by providing distracting information to modify her representation and consequent choice.

2.3. A 2×2 Quantum Decision Model

We next present a simplified model that we formulate in the terms of our experiment, that is a choice between two uncertain options with two attributes each. For a general and detailed exposition of the formal framework of quantum-like decision making see [23].

2.3.1. The Representation of a Choice Alternative

We have two animal protection projects Tiger Forever (TF) and Elephant Crisis Fund (ECF). The initial information is incomplete so their (utility) value is uncertain. Our decision-maker (DM) is endowed with a cognitive state which encapsulates the probability distribution for every possible state of the world. The notion of cognitive state is similar to the notion of belief. However, in contrast to beliefs, a cognitive state is not an (imperfect) image of the objective world. It is a mental construct, a representation of the objective world that evolves in a way that reflects the cognitive constraints that we focus on: namely that our DM cannot consider all perspectives (attributes) simultaneously. There exist perspectives that are not compatible in her mind. They are incompatible or Bohr complementary. The notion of Bohr complementarity is a central feature of Quantum Mechanics. It relates to properties of a physical system that cannot have definite values simulatneously. In quantum cognition, it relates to properties of the mental object (the represented choice item). When the object has a determinate value i.e., the individual is subjectively certain about e.g., the issue of urgency of a cause, her beliefs about an incompatible characteristics e.g., honesty of the NGO is necessarily mixed. As a consequence, the picture (representation) that arises depends on the order in which different pieces of information are processed.

The DM has an initial representation of the projects. To each project we associate a vector that captures the initial representation i.e., the cognitive state with respect to that project (hereafter we refer simply to *project-state* or simply *state*). We use Dirac's notation which allows easily connecting with the geometrical illustration, the initial states are denoted $|T\rangle$ and $|E\rangle$ with T for Tiger Forever and E for Elephants Crisis Fund. The two states are modelled as independent systems meaning that we assume that a measurement operation on one system has no impact on the other i.e., we have no entanglement. Each (represented) project is characterized by two properties (or characteristics) which are assumed incompatible with each other (in the DM mind). We call them Urgency (of the cause) and Honesty (of the NGO managing the project). The Urgency property (or perspective) is represented by a two dimensional space spanned by pure (subjective certainty) states $|U\rangle$, and $|U^{\perp}\rangle$ corresponding to the property of the project being Urgent respectively not-Urgent. The Honesty perspective is represented by an alternative basis of the same state space ($|H\rangle$, $|H^{\perp}\rangle$) corresponding to Honest respectively not-Honest NGO. The fact that the two bases span the same space is the geometrical expression of the (subjective) incompatibility of the two properties.

2.3.2. Preferences

Individual preferences are captured by the utility value attributed by the DM to the projects in the possible pure states e.g., $|U\rangle$ or $|U^{\perp}\rangle$. The utility of an uncertain state is calculated as a linear combination of those values. In addition, individual preferences are characterized by a "preferred perspective" corresponding to the (most) decision relevant characteristics of the item for the individual (e.g., the Urgency of the cause). When two or more perspectives are incompatible in her mind, the individual uses her preferred perspective to evaluate the expected utility of a project. This means that whereas our DM is capable of looking at alternative perspectives on the same item, when it comes to evaluation, she evaluates utility from one and the same perspective throughout the game. This secures that in any belief state, the utility value is uniquely defined while other incompatible perspectives affect choice through their impact on the belief state (see below). In the following, we assume that the preferred perspective is the same for the two projects.

This is illustrated in Figure 1. The two projects are represented each by two distinct states $|T\rangle$ and $|E\rangle$.

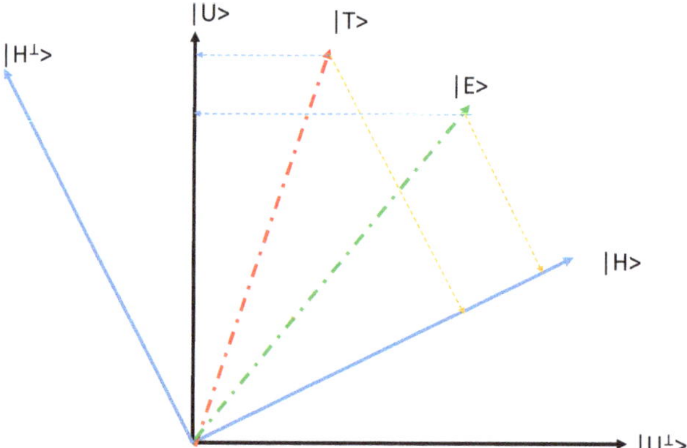

Figure 1. Project states.

The figure reads as follows. For a DM endowed with preferences that define Urgency (U) as her preferred perspective (a U-individual), the expected utility value of contributing 50 euros to the Elephant Crisis Fund in project-state $|E\rangle$ is denoted $u(ECF; |E\rangle, U)$. It depends on two things: 1. her beliefs (about the cause's urgency) encapsulated in state $|E\rangle$ and 2. her valuation of contributing to an urgent respectively non-urgent ECF project. We denote these values $x_U \in \mathbb{R}$ respectively $x_{U^{\perp}} \in \mathbb{R}$. It is useful to express an U- individual's preferences as $E_U = \begin{pmatrix} x_U & 0 \\ 0 & x_{U^\perp} \end{pmatrix}$. As shown in [38] the utility value of project ECF is given as follows

$$u(ECF; |E\rangle, U) = Tr(E_U, E) = x_U \cdot |\langle E|U\rangle|^2 + x_{U^\perp} \cdot |\langle E|U^\perp\rangle|^2. \tag{1}$$

Our DM is risk neutral: her expected utility of the project is as usual the utility associated with the possible states multiplied by the (subjective) probability for those states. So, for instance, with $x_U = 1$ and $x_{U^\perp} = 0$, the expected utility value of contributing to ECF when the individual has U-preferences is equal to her subjective probability that the elephant cause is urgent. That probability is calculated according to Born's rule which is the formula for calculating probability in a quantum setting. It corresponds to the square of the correlation coefficient $\langle E|U\rangle$; also called amplitude of probability. Graphically, the probability amplitudes are read off in the diagram as the orthogonal projection (yellow and blue thin doted lines) of vector $|E\rangle$ on the basis vectors ($|U\rangle, |U^\perp\rangle$, for a U-individual). Similarly, we have for an individual with H-preferences:

$$u(ECF; |E\rangle, H) = Tr(E_H, E) = x_H \cdot |\langle E|H\rangle|^2 + x_{H^\perp} \cdot |\langle E|H^\perp\rangle|^2 \tag{2}$$

where E_H the utility matrix is defined in the (preferred) $(H, H^\perp)$ basis: $E_H = \begin{pmatrix} x_H & 0 \\ 0 & x_{H^\perp} \end{pmatrix}$.
The corresponding expected utility values of the TF project for individual having U- and H-preferences respectively are:

$$\begin{aligned} u(TF; |T\rangle, U) &= Tr(T_U, T) = y_U \cdot |\langle T|U\rangle|^2 + y_{U^\perp} \cdot |\langle T|U^\perp\rangle|^2 \\ u(TF; |T\rangle, H) &= Tr(T_H, T) = y_T \cdot |\langle T|H\rangle|^2 + y_{T^\perp} \cdot |\langle T|H^\perp\rangle|^2 \end{aligned}$$

where $T_U = \begin{pmatrix} y_U & 0 \\ 0 & y_{U^\perp} \end{pmatrix}$ (as defined in the $(U, U^\perp)$ basis) and $T_H = \begin{pmatrix} y_H & 0 \\ 0 & y_{H^\perp} \end{pmatrix}$ (as defined in the $(H, H^\perp)$ basis) are the operator representing the utility value of choosing TF for a U-individual respectively a H-individual.

2.3.3. Choice

The individual makes her choice by comparing the expected utility of each project and selecting the one that yields the highest expected utility. For the sake of illustration take $x_U = y_U = 1$ and $x_{U^\perp} = y_{U^\perp} = 0$, reading directly from the figure we see that

$$u(TF; |T\rangle, U) = |\langle T|U\rangle|^2 > u(ECF; |E\rangle, U) = |\langle E|U\rangle|^2$$

and similarly setting $x_H = y_H = 1$ and $x_{H^\perp} = y_{H^\perp} = 0$

$$u(ECF; |E\rangle, H) = |\langle E|H\rangle|^2 > u(TF; |T\rangle, H) = |\langle T|H\rangle|^2$$

Which means that in our example a U-individual prefers to contribute to the TF project while a H-individual prefers to contribute to the ECF project.

2.3.4. Persuasion

Persuasion is about modifying the cognitive state, i.e., the (mental) project-states. This is achieved by means of an informational structure (IS) which we define as an operation that triggers the resolution of (subjective) uncertainty with respect to some aspect. This corresponds to complete (projective) measurements of the state. In a two-dimentional case, such measurements yield maximal (but not complete) knowledge. It is important to keep in mind that we are dealing with mental objects (represented projects). So, in our context, an IS can be a question that the individual puts to herself (alternatively an IS is an investigation of the outside world that determines whether the threat of extinction is real or not). The outcome is generally some level of conviction (subjective certainty).

In quantum persuasion, an IS is decomposed into two parts: a measurement device (MD) and an information channel (IC) (see [23] for details). An IC translates outcomes into signals. In the present context, we confine ourselves to trivial IC, where the signals are the outcomes of the MD. A MD is defined by a set of possible outcomes I, a collection of probabilities p_i to reach these outcomes where p_i depends on the (cognitive) project-state, $p_i = Tr(P_i E P_i)$ where $|E\rangle$ is the initial belief state (our project-state),the prior and P_i is the projector corresponding to outcome i. Upon obtaining outcome i, the belief-state transits (is updated) into $E_i = \frac{P_i E P_i}{p_i}$ according to Lüders' rule (a behavioral justification for this rule is provided in [38]). In line with the general theory, we focus on direct (or projective) measurements, that is, MDs whose outcomes transit the prior into a pure cognitive state, i.e., a state of full conviction (subjective certainty with respect to some aspect).

As in the standard persuasion problem, Sender chooses the MD. In our experiment Sender chooses the question put to Receiver. For instance "how urgent do you think it is to protect elephants from the threat of extinction". Such an MD is similar to a procedure that actualizes (rather than "elicit") the individual's beliefs about the severity of the threat. The distinction between eliciting and actualizing is that in the first case, it is assumed that the beliefs pre-existed the questioning, it is simply revealed. In contrast, actualizing means that the revealed beliefs were a potential among others which were made actual by the operation of questioning—they did not pre-exist. The statistical distribution of answers expresses the (mixed) beliefs. A crucial point that we emphasize here is that simply eliciting beliefs does not provide any *informational* justification for modifying those beliefs (project-state). In the classical context, belief elicitation has no impact. Yet, as we next shall see, as Sender asks such a question Receiver's cognitive state changes which is the signature of its intrinsic indeterminacy.

2.3.5. The Impact of Measurements

In the development, below we focus exclusively on introspective measurements, i.e., questions put to the decision-maker about the (represented) state of the world. This is in accordance with the experiment that follows. We distinguish between two types of measurements. Those that are compatible with each other, they correspond to commuting operations on the project-state. And those that are incompatible which correspond to non-commuting operations. In a similar way we speak of measurements that are compatible (incompatible) with the preferred perspective.

Compatible Measurements

The performance of a measurement of the (represented) projects in the individual's preference perspective corresponds to actualizing decision-relevant beliefs. The U-question: do you think that the cause is urgent YES/NO? is for a U-individual compatible with her preferred perspective. The question is formulated as a binary choice YES/NO, the initial mixed project state generates the probabilities for the responses.

In the classical (Bayesian) context this type of questioning is inconsequential (see below). This contrast with the quantum context where it modifies the project-state (beliefs). A compatible YES/NO question induces the 'collapse' of the (mixed) state onto one of the pure states. Consider a U-individual when questioned about her belief regarding the urgency of the elephant cause her prior $|E\rangle$ collapses onto $|E'\rangle = |U_E\rangle$ with probability $|\langle E|U_E\rangle|^2$ and onto $|E'\rangle = |U_E^\perp\rangle$ with probability $|\langle E|U_E^\perp\rangle|^2$ and similarly if questioned about the urgency of the Tiger cause $|T\rangle \rightarrow |T'\rangle = |U_T\rangle$ or $|T'\rangle = |U_T^\perp\rangle$ where the subscript informs about the project and are neglected when no confusion arises i.e., it is clear which project we talk about. The measurement of the two project-states (corresponding to ECF respectively TF) generates 4 possible combinations of project-states e.g., $(|E\rangle, |T\rangle)$ transits onto $(|U_E\rangle, |U_T\rangle)$ with a probabilities given by $|\langle E|U\rangle|^2 |\langle T|U\rangle|^2$.

We can now examine the impact of a measurement on the DM's choice in the graphical example above. Recall that in the absence of measurement our U-individual is selecting TF with probability 1 and ECF with probability 0. When the choice is preceded by the Urgency question, with probability $|\langle T|U^\perp\rangle|^2|\langle E|U\rangle|^2 > 0$ the resulting states are $(|U_E\rangle, |U_T^\perp\rangle)$. In this event, she selects ECF because $u(TF; |T'\rangle, U)_{T=U_T^\perp} = Tr(T_U U_T^\perp) = 0 < u(ECF; |E'\rangle, U)_{E=U_E} = Tr(E_U U_T) = 1$. So, we find that her choice behavior is affected by the question. Similarly, our H-individual will, after answering the compatible question, select TF with positive probability. Thus we have shown that even a "naive" question about the individual's decision relevant beliefs can induce a change in the expected revealed preferences i.e., we already have some "persuasion".

The impact of the mere actualization of beliefs underlines a distinction between the quantum and the classical framework. In the quantum world measurements generally change the state of the measured system (here the beliefs or project-states). This is an expression of the fundamental distinction with the classical world where it is assumed that reality preexists any measurement that merely reveals it. In the quantum world reality is contextual which means that measurements contribute in determining the state i.e., they do not reveal a preexisting state, they contribute in shaping that state. This is called contextuality (see [50] for a rich collection of contributions on contextuality). Another distinction with the classical case is the difference in the impact of compatible versus incompatible measurements as we show next.

Incompatible Measurement: Distraction

We now turn to distraction which we define as the actualization of beliefs with respect to features not directly relevant to decision-making i.e., belonging to a perspective that is incompatible with the preferred perspective. Below we depict the case when addressing a U-preference individual. Distraction corresponds to putting a H-question e.g., do you believe WWF (managing ECF) is honest YES/NO? As in the compatible case the question triggers the collapse of the project-state $|E\rangle$ in the basis corresponding to the question, here $(H, H^\perp)$. The state $|E\rangle$ transits into $|E'\rangle$ equal to either $|H_E\rangle$ or $|H_E^\perp\rangle$ and it does so with probability $|\langle E|H\rangle|^2$ and $|\langle E|H^\perp\rangle|^2$. And similarly for the H-question regarding TF (the NGO managing the Tiger project). We illustrate this in Figure 2 with the green lines for E the cognitive state representing ECF.

In contrast with the compatible measurement case, after having answered the incompatible question the resulting cognitive state does not allow the DM to evaluate the expected utility associated with the choice alternatives. She needs to project it back into her preferred perspective. The expected utility of ECF for a U-individual in initial project-state $|E\rangle$ subjected to the H-question, is obtained by considering a sequence of two non-commuting operations. First distraction, the state is projected onto the $(H, H^\perp)$ basis. Then the resulting state (H or $H^\perp$) is projected back onto the preferred basis $(U, U^\perp)$ in order to evaluate the project:

$$u(ECF; |E\rangle, U) = \left[|\langle E|H\rangle|^2|\langle H|U\rangle|^2 + |\langle E|H^\perp\rangle|^2|\langle H^\perp|U\rangle|^2\right]x_U +$$
$$\left[|\langle E|H\rangle|^2|\langle H|U^\perp\rangle|^2 + |\langle E|H^\perp\rangle|^2|\langle H^\perp|U^\perp\rangle|^2\right]x_{U^\perp}.$$

Example

Consider the following numerical example for a H-individual where we simplify the matter by assuming $|T\rangle = |E\rangle = |D\rangle$, that is the two projects are represented by the same project-state meaning that they are subjectively perceived as equally urgent and honest. Let this state in the H-perspective be

$$D = \begin{pmatrix} 4/5 & 2/5 \\ 2/5 & 1/5 \end{pmatrix}$$

Consider the following utility values $x_H = 8, x_{H^\perp} = 7, y_H = 10, y_{H^\perp} = 4$. Using the formula in (2) we obtain:

$$u(ECF; |D\rangle, H) = 4/5 \cdot 8 + 1/5 \cdot 7 = 39/5$$

And similarly

$$u(TF; |D\rangle, H) = 4/5 \cdot 10 + 1/5 \cdot 4 = 44/5$$

Which means that this H-individual chooses to donate to TF.

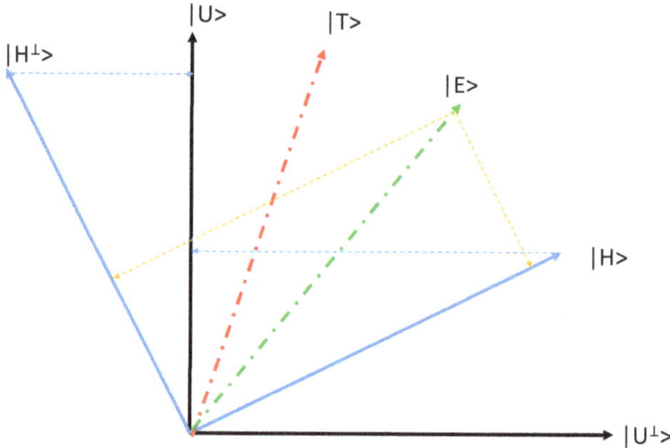

Figure 2. Distraction.

Let us now consider a distraction toward the Urgency perspective that we model for simplicity as a 45° rotation of the H-basis (which corresponds to the case when the pure states are statistically uncorrelated across perspectives) $U = \begin{pmatrix} 1/2 & 1/2 \\ 1/2 & 1/2 \end{pmatrix}$ and $U^\perp = \begin{pmatrix} 1/2 & -1/2 \\ -1/2 & 1/2 \end{pmatrix}$.

The distractive procedure is as follows: first the H-individual in project-state $|D\rangle$ is asked whether she thinks the Elephant respectively Tiger cause is Urgent or not Urgent which takes the state $|D\rangle$ onto $\left|U_{E(T)}\right\rangle$ or $\left|U^\perp_{E(T)}\right\rangle$. Then, our H-individual evaluates her expected utility value in the $(H, H^\perp)$ perspective. With a 45° rotation the computation simplifies greatly because whether distraction takes the states to U or $U^\perp$, the probability for H respectively $H^\perp$ is the same:

$$
\begin{aligned}
u(ECF; |D\rangle, H) &= \left[|\langle D|U\rangle|^2 + |\langle D|U^\perp\rangle|^2\right] 1/2 x_H + \left[|\langle D|U\rangle|^2 + |\langle D|U^\perp\rangle|^2\right] 1/2 x_{H^\perp} \\
&= 1/2 x_H + 1/2 x_{H^\perp} = 4 + 3.5 = 7.5
\end{aligned}
$$

and similarly

$$
\begin{aligned}
u(TF; |D\rangle, T) &= \left[|\langle D|U\rangle|^2 + |\langle T|U^\perp\rangle|^2\right] 1/2 y_H + \left[|\langle T|U\rangle|^2 + |\langle T|U^\perp\rangle|^2\right] 1/2 y_{H^\perp} \\
&= 1/2 y_H + 1/2 y_{H^\perp} = 5 + 2 = 7.
\end{aligned}
$$

So after the distraction our individual chooses to donate to the ECF project with probability 1 instead of TF in the absence of distraction. So we note that the impact of distraction can be a total reversal of the choice. This is in contrast with the impact of decision-relevant belief actualization (compatible measurement) which only triggers some partial reversal. For a general theoretical argument on the persuasion power of a distractive IS as compared with a compatible IS see [23].

Before moving to the experiment let us briefly remind ourselves of the classical subjective uncertainty approach in our example.

2.3.6. The Classical Uncertainty Approach

The classical uncertainty framework is nested in the quantum setting. It corresponds to the case when all properties of an item (perspectives) are compatible and therefore Lüder's rule for updating is equivalent to Bayesian updating. The individual can simultaneously considers Urgency and Honesty and combine them to obtain her expected utility value. Assuming a separable and additive utility function, we write

$$u(T) = \alpha\left(p_U^0(T)x_U + \left(1 - p_U^0(T)\right)x_{U\perp}\right) + (1 - \alpha)\left(p_T^0(T)x_H + \left(1 - p_T^0(T)\right)x_{H\perp}\right)$$

$$u(E) = \alpha\left(p_U^0(E)y_U + \left(1 - p_U^0(E)\right)y_{U\perp}\right) + (1 - \alpha)\left(p_T^0(E)y_T + \left(1 - p_T^0(E)\right)y_{T\perp}\right)$$

where $p_U^0(T)$ is the subjective probability in state $|T\rangle$ that the TF project is urgent (and $c(1 - p_U^0(T)c)$ that it is not urgent) and similarly for the other probabilities. The superscript refers to time $t = 0$ (initial beliefs). The α is the relative preference weight given to urgency $((1 - \alpha)$ the relative weight of honesty). An individual for whom honesty is determinant is an individual with $\alpha < 1/2$ and similarly for U-individuals ($\alpha \geq 1/2$).

Recall that the measurements that we consider are exclusively introspective i.e., no information appealing to the outside world is called upon. In other words, the questions correspond to eliciting Receiver's beliefs.

When asked "do you believe the tiger cause is urgent YES/NO. With probability $p_U^0(T)$ Receiver answers YES and with probability $\left(1 - p_U^0(T)\right)$ she answers NO. But her beliefs do not change, they remain mixed. Since beliefs are unchanged so is the expected utility from the two projects. As a consequence Receiver's choice is not affected by Sender's question. To put it differently, an introspective measurement has no persuasion power whatsoever in the classical context. The classical prediction contrasts starkly with the quantum model where an introspective measurement with respect to both compatible and incompatible perspectives has impact on decision-making. Those predictions appears more consistent with numerous experimental works that exhibit a significant impact of belief elicitation on decision-making see most recently [51]). In addition, as illustrated above, incompatible introspective measurements have the strongest potential to affect decision-making. TIt is precisely this prediction that we aim at testing with the next following experiment.

3. Experimental Design

Our main experiment features the choice to donate to either one of two projects concerned with the protection of endangered species. It uses the property of Bohr complementarity of mental perspectives. More precisely it relies on the hypothesis that two perspectives on the projects are incompatible in the mind of Receiver. The two perspectives that we consider are "the urgency of the cause" and "the honesty of the organization that manages the funds' (the terms "honesty" and "trustworthiness", or "trust", are used interchangeably). As a first step we provided experimental support for the incompatibility hypothesis. We know that when two properties are incompatible measuring them in different orders yields different outcomes. Therefore, we started with an experiment to check whether order matters for the response profile obtained. Note that even in Physics, there is no theoretical argument for establishing whether two properties are compatible or not. This must be done empirically.

3.1. Testing for the Incompatibility of Perspectives

At the time we conducted our study, the world was confronted with a severe refugee crisis in Myanmar. The situation actualized quite sharply the two perspectives we wanted to test. On the one hand, the urgency of the humanitarian crisis and, on the other hand, the uncertainty about the reliability/honesty of the NGOs on the ground.

We recruited 295 respondents through Amazon's Mechanical Turk, for which data quality has been confirmed by different studies (e.g., [52,53]). The respondents completed the short survey below on the website Typeform. They were paid $0.1 and spent on average 0:17 minutes to complete the survey.

The participants were first presented a screen with a short description of the situation of refugees in Myanmar including a mention of the main humanitarian NGO present in the field:

> *"About a million refugees (a majority of women and children) escaped persecution in Myanmar. Most of them fled to Bangladesh. The Bengali Red Crescent is the primary humanitarian organization that is providing help to the Rohingyas. They are in immediate need of drinkable water, food, shelter and first medical aid."*

They were then asked to evaluate the urgency of the cause and the honesty to the NGO on a scale from 1 ("Not urgent" or "Do not trust") to 5 ("Extremely urgent" or "Fully trust"). The order of presentation of the two questions was randomized so that half of participants responded to the urgency question before trust (U-T), and the other half conversely (T-U).

We prove the existence of order effects by showing that the responses are drawn from two different distributions. We do this using both a difference in means test (i.e., two-sample t-test with $t = -2.54$ and p-value $= 0.011$) and a nonparametric test of the two sample distributions (i.e., two-sample Kolmogorov-Smirnov test with $D = 0.11$ and p-value $= 0.047$) in R.

The results are consistent with the hypothesis that the two perspectives are incompatible in the mind of people. As well-known there exist other theories for order effects. We observe that the value of the responses to the Trust question tend to be lower when that question comes first (i.e., T-U), whereas the responses to the Urgency question tends to have a higher value when that question comes first (i.e., U-T). Therefore, we can reject both the hypothesis of a recency biais and that of a primacy bias. This strengthens our quantum interpretation. We next proceed to the main experiment using those two perspectives.

3.2. Main Experiment

1253 participants completed the survey on the website Typeform, they were recruited through Amazon's Mechanical Turk. They were paid either $1 to $0.75 depending on the condition.

The participants were divided into three groups. Two treatment groups and a control group as explained below. All three groups were presented a screen with an introductory message, informing them that the questionnaire is part of a research project on quantum cognition and that they will contribute in deciding which one of two NGOs projects will receive a €50 donation. The decision will be made by randomly selecting a respondent and implementing his or her decision. Presumably, this created an incentive to respond truthfully. The respondents were next asked to click on a button that randomly assigned them to a specific condition. In all conditions, participants were shown a short text about the situation of elephants respectively tigers and of ongoing actions of two NGOs working for their protection the Elephant Crisis Fund (ECF) and Tiger Forever (TF). The order of presentation of the text was reversed for half of the subjects. This aimed at isolating order effects not relevant to our main point. The screen displayed the following two texts:

> *"Elephant crisis fund: A virulent wave of poaching is on-going with an elephant killed for its tusks every 15 min. The current population is estimated to around 700,000 elephants in the wild. Driving the killing is international ivory trade that thrives on poverty, corruption, and greed. But there is hope. The Elephant Crisis Fund closely linked to World Wildlife Fund (WWF) exists to encourage collaboration, and deliver rapid impact on the ground to stop the killing, the trafficking, and the demand for ivory."*

"Tiger Forever: Tigers are illegally killed for their pelts and body parts used in traditional Asian medicines. They are also seen as threats to human communities. They suffer from large scale habitat loss due to human population growth and expansion. Tiger Forever was founded 2006 with the goal of reversing global tiger decline. It is active in 17 sites with Non-Governmental Organizations (NGOs) and government partners. The sites host about 2260 tigers or 70% of the total world's tiger population"

It is worth mentioning that the descriptions were formulated so as to slightly suggest that the elephants' NGO (EFC) could be perceived as more trustworthy (because of its link with of WWF, a well-known NGO). In contrast, the text about tigers suggested a higher level of urgency (the absolute number of remaining tigers is significantly lower than the number of remaining elephants). Thereafter, all respondents were confronted with a choice:

"When considering donating money in support of a project to protect endangered species, different aspects may be relevant to your choice. Let us know what counts most to you:

-The urgency of the cause: among the many important issues in today's world, does the cause you consider belong to those that deserve urgent action? or

-The honesty of the organization to which you donate: do you trust the organization managing the project to be reliable; i.e., do you trust the money will be used as advertised rather than diverted."

The objective was to elicit an element of their preferences namely their preferred perspective, see Section 2. The rest of the questionnaire depended on which one of the three groups the participants belonged to.

In the control condition (baseline), they were next asked whether or not they wanted to read the first descriptions again or if they wanted to make their final decision i.e., to make their choice between supporting the Elephant Crisis Fund or Tiger Forever both represented by an image of an adult elephant respectively adult tiger (presented in random order on the same screen).

In the first treatment condition, the respondents were redirected to a screen with general information compatible with the aspect they indicated as determinant to their choice when making a donation. Importantly, the information did not directly or indirectly favor or disfavor any of the two projects. The information was aimed at triggering a measurement as they were invited to determine themselves with respect to which of the species was most urgent to save respectively which NGO was most trustworthy. We below return to the role and expected impact of the general information screens. Those who cared most honesty saw a screen with the following text:

"Did you know that most Elephant and Tiger projects are run by Non-Governmental Organizations (NGOs)? But NGOs are not always honest! NGOs operating in countries with endemic corruption face particular risks. NGOs are created by enthusiastic benevolent citizens who often lack proper competence to manage both internal and external risks. Numerous scandals have shown how even long standing NGOs had been captured by less scrupulous people to serve their own interest. So a reasonable concern is whether Tiger Forever respectively Elephant Crisis Fund deserves our trust."

Those who cared most for urgency saw:

"Did you know that global wildlife populations have declined 58% since 1970, primarily due to habitat destruction, over-hunting and pollution. It is urgent to reverse the decline! "For the first time since the demise of the dinosaurs 65 million years ago, we face a global mass extinction of wildlife. We ignore the decline of other species at our peril—for they are the barometer that reveals our impact on the world that sustains us." —Mike Barrett, director of science and policy at WWF's UK branch. A reasonable concern is how urgent protecting tigers or elephants actually is."

Thereafter the respondents were offered the opportunity to read again the descriptions before making their image choice between ECF and TF.

In the main treatment condition (distraction), participants were redirected to a screen with general information on the aspect they did not select as determinant to their choice; this is what we call a distraction. So, those who selected honesty (resp. urgency) saw the screen on global wildlife decline (resp. NGO's scandals). Thereafter, the respondents were offered the opportunity to read again the initial project description before making their image choice.

Finally, information about their age, gender, education and habits of donation to NGOs was collected before the thank-you message ending of the experiment.

Before presenting the results, we wish to address a feature of the experimental design absent from the theoretical model.

The General Information Screens

First, we note that the theoretical model does not account for anything like a general information screen. The connection with the model is with the questions that follow the general information. Indeed, general information plays no role for persuasion since it conveys no new data on the relative urgency or honesty of the specific projects. Therefore it should not affect the choice between the two projects. So, what is the role of those screens?

Our justification for the general information screens is to be found in the quantum approach to cognition. Quantum cognition recognizes that people consider project from different perspectives some of which may be incompatible but not mutually exclusive. They are Bohr complementary which implies that the questions related to those perspectives do not commute i.e., order matters. This in turn is an expression of the fact that the cognitive state is modified by responding to a question. Our intuition is that there can be some inertia. Consider a person who declared that Honesty is her priority which we interpret as her being in the Honesty perspective. If you abruptly ask whether the elephant cause is urgent, she might not make the effort to switch perspective in order to respond faithfully. In contrast if you softly accompany her into the switch with an engaging short text, she will find herself capable of responding truthfully without particular effort.

Hence the point with those screens is to accompany the change in perspective. Clearly, that is only justified in the distraction treatment, but for the sake of symmetry, we have a similar screen in the treatment where no change of perspective is required.

Next, one may wonder why we do not, in the experiment, simply ask people for their beliefs e.g., do you trust that NGO? But instead we suggest a questioning: "So a reasonable concern is whether Tiger Forever respectively Elephant Crisis Fund deserves our trust". The reason is that we wanted to avoid that the response would influence the respondent beyond the impact under investigation. Additional impact can be expected because of perceived dissonance. Assume the individual cares for the urgency of the cause and since there are only 2700 tigers left, so she is most likely to choose TF. If she is explicitly asked whether she believes that the TF NGO is honest and decides that she does not trust them much, then it becomes psychologically difficult to select TF. Since we do not put an explicit question but use a general text to induce the measurement, people are expected to be less likely to perceive dissonance and choose more spontaneously. These precautions are among the difficult decisions we have to make in quantum cognition when trying to exhibit quantum effect in behavior. Individuals are, in contrast with particles, thinking systems endowed with, among other things, a drive toward consistency which can interfere with the intrinsic indeterminacy of preferences (see e.g., [26,54]) and Discussion in Section 5).

3.3. Theoretical Predictions

Before getting into the results and their interpretation, let us remind of the main theoretical predictions:

- The Bayesian model predicts no effect in both treatment groups.
- The quantum model predicts that the distraction treatment group should exhibit a significantly different allocation of responses compared with the control group's choice profile. It also predicts some milder impact of the question in the compatible

treatment compared with the control group. It should be emphasized that since we lack information about the correlation coefficients between the two perspectives and the utility values, we do not have quantitative predictions. Generally, the less correlated two perspectives (in the example of Section 2 they were fully uncorrelated) the larger the expected impact in terms of switching the choice for given utility values.

4. Results

4.1. Descriptive Statistics

Data were processed, cleaned and analyzed with statistical software R. The number of acceptable observations was 1253 (114 participants were removed from the data due to a technical error which created a risk that some participants might have responded twice). 58.5% of all respondents were male, the average age was 35.6 years and the average education level was undergraduate. Overall, 71.1 % of the participants declared that the Honesty of the NGO rather than the Urgency of the cause is what counts most to their choice. Across the three conditions, 54.4% chose to support with their donation the Elephants Crisis Fund (ECF) and 45.6 the Tiger Forever project (TF). Looking into the different treatment groups, we find that 59% of the respondents in the control condition chose ECF and 54% in the compatible information treatment group. In contrast, in the distraction treatment group, only 47% chose ECF. Conditional on revealed preferences, 50% of the respondents who valued Urgency most chose to support TF, whereas 56% of those who valued Honesty most chose to support ECF. Overall, 87.8% made their final decision without reading the description of the projects a second time. They spent, on average, 1:33 minutes to complete the experiment.

We divided up the respondents into a number of subgroups based on their preferences and the detailed treatment they received. Figure 3 represents the number of participants who chose TF respectively ECF conditional on their preference (Honesty vs. Urgency) and the order of the presentation they have been exposed to (ECF-TF vs. TF-ECF). Note that for all conditions but "Honesty-ET", a majority of participants chose ECF in the control condition and TF in the incompatible condition. This is particularly striking for "Honesty-TE" and "Urgency-ET". At first glance, we find a clear reversal in three of the subgroups.

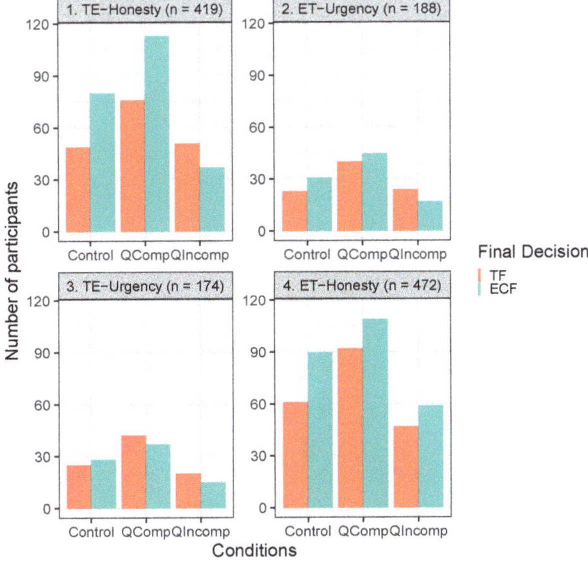

Figure 3. Descriptive Histograms.

4.1.1. Data Analysis

General Results

The first set of results displayed in Table 1's first column establishes that distraction—i.e., the question related to the non-determinant perspective - has a statistically significant impact on the final choice ($p = 0.005$). This result stands across different specifications see Table A1 in the Appendix A). In particular, it appears that everything else being constant, the predicted probability of choosing ECF is 11.1% lower for an individual in the incompatible condition than for an individual in the control condition. By contrast, there is no statistically significant impact of the compatible question on the final decision ($p = 0.173$). This result is also persistent over alternative specifications. Note however that the effect of the compatible question on the final choice is nonzero. We come back to this later on.

Not surprisingly, there is a statistically significant impact ($p = 0.046$) on the final choice of the declared determinant—i.e., Honesty versus Urgency, which captures an element of preferences. Here again, the influence is robust to alternative specifications (see Table A1 in the Appendix A). More precisely, the predicted probability of choosing ECF is 6.33% higher for an individual claiming that Honesty is determinant than for someone who reported Urgency as determinant to her choice. By contrast there is no statistically significant impact ($p = 0.7$) of the order of presentation of the project descriptions on the final decision.

The correlation between covariates were not bigger than -0.13 (between Age and Male)—hence putting aside potential issues of multicollinearity. In particular, regressing on revealed preferences (i.e., choice between Honesty and Urgency) shows no relationship with the order of presentation of the descriptions (ECF-TF and TF-ECF). Interestingly, none of the variables significantly affected the revealed preferences.

Table 1. Logit Regressions on Final Choice (ECF)—Coefficients transformed.

| | *Dependent Variable:* | | | | | | | | |
| | **Decision_ECF** | | | | | | | | |
	General	Hon	Urg	ET	TE	ET-Hon	TE-Hon	ET-Urg	TE-Urg
Compatible condition	-0.170 $t = -1.363$	-0.142 $t = -0.938$	-0.231 $t = -1.030$	-0.181 $t = -1.063$	-0.175 $t = -0.958$	-0.164 $t = -0.807$	-0.154 $t = -0.683$	-0.216 $t = -0.669$	-0.263 $t = -0.833$
Incompatible condition	-0.363 *** $t = -2.791$	-0.336 ** $t = -2.152$	-0.455 * $t = -1.953$	-0.249 $t = -1.307$	-0.488 *** $t = -2.763$	-0.142 $t = -0.591$	-0.523 ** $t = -2.563$	-0.510 * $t = -1.658$	-0.404 $t = -1.129$
Honesty	0.294 ** $t = 1.998$			0.196 $t = 1.007$	0.407 * $t = 1.809$				
Order (ECF-TF)	0.046 $t = 0.383$	-0.014 $t = -0.101$	0.209 $t = 0.868$						
Reread Descriptions	-0.088 $t = -0.502$	-0.178 $t = -0.941$	0.341 $t = 0.720$	-0.057 $t = -0.225$	-0.117 $t = -0.467$	-0.139 $t = -0.508$	-0.209 $t = -0.777$	0.374 $t = 0.566$	0.241 $t = 0.354$
Age	0.005 $t = 0.894$	0.008 $t = 1.198$	-0.002 $t = -0.186$	0.006 $t = 0.826$	0.003 $t = 0.394$	0.002 $t = 0.227$	0.015 $t = 1.455$	0.019 $t = 1.245$	-0.022 $t = -1.513$
Male	-0.172 $t = -1.566$	-0.128 $t = -0.952$	-0.262 $t = -1.348$	-0.067 $t = -0.414$	-0.274 * $t = -1.836$	-0.086 $t = -0.447$	-0.184 $t = -0.968$	-0.119 $t = -0.393$	-0.390 $t = -1.518$
Education	0.021 $t = 0.252$	0.006 $t = 0.064$	0.072 $t = 0.450$	0.002 $t = 0.019$	0.026 $t = 0.210$	-0.011 $t = -0.085$	0.012 $t = 0.080$	0.032 $t = 0.147$	0.116 $t = 0.479$
NGO	0.006 $t = 0.051$	0.038 $t = 0.259$	-0.062 $t = -0.287$	0.190 $t = 1.058$	-0.187 $t = -1.154$	0.356 $t = 1.554$	-0.272 $t = -1.446$	-0.167 $t = -0.589$	0.137 $t = 0.392$
Constant	0.068 $t = 0.216$	0.251 $t = 0.617$	0.317 $t = 0.534$	-0.028 $t = -0.068$	0.303 $t = 0.597$	0.292 $t = 0.528$	0.253 $t = 0.419$	-0.213 $t = -0.329$	1.559 $t = 1.240$
Observations	1211	864	347	638	573	458	406	180	167
Log Likelihood	-825.591	-586.663	-237.452	-436.394	-385.940	-311.893	-269.413	-122.258	-112.568
Akaike Inf. Crit.	1671.183	1191.325	492.904	890.788	789.879	639.786	554.825	260.515	241.136

Note: * $p < 0.1$; ** $p < 0.05$; *** $p < 0.01$. Coefficients are transformed as : $exp(\beta) - 1$.

Advanced Results

As shown by Table 1, we find that the distraction effect is not homogeneous across subgroups. First, we find that the distraction effect was stronger in the Urgency subgroup than in Honesty subgroup: for Urgency-individuals, the predicted probability of choosing ECF in the incompatible condition is 14.93% lower compared to the control condition ($p = 0.051$); for Honesty-individuals, it is 10.05% lower ($p = 0.031$). Next, it appears that the impact of distraction is most pronounced for those who were presented the Tiger Forever project first and Elephant Crisis Fund last (TE subgroup; see Table 1). Distraction statistically significantly affected the final choice in that group (corresponding to 50% of the respondents). In fact, TE-participants in the incompatible condition had a predicted probability of choosing ECF that was 16.3% lower than TE-participants in the control condition ($p = 0.006$; see Table 2).

Table 2. Predicted Probability Difference between Incompatible and Control condition.

	General	Hon	Urg	ET	TE	ET_Hon	TE_Hon	ET_Urg	TE_Urg
Difference	−0.111	−0.101	−0.149	−0.070	−0.163	−0.037	−0.179	−0.174	−0.125
p-value	0.005	0.031	0.051	0.191	0.006	0.555	0.010	0.097	0.259

When combining the TE presentation order with the Honesty subgroup: the difference in predicted probability to choose ECF between the incompatible condition and the control condition is 17.9% ($p = 0.01$). For ET-Urgency a 17.4% change in probability can be seen (cf. Figure 3, Tables 2 and 1), even though the effect fails to reach statistical significance at the 5% level ($p = 0.097$). Note however that group only constitutes around 15% of the sample, while the TE-Honesty subgroup represents around 34% of the data. For the other two subgroups, distraction had no statistical significant impact ($p = 0.555$), but a switch of 12.5% may still be noticed for TE-Urg.

4.1.2. Interpretation

First, we note that the significance of preferences (i.e., the answer to "what is determinant to your choice") for the final choice combined with the fact that a majority of participants who chose Honesty also chose ECF regardless of their condition, suggests that the initial texts were generally well-understood. As explained earlier, the description of the Elephant project was designed to suggest more trust to the NGO managing the project and the description of the Tiger project to suggest a higher level of urgency.

The general results show with no ambiguity that the question triggered by the incompatible information (distraction) had a significant impact on the final choice. It induced some extent of switch as compared to both the control group and the compatible information group. Interestingly, the switch does not reflect the thematic content of the screens. This is consistent with the fact that the information in the screens did not favor any one of the projects. The quantum model provides an explanation for why Urgency individuals made aware of corruption problems reduced their support for ECF (presumably managed by the more reliable WWF). Distraction can induce such change. This is the case for example when the two perspectives (Urgency and Honesty) are uncorrelated (45° rotation as in the example) for a class of project-states and preferences. And it does seem reasonable to expect no or minimal correlation between Urgency of the cause and the Honesty of the NGO (in people's mind). The fact that the compatible information had no statistically significant impact supports the thesis that being simply exposed to a general information screen does not affect the choice. Instead it is only when the appending question induces a change in perspective that something happens.

We found significant variations between subgroups. First, we could exhibit a distinction in the reaction to distraction depending on preferences alone. On average, Urgency-individuals have been more sensitive to distraction than Honesty-individuals. This could be explained by a the fact that Urgency people tend to be more passionate about the situation. A passionate individual may feature a more pronounced quantum-like working of the mind because she is expected to be less constrained by the rational mind (see below for further arguments).

More intriguing is the fact that when combining preference and the order of presentation, we find that individuals whose preferences are congruent with the last presented project (TE-Hon and ET-Urg) tend to be more sensitive to distraction. The order of presentation of the projects is an element of the "preparation procedure '(in QM, the state of a quantum system is determined by a suitable preparation procedure). One possible explanation is that "congruent respondents" are more manipulable because both beliefs and preferences are indeterminate. Although this paper focuses on the indeterminacy of beliefs, both beliefs and preferences are mental objects that we expect can exhibit quantum-like properties. Indeed a number of works in quantum cognition address preference indeterminacy (see for instance [55]). As we discuss in the next section the rational mind tends to constrain the quantum-like working of the mind. We can thus conjecture that "congruent respondents" include respondents for whom the rational mind was less constraining. Their preferences were partly determined by the information received just before they had to respond to "what is most important for you?". This line of interpretation goes outside of our quantum model which focuses on beliefs indeterminacy however. It suggests that future research in quantum cognition should address both determinants of decision-making simultaneously.

Even when looking more closely at the results, we find no statistically significant impact of the compatible question. This is consistent with the Bayesian model because no information relevant to the choice between the two projects is provided. Note however that, despite the lack of statistical significance, the effect in the compatible condition is nonzero. We note that in the ET-Honesty subgroup is close to the effect of the incompatible condition. We recall that the theoretical model predicts some mild impact on the belief state. An U-individual who chooses TF on the basis of mixed beliefs will choose ECF with some probability if she is forced to decide for herself whether the cause is urgent YES or NO, prior to decision (see Section 2.3.4). The same holds for those who choose ECF while holding mixed beliefs. We conjecture that, at the sample level these effects also counter-balance each other so the overall impact is not statistically significant.

The time for responding to the whole questionnaire was between 1 and 3 min which is rather short. We interpret this feature as an evidence that the quantum working of the mind could be part of what Nobel prize Kahneman calls System 1—the fast, non-rational reasoning [12]: no new information of relevance for the choice was provided yet decision-making was affected. The respondent did not take time to reflect, they reacted spontaneously to the distraction. Recall that we do not elicit their preferences for the projects but only for what is determinant in a class of situations. That choice in our experiment was made to minimize interference from the rational mind. Nevertheless, we found that those determinants were highly correlated with the final choice both in the control and compatible information groups. The significance of the impact of distraction distraction results suggest that as we had conjectured respondents were not aware of the correlation (and the logic behind it). Therefore, they were not confronted with a (conscious) cognitive dissonance when the distraction changed their focus and eventually affected their decision. In the same line of thoughts the respondents overwhelmingly passed the chance to reassess their understanding of the project before making their choice. Only 12% used the opportunity re-read before making their choice.

An interesting finding is that the results are fully independent of population variables which supports the hypothesis that the quantum-like structure is a general regularity of the human mind.

5. Concluding Remarks and Discussion

In this paper, we have proposed an explanation of the manipulability of people's decision-making based on the intrinsic indeterminacy of the individual's subjective representation of the world. We first developed a simple quantum model of choice between two uncertain alternatives. Compared with the classical approach, the main distinction is in the modelisation of uncertainty. Where the classical approach relies on a single integrated representation on the world, the quantum-like modelling of uncertainty allows for a multiplicity of equally valid but subjectively incompatible perspectives on the world which is the expression of the intrinsic indeterminacy of mental objects. We show how an indeterminate representation of the world can be exploited to manipulate a decision-maker by a Sender who simply asks questions. Our focus has been on introspective questions, that is question about beliefs that bring no new information from the outside world. This allows establishing a clear distinction between the classical model's predictions and the quantum one. In particular, we show with an example that the quantum model predicts that distractive questions have strong persuasion power when the classical model predicts no impact of such questions at all.

We provided a first empirical test of that prediction in an experiment where individuals choose between supporting either one of two projects to save elephants respectively tigers. In the experiment that we performed, the change of focus or of narratives brought about by the distractive question was shown to statistically significantly affect revealed preferences for the projects. This central result is in accordance with the predictions of the quantum model when dealing with two incompatible perspectives here Urgency and Honesty. Looking closer, we find some significant differences in reaction between subgroups, with some reacting very strongly and others much less so. While this calls for further investigation, we find that this first experimental test was successful in providing some support for the hypothesis that the manipulability of people may have its roots in the indeterminacy of their subjective representation of the world.

In the real world however, a cause can simultaneously be urgent and the NGO supporting the project dishonest. There is thus a discrepancy between the properties of the true classical objects (the projects) and the properties of their representation, the mental objects (project-states). When Receiver processes information about a classical object as if it was a quantum system, she is mistaken. But as amply evidenced in Kahneman's best selling book "Thinking Fast and Slow", information processing is not always disciplined by (Bayesian) rational thinking when the brain operates quickly. The two-system approach does also open the way for manipulation because when the individual thinks fast she makes mistakes which could be exploited. Our view is that the quantum approach rather than being an alternative to most behavioral explanations, provides a rigorous foundations to a number of them. The interpretation that arises from its structure can however be different. Quantum cognition proposes that all forms of thinking are contextual due to the intrinsic indeterminacy of mental objects including beliefs and preferences. Conscious thinking may however interfere and constrain contextuality. Galberti [44] relies on a similar argument to explain Receiver's resistance to change worldview. The reason is that individuals have a resistance to changing their mind without a "good reason" due to drive toward consistency. This drive needs not be related to true rationality however but instead to an entrenched attachment to a stable identity or ego. The existence of a stable identity has been questioned by numerous experimental results (see e.g., self-perception theory and [56]). Those studies are consistent with a contextual and thus unstable identity [56]. As in the two-system approach the extent of conscious thinking matters. This is because the drive toward maintaining a coherent ego is more effectual when the individual is conscious about her instability. As argued in [55] cognitive dissonance and its resolution is an expression of that drive in face of instability (arising from intrinsic indeterminacy). We close this short discussion by suggesting that the quantum-like nature of mental objects needs not reflect a cognitive failure but would be the expression of the intrinsic indeterminacy (contextuality) of *human reality*. The question of rationality in such a context deserves further investigation.

Finally, we recognize that quantum cognition experiments cannot have the same degree of precision of physical experiments which prevents making and testing quantitative predictions. To a large part, this is because it is (today) impossible to fully characterize the state of a cognitive system which is incommensurably more complex that of an atomic particle. Nevertheless, our experimental exercise shows that it may be useful to test some theoretical predictions in contrast with standard classical (Bayesian) ones.

Author Contributions: Theory: A.L.-M.; Methodology: A.L.-M., A.C.; Data analysis: A.C.; Writing A.L.-M., A.C. Both authors have read and agreed to the published version of the manuscript.

Funding: This research was funded by the Paris School of Economics' small grants.

Institutional Review Board Statement: Not requested by funding agency.

Informed Consent Statement: Informed consent was obtained from all subjects involved in the study.

Data Availability Statement: The data presented in this study are available on request from the corresponding author.

Acknowledgments: We would like to thank seminar participants at Paris School of Economics and at the QI2028 symposium for their enriching comments as well as Jerome Busemeyer for a very valuable suggestions on the design of the experiment.

Conflicts of Interest: The authors declare no conflict of interest.

Appendix A

Table A1. Alternative Specifications.

	Dependent Variable:			
	Decision_ECF			
	(1)	**(2)**	**(3)**	**(4)**
QComp	−0.176	−0.170	−0.172	−0.186
	(0.134)	(0.134)	(0.134)	(0.137)
QIncomp	−0.475 ***	−0.476 ***	−0.477 ***	−0.451 ***
	(0.160)	(0.160)	(0.160)	(0.162)
Honesty		0.267 **	0.356 *	0.258 **
		(0.128)	(0.185)	(0.129)
ET		0.039	0.160	0.045
		(0.116)	(0.216)	(0.117)
Honesty:ET			−0.171	
			(0.256)	
Reread_Descriptions				−0.093
				(0.185)
Age				0.005
				(0.006)
Male				−0.189
				(0.121)
Education				0.021
				(0.083)
NGO				0.006
				(0.121)
Constant	0.371 ***	0.158	0.096	0.066
	(0.103)	(0.151)	(0.177)	(0.306)
Observations	1211	1211	1211	1211
Log Likelihood	−829.852	−827.610	−827.387	−825.591
Akaike Inf. Crit.	1665.704	1665.219	1666.775	1671.183
Bayesian Inf. Crit.	1681.002	1690.715	1697.370	1722.175

Note: * $p < 0.1$; ** $p < 0.05$; *** $p < 0.01$

References

1. P & G: Thank You, Mom | Wieden+Kennedy. Available online: https://www.wk.com/work/p-and-g-thank-you-mom/ (accessed on 16 January 2021).
2. Packard, V. *The Hidden Persuaders*. McKay: New York, NY, USA, 1957.
3. Cialdini, R.B.; Cialdini, R.B. *Influence: The Psychology of Persuasion*; Collins: New York, NY, USA, 2007; p. 55.
4. Petty, R.E.; Cacioppo, J.T. *Communication and Persuasion: Central and Peripheral Routes to Attitude Change*; Springer Science and Business Media: Berlin, Germany, 2012.
5. Festinger, L.; Maccoby, N. On resistance to persuasive communications. *J. Abnorm. Soc.* **1964**, *68*, 359. [CrossRef] [PubMed]
6. Baron, R.S.; Baron, P.H.; Miller, N. The relation between distraction and persuasion. *Psychol. Bull.* **1973**, *80*, 310. [CrossRef]
7. Petty, R.E.; Wells, G.L.; Brock, T.C. Distraction can enhance or reduce yielding to propaganda: Thought disruption versus effort justification. *J. Personal. Soc. Psychol.* **1976**, *34*, 874. [CrossRef]
8. Petty, R.E.; Cacioppo, J.T. The elaboration likelihood model of persuasion. In *Communication and Persuasion*; Springer: New York, NY, USA, 1986; pp. 1–24.
9. DellaVigna, S.; Gentzkow, M. Persuasion: Empirical evidence. *Annu. Rev. Econ.* **2010**, *2*, 643–669. [CrossRef]
10. Dudukovic, N.M.; DuBrow, S.; Wagner, A.D. Attention during memory retrieval enhances future remembering. *Mem. Cogn.* **2009**, *37*, 953–961. [CrossRef] [PubMed]
11. Fernandes, M.A.; Moscovitch, M. Divided attention and memory: Evidence of substantial interference effects at retrieval and encoding. *J. Exp. Psychol. Gen.* **2000**, *129*, 155. [CrossRef]
12. Kahneman, D. *Thinking Fast and Slow*, 1st ed.; Farrar, Straus and Giroux: New York, NY, USA, 2011.
13. Akerlof, G.; Shiller, R. *Phishing for Phools—The Economics of Manipulation and Deception*; Princeton University Press: Princeton, NJ, USA, 2015.
14. Tversky, A.; Kahneman, D. The framing of decisions and the psychology of choice. *Science* **1981**, *211*, 453–458. [CrossRef]
15. Chong, D.; Druckman, J. Framing Theory. *Annu. Rev. Polit. Sci.* **2007**, *10*, 103–126. [CrossRef]
16. Dehaene, S.; Naccache, L.; Le Clec'H, G.; Koechlin, E.; Mueller, M.; Dehaene-Lambertz, G.; Le Bihan, D. Imaging unconscious semantic priming. *Nature* **1998**, *395*, 597–600. [CrossRef]
17. Taylor, S.E. The Availability Bias in Social Perception and Interaction. In *Judgment under Uncertainty: Heuristics and Biases*; Kahneman, D., Slovic, P., Tversky, A., Eds.; Cambridge University Press: New York, NY, USA, 1982.
18. Latham, G. Unanswered questions and new directions for future research on priming goals in the subconscious. *Acad. Manag. Discov.* **2019**, *5*, 111–113. [CrossRef]
19. Bargh, J.A. The historical origins of priming as the preparation of behavioral responses: Unconscious carryover and contextual influences of real-world importance. *Soc. Cogn.* **2014**, *32*, 209–224. [CrossRef]
20. Bargh, J.A. *Before You Know It: The Unconscious Reasons We Do What We Do*; Simon & Schuster: New York, NY, USA, 2017.
21. Dijksterhuis, A.; Aarts, H. Goals, attention, and (un) consciousness. *Annu. Rev. Psychol.* **2010**, *61*, 467–490. [CrossRef] [PubMed]
22. Dubois, F.; Lambert-Mogiliansky, A. Our (represented) world and quantum-like object. In *Contextuality in Quantum Physics and Psychology*; Advanced Series in Mathematical Psychology; Dzhafarov, E., Scott, J., Ru, Z., Cervantes, V., Eds.; World Scientific: Singapore, 2016; Volume 6, pp. 367–387.
23. Danilov, V.I.; Lambert-Mogiliansky, A. Targeting in Persuasion Problems. *J. Math. Econ.* **2018**, *78*, 142–149. [CrossRef]
24. Benjamin, D.J. Errors in probabilistic reasoning and judgment biases. In *Handbook of Behavioral Economics—Foundations and Applications 2*; Bernheim, B.D., DellaVigna, S., Laibson, D., Eds.; North-Holland: Amsterdam, The Netherlands, 2019; Chapter 2, pp. 69–186.
25. Camerer, C. Bounded rationality in individual decision making. *Exp. Econ.* **1998**, *1*, 163–183. [CrossRef]
26. Edwards, W. Conservatism in human information processing. In *Judgment under Uncertainty: Heuristics and Biases*; Kahneman, D., Slovic, P., Tversky, A., Eds.; Cambridge University Press: Cambridge, UK, 1982; pp. 359–369.
27. Grether, D.M. Testing Bayes rule and the representativeness heuristic: Some experimental evidence. *J. Econ. Behav. Organ.* **1992**, *17*, 31–57. [CrossRef]
28. Tversky, A.; Kahneman, D. Extensional versus intuitive reasoning: The conjunction fallacy in probability judgment. *Psychol. Rev.* **1983**, *90*, 293. [CrossRef]
29. Zizzo, D.J.; Stolarz-Fantino, S.; Wen, J.; Fantino, E. A violation of the monotonicity axiom: Experimental evidence on the conjunction fallacy. *J. Econ. Behav. Organ.* **2000**, *41*, 263–276. [CrossRef]
30. Bruza, P.; Busemeyer, J.R. *Quantum Cognition and Decision-Making*; Cambridge University Press: Cambridge, UK, 2012.
31. Haven, E.; Khrennikov, A. *The Palgrave Handbook of Quantum Models in Social Science*; Macmillan Publishers Ltd.: London, UK, 2017; pp. 1–17.
32. Khrennikov, A.; Basieva, I.; Dzhafarov, E.N.; Busemeyer, J.R. Quantum models for psychological measurements: An unsolved problem. *PLoS ONE* **2014**, *9*, e110909. [CrossRef]
33. Ozawa, M.; Khrennikov, A. Application of theory of quantum instruments to psychology: Combination of question order effect with response replicability effect. *Entropy* **2020**, *22*, 37. [CrossRef]
34. Bagarello, F.; Basieva, I.; Khrennikov, A. Quantum field inspired model of decision making: Asymptotic stabilization of belief state via interaction with surrounding mental environment. *J. Math. Psychol.* **2018**, *82*, 159–168. [CrossRef]

35. Basieva, I.; Cervantes, V.H.; Dzhafarov, E.N.; Khrennikov, A. True Contextuality Beats Direct Influences in Human Decision Making. *J. Exp. Psychol. Gen.* **2019**, *148*, 1925–1937. [CrossRef] [PubMed]
36. Aerts, D.; Haven, E.; Sozzo, S. A proposal to extend expected utility in a quantum probabilistic framework. *Econ. Theory* **2018**, *65*, 1079–1109. [CrossRef]
37. Danilov, V.I.; Lambert-Mogiliansky, A. Expected Utility under Non-classical Uncertainty. *Theory Decis.* **2010**, *68*, 25–47. [CrossRef]
38. Danilov, V.I.; Lambert-Mogiliansky, A.; Vergopoulos, V. Dynamic consistency of expected utility under non-classical (quantum) uncertainty. *Theory Decis.* **2018**, *84*, 645–670. [CrossRef]
39. Danilov, V.I.; Lambert-Mogiliansky, A. Preparing a (quantum) belief system. *Theor. Comput. Sci.* **2018**, *752*, 97–103. [CrossRef]
40. Cohen-Tannoudji, C.; Diu, B.; Laloë, F. *Mécanique Quantique*; Hermann, EDP Sciences: Paris, France, 2000.
41. Kamenica, E.; Gentzkow, M. Bayesian Persuasion. *Am. Econ. Rev.* **2011**, *101*, 2590–2615. [CrossRef]
42. Bloedel, A.W.; Segal, I.R. Persuasion with Rational Inattention. Available online: https://ssrn.com/abstract=3164033 (accessed on 16 April 2018).
43. Lipnowski, E.; Mathevet, L. Disclosure to a psychological audience. *Am. Econ. J. Microeconom.* **2018**, *10*, 67–93. [CrossRef]
44. Galperti, S. Persuasion: The Art of Changing Worldviews. *Am. Econ. Rev.* **2019**, *109*, 996–1031. [CrossRef]
45. De Clippel, G.; Zhang, Z. *Non-Bayesian Persuasion*; Working Paper; Brown University: Providence, RI, USA, 2020.
46. Broekaert, J.B.; Busemeyer, J.R.; Pothos, E.M. The disjunction effect in two-stage simulated gambles. An experimental study and comparison of a heuristic logistic, Markov and quantum-like model. *Cogn. Psychol.* **2020**, *117*, 101262. [CrossRef]
47. Busemeyer, J.R.; Wang, Z.; Shiffrin, R.S. Bayesian model comparison favors quantum over standard decision theory account of dynamic inconsistency. *Decision* **2015**, *2*, 1–12. [CrossRef]
48. Denolf, J.; Martínez-Martínez, I.; Josephy, H.; Barque-Duran, A. A quantum-like model for complementarity of preferences and beliefs in dilemma games. *J. Math. Psychol.* **2016**, *78*, 96–106. [CrossRef]
49. Moreira, C.; Wichert, A. Are quantum-like Bayesian networks more powerful than classical Bayesian networks? *J. Math. Psychol.* **2018**, *82*, 73–83. [CrossRef]
50. Dzhafarov, E.; Scott, J.; Ru, Z.; Cervantes, V. (Eds.) *Contextuality from Quantum Physics to Psychology*; Advanced Series in Mathematical Psychology; World Scientific: Singapore, 2016; Volume 6, ISBN 978–981-4730-62-4.
51. Wang, Z; Busemeyer, J.R.; DeBuys, B. Beliefs, action and rationality in strategical decisions. *Top. Cogn. Sci.* **2020**, in press.
52. Bartneck, C.; Duenser, A.; Moltchanova, E.; Zawieska, K. Comparing the similarity of responses received from studies in Amazon's Mechanical Turk to studies conducted online and with direct recruitment. *PLoS ONE* **2015**, *10*, e0121595. [CrossRef] [PubMed]
53. Kees, J.; Berry, C.; Burton, S.; Sheehan, K. An analysis of data quality: Professional panels, student subject pools, and Amazon's Mechanical Turk. *J. Advert.* **2017**, *46*, 141–155. [CrossRef]
54. Festinger, L. *A Theory of Cognitive Dissonance*; Stanford University Press: Stanford, CA, USA, 1957; Volume 2.
55. Lambert-Mogiliansky A.; Zamir S.; Zwirn H. Type-Indeterminacy a Model of the KT (Kahnemann and Tversky) Man. *J. Math. Psychol.* **2009**, *53*, 349–361. [CrossRef]
56. Lambert-Mogiliansky A. Quantum Type Indeterminacy in Dynamic Decision-Making: Self-Control through Identity Management. *Games* **2012**, *3*, 97–118. [CrossRef]

Article

Optical-Cavity-Induced Current

Garret Moddel *, Ayendra Weerakkody, David Doroski and Dylan Bartusiak

Department of Electrical, Computer, and Energy Engineering, University of Colorado,
Boulder, CO 80309-0425, USA; Don.Weerakkody@Colorado.EDU (A.W.); doroski@Colorado.EDU (D.D.);
Dylan.Bartusiak@colorado.edu (D.B.)
* Correspondence: moddel@colorado.edu

Abstract: The formation of a submicron optical cavity on one side of a metal–insulator–metal (MIM) tunneling device induces a measurable electrical current between the two metal layers with no applied voltage. Reducing the cavity thickness increases the measured current. Eight types of tests were carried out to determine whether the output could be due to experimental artifacts. All gave negative results, supporting the conclusion that the observed electrical output is genuinely produced by the device. We interpret the results as being due to the suppression of vacuum optical modes by the optical cavity on one side of the MIM device, which upsets a balance in the injection of electrons excited by zero-point fluctuations. This interpretation is in accord with observed changes in the electrical output as other device parameters are varied. A feature of the MIM devices is their femtosecond-fast transport and scattering times for hot charge carriers. The fast capture in these devices is consistent with a model in which an energy ΔE may be accessed from zero-point fluctuations for a time Δt, following a $\Delta E \Delta t$ uncertainty-principle-like relation governing the process.

Keywords: MIM diode; metal–insulator–metal diode; photoinjection; internal photoemission; vacuum fluctuations; Casimir effect; zero-point fluctuations; geometrical asymmetry

Citation: Moddel, G.; Weerakkody, A.; Doroski, D.; Bartusiak, D. Optical-Cavity-Induced Current. *Symmetry* **2021**, *13*, 517. https://doi.org/10.3390/sym13030517

Academic Editors: Christophe Humbert and G. Jordan Maclay

Received: 7 January 2021
Accepted: 16 March 2021
Published: 22 March 2021

Publisher's Note: MDPI stays neutral with regard to jurisdictional claims in published maps and institutional affiliations.

1. Introduction

Metal–insulator–metal (MIM) tunnel diodes have been used to provide rectification and nonlinearity [1–3] for a variety of applications. The insulator forms a barrier that charge carriers—electrons or holes—must cross to provide current when a voltage is applied across the device. In addition, current can be produced by the direct absorption of light on one of the metal surfaces of an MIM sandwich structure, which generates the hot carriers that cross the metal and are injected into the insulator. This internal photoemission [4,5] is also called photoinjection. For current to be provided, the metal layer must be thinner than the hot-carrier mean-free path length so that the carriers can cross it without being scattered. Once they reach the insulator, the hot carriers must have sufficient energy to surmount the energy barrier at the interface and traverse the insulator ballistically above its conduction band edge, or alternatively they can tunnel through the insulator. Thinner insulators favor tunneling [4,6]. After entering the metal base electrode on the other side, the hot carriers are scattered and captured.

For over two decades, our lab has designed and fabricated MIM diodes for ultrahigh speed rectification [3]. We have found that incorporating a thin optical resonator used as an optical cavity on one side of the MIM structure induces a reduction in the device conductivity measured over a range of several tenths of a volt [7]. At lower, submillivolt voltages, a persistent induced current and voltage is evident, which we report here. We describe observed trends in the current output as the thicknesses of the optical and electrical layers in the device are varied. After showing these trends, we present the results of a range of tests that are carried out to check whether the results could be due to some sort of an experimental artifact. Finally, to explain what could produce the observed output, we present a conjectural photoinjection model in which hot carriers are excited by the quantum vacuum field.

79

The devices consist of thin optical cavities deposited over MIM structures, as depicted in Figure 1. The cavity thickness is in the range of tens of nanometers up to approximately 1 μm, which results in a cavity optical mode density with a wavelength dependence described by an Airy function [8]. This cavity largely suppresses wavelengths longer than twice the cavity thickness multiplied by the refractive index of the transparent dielectric. For our devices, the resulting wavelength cutoff, above which modes are suppressed, varies from the near-infrared (NIR) through the near-ultraviolet, depending on the cavity thickness. The MIM structures include a nanometer-thick insulator to form the barrier. The upper electrode is sufficiently thin to allow hot carriers that are photoexcited on the optical cavity side of the electrode to penetrate the electrode and reach the insulator without being scattered.

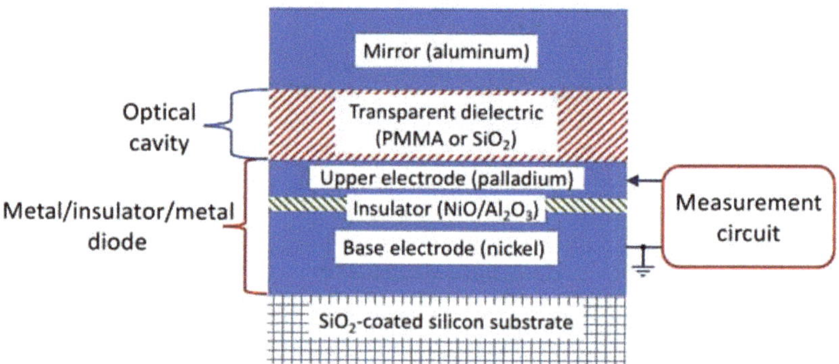

Figure 1. Device cross section, showing a metal–insulator–metal (MIM) structure adjoining an optical cavity. The electrical characteristics of the device are measured between the two metal layers of the MIM structure, where the polarity of the upper electrode voltage is with reference to the base electrode, which is defined as ground. Positive current is defined to be in the direction of the arrow.

2. Materials and Methods

2.1. Device Fabrication

Two different processes were used to form the devices. Submicron devices were fabricated using a germanium shadow-mask (GSM) process [9,10]. Using a deep-ultraviolet stepper, a 250 nm wide germanium bridge is formed over an SiO$_2$-coated surface of a silicon wafer, as depicted in Figure 2a. First the nickel base electrode is evaporated under the bridge from one side. This is followed by native NiO$_x$ growth at room temperature, and then by conformal Al$_2$O$_3$ deposited by sputtering. After the insulator is formed, the palladium upper electrode is evaporated from the opposite side. The resulting overlap of the two metals forms an ellipse with an area of 0.02 ± 0.006 μm^2, as shown in Figure 2b. After the germanium bridge is removed, a transparent dielectric, i.e., spun-on polymethyl methacrylate (PMMA) or sputtered SiO$_2$, is deposited to form an optical cavity over the MIM structure. The dielectric layer is then coated with an aluminum mirror. In addition to providing a reduced density of optical modes, the optical cavity encapsulates and stabilizes the MIM structure, blocking further oxidation. It should be noted that the use of PMMA to support the germanium bridge during fabrication, as shown in Figure 2a, is independent of whether PMMA is also used in a subsequent layer to form an optical cavity.

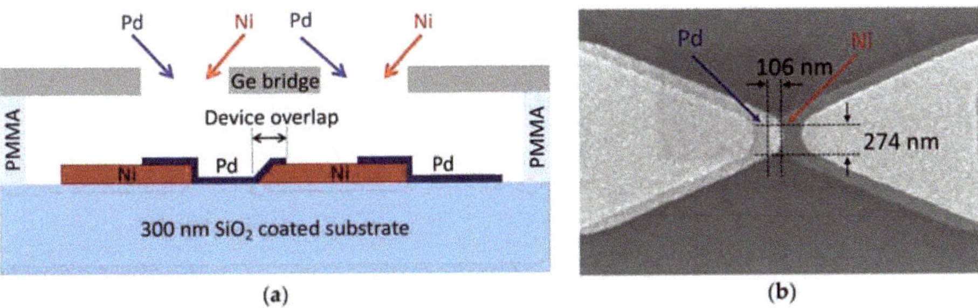

Figure 2. Germanium shadow mask (GSM) device fabrication. (**a**) Depiction of fabrication process, showing a cross-sectional view of materials deposited under a germanium bridge. The NiO$_x$ and Al$_2$O$_3$ insulating layer formed over the Ni layer is not shown. The active area of the device is formed in the overlap region. (**b**) A scanning electron microscope (SEM) image of a completed device, with an overlap area of 0.02 ± 0.006 μm^2. The Ni and Pd-coated regions are indicated; the lightest regions, in the center and at the left and right-hand sides, are coated with both Ni and Pd layers with the insulator layer between them.

Additional devices, having larger areas, were fabricated using standard photolithographic techniques. The top view of one of these devices is shown in Figure 3.

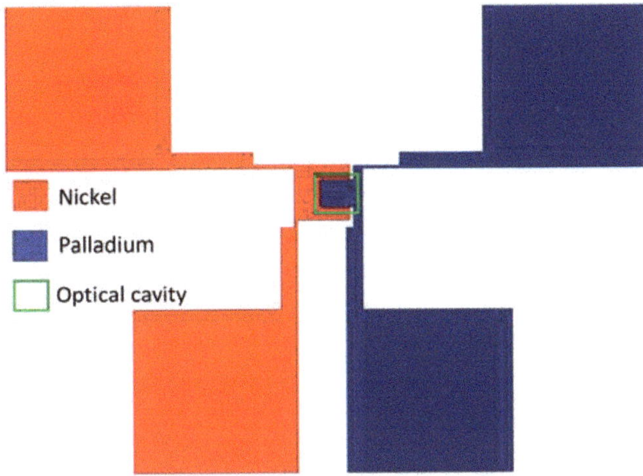

Figure 3. Photolithographic device. Top view of devices formed by the standard photolithographic technique. The overlap of the pallidum upper electrode, shown to the right, with the nickel lower electrode, shown to the left, forms active square regions with edge lengths between 5 and 100 μm.

The thicknesses and refractive indices of most of the dielectric layers were determined by UV–Vis–NIR variable angle spectroscopic ellipsometry (VASE) measurements. The measured refractive indices for the spun-on PMMA and deposited SiO$_2$ are 1.52 and 1.49, respectively, at a wavelength of 300 nm. The Al$_2$O$_3$ thickness values were measured by VASE on silicon witness samples that were placed in the sputtering system along with the devices. For the photolithographic devices, we also measured the native NiO$_x$ thickness by VASE and found it to be 2.3 nm. In addition, a monolayer (0.4 nm) of photoresist remained over the NiO$_x$ layer in the photolithographic devices. The reason for this is that after depositing the insulator and patterning the upper electrode using a liftoff process, we could not use the standard oxygen plasma to clean residual photoresist off the insulator surface without further oxidizing the insulator, with the result that the devices become too resistive. For GSM devices the patterning was accomplished by the shadow-mask deposition depicted in Figure 2a, and so no photoresist was required. Because the thickness of the native NiO_x insulator in the GSM MIM structures could not be measured directly, we determined its effective thickness from electrical measurements and simulations of Ni/native NiO_x/Pd structures. We extracted a NiO_x thicknesses of 0.6–1 nm for effective barrier heights in the range of 0.06–0.08 eV [11]. The barrier height was calculated by performing a Fowler–Nordheim analysis on a low resistance (~100 Ω) device with nonlinear current–voltage characteristics. The native NiO_x thickness for GSM structures is smaller than that for the photolithographic devices because of the higher processing temperatures for the photolithographic devices, and also possibly because the junctions in GSM structures were partially protected by the germanium bridge. The total effective insulator thickness for the Al_2O_3/NiO_x combination is the sum of the thickness values for each layer. Thickness values for the Ni base electrodes are 38 nm for the GSM devices and 50 nm for the photolithographic devices. The aluminum mirror is 150 nm thick for all devices. The thickness values for the other layers in the devices for each figure are provided in Table 1.

Table 1. Device parameters for each figure.

Figure	Area	NiO_x	Al_2O_3	Pd Upper Electrode	Transparent Dielectric	
	(μm^2)	(nm)	(nm)	(nm)	Values (nm)	Material
4 (a)	0.02		1.3	8.3	33–1100	PMMA
4 (b)	0.02	1	0.9	8.3	33–1100	PMMA or SiO_2
5 (a) [1]	10,000	2.3 + 0.4 resist	2.3	8.7–24	11	SiO_2
5 (b) [1]	625	2.3 + 0.4 resist	0.7–1.5	15	11	SiO_2
6 (a)	0.02	1	0.7	8.3	35	PMMA
6 (b)	25–10,000	2.3 + 0.4 resist	2.3	12	11	SiO_2
7 (b)	0.02 × 16	1	0.9	15.6	107	PMMA
8	0.02	1	0.9 0.7	8.3	36 50	PMMA SiO_2
9	0.02	1	0.7	8.7	–	PMMA
10	0.02	1	05	8.7	33	PMMA
11	0.02	1	0.7	8.7	35	PMMA

[1] Photolithographic devices; all other devices used the GSM fabrication process.

2.2. Device Measurement

Once the MIM structures were fabricated, we carried out current–voltage ($I(V)$) measurements at room temperature using a four-point probe configuration to circumvent the effects of lead resistance. A high precision Keithley 2612 source meter (calibrated to NIST standards) was used to source either voltage or current across two pads, and an HP 3478A digital multimeter (DMM) was used to measure the voltage drop across the MIM junctions. Although the standard technique is to source a voltage and measure the current using the source meter, this can result in erroneous offsets for low resistance devices, e.g., for currents on the order of 1 nA through a device having a resistance less than 1 MΩ. Therefore, we carried out some of the measurements, particularly for low-resistance devices, by sourcing the current ($\pm 0.06\% \pm 100$ pA accuracy) and measuring the voltage. To eliminate any effects due to thermoelectric potentials resulting from a temperature difference between the source meter and the probes, we used a current reversal method [12]. Following this method, we performed two measurements with currents of opposite polarity, i.e., one when the base electrode was grounded and another when the upper electrode was grounded, and then we subtracted the difference in the currents to yield the final value.

In fabricating and testing tens of thousands of MIM devices, we generally found a wide range of resistances for nominally the same fabrication conditions due to slight uncontrollable variations in the insulator thickness [13], which is fewer than 10 lattice constants thick. In most cases, the measurement results presented are averages across each wafer chip, with error bars showing the standard deviation.

3. Results

3.1. Electrical Response Measurements

The electrical response of MIM devices having an adjoining optical cavity is shown in Figure 4. In Figure 4a, the $I(V)$ curves extend into the second quadrant and therefore exhibit a positive power output. For linear $I(V)$ characteristics, the maximum power is $|I_{SC}V_{OC}|/4$, where I_{SC} and V_{OC} are the short-circuit current and open-circuit voltage, respectively. For the device with a 33 nm thick cavity, the maximum power is 1.4 pW in a $0.02~\mu m^2$ area (see the SEM image of Figure 2b). The short-circuit current increases when decreasing cavity thickness, as shown in Figure 4b for two different cavity dielectrics. This increase in current with decreasing cavity thickness corresponds to an increasing range of suppressed optical modes with decreasing thickness, as described in the Introduction.

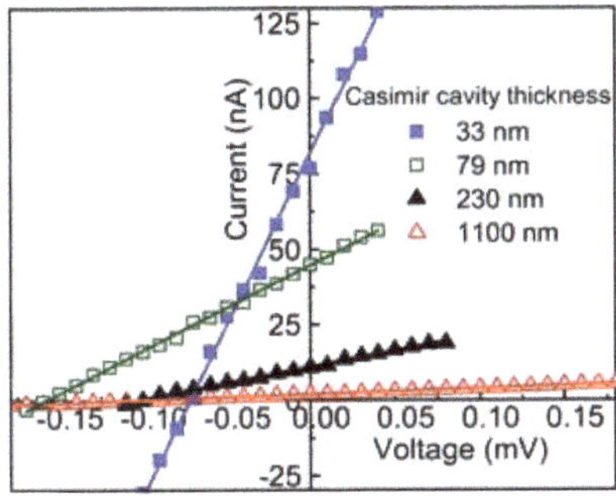

(a)

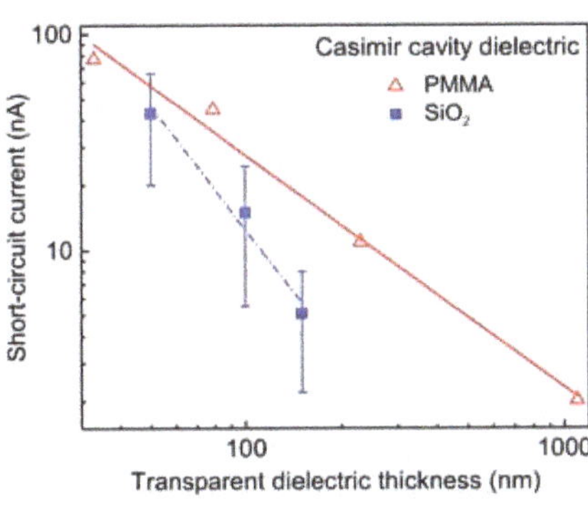

(b)

Figure 4. Electrical response as a function of cavity thickness. (**a**) Current as a function of voltage for different polymethyl methacrylate (PMMA) cavity thicknesses. (**b**) Short-circuit current as a function of cavity thickness for PMMA and SiO_2–filled cavities.

To understand the current-producing mechanism, we carried out tests to determine whether there is evidence that it involves hot charge carriers generated from optical fields in the optical cavity. These carriers, generated in the Pd upper electrode near the interface to the transparent dielectric cavity shown in Figure 1, could be injected into the insulator if they could traverse the Pd without being scattered. This photoinjection current should decrease when increasing the upper electrode thickness because of the increased scattering of the hot carriers before they reach the insulator, which results in excited carriers not contributing to the current. The hot electron mean-free path length in metals at room temperature is on the order of 10 nm [14]. On the other hand, when the upper electrode thicknesses is below the absorption depth of Pd, which is 10 nm for 0.4 μm radiation [15], the photoinjection current would be expected to decrease when decreasing the electrode thickness because the rate of carrier excitation is reduced. Figure 5a shows the short circuit current as a function of upper electrode thickness. This current does, in fact, decrease with increasing thickness, and also decreases for the thickness below 10 nm, peaking for a Pd thickness of approximately 12 nm.

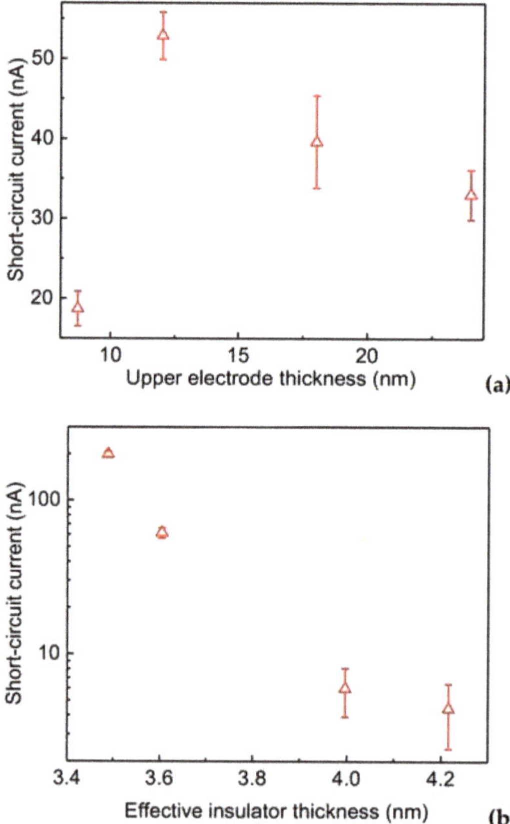

Figure 5. Tests for photoinjection of hot carriers through the Pd upper electrode and through the insulator. (**a**) Short-circuit current as a function of upper electrode Pd thickness. (**b**) Short-circuit current as a function of effective insulator thickness. Both trends are consistent with hot-carrier photoinjection from optical fields in the optical cavity.

To be collected, these hot carriers must traverse the insulator either ballistically or via tunneling. The ballistic transport is limited by the mean-free path length in the insulator, on the order of several nanometers [16]. The tunneling probability decreases exponentially with insulator thickness [17]. In either case, the short-circuit current would be expected to decrease with increasing insulator thickness if the current is due to charge injection through the insulator. This trend is observed by the data of Figure 5b.

Although the devices of Figure 4 were fabricated using the GSM process and the devices of Figure 5 used standard photolithography, devices fabricated by both processes exhibited similar trends. The GSM process allowed for much shorter fabrication times and for the absence of residual photoresist (described in Section 2.2), but it did not allow for large device areas or for varying the device area.

3.2. Testing for Experimental Artifacts

3.2.1. Stability over Time

We carried out a series of experiments to test whether the results presented above might be due to some sort of experimental artifact rather than a genuine electrical response from these structures. One concern is whether the observed current is a transient or hysteresis effect, possibly due to charging, as opposed to being a stable output from the devices. If, for example, one charge was trapped for every Al_2O_3 molecule (having a lattice constant of 0.5 nm) in a 2.5-nm thick layer of area 0.02 μm^2, depleting that charge could produce a current of 20 nA for 3.2 μs. To test for this, we measured the short-circuit current continuously over a period of four hours. The data, provided in Figure 6a, shows no change over time.

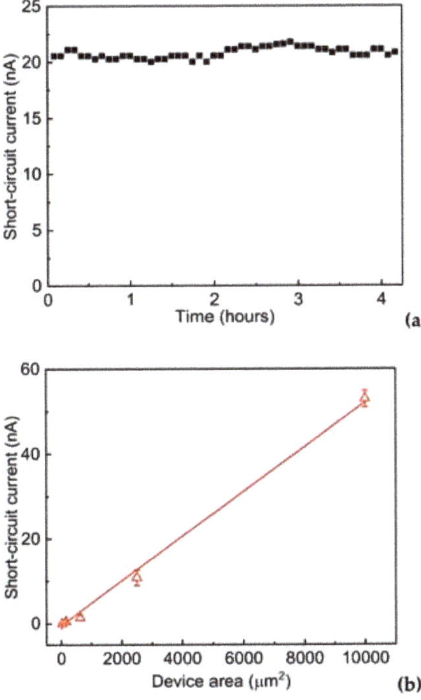

Figure 6. Two tests to check whether the measured current could be an experimental artifact. (a) Short-circuit current as a function of time over a period of four hours for a GSM device. (b) Short-circuit current as a function of active device area, as defined by the upper electrode area, for devices fabricated using the photolithography process.

3.2.2. Area Dependence

Another concern is whether the output current is collected from just the active area covered by the upper electrode, or whether it is due to some other effect that would not scale with the active area. To test for this, we fabricated devices with a range of areas, in which the overlap shown in Figure 3 was varied. The results are shown in Figure 6b. The current scales linearly with the active area, supporting the conclusion that the source for the current is the active area.

3.2.3. Array Dependence

In addition to scaling with area, the short-circuit current should scale with the number of devices in parallel. Similarly, the open-circuit voltage should scale with the number of devices in series. If, for example, the output was the result of thermoelectric effects at the contacts, it would not scale with the number of devices. We tested for this possibility by fabricating and measuring two types of 4 × 4 arrays. One is a staggered array, schematically depicted in Figure 7a. It is a combination of series and parallel connections designed to reduce the effect of defective individual devices. The other 4 × 4 array is a series–parallel array consisting of four parallel sets of four devices in series. The results are shown in Figure 7b. The array currents and voltages are approximately four times those of the single devices, supporting the validity of the measured results.

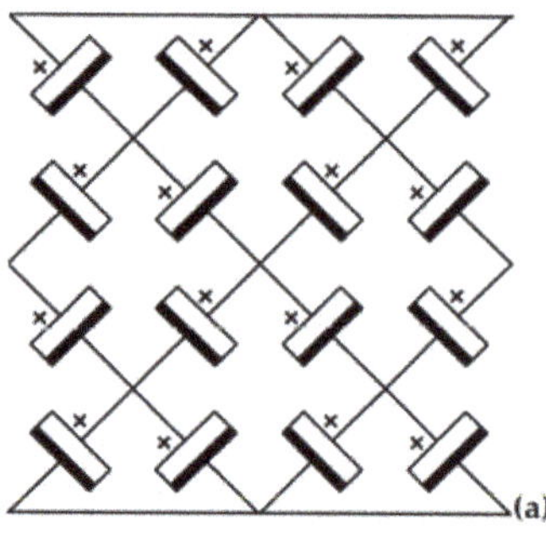

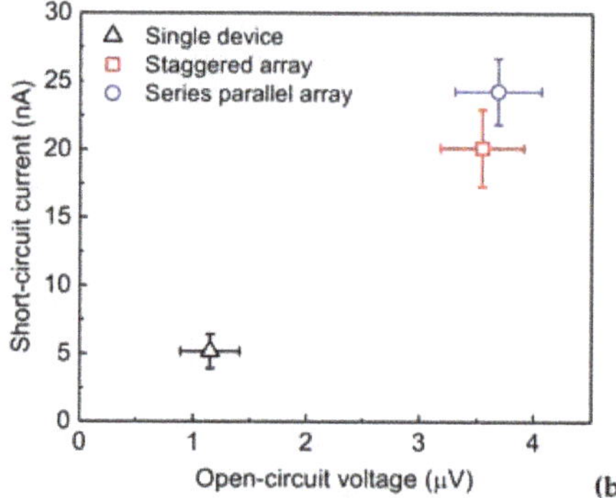

Figure 7. Measurements of device arrays. (**a**) Schematic representation of a 4 × 4 staggered array, where the circuit symbols represent cavity/electronic-device elements; the measured total short-circuit current between the top and bottom bars and the open-circuit voltage are each four times that for a single element. (**b**) Measured short-circuit current and open-circuit voltage for single devices, staggered 4 × 4 arrays, and series–parallel arrays consisting of four parallel sets of four devices in series.

3.2.4. Processing Dependence

From experience fabricating and testing many thousands of MIM structures, we are confident that it is not the MIM structure alone that produces the observed electrical output. It is conceivable, however, that it is the additional processing of the MIM structures to form the adjoining cavity, and not the cavity itself, that gives rise to the output. To check for that, we measured devices at different stages of cavity formation. Figure 8 shows the short-circuit current from MIM devices at these different stages. The first stage is for an as-built MIM structure. Only a negligible current is produced. We then annealed the MIM structure at 180 °C for 15 min to replicate the temperature cycle that it would undergo during the process in which the mirror was defined. Again, no significant current was evident. We then deposited the cavity dielectric, i.e., PMMA in one case and SiO_2 in the other, and we saw no change in the current. Finally, after depositing the aluminum mirror, the current jumped to the values that we observed in the completed devices. This makes it clear that it is the cavity with the mirror, and not just the device processing, that yields a device producing the observed electrical output.

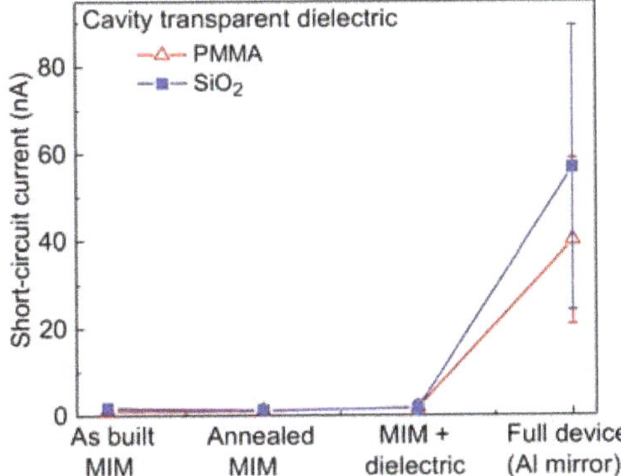

Figure 8. Effect of cavity formation on short-circuit current. The anneal was carried out at 180 °C for 15 min to replicate the mirror processing temperature cycle. Only a completed device produces a significant current.

The characteristics remain stable over time; for instance, for the full devices of Figure 8, after approximately six months the PMMA-cavity device output degraded by less than 10% and the SiO_2-cavity device output degraded by less than 20%.

3.2.5. Current Leakage through the Cavity

A possible source for the observed currents and voltages produced between the MIM electrodes might be leakage currents through the transparent dielectric from the mirror, which somehow picks up anomalous voltages. For this to be the case, resistance between the mirror and the MIM upper electrode would have to be on the order of or less than the resistance between the MIM electrodes. To test for this, we measured the relevant resistances in some completed devices. The results, given in Figure 9, show that the observed currents could not be due to leakage from the mirror.

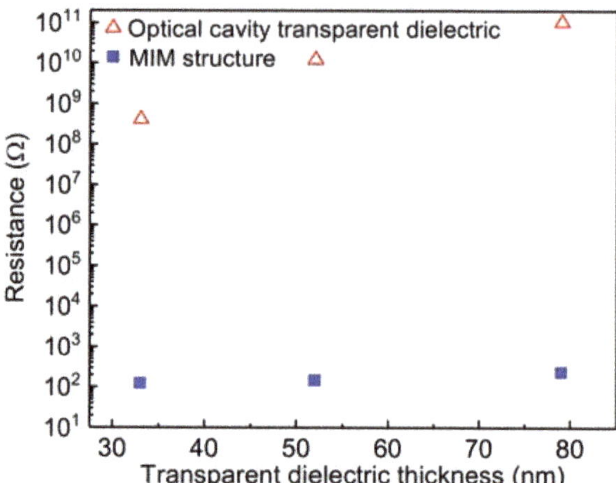

Figure 9. Comparison of resistance values across the optical cavity transparent dielectrics and across MIM structures. Because of the much higher resistance between the mirror and upper electrode than between the upper and the base electrodes (shown in Figure 1), the currents observed between the electrodes could not be the result of leakage from the mirror.

3.2.6. Electromagnetic Pickup

Another potential source of anomalous currents and voltage is ambient electromagnetic radiation. Electromagnetic pickup might occur somewhere in the device and result in current through the MIM structure, which would rectify it to produce the $I(V)$ characteristics shown in Figure 4a. To avoid such rectification, we designed the MIM devices to have low barrier heights and consequently linear $I(V)$ characteristics, as is evident from the lack of curvature in the data of Figure 4a. Still, a slight nonlinearity might rectify picked up signals. To test for that, we carried out measurements in three different environments: (i) the usual measurement conditions using a probed wafer chip mounted onto a measurement stage exposed to ambient fields, (ii) inside a mu-metal box, which blocks low frequency electromagnetic radiation, and (iii) inside an aluminum box, which blocks higher frequency radiation. The results, given in Figure 10, show that the current–voltage characteristics are the same for all three measurements. While these environments do not totally block all ambient radiation, the fact that the three measurements give the same results make it highly unlikely that electromagnetic pickup is the source for the electrical outputs that we observed.

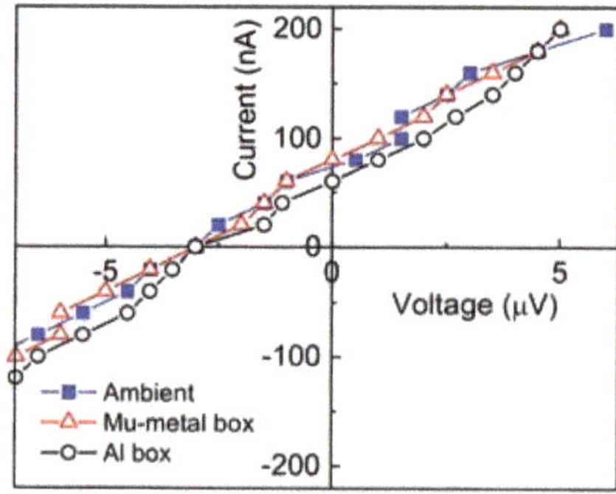

Figure 10. Effect of blocking ambient electromagnetic radiation during measurement of current-voltage characteristics. The characteristics measured under open ambient conditions did not change when the device was place in mu-metal or aluminum boxes.

3.2.7. Thermoelectric Effects on Electrodes

A common source of errors in low-voltage measurements is thermoelectric effects. One type is generated from a temperature difference between electrodes of different types at the device and at the measurement electronics. This voltage can be cancelled using the voltage reversal method [12] during measurement, as described in Section 2.1. We used this method consistently in our measurements. Another experiment that would indicate whether this type of thermoelectric effect could be the source of the electrical output is the measurement of device arrays. As the devices on a single substrate and at a uniform temperature are linked together, the thermoelectric voltage measured at the electronics would not change, and so the voltage would not scale with the number of devices in series. As shown in Figure 7b, the measured voltage does scale with the number of devices in series, and so it is not due to such a thermoelectric effect. Both the voltage reversal method and the device array results show that this type of thermoelectric voltage does not affect the results.

3.2.8. Thermoelectric Effects on Devices

Another potential source of thermoelectric effects is temperature differences within the sample itself. If the upper electrode was at a slightly different temperature than the lower electrode, this would generate a thermoelectric voltage. Such a temperature difference would be unlikely because of the much greater thermal conductance across the thin insulator than between the upper part of the device, shown in Figure 1, and the surrounding air. To be sure that this sort of thermoelectric effect is not the source of the measured electrical output, we carried out measurements to test for effects from such a temperature difference. The wafer chip containing the device is held tightly onto the metal measurement stage with a vacuum chuck. We varied the temperature of the measurement stage while the ambient temperature remained constant, and measured the short-circuit current and open-circuit voltage from the device. If there was a difference in temperatures between the upper and lower MIM electrodes that gave rise to a thermoelectric voltage, such a test would shift it or reverse its polarity. No change in the output voltage or current was observed, as shown in Figure 11, providing evidence that a temperature difference is not the source of the measured electrical output.

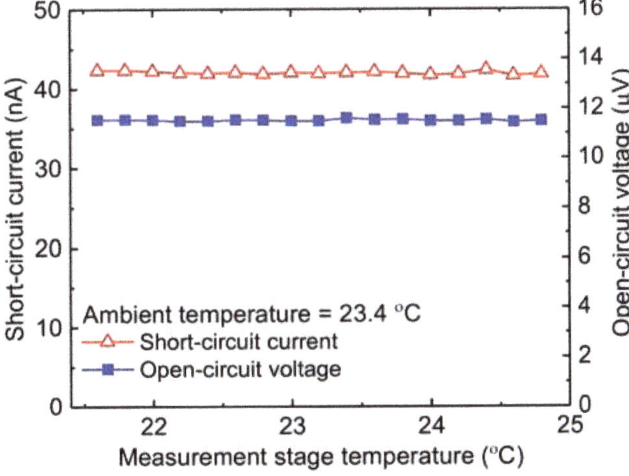

Figure 11. Test for possible thermoelectric effects. The electrical output is measured as the difference between the substrate and ambient temperatures is varied. The measured short-circuit current and open-circuit voltage does not vary, providing evidence that such a temperature difference is not the source of the measured electrical output.

The data shown here are not flukes observed in rare devices. The trends reported in this paper were replicated in over 1000 MIM-based devices produced in 21 different batches. Virtually all the devices with working (nonshorted) MIM structures exhibited the type of electrical characteristics shown in Figure 4a.

Based on all these checks for experimental artifacts, it appears that the measured electrical characteristics of the photoinjector cavity devices are, in fact, real and due to the devices themselves.

4. Discussion

The data show than when an MIM structure adjoins a thin optical cavity, electrical power is produced. An extensive range of tests show that real power is provided, and the results are not due to experimental artifacts. We consider a possible mechanism for the electrical characteristics based upon the experimental observations.

To produce current, the upper electrode in the MIM device must be thin, less than or on the order of a mean-free path length for ballistic charge carriers, consistent with Figure 5a. This suggests that the carriers are photoinjected from the cavity side of the MIM device. The observation that the current decreases with increasing insulator thickness, shown in Figure 5b, is consistent with the charge traversing the insulator by surmounting the metal/insulator barrier or tunneling through it.

The current increases with decreasing cavity thickness, as shown in Figure 4b. Optical cavities largely suppress wavelengths greater than twice the cavity spacing, such that the suppressed band extends to shorter wavelengths for thinner cavities. Therefore, the increase in current corresponds to an increasingly wide band of suppressed optical modes in the cavity. The source of these optical modes could be the quantum vacuum field, which gives rise to the Casimir force [18–21], the Lamb shift [22], and other physical effects [23]. It was argued that the use of energy from the vacuum field does not violate fundamental laws of thermodynamics [24].

In what follows, we present an operational model consistent with the observations to provide an interpretation of the results until a rigorous theoretical explanation is developed. The energy density of the quantum vacuum field varies with frequency cubed [23], and therefore the energy density of the suppressed cavity modes would vary with the reciprocal of the cavity thickness cubed. At first blush, one might expect to see this cubic dependence in Figure 4b. However, a multiplicity of other frequency-dependent processes could obscure this cubic dependence. They include (i) the dependence of photoinjection yield on photon energy, as described by extensions of Fowler's theory of photoemission [25]; (ii) limitations in the transport of high energy carriers through the Pd upper electrode due to the interband transition threshold [26]; (iii) the mirror energy-dependent reflectivity; (iv) the energy-dependent absorptivity of the transparent dielectric; and (v) the energy dependence of hot-carrier scattering [27]. A quantitative fit to the data of Figure 4b would require an extensive investigation of each of the energy-dependent mechanisms involved in producing the current. Despite this multiplicity of effects, the overall increase in current with decreasing cavity thickness is qualitatively consistent with what would be expected for the quantum vacuum field as the source for the hot-carrier excitation.

The hot carriers could be electrons, holes, or a combination of the two. The barrier heights are the main factor that determines which one dominates. The effective barrier heights for electrons between the Pd upper electrode and the NiO_x and Al_2O_3 insulators are approximately 0.2 eV and 0.3 eV, respectively, whereas for holes, the respective barrier heights are 3.2 eV and 5.9 eV [6]. For the materials in the devices reported here, the higher barriers for holes are consistent with electrons being the dominant carriers. As described in the introduction, the charge transport through the insulator could be ballistic or via tunneling. For an insulator thickness below 4 nm, which is the case for the devices reported here, the dominant mechanism is tunneling [6].

The measured current from the devices is positive, i.e., in the direction of the arrow shown in the measurement circuit of Figure 1. This corresponds to a net current of electrons flowing from the nickel base electrode through the insulator to the palladium upper electrode. This direction of the current can be understood in terms of the three current components shown in Figure 12.

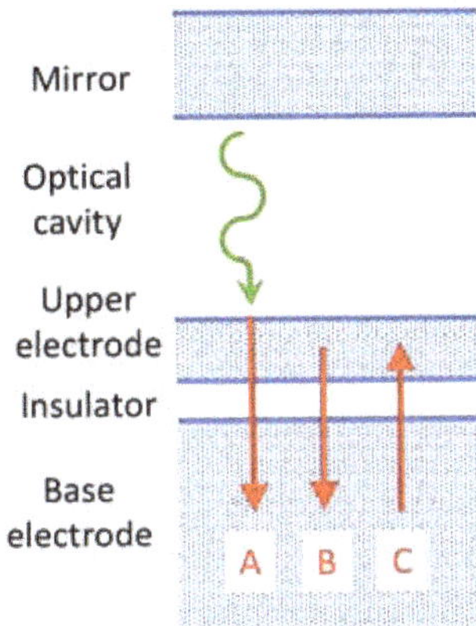

Figure 12. Cross section of the photoinjector device, showing optically generated electron current component **A**, and internally generated components **B** and **C**.

1. We consider first the MIM device in the absence of the optical cavity and the mirror. Component A is produced by free-space ambient optical modes impinging on the upper electrode, where they excite hot electrons. These hot electrons are injected into the insulator and then are absorbed in the base electrode. Component B is due to electrons excited within the upper electrode, e.g., from plasmonic zero-point fluctuations [28]. The electrons are injected into the insulator and then absorbed in the base electrode. Component C, in the opposite direction, is due to electrons that are excited by fluctuations within the base electrode, injected into the insulator, and then absorbed in the upper electrode. There is no optically excited current component of electrons from the base electrode to the upper electrode because the base electrode is thicker than the electron mean-free path length, and so any electrons excited at the outer (lower) surface of the electrode are scattered before reaching the insulator. In equilibrium, the net current is zero, and component C is balanced by the sum of components A and B.

2. We now consider the MIM device in the presence of an adjoining optical cavity and the mirror. The addition of the adjoining structure upsets the balance in current components. Because the cavity reduces the density of the optical modes impinging on the upper electrode, component A is reduced while components B and C remain unchanged. This results in a net electron current from the base electrode to the upper electrode, which is consistent with our observations.

To understand the enhancement of the measured current with varying thicknesses shown in Figure 5a,b we again consider the current components of Figure 12. As discussed with respect to Figure 5a, decreasing the upper electrode thickness from 24 nm down to the absorption depth of ~10 nm, allows an increasing fraction of the photoexcited electrons to traverse the upper electrode without being scattered, with the result that they can produce measurable current. As discussed with respect to Figure 5b, decreasing the insulator thickness increases the fraction of photoexcited electrons that can traverse the insulator and produce measurable current. In both cases, decreasing the thickness of the layers tends to increase the proportion of incoming photons that excite electrons, which can then be injected. The effect is to enhance the photoinjection yield, which increases electron current component A. To understand how the increased photoinjection yield can result in a greater electron current in the direction opposite to the photoinjection when an adjoining cavity is added, we consider the process first without and then with the cavity.

1. In examining the balance of current components when the MIM device is not perturbed by the presence of the optical cavity and mirror, we consider first the device in the absence of the cavity structure. To maintain the zero net current, this increase in photoinjection yield with changing layer thickness must be balanced by a decrease in the internally generated component B or an increase in the internally generated component C. These changes in components B and C result from the thickness changes and are independent of whether there is an adjoining optical cavity or not.

2. We once again consider the MIM device in the presence of an adjoining optical cavity and mirror. With the reintroduction of this adjoining structure, the enhanced photoinjection yield represented by component A now leads to a greater suppression of component A. Because of the greater reduction in component A, the net electron current from the base electrode to the upper electrode is enhanced. Thus, increasing the photoinjection yield leads to an enhanced current that is induced by the presence of an adjoining optical cavity, in the direction opposite to the photoinjection current.

There are additional current components in play beyond those indicated by the three arrows in Figure 12, but they are not expected to add significantly to the current.

- Additional components result from the fact that the insulator itself forms a very thin optical cavity. This cavity is symmetric with respect to the MIM structure itself, as opposed to the optical cavity shown in the figure, which is to one side of the MIM structure. Because there is no longer a cavity having a reduced density of vacuum modes on one side of the tunneling region, the current components in each direction balance each other out, resulting in no net current. This is consistent with the observation that MIM structures without the adjoining optical cavity do not produce a current, as shown in Figure 8.

- Another component results from the upper electrode being slightly transparent. As a result, a small fraction of the optical radiation from the optical cavity impinges on the lower electrode and produces hot carriers. Because the optical transmission through the upper electrode is small, this produces only negligible effects.

- Additional effects, such as those from the surface plasmon modes in the cavity [29], cannot be ruled out.

There is a particular characteristic of MIM structures that adjoin the optical cavities that may be key in allowing the observed current and voltage outputs to be induced. This characteristic is the time required for hot-charge carriers to traverse the combination of the upper electrode and the thin insulator, followed by capture in the base electrode. The entire process can be very fast. The hot-carrier velocity in the metal is at least the Fermi velocity of 10^6 m/s [30,31]. For a metal thickness of approximately 10 nm, the resulting transit time is less than 10 fs. In the insulator, which is even thinner, if the carriers travel ballistically, the velocity is 10^6 m/s [32]. This results in a roughly 1 fs transit time through the insulator. This is on the order of the same time that is required for tunneling [33], which appears to be the dominant transport mechanism, as discussed above. Finally, the hot carriers scatter inelastically in the base electrode, with a lifetime of at most 10 fs. The combination of hot-carrier transport and scattering takes place in the order of 10 fs.

We speculate that the reason the femtosecond transit and capture gives rise to the observed electrical output has to do with an uncertainty-principle-like relation that governs the process. It has been argued that an amount of energy ΔE may be borrowed from the quantum vacuum field for a time Δt [34–36], although that has yet to be supported by experiments [37]. For $\Delta E \Delta t \sim \hbar/2$, hot electrons from 1 eV excitations would be available for 0.3 fs. For a transit and capture that is longer than that, a fraction of the hot electrons would be available. MIM devices adjoining optical cavities could capture photoinjected electrons sufficiently quickly and irreversibly, which would give rise to the observed currents.

5. Conclusions

In metal–insulator–metal (MIM) tunnel devices adjoining thin optical cavities, we consistently observed a small output current and voltage, as reproduced in over 1000 devices produced in 21 batches. When the cavities were made thinner, which corresponds to suppressing a wider range of optical modes, the current increased. The output scaled with number of devices in parallel and series, and the current scaled with the device area. Changing the layer thicknesses in the MIM structure resulted in changes in the current that are consistent with modifying the suppression of hot electron photoinjection from the side of the MIM structure adjoining the optical cavity.

We carried out a set of tests to determine whether the measured electrical output could be the result of some sort of experimental artifact. The results support the conclusion that the source of the electrical output is not due to measurement offsets or errors, transient stored charge, characteristics of the structure not related to the optical cavity, electromagnetic pick-up, electrical leakage through the optical cavity, or thermoelectric effects in the electrodes or in the device itself. All evidence is that the device itself produces the measured outputs.

The observations are consistent with the optical cavity upsetting an equilibrium balance of currents in the MIM structures. We posit that quantum fluctuations excite the observed currents. If access to such excitation is limited by a $\Delta E \Delta t$ uncertainty-principle-like relation, the available energy ΔE would be accessible for a very short time Δt. The ultra-fast charge transport and capture in MIM devices is compatible with such a short time requirement.

Author Contributions: G.M. conceived the device concept, directed the research, and wrote the manuscript; A.W. supervised the fabrication and testing, and carried out much of the GSM device fabrication; D.D. carried out the fabrication of the initial devices and layer depositions; and D.B. carried out the design and fabrication of the photolithographic devices. All authors have read and agreed to the published version of the manuscript.

Funding: Much of the work was supported by The Denver Foundation, which we gratefully acknowledge.

Institutional Review: Not applicable.

Informed Consent Statement: Not applicable.

Data Availability Statement: All data analyzed for this study are included in this published article.

Acknowledgments: We greatly appreciate the insightful comments on our results from P.C.W. Davies, E.A. Cornell, J. Maclay, and M.J. Estes, and M. McConnell and R. Cantwell for help in verifying and improving the measurements, and additional comments on the manuscript from B. Pelz and J. Stearns.

Conflicts of Interest: The authors declare no conflict of interest.

References

1. Sanchez, A.; Davis, C.F., Jr.; Liu, K.C.; Javan, A. The MOM Tunneling Diode: Theoretical Estimate of Its Performance at Microwave and Infrared Frequencies. *J. Appl. Phys.* **1978**, *49*, 5270–5277. [CrossRef]
2. Fumeaux, C.; Herrmann, W.; Kneubuhl, F.K.; Rothuizen, H. Nanometer Thin-Film Ni-NiO-Ni Diodes for Detection and Mixing of 30 THz Radiation. *Infrared Phys. Technol.* **1998**, *39*, 123–183. [CrossRef]
3. Grover, S.; Moddel, G. Engineering the Current-Voltage Characteristics of Metal-Insulator-Metal Diodes Using Double-Insulator Tunnel Barriers. *Solid-State Electron.* **2012**, *67*, 94–99. [CrossRef]
4. Kadlec, J.; Gundlach, K.H. Results and Problems of Internal Photoemission in Sandwich Structures. *Phys. Status Solidi (A)* **1976**, *37*, 11–28. [CrossRef]
5. Kovacs, D.A.; Winter, J.; Meyer, S.; Wucher, A.; Diesing, D. Photo and Particle Induced Transport of Excited Carriers in Thin Film Tunnel Junctions. *Phys. Rev. B—Condens. Matter Mater. Phys.* **2007**, *76*. [CrossRef]
6. Simmons, J.G. Potential Barriers and Emission-limited Current Flow between Closely Spaced Parallel Metal Electrodes. *J. Appl. Phys.* **1964**, *35*, 2472–2481. [CrossRef]
7. Moddel, G.; Weerakkody, A.; Doroski, D.; Bartusiak, D. Optical-Cavity-Induced Conductance Change. **2021**. Unpublished work.
8. Lambrecht, A.; Jaekel, M.-T.; Reynaud, S. The Casimir Force for Passive Mirrors. *Phys. Lett. A* **1997**, *225*, 188–194. [CrossRef]

9.   Pelz, B.; Moddel, G. Demonstration of Distributed Capacitance Compensation in a Metal-Insulator-Metal Infrared Rectenna Incorporating a Traveling-Wave Diode. *J. Appl. Phys.* **2019**, *125*. [CrossRef]

10. Hobbs, P.C.D.; Laibowitz, R.B.; Libsch, F.R. Ni–NiO–Ni Tunnel Junctions for Terahertz and Infrared Detection. *Appl. Opt.* **2005**, *44*, 6813–6822. [CrossRef]

11. Weerakkody, A.; Belkadi, A.; Moddel, G. Nonstoichiometric Nanolayered Ni/NiO/Al2O3/CrAu Metal–Insulator–Metal Infrared Rectenna. *ACS Appl. Nano Mater.* **2021**. [CrossRef]

12. Keithley. *Low Level Measurements Handbook Precision DC Current, Voltage, and Resistance Measurements Low Level Measurements Handbook*, 7th ed.; Keithley Instruments Inc.: Cleveland, OH, USA, 2013.

13. Herner, S.B.; Belkadi, A.; Weerakkody, A.; Pelz, B.; Moddel, G. Responsivity-Resistance Relationship in MIIM Diodes. *IEEE J. Photovolt.* **2018**, *8*, 499–504. [CrossRef]

14. Gall, D. Electron Mean Free Path in Elemental Metals. *J. Appl. Phys.* **2016**, *119*. [CrossRef]

15. Borghesi, A.; Guizzetti, G. Palladium. In *Handbook of Optical Constants of Solids*; Palik, E.D., Ed.; Academic: New York, NY, USA, 1991; Volume II.

16. Nelson, O.L.; Anderson, D.E. Hot-Electron Transfer through Thin-Film Al–Al2O3 Triodes. *J. Appl. Phys.* **1966**, *37*, 66–76. [CrossRef]

17. Simmons, J.G. Generalized Formula for the Electric Tunnel Effect between Similar Electrodes Separated by a Thin Insulating Film. *J. Appl. Phys.* **1963**, *34*, 1793–1803. [CrossRef]

18. Casimir, H.B.G. On the Attraction between Two Perfectly Conducting Plates. In Proceedings of the Koninklijke Nederlandse Akademie van Wetenschappen, 29 May 1948; Volume 51, p. 793.

19. Lamoreaux, S.K. Demonstration of the Casimir Force in the 0.6 to 6 μ m Range. *Phys. Rev. Lett.* **1997**, *78*, 5. [CrossRef]

20. Woods, L.M.; Krüger, M.; Dodonov, V. Perspective on Some Recent and Future Developments in Casimir Interactions. *Appl. Sci.* **2021**, *11*, 293. [CrossRef]

21. Woods, L.M.; Dalvit, D.A.R.; Tkatchenko, A.; Rodriguez-Lopez, P.; Rodriguez, A.W.; Podgornik, R. Materials Perspective on Casimir and van Der Waals Interactions. *Rev. Mod. Phys.* **2016**, *88*. [CrossRef]

22. Lamb, W.E., Jr.; Retherford, R.C. Fine Structure of the Hydrogen Atom by a Microwave Method. *Phys. Rev.* **1947**, *72*, 241. [CrossRef]

23. Milonni, P.W. *The Quantum Vacuum: An Introduction to Quantum Electrodynamics*; Academic press: Boston, MA, USA, 1994; ISBN 0080571492.

24. Cole, D.C.; Puthoff, H.E. Extracting Energy and Heat from the Vacuum. *Phys. Rev. E* **1993**, *48*, 1562–1565. [CrossRef]

25. Chalabi, H.; Schoen, D.; Brongersma, M.L. Hot-Electron Photodetection with a Plasmonic Nanostripe Antenna. *Nano Lett.* **2014**, *14*, 1374–1380. [CrossRef]

26. Tagliabue, G.; Jermyn, A.S.; Sundararaman, R.; Welch, A.J.; DuChene, J.S.; Pala, R.; Davoyan, A.R.; Narang, P.; Atwater, H.A. Quantifying the Role of Surface Plasmon Excitation and Hot Carrier Transport in Plasmonic Devices. *Nat. Commun.* **2018**, *9*. [CrossRef]

27. Ludeke, R.; Bauer, A.; Cartier, E. Hot Electron Transport through Metal–Oxide–Semiconductor Structures Studied by Ballistic Electron Emission Spectroscopy. *J. Vac. Sci. Technol. B Microelectron. Nanometer Struct. Process. Meas. Phenom.* **1995**, *13*, 1830–1840. [CrossRef]

28. Rivera, N.; Wong, L.J.; Joannopoulos, J.D.; Soljačić, M.; Kaminer, I. Light Emission Based on Nanophotonic Vacuum Forces. *Nat. Phys.* **2019**, *15*, 1284–1289. [CrossRef]

29. Intravaia, F.; Henkel, C.; Lambrecht, A. Role of Surface Plasmons in the Casimir Effect. *Phys. Rev. A At. Mol. Opt. Phys.* **2007**, *76*, 1–11. [CrossRef]

30. Knoesel, E.; Hotzel, A.; Wolf, M. Ultrafast Dynamics of Hot Electrons and Holes in Copper: Excitation, Energy Relaxation, and Transport Effects. *Phys. Rev. B* **1998**, *57*, 12812. [CrossRef]

31. Williams, R. Injection by Internal Photoemission. In *Williams, Richard. "Injection by Internal Photoemission." Semiconductors and Semimetals*; Academic Press: Amsterdam, The Netherlands, 1970; pp. 97–139.

32. Heiblum, M.; Fischetti, M. Ballistic Hot-Electron Transistors. *IBM J. Res. Dev.* **1990**, *34*, 530–549. [CrossRef]

33. Nguyen, H.Q.; Cutler, P.H.; Feuchtwang, T.E.; Huang, Z.-H.; Kuk, Y.; Silverman, P.J.; Lucas, A.A.; Sullivan, T.E. Mechanisms of Current Rectification in an STM Tunnel Junction and the Measurement of an Operational Tunneling Time. *IEEE Trans. Electron Devices* **1989**, *36*, 2671–2678. [CrossRef]

34. Ford, L.H. Constraints on Negative-Energy Fluxes. *Phys. Rev. D* **1991**, *43*, 3972. [CrossRef]

35. Davies, P.C.W.; Ottewill, A.C. Detection of Negative Energy: 4-Dimensional Examples. *Phys. Rev. D* **2002**, *65*. [CrossRef]

36. Huang, H.; Ford, L.H. Quantum Electric Field Fluctuations and Potential Scattering. *Phys. Rev. D Part. Fieldsgravitation Cosmol.* **2015**, *91*, 1–8. [CrossRef]

37. Maclay, G.J.; Davis, E.W. Testing a Quantum Inequality with a Meta-Analysis of Data for Squeezed Light. *Found. Phys.* **2019**, *49*, 797–815. [CrossRef]

Article

Algebraic DVR Approaches Applied to Describe the Stark Effect

Marisol Bermúdez-Montaña [1,2], **Marisol Rodríguez-Arcos** [1], **Renato Lemus** [1,*], **José M. Arias** [3,4], **Joaquín Gómez-Camacho** [3,5] and **Emilio Orgaz** [2]

[1] Instituto de Ciencias Nucleares, Universidad Nacional Autónoma de México, Apartado Postal 70-543, CDMX 04510, Mexico; marisol.bermudez@nucleares.unam.mx (M.B.-M.); marisol_ra@ciencias.unam.mx (M.R.-A.)

[2] Facultad de Química, Universidad Nacional Autónoma de México, Apartado Postal 70-543, CDMX 04510, Mexico; emilio.Orgaz@unam.mx

[3] Departamento de Física Atómica, Molecular y Nuclear, Facultad de Física, Universidad de Sevilla, Apartado 1065, 41080 Sevilla, Spain; ariasc@us.es (J.M.A.); gomez@us.es (J.G.-C.)

[4] Instituto Carlos I (iCI) de Física Teórica y Computacional, Universidad de Sevilla, Apartado 1065, 41080 Sevilla, Spain

[5] CN de Aceleradores (U. Sevilla, J. Andalucía, CSIC), 41092 Sevilla, Spain

* Correspondence: renato@nucleares.unam.mx

Received: 05 August 2020; Accepted: 30 August 2020; Published: 19 October 2020

Abstract: Two algebraic approaches based on a discrete variable representation are introduced and applied to describe the Stark effect in the non-relativistic Hydrogen atom. One approach consists of considering an algebraic representation of a cutoff 3D harmonic oscillator where the matrix representation of the operators r^2 and p^2 are diagonalized to define useful bases to obtain the matrix representation of the Hamiltonian in a simple form in terms of diagonal matrices. The second approach is based on the $U(4)$ dynamical algebra which consists of the addition of a scalar boson to the 3D harmonic oscillator space keeping constant the total number of bosons. This allows the kets associated with the different subgroup chains to be linked to energy, coordinate and momentum representations, whose involved branching rules define the discrete variable representation. Both methods, although originating from the harmonic oscillator basis, provide different convergence tests due to the fact that the associated discrete bases turn out to be different. These approaches provide powerful tools to obtain the matrix representation of 3D general Hamiltonians in a simple form. In particular, the Hydrogen atom interacting with a static electric field is described. To accomplish this task, the diagonalization of the exact matrix representation of the Hamiltonian is carried out. Particular attention is paid to the subspaces associated with the quantum numbers $n = 2, 3$ with $m = 0$.

Keywords: algebraic approach; DVR method; coulombic potential; stark effect

---

1. Introduction

A discrete variable representation approach is based on the search of a discrete basis in terms of which any function of the coordinates is diagonal. The discrete variable representation methods (DVR) in configuration space were developed with some variants since the 1960s, but systematically widely used during the 1980s with different names: discrete-variable representation method [1–4], quadrature discretization method [5,6], configuration localized states (CLS) approach [7], and Lagrange mesh method (LMM) [8–11], whose similarities and differences are discussed in Ref. [11]. The basic ingredient of these methods is the use of orthogonal polynomials and their associated grids from Gaussian quadratures.

In this contribution, we introduce two algebraic DVR methods where the discrete bases are obtained using purely algebraic means without any explicit reference to polynomials. This is accomplished through the diagonalization of the matrix representations associated with the coordinates and momenta. From this perspective, the recently-proposed unitary group approach ($U(4)$-UGA) belongs to an algebraic DVR method [12–16]. In this approach, the discretization provided by the zeros of the polynomials in configuration space is substituted by the branching rules involved in the subgroup chains embedded in the dynamical group. An alternative approach to establish an algebraic DVR method consists of taking a cutoff 3D harmonic oscillator in Fock space and diagonalize the matrix representation of the squares of the coordinates and momenta to obtain two discrete bases (HO-DVR approach), which in turn are used to diagonalize the matrix representation of the Hamiltonian in a simple manner. Both methods are based on a harmonic oscillator basis, albeit they provide different convergence behavior due to the fact that the provided discrete bases are distinct.

In this contribution we present the HO-DVR approach as well as the salient features of the $U(4)$-UGA, both applied to the Coulombic potential. We first show the confidence of our approaches to reproduce the analytical wave functions for the non relativistic Hydrogen atom. Both approaches are able to arrive at the same level of accuracy with the appropriate basis dimension. In order to explore the capabilities and differences between these approaches, the Stark effect is studied by adding the potential energy for the electron in an homogeneous electric field. Our treatment involves a complete dipole matrix representation contemplating the possibility of the action of fields beyond the perturbation region. Since the conservation of the angular momentum is broken, the calculations become particularly heavy. This fact forces us to constraint the basis to a limited angular momentum. With the proposal of these DVR algebraic approaches, we do not intend to improve the variational approaches used to describe the Stark effect, but rather for showing a simple way to deal with this problem as well as to evaluate the differences between them.

Our contribution is organized as follows. In Section 2 the salient features of both the algebraic HO-DVR method and the $U(4)$-UGA are presented. Section 3 is devoted to applying the methods to describe the isolated non-relativistic Hydrogen atom. To this end, both energies and wave functions are tested. In Section 4 the Stark effect is described, paying attention to the bound states originated from the subspaces with $n = 2, 3$. Finally, in Section 5 the summary and our conclusions are presented.

2. Algebraic DVR Methods in 1D Systems

In this section, we present the salient features of the HO-DVR method as well as of the $U(4)$-UGA. Since both approaches are based on the 3D harmonic oscillator we start presenting the main features associated with its algebraic solutions.

2.1. 3D Harmonic Oscillator

We start considering the isotropic 3D harmonic oscillator with reduced mass m and frequency ω. The Hamiltonian is given by

$$\hat{H}_{\text{CS}}^{\text{HO}} = \frac{1}{2m}p^2 + \frac{m\omega^2}{2}r^2. \tag{1}$$

The solutions in configuration space are well known, but this is not the case in Fock space [17,18]. The Hamiltonian, as well as the eigenfunctions can be translated into the Fock space through the introduction of the bosonic creation $a_\mu^\dagger$ and annihilation a_μ operators with commutation relations

$$[a_\mu, a_\nu^\dagger] = \delta_{\mu\nu}; \quad [a_\mu^\dagger, a_\nu^\dagger] = [a_\mu, a_\nu] = 0. \tag{2}$$

The reflection and rotation properties of $a_\mu^\dagger$ are summarized by their transformation under a rotation $\hat{R}(\alpha, \beta, \gamma)$:

$$\hat{R}^{-1}a_\mu^\dagger\hat{R} = \sum_{\mu'} D_{\mu'\mu}^{(1)}(\alpha, \beta, \gamma)a_{\mu'}^\dagger, \tag{3}$$

characterized by the three Euler angles (α, β, γ). The annihilation operators a_μ, however, do not transform as the components of a spherical tensor operator, but the modified operator

$$\tilde{a}_\mu = (-1)^{1-\mu} a_{-\mu} \tag{4}$$

do satisfy the transformation property (3). In this way the the operators $a_\mu^\dagger$ and $\tilde{a}_\mu$ can be coupled to definite angular momentum λ and projection m

$$[a^\dagger \times \tilde{a}]_m^{(\lambda)} = \sum_{\mu',\mu} \langle 1\mu 1\mu' | \lambda m \rangle a_\mu^\dagger \tilde{a}_{\mu'}, \tag{5}$$

using the standard notations $\times$ for the angular momentum coupling and the Clebsh–Gordan coefficients. These operators in terms of the spherical coordinate and momentum components are given by

$$a_\mu^\dagger = \frac{1}{\sqrt{2}} \left(\frac{1}{d} q_\mu - i \frac{d}{\hbar} (-1)^\mu p_{-\mu} \right); \tag{6}$$

$$\tilde{a}_\mu = -\frac{1}{\sqrt{2}} \left(\frac{1}{d} q_\mu + i \frac{d}{\hbar} (-1)^\mu p_{-\mu} \right), \tag{7}$$

with commutation relations

$$[\tilde{a}_\mu, a_\nu^\dagger] = (-1)^{1-\mu} \delta_{-\mu\nu}; \quad [a_\mu^\dagger, a_\nu^\dagger] = [\tilde{a}_\mu, \tilde{a}_\nu] = 0, \tag{8}$$

where $d = \sqrt{\hbar/m\omega}$ has been introduced as unit length, q_μ ($\mu = 1, 0, -1$) are the coordinate components with $r^2 = \sum_\mu (-1)^\mu q_\mu q_{-\mu}$ and corresponding momenta $p_\mu = -i\hbar \frac{\partial}{\partial q_\mu}$. The solutions of the eigensystem associated with (1) in Fock space are given by the kets [17,18]

$$|nlm\rangle = B_{nl} \, (\mathbf{a}^\dagger \cdot \mathbf{a}^\dagger)^{(n-l)/2} \mathcal{Y}_m^l(\mathbf{a}^\dagger)|0\rangle, \tag{9}$$

characterized by the eigensystem

$$\hat{n}|nlm\rangle = n|nlm\rangle, \tag{10}$$
$$\hat{L}^2|nlm\rangle = l(l+1)|nlm\rangle, \tag{11}$$
$$\hat{L}_z|nlm\rangle = m|nlm\rangle, \tag{12}$$

where $\mathcal{Y}_m^l(\mathbf{a}^\dagger)$ are the solid spherical harmonics defined by

$$\mathcal{Y}_m^l(\mathbf{a}^\dagger) = 2^{-l/2}(\mathbf{a}^\dagger \cdot \mathbf{a}^\dagger)^{l/2} Y_m^l(\mathbf{a}^\dagger) \tag{13}$$

in terms of the spherical harmonics $Y_m^l(\mathbf{a}^\dagger)$ with normalization [17,18]

$$B_{nl} = (-1)^{(n-l)/2} \sqrt{\frac{4\pi}{(n-l)!!(n+l+1)!!}}, \tag{14}$$

and eigenvalues

$$E_n^{\mathrm{HO}} = \hbar\omega(n + 3/2), \tag{15}$$

with frequency $\omega = \sqrt{k/m}$ in terms of the force constant k and number operator

$$\hat{n} = \sum_\mu a_\mu^\dagger a_\mu = \sqrt{3}[\mathbf{a}^\dagger \times \tilde{\mathbf{a}}]_0^{(0)}, \tag{16}$$

while the notation $\mathbf{a}^{\dagger} \cdot \mathbf{a}^{\dagger}$ stands for the dot product defined by

$$\mathbf{a}^{\dagger} \cdot \mathbf{a}^{\dagger} = \sqrt{3}[\mathbf{a}^{\dagger} \times \mathbf{a}^{\dagger}]_0^{(0)} = \sum_{\mu}(-1)^{\mu} a_{\mu}^{\dagger} a_{-\mu}^{\dagger}, \tag{17}$$

with the realization of the angular momentum components

$$\hat{L}_{\mu} = \sqrt{2}[a^{\dagger} \times \tilde{a}]_{\mu}^{(1)}. \tag{18}$$

This summary for the harmonic oscillator paves the way to introduce the algebraic DVR methods.

2.2. HO-DVR Approach

The HO-DVR approach consists of establishing discrete bases where functions of the radial coordinate and the corresponding momentum are diagonal. To accomplish this task we should consider the algebraic realization of the variables involved in Equation (1). Based on the relations (6) and (7), the squared of the radius and momenta in Fock space takes the form

$$r^2 = (\hat{n} + 3/2) + \frac{1}{2}(\mathbf{a}^{\dagger} \cdot \mathbf{a}^{\dagger} + \mathbf{a} \cdot \mathbf{a}), \tag{19}$$

$$p^2 = (\hat{n} + 3/2) - \frac{1}{2}(\mathbf{a}^{\dagger} \cdot \mathbf{a}^{\dagger} + \mathbf{a} \cdot \mathbf{a}), \tag{20}$$

Since both operators (19) and (20) preserve the angular momentum we can obtain their matrix representation for a given l and m in the N-dimensional space defined by

$$\mathcal{L}_N^l = \{|nlm\rangle, n = 0, 1, \ldots, N - 1\}, \tag{21}$$

with the restrictions $l = n, n - 2, \ldots, 1$ or 0, and the usual reduction $SO(3) \supset SO(2)$ given by $-l \leq m \leq l$. It is clear that the N-dimensional space (21) is not complete, but it establishes a space to carry our the calculations in an approximate way. Explicitly the matrix elements of the operators (19) and (20) in dimensionless units turn out to be

$$\langle nlm|r^2|nlm\rangle = \langle nlm|p^2|nlm\rangle,$$
$$= n + \frac{3}{2}, \tag{22}$$
$$\langle n+2, lm|r^2|nlm\rangle = -\langle n+2, lm|p^2|nlm\rangle$$
$$= \frac{1}{2}\sqrt{(n+l+2)(n-l+2)}. \tag{23}$$

The numerical diagonalizations of the representation matrix of these operators provide a discrete set of eigenvectors

$$|lm, r_i^2\rangle = \sum_{n=0}^{N-1} \langle nlm|lm, r_i^2\rangle |nlm\rangle, \tag{24}$$

$$|lm, p_i^2\rangle = \sum_{n=0}^{N-1} \langle nlm|lm, p_i^2\rangle |nlm\rangle, \tag{25}$$

which provide the radial coordinate and momentum representations characterized by

$$r|lm, r_i^2\rangle = \sqrt{r_i^2}\,|lm, r_i^2\rangle, \tag{26}$$

$$p|lm, p_i^2\rangle = \sqrt{p_i^2}|lm, p_i^2\rangle. \tag{27}$$

Because of these properties, the basis $|lm, r_i^2\rangle$ is identified with the position representation, while $|lm, p_i^2\rangle$ stands for the momentum representation. In addition since the Hamiltonian (15) is diagonal in the basis (9) we refer the $U(3)$ basis as the energy representation. As a consequence, for any functions $V(r)$ and $G(p)$ we have for their matrix elements

$$\langle lm, r_j^2 | V(r) | lm, r_i^2 \rangle = V\left(\sqrt{r_i^2}\right)\delta_{ij},\tag{28}$$

$$\langle lm, p_j^2 | G(p) | lm, p_i^2 \rangle = G\left(\sqrt{p_i^2}\right)\delta_{ij}.\tag{29}$$

The eigenvectors (24) and (25) define the transformation coefficients

$$\mathbf{W} = ||\langle nlm | lm, p_i^2 \rangle||,\tag{30}$$
$$\mathbf{T} = ||\langle nlm | lm, r_i^2 \rangle||,\tag{31}$$

while the properties (28) and (29) lead to a DVR method where the matrix representation of the general 3D Hamiltonian depending on the radial coordinate

$$\hat{H} = \frac{p^2}{2m} + V(r),\tag{32}$$

takes the following matrix form in the energy representation:

$$\mathbf{H} = \mathbf{W}^\dagger \mathbf{\Lambda}^{(p)} \mathbf{W} + \mathbf{T}^\dagger \mathbf{\Lambda}^{(r)} \mathbf{T},\tag{33}$$

where $\mathbf{\Lambda}^{(p)}$ is the diagonal contribution of the kinetic energy in the momentum representation (25):

$$||\mathbf{\Lambda}^{(p)}|| = ||\langle lm, p_j^2 | \frac{p^2}{2m} | lm, p_i^2 \rangle|| = \frac{p_i^2}{2m}\delta_{ij},\tag{34}$$

where the notation for the mass m should not be confused with the angular momentum projection m. In addition the matrix $\mathbf{\Lambda}^{(r)}$ corresponds to the diagonal matrix of the potential $V(r)$ in the position representation:

$$||\mathbf{\Lambda}^{(r)}|| = ||\langle lm, r_j^2 | \left[V\left(\sqrt{r^2}\right)\right] | lm, r_i^2 \rangle|| = \left[V\left(\sqrt{r_i^2}\right)\right]\delta_{ij}.\tag{35}$$

Both transformation coefficients $\mathbf{W}$ and $\mathbf{T}$ are numerically calculated only once for a given l and m and space dimension N, providing the methodology for any potential in the framework of the HO-DVR method. However since the basis corresponds to the harmonic oscillator functions, adding and subtracting the quadratic potential makes possible to reformulate the Equation (33) in terms of only the transformation coefficients $\mathbf{T}$ in the form:

$$\mathbf{H} = \mathbf{\Lambda}^{(E)} + \kappa \mathbf{T}^\dagger \mathbf{\Lambda}'^{(r)} \mathbf{T},\tag{36}$$

where for a given l and m, $\mathbf{\Lambda}^{(E)}$ is the diagonal matrix in the energy representation

$$||\mathbf{\Lambda}^{(n)}|| = ||\langle n'lm | \hbar\omega(\hat{n} + 3/2) | nlm \rangle|| = \hbar\omega(n + 3/2)\delta_{n'n},\tag{37}$$

while in the position representation

$$||\mathbf{\Lambda}'^{(r)}|| = -\frac{m\omega^2}{2}r_i^2\delta_{ij} + V\left(\sqrt{r_i^2}\right)\delta_{ij}.\tag{38}$$

In Equation (36), κ stands for a control parameter in the interval $[0,1]$, whose variation allows the potential to be transformed from a harmonic oscillator ($\kappa = 0$) to the arbitrary potential $V(\sqrt{r^2})$ ($\kappa = 1$).

The methodology of this approach is simple. We diagonalize both matrix representations (22) and (23) to obtain the coefficients (30) and (31). Then we proceed to diagonalize in the energy representation the Hamiltonian (36) associated with the potential we are interested in for a given angular momentum and projection.

2.3. SU(4)-UGA

This approach starts by establishing the $U(4)$ dynamical group for 3D systems by adding a scalar boson $s^\dagger(s)$ to the space of the same bosonic operators $a_\mu^\dagger(\tilde{a}_\mu)$ previously introduced carrying angular momentum $l = 1$ in such a way that the bilinear products satisfy the commutation relations associated with the generators of the $U(4)$ group [19]:

$$\hat{n}_a = \sqrt{3}[a \times \tilde{a}]_0^{(0)}; \qquad \hat{n}_s = s^\dagger s, \tag{39}$$

$$\hat{L}_\mu = \sqrt{2}[a \times \tilde{a}]_\mu^{(1)}; \qquad \hat{D}_\mu = a_\mu^\dagger s - s^\dagger \tilde{a}_\mu, \tag{40}$$

$$\hat{Q}_\mu = [a \times \tilde{a}]_\mu^{(2)}; \qquad \hat{R}_\mu = i(a_\mu^\dagger s + s^\dagger \tilde{a}_\mu), \tag{41}$$

with the constraint that the total number of bosons $\hat{N} = \hat{n}_a + s^\dagger s$ is constant. Hence, the associated quantum number N establishes the dimension of the dynamical space. From the dynamical group $U(4)$, three subgroup chains can be obtained, preserving the angular momentum [18,19]:

$$U(4) \supset U(3) \supset SO(3) \supset SO(2), \tag{42}$$

$$U(4) \supset SO(4) \supset SO(3) \supset SO(2), \tag{43}$$

$$U(4) \supset \overline{SO}(4) \supset SO(3) \supset SO(2). \tag{44}$$

Each chain provides a basis with a definitive meaning, as will become clear shortly. A particular chain, (42), leads to the basis $|[N]n_a LM\rangle$ defined by the operator equations

$$\hat{N}|[N]n_a LM\rangle = N|[N]n_a LM\rangle, \tag{45}$$

$$\hat{n}_a|[N]n_a LM\rangle = n_a|[N]n_a LM\rangle, \tag{46}$$

$$\hat{L}^2|[N]n_a LM\rangle = L(L+1)|[N]n_a LM\rangle, \tag{47}$$

$$\hat{L}_z|[N]n_a LM\rangle = M|[N]n_a LM\rangle. \tag{48}$$

We now turn our attention to the connection with configuration space. Using the approach recently proposed [20] with a mapping to the 3D harmonic oscillator basis

$$|[N]n_a LM\rangle \rightarrow |nlm\rangle. \tag{49}$$

It is possible to show that the $U(4)$ algebraic realization $\hat{Q}_\mu$ and $\hat{P}_{-\mu}$ for the coordinates and momenta, respectively, take the form [16]

$$\hat{Q}_\mu = \frac{d}{\sqrt{2N}}\hat{D}_\mu, \tag{50}$$

$$\hat{P}_{-\mu} = -(-)^{1-\mu}\frac{\hbar}{d}\frac{1}{\sqrt{2N}}\hat{R}_\mu. \tag{51}$$

The mapping

$$q_\mu \rightarrow \hat{Q}_\mu \qquad p_\mu \rightarrow \hat{P}_\mu \tag{52}$$

allows the algebraic Hamiltonian (1) to be obtained in the $U(4)$ space:

$$\hat{H}^{U(4)}_{alg} = \frac{1}{2m}\,\mathcal{P}^2 + \frac{m\omega^2}{2}\,\mathcal{Q}^2. \tag{53}$$

Notice that the operators $\mathcal{P}^2$ and $\mathcal{Q}^2$ are not related in a straightforward way with the physical momentum p^2 and coordinate q^2, but their matrix elements coincide in the N large limit:

$$\lim_{N\to\infty} ||\langle Nn'_a LM|\mathcal{P}^2|Nn_a LM\rangle|| = ||\langle n'lm|p^2|nlm\rangle||, \tag{54}$$

with similar equation for the coordinate.

Taking into account (50) and (51), the Hamiltonian (53) is recast in the form:

$$\hat{H}^{U(4)}_{alg} = \frac{\hbar\omega}{4}\left[\frac{\hat{R}^2 + \hat{D}^2}{N}\right]. \tag{55}$$

The second order Casimir operators of $U(4)$ and $U(3)$ are given by [18]

$$\hat{C}_2[U(4)] = \frac{1}{3}\hat{n}_a^2 + \frac{1}{2}(\hat{L}^2 + \hat{D}^2 + \hat{R}^2) + \hat{Q}^2 + (\hat{N} - \hat{n}_a)^2, \tag{56}$$

$$\hat{C}_2[U(3)] = \frac{1}{3}\hat{n}_a^2 + \frac{1}{2}\hat{L}^2 + \hat{Q}^2, \tag{57}$$

and in the symmetrical space of identical bosons the following replacements are valid [18]

$$C_2[U(4)] \to \hat{N}(\hat{N} + 3) \tag{58}$$

$$C_2[U(3)] \to \hat{n}_a(\hat{n}_a + 2), \tag{59}$$

which allows (55) to be transformed into

$$\hat{H}^{U(4)}_{alg} = \hbar\omega\left[\left(1 - \frac{1}{N}\right)\hat{n}_a + \frac{3}{2} - \frac{\hat{n}_a^2}{N}\right]. \tag{60}$$

Here we have taken into account that the operator $\hat{N}$ is diagonal in any basis and consequently it can be substituted by its eigenvalue N. This result leads to the identification of the $U(3)$ chain (42) as the energy representation with eigenkets

$$\hat{n}_a|[N]n_a LM\rangle = n_a|[N]n_a LM\rangle, \tag{61}$$

and branching rules $n_a = 0, 1, \ldots, N$ and $L = n_a, n_a - 2, \ldots, 1$ or 0.

Let us now come back to chains (43, 44). Since the Casimir operators characterizing the $SO(4)$ and $\overline{SO}(4)$ groups are $\hat{W}^2 = \hat{D}^2 + \hat{L}^2$ and $\hat{\bar{W}}^2 = \hat{R}^2 + \hat{L}^2$ respectively, Equation (50) and (51) implies that chain (43) provides the coordinate representation while chain (44) gives the momentum representation. The corresponding eigenvectors satisfy

$$\hat{W}^2|[N]\zeta LM\rangle = \zeta(\zeta + 2)|[N]\zeta LM\rangle, \tag{62}$$

$$\hat{\bar{W}}^2|[N]\bar{\zeta} LM\rangle = \bar{\zeta}(\bar{\zeta} + 2)|[N]\bar{\zeta} LM\rangle, \tag{63}$$

with branching rules $\zeta, \bar{\zeta} = N, N - 2, \ldots, 1$ or 0, and $L = 0, 1, \ldots, \zeta(\bar{\zeta})$. The importance of these transformations is twofold. On one hand, we have the mapping (49) and consequently it is possible to project any state to the position representation as an expansion of harmonic oscillator functions. On the other hand, an arbitrary potential depending on $\sqrt{\hat{Q}^2}$ takes a diagonal form in the basis (62), providing a direct way to obtain its matrix representation in the energy representation through a similarity transformation.

Let us now consider a general situation of a 3D Hamiltonian of the form (32). The corresponding algebraic Hamiltonian is obtained by applying the anharmonization procedure (52):

$$\hat{H}_{alg}^{U(4)} = \hat{H}\Big|_{q \to \hat{Q}, p \to \hat{\mathcal{P}}} = \frac{1}{2m}\hat{\mathcal{P}}^2 + \hat{V}(\sqrt{\hat{Q}^2}). \tag{64}$$

A practical convenient form for this Hamiltonian is obtained by adding and subtracting a quadratic term in such a way that the harmonic oscillator term (1) is identified:

$$\hat{H}_{alg}^{U(4)} = \hbar\omega\left[\left(1 - \frac{1}{N}\right)\hat{n}_a + \frac{3}{2} - \frac{\hat{n}_a^2}{N}\right] + \kappa\hat{V}'(\sqrt{\hat{Q}^2}). \tag{65}$$

where

$$V'(\hat{Q}^2) = -\frac{m\omega^2}{2}\hat{Q}^2 + \hat{V}(\sqrt{\hat{Q}^2}). \tag{66}$$

Taking into account the scalar character of $\hat{Q}^2$, we have its following matrix elements in the coordinate representation:

$$\langle [N]\zeta'LM|\hat{Q}^2|[N]\zeta LM\rangle = \frac{d^2}{2}\frac{[\zeta(\zeta+2) - L(L+1)]}{N}\delta_{\zeta',\zeta}. \tag{67}$$

Hence, the matrix elements of the Hamiltonian (for a given L and M) in the energy representation take the general form

$$\mathbf{H}^{(E)} = \mathbf{\Lambda}^{(E)} + \kappa\mathbf{T}^\dagger\mathbf{\Lambda}^{(Q)}\mathbf{T}, \tag{68}$$

where $\mathbf{\Lambda}^{(E)}$ is the diagonal contribution of the deformed harmonic oscillator

$$||\mathbf{\Lambda}^{(E)}|| = \hbar\omega\left[\left(1 - \frac{1}{N}\right)n_a + \frac{3}{2} - \frac{n_a^2}{N}\right]\delta_{n_a,n_a'}, \tag{69}$$

while $\mathbf{\Lambda}^{(Q)}$ is the diagonal matrix of the potential $V'(\sqrt{\hat{Q}^2})$ in the position representation:

$$||\mathbf{\Lambda}^{(Q)}|| = \hbar\omega\left[-\frac{1}{2}\frac{\xi(\zeta,L)^2}{2N} + \frac{1}{\hbar\omega}V\left(d\frac{\xi(\zeta,L)}{\sqrt{2N}}\right)\right]\delta_{\zeta,\zeta'}, \tag{70}$$

with $\xi(\zeta,L) = \sqrt{\zeta(\zeta+2) - L(L+1)}$. The $\mathbf{T}$ matrix stands for the transformation brackets: $\mathbf{T} = ||\langle[N]\zeta LM|[N]n_a LM\rangle||$, which are obtained by diagonalizing the matrix representation of the Casimir operators $\hat{W}^2$ given by

$$\langle[N]n_a LM|\hat{W}^2|[N];n_a LM\rangle =$$
$$N(2n_a + 3) - 2n_a(n_a + 1) + L(L+1),$$
$$\langle[N]n_a + 2, LM|\hat{W}^2|[N];n_a LM\rangle =$$
$$-\sqrt{(N - n_a - 1)(N - n_a)(n_a - L + 2)(n_a + L + 3)},$$

or in analytic form [18,21]. This matrix representation of $\hat{W}^2$ is tridiagonal and its diagonalization to provide $\mathbf{T}$ is calculated once and for all for a given N, while the diagonal matrix $\mathbf{\Lambda}^{(Q)}$ is modified in accordance with the specific potential. The contribution (69) involves accidental degeneracy due to the quadratic contribution $\hat{n}_a^2$. This degeneracy manifests in the full Hamiltonian (68) by the appearance of degeneracy for the bound states in the N large limit. This degeneracy is removed by substituting the diagonal term (69) by [16]

$$\left[\left(1-\frac{1}{N}\right)\hat{n}_a + \frac{3}{2} - \delta\frac{\hat{n}_a^2}{N}\right],\tag{71}$$

with $\delta = 1$ for $n_a < N/2$ and $\delta = 0$ for $n_a \geq N/2$.

The expression (68) contains the main features of our approach. The physical information characterizing the system is given by the diagonal matrix $\Lambda^{(\mathcal{Q})}$ easily generated and given by (70), while the calculation of the transformation brackets allows any system to be analyzed in simple form using the standard tools of matrix diagonalization. It is important to notice that this approach is intended to be considered in the N large limit [16]. This is in contrast to the previous analysis of the algebraic models where $N \to \infty$ represents the classical limit [22].

3. Coulombic Potential: Hydrogen Atom

Because of its importance in dealing with molecular systems the Coulombic potential deserves special attention. For this reason we shall consider the non-relativistic Hydrogen atom together with the effect of an external electric field to show the advantages of the proposed approaches. We start considering the Hamiltonian of the Hydrogen atom.

Following M. Moshinsky, [17], in energy units of the first Bohr orbit $E_B = \frac{1}{2}me^4/\hbar^2$, the Hamiltonian takes the form

$$\frac{\hat{H}}{E_B} = \frac{1}{2mE_B}p^2 - \sqrt{\frac{2\hbar^2}{mE_B}}\frac{1}{r}.\tag{72}$$

Here, the mass associated with the harmonic oscillator is identified with the reduced mass of the Hydrogen system. Adding and subtracting the harmonic oscillator potential we obtain

$$\frac{\hat{H}}{E_B} = \epsilon^2\hat{H}^{\text{HO}} - \left[\frac{\epsilon^2}{2}\vec{r}^2 + \epsilon\sqrt{2}\frac{1}{\vec{r}}\right],\tag{73}$$

where $\vec{r}$ is assumed to be in units of $d = \sqrt{\hbar/m\omega}$, the harmonic oscillator Hamiltonian $\hat{H}^{\text{HO}}$ in units of $\hbar\omega$ and ϵ a dimensionless parameter given by

$$\epsilon = \sqrt{\frac{\hbar\omega}{E_B}},\tag{74}$$

with ω a variational parameter to be determined.

Let us now establish, on one hand, the application of the HO-DVR approach to this problem. The matrix representation of the Hamiltonian in units of $\hbar\omega$ takes the form

$$\mathbf{H} = \epsilon^2\Lambda^{(n)} - \kappa\mathbf{T}^{\dagger}\Lambda'^{(r)}\mathbf{T},\tag{75}$$

where the diagonal matrix $\Lambda^{(n)}$ in the energy representation is given by

$$||\Lambda^{(n)}|| = (n+3/2)\delta_{n'n},\tag{76}$$

while for the diagonal matrix $\Lambda'^{(r)}$ in the position representation we have

$$||\Lambda'^{(r)}|| = \left[\frac{\epsilon^2}{2}\vec{r}_i^2 + \epsilon\sqrt{2}\frac{1}{\sqrt{\vec{r}_i^2}}\right]\delta_{ij}.\tag{77}$$

On the other hand, in the framework of the $U(4)$-DVR, we introduce the realization (50) and (51) into (73) to obtain

$$\frac{\hat{H}}{E_B} = \epsilon^2\left[\left(1-\frac{1}{N}\right)\hat{n}_a + \frac{3}{2} - \frac{\hat{n}_a^2}{N}\right] - \left[\frac{1}{2}\epsilon^2\left(\frac{\hat{D}^2}{2N}\right) + \epsilon\frac{2\sqrt{N}}{\sqrt{\hat{D}^2}}\right]. \tag{78}$$

The corresponding matrix representation takes the form

$$\mathbf{H}^{(E)} = \mathbf{\Lambda}^{(E)} - \kappa\mathbf{T}^\dagger\mathbf{\Lambda}^{(Q)}\mathbf{T}, \tag{79}$$

where $\mathbf{\Lambda}^{(E)}$ is the diagonal contribution of the deformed harmonic oscillator

$$||\mathbf{\Lambda}^{(E)}|| = \epsilon^2\left[\left(1-\frac{1}{N}\right)n_a + \frac{3}{2} - \frac{n_a^2}{N}\right]\delta_{n_a',n_a}, \tag{80}$$

while $\mathbf{\Lambda}^{(Q)}$ is the diagonal matrix containing the potential with the form

$$||\mathbf{\Lambda}^{(Q)}|| = \left[\frac{1}{2}\epsilon^2\left(\frac{1}{2N}\right)\xi(\zeta,L)^2 + \frac{\epsilon\,2\sqrt{N}}{\xi(\zeta,L)}\right]\delta_{\zeta,\zeta'}, \tag{81}$$

where the preservation of the angular momentum has been considered implicitly. It is importance to notice that because of the branching rule $\zeta = N, N-2, \ldots, 1$ or 0, the number of bosons N must be taken to be odd in order to avoid a singularity (obtained when $L = \zeta = 0$).

The diagonalization of the Hamiltonian matrices (75) and (79) lead to the correlation diagram displayed in Figure 1 as a function of κ. The parameter (74) turns out to be $\epsilon = 1$. Although the correlations diagram is the same for both methods, the corresponding basis dimension is different in order to reach convergence. As noticed, the levels corresponding to the principal quantum number in the Hydrogen atom converge to the well known n^2 degeneracy, although to simplify the figure we have only included angular momenta up to 2. The degeneracy $2L + 1 = 2l + 1$ corresponding to the projection (M, m) should be assumed but not shown because of the (M, m)-independence of the Hamiltonian. The states above the zero energy represent a discretization of the continuum. This kind of the description of the continuum is also present in the Morse and Pöeschl–Teller potentials [23–25].

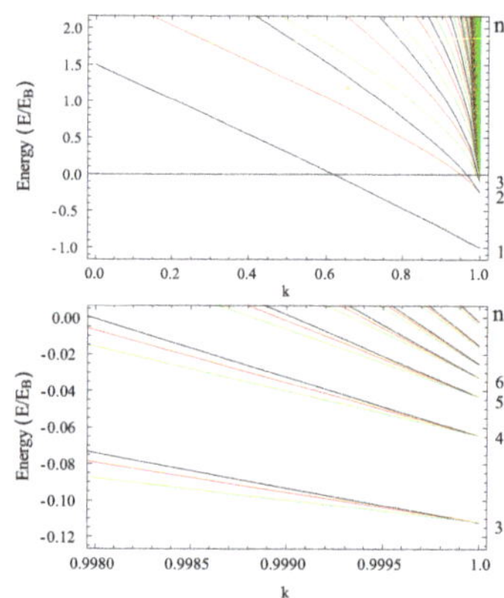

Figure 1. Correlation diagram from the harmonic oscillator ($\kappa = 0$) to the Coulomb potential ($\kappa = 1$) corresponding to the Hamiltonians (36) and (68), with (77) and (81), respectively. At the right of the figures the principal quantum number for the electron in the H atom, n, is indicated. To simplify only the levels of the Hydrogen system with $L, l = 0, 1, 2$ are included. In the bottom panel a zoom is displayed in order to show the convergence of the levels with different angular momenta. The top level in each group corresponds to angular momentum $L, l = 0$. The parameters were taken to be $\epsilon = 1$ and $N = 4001$ for the $U(4)$-unitary group approach (UGA), while $N = 2001$ for the HO-DVR method.

In order to explicitly show the energy convergence, the comparison between the exact energy levels $E_n = -1/n^2$ and the calculated ones are displayed in Table 1 for the HO-DVR method and in Table 2 for the $U(4)$-UGA. The convergence is similar, but while in the HO-DVR it is reached for $N = 2001$, the $U(4)$-UGA needed a number of bosons $N = 4001$. This difference may be explained by the accidental degeneracy involved in the $U(4)$-UGA but also because the bases are different. In both cases, the space dimension is large enough to obtain a reasonable convergence to the exact values. In general, the convergence can be measured through the fidelity F or overlap of eigenstates involving different parameters N [26]. Along this venue we have calculated the energies for number of bosons $N \pm 50$, registering the energy difference as the error displayed in Table 3 for the HO-DVR method and in Table 4 for the $U(4)$-UGA. It is interesting that the uniformity of the errors in the latter case. On the other hand, we know the analytical solutions and consequently the wave functions can be directly compared.

Table 1. Exact bond energies compared with the energies provided by the HO-DVR method, the basis dimension was taken to be $N = 2001$.

	Basis Dimension N		
	2001		
E_n	$L = 0$	$L = 1$	$L = 2$
$-1.$	-0.9969		
-0.25	-0.2496	-0.2500	
-0.1111	-0.1111	-0.1111	-0.1111
-0.0625	-0.0625	-0.0625	-0.0625
-0.04	-0.0400	-0.0400	-0.0400
-0.0278	-0.0275	-0.0276	-0.0276
-0.0204	-0.0173	-0.0175	-0.0180

Table 2. Exact bond energies compared with the energies provided by the diagonalization of (68) taking a total number of bosons $N = 4001$ with the $U(4)$-UGA method.

	Total Number of Bosons N		
	4001		
E_n	$L = 0$	$L = 1$	$L = 2$
$-1.$	-1.00330		
-0.25	-0.25088	-0.25039	
-0.1111	-0.11237	-0.11222	-0.11220
-0.0625	-0.06463	-0.06457	-0.06455
-0.0400	-0.04344	-0.04340	-0.04336
-0.0278	-0.0330	-0.0330	-0.0330
-0.0204	-0.02541	-0.02555	-0.02586

Table 3. Errors of the energies provided by the HO-DVR method. Errors were calculated considering the difference of energies $|E(N = 2051) - E(N = 1951)|$.

Error for $L = 0$	Error for $L = 1$	Error for $L = 2$
0.0001		
0.00002	9×10^{-9}	
5×10^{-6}	3×10^{-9}	8×10^{-13}
2×10^{-6}	1×10^{-9}	1×10^{-11}
2×10^{-6}	5×10^{-7}	3×10^{-7}
0.00008	0.00007	0.00005
0.0006	0.0005	0.0005

Table 4. Errors of the energies provided by the $U(4)$-UGA method. Errors were calculated considering the difference of energies $|E(N = 4051) - E(N = 3951)|$.

Error for $L = 0$	Error for $L = 1$	Error for $L = 2$
0.00007		
0.00002	0.00001	
0.00003	0.00003	0.00003
0.00005	0.00005	0.00005
0.00009	0.00009	0.00009
0.0001	0.0001	0.0001
0.00008	0.00006	0.00003

In our approach the α-th eigenstate in the HO-DVR method takes the form

$$|\psi^N_{\alpha,lm}\rangle = \sum_{n=0}^{N} \langle nlm|\psi^N_{\alpha,lm}\rangle \, |nlm\rangle, \tag{82}$$

while for the $U(4)$-UGA

$$|\psi^N_{\alpha,LM}\rangle = \sum_{n_a=0}^{N} \langle [N]n_a LM|\psi^N_{\alpha,LM}\rangle \, |[N]n_a LM\rangle. \tag{83}$$

From these expressions we obtain the corresponding position representation for the radial contribution of the wave function. Integrating over the angular coordinates and applying the coordinate projection we obtain $\langle r|\psi^N_{\alpha L}\rangle$ or $\langle r|\psi^N_{\alpha l}\rangle$ with

$$\langle r|nl\rangle = \langle r|[N]n_a L\rangle = A_{nl} r^l e^{r^2/2} L^{l+1/2}_{(n-l)/2}(r^2), \tag{84}$$

where $A_{nl} = \sqrt{(2^{l+2}(n-l)!!)/(\sqrt{\pi}(n+l+1)!!)}$. The projections $\langle r|\psi^N_{\alpha L}\rangle$ or $\langle r|\psi^N_{\alpha l}\rangle$ can be compared to the exact analytical functions $R_{nL}(r)$. Here the variable r is given in the distance unit d. Since $\epsilon = 1$ we have that $\hbar\omega = E_B$, a result that determines the distance units in the MKS units system. In Figure 2 the position projection of the eigenstates (82) and (83) together with the exact solutions for the $1s, 2s$ and $3s$ orbitals are displayed, while in Figure 3 the corresponding $2p, 3p$ and $3d$ orbitals are shown. Beyond the intrinsic convergence ($N = 2001$ for the HO-DVR method and $N = 4001$ for the U(4)-UGA) provided by the harmonic oscillator basis, we notice that the wave functions are very well described, a feature that proves the validity of both approaches for the free Hydrogen atom.

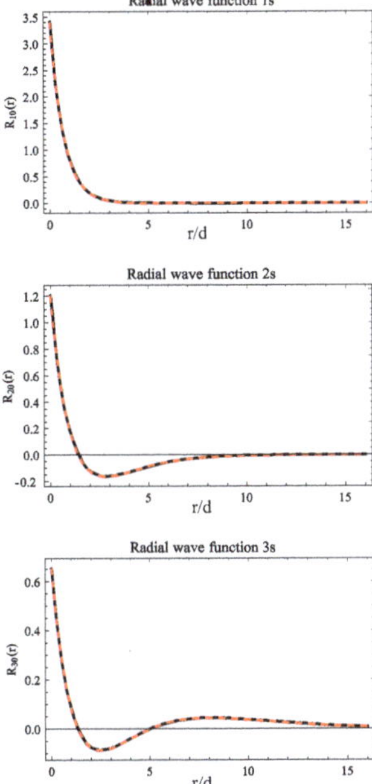

Figure 2. Comparison between exact and calculated radial wave functions $R_{n0}(r)$; $n = 1, 2, 3$ for the Hydrogen atom using (75) and (79) with parameters $N = 2001$ and $N = 4001$ respectively, with $\epsilon = 1$. The dash lines correspond to the exact wave functions.

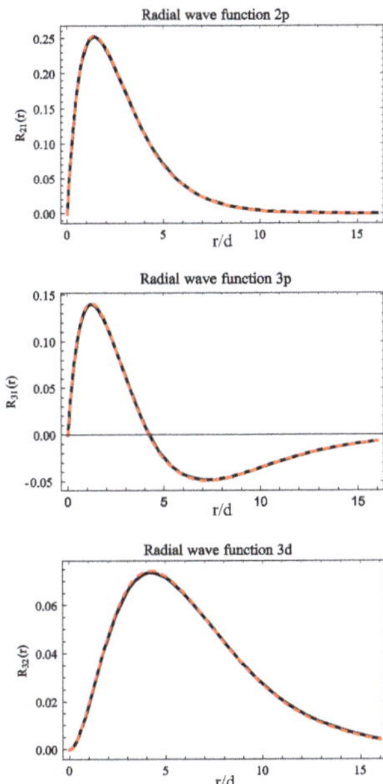

Figure 3. Comparison between exact and calculated radial wave functions $R_{21}(r)$, $R_{31}(r)$, $R_{32}(r)$ for the Hydrogen atom using (75) and (79) with parameters $N = 2001$ and $N = 4001$ respectively, with $\epsilon = 1$. The dash lines correspond to the exact wave functions.

4. Stark Effect

In this section, we study the effect of an static external electric field on the spectrum of the Hydrogen atom. The electric field interacts through the dipole producing a splitting of the spectral lines called Stark effect, since this phenomenon was experimentally shown by Stark in 1913 [27]. There is a vast amount of literature on the Stark effect in the Hydrogen atom. This problem has been treated in configuration space using semi-classical methods [28–30], as well as variational approaches [31,32], but also algebraically in an elegant way. Since it is impossible to mention all the contributions to this problem we confine our discussion to some relevant works to put in context our algebraic method. We do not pretend that our method represents an improvement to the previous descriptions. Instead we are interested in showing that it is possible to tackle this problem along the venue worked out by *M. Moshinsky*, where the harmonic oscillator was the basis to solve problems from atoms to quarks [17].

Algebraic methods applied to the Hydrogen atom may be considered to have originated from the identification of the $SO(4)$ group as the symmetry group of the non-relativistic Hydrogen atom for the bound states and $SO(3,1)$ for the continuum [33–39]. Based upon the $SO(4)$ symmetry group the Stark effect for both static and time-dependent electric fields has been described from different points of view, including semiclassical methods taking into account the crucial result that the Runge–Lenz vector is proportional to the coordinate and the energy [40–43]. In any case, these treatments are confined to the subspace characterized for a given principal quantum number n since the generators of the group commute with the Hamiltonian, a fact that constraints the descriptions to perturbative schemes. In addition the natural basis to deal with this problem, called the Stark basis $|n_1 n_2 m\rangle$, is based on the parabolic coordinates where the Schrödinger equation is separable. In this context an important role is played by the transformation brackets $\langle n_1 n_2 m | nlm \rangle$ connecting to the states $|nlm\rangle$ with good angular momentum [44].

A full algebraic treatment for the Stark effect needs the introduction of the $SO(4,2)$ dynamical group [38,45–48] or alternative groups [49]. The $SO(4,2)$ group can be identified by the merging of the $SO(2,1)$ and $SO(4)$ groups [50]. The $SO(2,1)$ group, or alternatively, the $SU(1,1)$ group deals with the radial contributions that shift the principal quantum numbers [50–52]. Even though the $SO(4,2)$ group allows the dipole matrix elements to be calculated, the use of the dynamical group tends to perturbation treatments [53,54]. In contrast, our algebraic approach is not based on the use of the dynamical group, instead, our approach falls in the framework of a discrete variable representation approach, whose salient feature consists in its simplicity as we next present.

In the previous section, we have shown that both algebraic DVR approaches are able to describe the free Hydrogen atom, although with a different convergence degree. The result is that the HO-DVR method needs half of the number of the basis functions, albeit with somewhat less uniformity of errors. In this section, the interaction with a static electric field is introduced whose better description is tempting to be given by the HO-DVR approach. The result, however, is that the $SU(4)$-UGA is more suitable to describe the Stark Effect given the dimensions taken into account for the free Hydrogen atom, as we show next.

The Hydrogen atom under the action of the electric field takes the form

$$\hat{H}^S = \hat{H} + \hat{V}(E), \tag{85}$$

where $\hat{H}$ represents the unperturbed Hydrogen atom (72) and $\hat{V}(E)$ is the potential energy associated with the interaction of the electron with the external electric field $|\mathbf{E}|$ along the z direction

$$\hat{V}(E) = -e|\mathbf{E}|z, \tag{86}$$

and in the convenient units previously used

$$\frac{\hat{V}(E)}{E_B} = -\frac{\bar{z}}{\epsilon} 2\sqrt{2}|\bar{\mathbf{E}}|, \tag{87}$$

with $\bar{z}$ in units of d and $\bar{E}$ in atomic units defined by the electric field of the nucleus given by $e/a_0^2 \approx 5 \times 10^9$ V/cm.

Let us now consider the realization of the potential $\hat{V}(E)$ in accordance with the two algebraic DVR methods. According to the relations (6) and (7) the z component is given by

$$\bar{z} = \frac{1}{2}(a_0^\dagger + a_0).\tag{88}$$

Hence, in the HO-DVR method we just calculate the matrix elements of the operator $a_0^\dagger$ in the basis (9). We obtain

$$\langle n+1, l+1, m|a_0^\dagger|nlm\rangle = \sqrt{(n+l+3)}\sqrt{\frac{(l+1)}{(2l+3)}}\sqrt{\frac{(l+m+1)(l-m+1)}{(2l+1)(l+1)}},\tag{89}$$

$$\langle n+1, l-1, m|a_0^\dagger|nlm\rangle = -\sqrt{(n-l+2)}\sqrt{\frac{l}{(2l-1)}}\sqrt{\frac{(l+m)(l-m)}{l(2l+1)}}.\tag{90}$$

On the other hand, in the $U(4)$-UGA method, taking into account (50) and (51) with the correspondence $\bar{z} \to \hat{Q}/d$, we have

$$\bar{z} = \frac{1}{\sqrt{2N}}\hat{D}_0.\tag{91}$$

Using the Wigner–Eckart theorem, its matrix elements take the form

$$\langle [N]n'_a L'M|\hat{D}|[N]n_a LM\rangle = \langle LM; 10|L'M\rangle \langle [N]n'_a L'||\hat{D}||[N]n_a L\rangle,\tag{92}$$

with the reduced matrix elements

$$\langle [N]n_a - 1L - 1\rangle||\hat{D}||[N]n_a L\rangle = -\sqrt{\frac{(N - n_a + 1)(n_a + L + 1)L}{2L - 1}},\tag{93}$$

$$\langle [N]n_a - 1L + 1\rangle||\hat{D}||[N]n_a L\rangle = -\sqrt{\frac{(N - n_a + 1)(n_a - L)(L + 1)}{2L + 3}},\tag{94}$$

$$\langle [N]n_a + 1L - 1\rangle||\hat{D}||[N]n_a L\rangle = \sqrt{\frac{(N - n_a)(n_a - L + 2)L}{2L - 1}},\tag{95}$$

$$\langle [N]n_a + 1L + 1\rangle||\hat{D}||[N]n_a L\rangle = \sqrt{\frac{(N - n_a)(n_a + L + 3)(L + 1)}{2L + 3}}.\tag{96}$$

Taking into account these technical ingredients, we carried out the diagonalization of the Hamiltonian (85) in the framework of both methods.

In Figure 4 the energy levels associated with the subspace $n = 2$ as a function of the electrical field are depicted. In order to obtain these results, a cutoff of the dimension space determined by the feasibility of the calculations was necessary as well as a constraint in the involved angular momenta. The black solid circles correspond to the spectrum obtained using the $U(4)$-UGA, where the number of boson was taken to be $N = 3501$ up to $L = 16$, while the blue triangles correspond to HO-DVR method taking $N = 3001$ up to $l = 11$. The continuous lines correspond to the energy-splitting obtained by the variational approach described in Ref. [32], where an accurate variational method is presented for the $n = 2-5$ shells for different $m's$ including a display of the wave functions. Since, for each approach, the convergence of the energy levels for the null field is different the continuous lines were appropriately located to follow the splitting trend. There are two remarkable facts to be highlighted. First, the splitting is correctly described by the $SU(4)$-UGA, while the HO-DVR method tends to overestimate the splitting as the field intensity increases. This behavior is explained by the different basis dimensions. As a second point, we notice that for small-field intensity the splitting is rather flattened, showing an evident deviation from the linearity predicted by perturbation theory. In order to clearly see this effect a zoom of Figure 4 is depicted in Figure 5 for low field intensities in the interval $[0, 10^{-4}]$. Both approaches present this anomalous behavior, which we believe corresponds to a lack of convergence. To make clear that this is in fact the case, a similar plot is depicted in Figure 6 for the HO-DVR approach for different boson numbers. The black triangles correspond to $N = 3001$, while the red ones to $N = 1501$. From these plots, it becomes clear that as the basis dimension increases the splittings approach to the correct results.

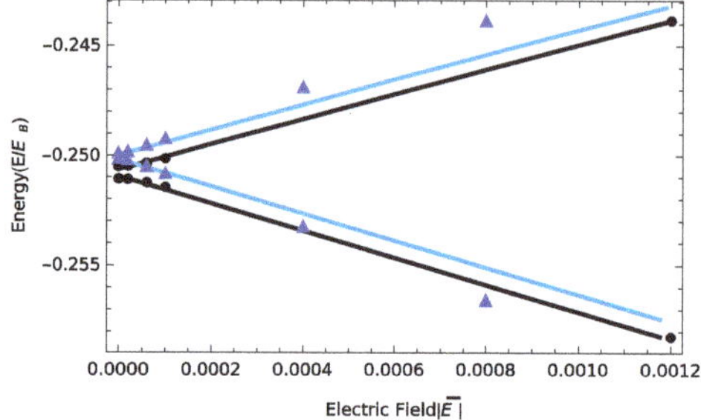

Figure 4. Effect of the electric field effect over the subspace $n = 2$ with $m = M = 0$ in the Hydrogen atom. The black solid circles correspond to spectrum obtained with the $U(4)$-UGA taking $N = 3501$ up to $L = 16$, while the blue triangles correspond to HO-DVR method taking $N = 3001$ up to $l = 11$. Since the calculations were carried out independently, the plotted points do not coincide. The continuous line correspond to the variational approach described in Ref. [32]. $|\bar{E}|$ is the electric field in atomic units.

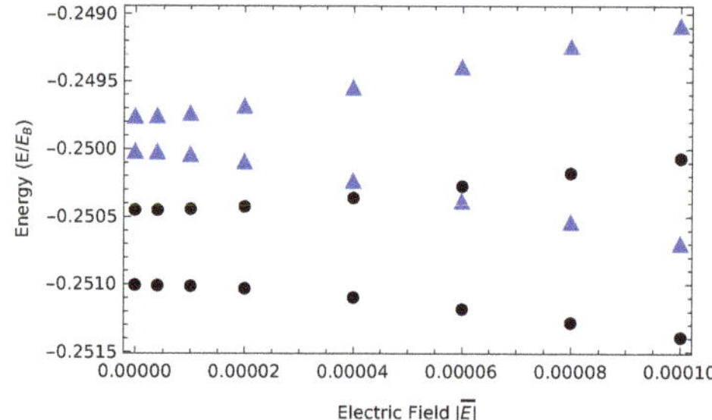

Figure 5. Electric field effect over the $n = 2$ levels of Hydrogen atom for $M = m = 0$. The black solid circles correspond to the results using the $U(4)$-UGA method taking $N = 3501$ and $L = 16$. The blue triangles correspond to HO-DVR method taking $N = 3001$ and $l = 11$.

It is worth mentioning the complexity of these calculations due to the spherical symmetry breaking. Due to our computational limitations, we are forced to fix a maximum value for the angular momentum. Without this approximation, we would be dealing with a $(3,067,752)^2$ dimension matrices, which are impossible to calculate given our computational capacity. However, the restriction of the angular momentum components has physical justification in the sense that not all the functions with a given projection are expected to be mixed. Indeed for the $n = 2$ and $n = 3$ levels the mixing of states with different angular momenta are not expected to be so large (may be up to $L, l = 5$), but in this work we have included the maximum angular momenta that allows our calculations to be carried out.

In order to see whether the previous behavior for the p-space is reproduced for other subspaces, in Figure 7 we display the effect of the electric field provided by both algebraic DVR methods over the subspace with $n = 3$ for $M = m = 0$. The same trend is obtained given the comparison with the exact result given in Ref. [32]. While the $SU(4)$-UGA reproduces the correct splitting, the HO-DVR method overestimates it.

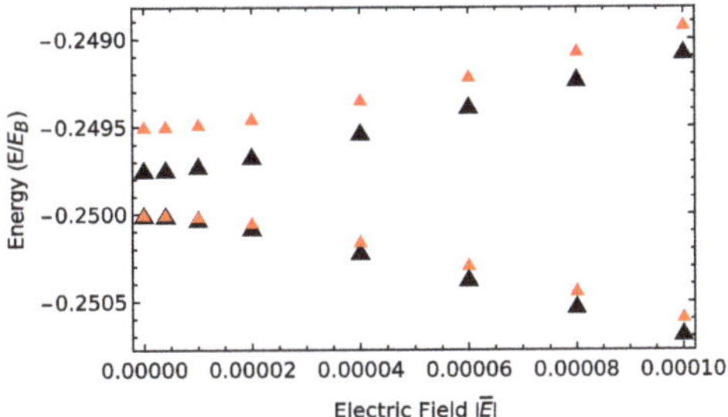

Figure 6. Zoom of the electric field effect over the subspace $n = 2$ with $m = M = 0$ in the Hydrogen atom provided by the HO-DVR method for $N = 1501$ (red triangles) and $N = 3001$ (black triangles).

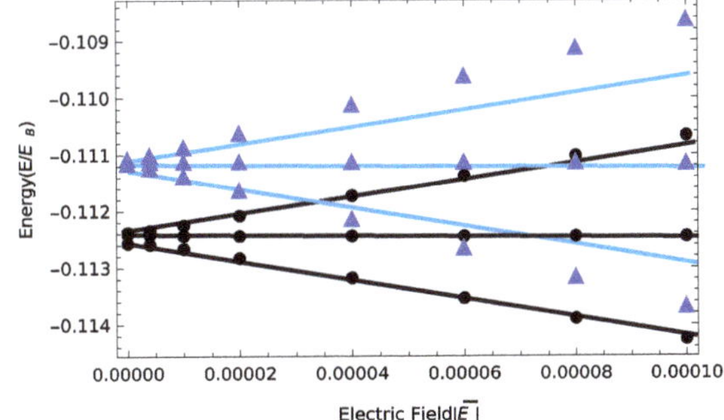

Figure 7. Electric field effect over the subspace characterized by the principal quantum number $n = 3$ with $M = m = 0$ in the Hydrogen atom. The solid black dots correspond to the $U(4)$-UGA method taking $N = 3501$ up to $L = 16$, while the blue triangles correspond to the HO-DVR method taking $N = 3001$ up to $l = 11$. The continuous lines correspond to the exact results provided by Ref. [32].

Finally it is convenient to recall the usual treatment of the Stark effect using perturbation theory. For the first excited state ($n = 2$) the fourfold degeneracy must be born in mind. In this context, since the angular momentum projection is preserved and z is odd, only the states $|\Psi_{200}\rangle$ and $|\Psi_{210}\rangle$ interact through (86). The rest of the matrix element vanish leading to the eigenvalues for $M = m = 0$ [55,56]:

$$E_{\pm} = E_2 \pm 6|\bar{\mathbf{E}}|, \tag{97}$$

where the energy is intended to be in units of E_B and the electric field in atomic units. The magnitude of the splitting of the levels is proportional to the electric field strength, giving rise to the linear Stark effect. The straight lines provided by the equation (97) coincide with the exact values displayed in Figure 4. Deviation from linearity manifests around field intensities of 0.004 au.

5. Conclusions

In this work, we have applied two algebraic discrete variable representation approaches to describe the non-relativistic Hydrogen atom together with the Sark effect. The selection of the Coulombic potential to test our approaches is twofold. On one hand, its study is compulsory to consider more complex situations like molecular systems. On the other hand it is reasonable to evaluate the feasibility of these approaches to take into account additional interactions breaking the symmetry, in particular for the $SU(4)$-UGA where the condition (71) is imposed to remove the intrinsic degeneracy associated with the deformed harmonic oscillator (60).

The two approaches we have proposed may be, in principle, applied to general potentials to obtain the solutions of the time-independent Schrödinger equation. In both approaches, the matrix representation of the Hamiltonian given by either (36) or (79) is given in terms of two diagonal matrices together with the transformation brackets. The calculation of the transformation brackets is independent of the potential and hence they are calculated only once for a given space dimension. The diagonal matrices $\Lambda^{(E)}$ corresponding to the simple or deformed harmonic oscillator is trivially generated and fixed for any system, while $\Lambda^{(r)}$ and $\Lambda^{(Q)}$ are diagonal in the coordinate representation which are also easily constructed. Hence, these approaches represent powerful simple alternatives to study systems where the analytical solutions of the Schrödinger equations are either not known or difficult to obtain by integration methods. Even considering moderate values of space dimension, the approach provides solutions reflecting the main salient features of the system, a situation very useful when the solution cannot be obtained by conventional methods. Hence, this approach may be useful in catastrophe theory where a small change in the original stable potential changes the topology of the problem [57], as well as in the study of quantum phase transitions [58], but also provides new ways to obtain Franck–Condon factors between different potentials since in our approach a common basis of harmonic oscillators are used [59].

We have shown the feasibility of both approaches, the HO-DVR method and U(4)-UGA to obtain the solutions of the 3D Coulombic potential. Although both algebraic methods provide the solutions, the HO-DVR approach proves to show a faster convergence, a fact that makes it the best candidate to study this system. A quite different situation is manifested when the dipole interaction is taken into account to describe the Stark effect, which is particularly difficult due to the spherical symmetry breaking. The full dimension of the matrices to be diagonalized is too large to be consider in realistic terms and consequently, we were forced to cutoff the included angular momenta. With this approximation, the calculations become feasible with reasonable convergence. Fixing the total number of bosons obtained to describe the free Hydrogen atom, the $SU(4)$-UGA turns out to be the suitable method to describe the energy splittings, while the HO-DVR method overestimates the Stark effect. Although both approaches are based on the harmonic oscillator basis, the localized functions associated with the discrete variable representation basis are different, a fact that makes the difference between these approaches. We have thus obtained the unexpected remarkable result that a preference of a model based on the convergence in the zeroth-order Hamiltonian does not imply the same preference when a symmetry breaking interaction is included.

The two presented algebraic approaches provide alternatives to obtain the solutions of 3D systems for any potential where the harmonic oscillator functions represent a good basis. Even though for the Coulombic potential the harmonic oscillator is not by far the best basis, our methods provide reasonable results in a simple way. This opens the possibility of studying systems with different algebraic DVR methods where the exact solutions are not known, providing the possibility of extracting properties by purely algebraic means using the standard tools of matrix diagonalization.

Author Contributions: M.R.-A., M.B.-M.: investigation, software, methodology, reviewing, discussion. R.L.: conceptualization, supervision, investigation, original draft preparation, discussion. J.M.A., J.G.-C.: conceptualization, investigation, writing, discussion. E.O.: investigation, discussion. All authors have read and agreed to the published version of the manuscript.

Funding: This research received no external funding.

Acknowledgments: This work is partially supported by DGAPA-UNAM, Mexico, under project IN-212020, by the Consejería de Economía, Conocimiento, Empresas y Universidad de la Junta de Andalucía (Spain) under Group FQM-160, by the Spanish Ministerio de Ciencia e Innovaciíon, ref. FIS2017-88410-P, RTI2018-098117-B-C21 and PID2019-104002GB-C22, and by the European Commission, ref. H2020-INFRAIA-2014-2015 (ENSAR2). First author is grateful to DGAPA-UNAM for the postdoctoral scholarship (Facultad de Química). Second author is also grateful for the scholarship (Posgrado en Ciencia e Ingeniería de Materiales) provided by CONACyT, México.

References

1. Light, J.C.; Hamilton, I.P.; Lill, J.V. Generalized discrete variable approximation in quantum mechanics. *J. Chem. Phys.* **1985**, *82*, 1400. [CrossRef]
2. Light, J.C.; Carrington, T., Jr. Discrete-variable representations and their utilization. *Adv. Chem. Phys.* **2000**, *114*, 263.
3. Littlejohn, R.G.; Cargo, M.; Carrington; Mitchell, K.A.; Poirier, B. A general framework for discrete variable representation basis sets. *J. Chem. Phys.* **2002**, *116*, 8691. [CrossRef]
4. Wang, X.-G.; Carrington, T. J. A discrete variable representation method for studying the rovibrational quantum dynamics of molecules with more than three atoms. *Chem. Phys.* **2009**, *130*, 094101. [CrossRef] [PubMed]
5. Shizgal, B.; Blackmore, R. A discrete ordinate method of solution of linear boundary value and eigenvalue problems. *J. Comput. Phys.* **1984**, *55*, 313. [CrossRef]
6. Shizgal, B.D.; Chen, H. The quadrature discretization method (QDM) in the solution of the Schrödinger equation with nonclassical basis functions. *J. Chem. Phys.* **1996**, *104*, 4137. [CrossRef]
7. Pérez-Bernal, F.; Arias, J.M.; Carvajal, M.; Gómez-Camacho, J. Configuration localized wave functions: General formalism and applications to vibrational spectroscopy of diatomic molecules. *Phys. Rev. A* **2000**, *61*, 042504. [CrossRef]
8. Baye, D.; Heenen, P.-H. Generalised meshes for quantum mechanical problems. *J. Phys. A* **1986**, *19*, 2041. [CrossRef]
9. Vincke, M.; Malegat, L.; Baye, D. Regularization of singularities in Lagrange-mesh calculations. *J. Phys. B At. Mol. Opt. Phys.* **1993**, *26*, 811. [CrossRef]
10. Baye, D. Lagrange-mesh method for quantum-mechanical problems. *Phys. Status Solidi B* **2006**, *243*, 1095. [CrossRef]
11. Baye, D. The Lagrange-mesh method. *Phys. Rep.* **2015**, *565*, 1. [CrossRef]
12. Estévez-Fregoso, M.M.; Lemus, R. Connection between the $su(3)$ algebraic and configuration spaces: Bending modes of linear molecules. *Mol. Phys.* **2018**, *116*, 2374. [CrossRef]
13. Rodríguez-Arcos, M.; Lemus, R. Unitary group approach for effective potentials in 2D systems: Application to carbon suboxide C_3O_2. *Chem. Phys. Lett.* **2018**, *713*, 266. [CrossRef]
14. Lemus, R. Unitary group approach for effective molecular potentials: 1D systems. *Mol. Phys.* **2019**, *117*, 167. [CrossRef]
15. Lemus, R. A simple approach to solve the time independent Schrödinger equation for 1D systems. *J. Phys. Commun.* **2019**, *3*, 025012. [CrossRef]
16. Rodríguez-Arcos, M.; Lemus, R.; Arias, J.M.; Gómez-Camacho, J. Unitary group approach to describe interatomic potentials in 3D systems. *Mol. Phys.* **2019**, *118*, e1662959. [CrossRef]
17. Moshinsky, M. *The Harmonic Oscillator in Modern Physics: From Atoms to Quarks*; Gordon and Breach: New York, NY, USA, 1969.
18. Frank, A.; Van Isacker, P. *Algebraic Methods in Molecular and Nuclear Structure Physics*; Wiley and Sons: New York, NY, USA, 1994.
19. Iachello, F.; Levine, R.D. *Algebraic Theory of Molecules*; Oxford University Press: Oxford, UK, 1995.
20. Santiago, R.D.; Arias, J.M.; Gómez-Camacho, J.; Lemus, R. An approach to establish the connection between configuration and $su(n+1)$ algebraic spaces in molecular physics: application to ammonia. *Mol. Phys.* **2017**, *115*, 3206. [CrossRef]
21. Santopinto, E.; Bijker, R.; Iachello, F. Transformation brackets between U(ν+ 1)$\supset$ U(ν)$\supset$ SO(ν) and U(ν+ 1)$\supset$ SO(ν+ 1)$\supset$ SO(ν). *J. Math. Phys.* **1996**, *37*, 2674. [CrossRef]
22. Cejnar, P.; Jolie, J. Quantum phase transitions in the interacting boson model. *Prog. Part. Nucl. Phys.* **2009**, *62*, 210. [CrossRef]
23. Arias, J.M.; Camacho, J.G.; Lemus, R. An $su(1,1)$ dynamical algebra for the Pöschl–Teller potential. *J. Phys. A Math. Gen.* **2004**, *37*, 877. [CrossRef]
24. Lemus, R.; Arias, J.M.; Gómez-Camacho, J. An $su(1,1)$ dynamical algebra for the Morse potential. *J. Phys. A Math. Gen.* **2004**, *37*, 1805. [CrossRef]
25. Álvarez-Bajo, O.; Arias, J.M.; Gómez-Camacho, J.; Lemus, R. An approach to the study of the continuum effects in systems of interacting Morse oscillators. *Mol. Phys.* **2008**, *106*, 1275. [CrossRef]

26. Bermúdez-Montaña, M.; Lemus, R.; Castaños, O. Polyad breaking phenomenon associated with a local-to-normal mode transition and suitability to estimate force constants. *Mol. Phys.* **2017**, *115*, 3076. [CrossRef]

27. Condon, E.; Shortley, G. *The Theory of Atomic Spectra*; Cambridge University Press: Cambridge, UK, 1951.

28. Harmin, D. Hydrogenic Stark effect: Properties of the wave functions. *Phys. Rev. A* **1981**, *24*, 2491. [CrossRef]

29. Harmin, D. Theory of the Stark effect. *Phys. Rev. A* **1982**, *26*, 2656. [CrossRef]

30. Rice, M.H.; Good, R.H., Jr. Stark effect in Hydrogen. *J. Opt. Soc. Am.* **1962**, *52*, 239. [CrossRef]

31. Fernández, F.M. Direct calculation of Stark resonances in Hydrogen. *Phys. Rev. A* **1996**, *54*, 1206. [CrossRef]

32. Fernández-Menchero, L.; Summers, H.P. Stark effect in neutral Hydrogen by direct integration of the Hamiltonian in parabolic coordinates. *Phys. Rev. A.* **2013**, *88*, 022509. [CrossRef]

33. Lenz, W. Über den Bewegungsverlauf und die Quantenzustände der gestörten Keplerbewegung. *Z. Phys.* **1924**, *24*, 197. [CrossRef]

34. Pauli, W. On the spectrum of the Hydrogen from the standpoint of the new Quantum Mechanics. *Z. Phys.* **1926**, *36*, 336. [CrossRef]

35. Fock, V. 2. Phys. 98 145 Bargmann V 1936. *Z. Phys.* **1935**, *98*, 145. [CrossRef]

36. Bargman, V. Zur theorie des Wasserstoffatoms. *Z. Phys.* **1936**, *99*, 576. [CrossRef]

37. Schiff, L.I. *Quantum Mechanics*; McGraw-Hill: New York, NY, USA,1968.

38. Wybourne, B.G. *Classical Groups for Physicists*; John Wiley: New York, NY, USA, 1974.

39. Majumbdar, S.D.; Basu, D. $O(3,1)$ symmetry of the Hydrogen atom. *J. Phys. A* **1974**, *7*, 787. [CrossRef]

40. Flamand, G. The Solution of a Stark-Effect Model as a Dynamical Group Calculation. *J. Math. Phys.* **1966**, *7*, 1924. [CrossRef]

41. Bakshi, P.; Kalman, G.; Cohn, A. hydrogenic Stark-Zeeman Spectra for Combined Static and Dynamic Fields. *Phys. Rev. Lett.* **1973**, *31*, 1576. [CrossRef]

42. Gigosos, M.A.; Fraile, J.; Torres, F. Hydrogen Stark profiles: A simulation-oriented mathematical simplification. *Phys. Rev. A* **1985**, *31*, 3509. [CrossRef]

43. Demkov, Y.N.; Monozon, B.S.; Ostrovskii, V.N. Energy levels of a Hydrogen atom in crossed electric and magnetic fields. *Sov. Phys. JETP* **1970**, *30*, 775.

44. Hughes, J.W.B. Stark states and $O(4)$ symmetry of hydrogenic atoms. *Proc. Phys. Soc.* **1967**, *91*, 810. [CrossRef]

45. Barut, A.O.; Kleinert, H. Transition probabilities of the Hydrogen atom from noncompact dynamical groups. *Phys. Rev. A* **1967**, *156*, 1541. [CrossRef]

46. Fronsdal, C. Infinite multiplets and the Hydrogen atom. *Phys. Rev.* **1967**, *156*, 1665. [CrossRef]

47. Kleinert, H. Group Dynamics of Elementary Particles. *Fortsch. Phys.* **1968**, *16*, 1. [CrossRef]

48. Bednar, M. Algebraic treatment of quantum-mechanical models with modified Coulomb potentials. *Ann. Phys.* **1973**, *75*, 305. [CrossRef]

49. Hughes, J.W.B.; Yadagar, J. Theory of laser-induced inelastic collisions. *J. Phys. A.* **1976**, *9*, 1569. [CrossRef]

50. Adams, B.G.; Cizek, J.; Paldus, J. Representation theory of $so(4,2)$ for the perturbation treatment of Hydrogenic-type hamiltonians by algebraic methods. *Int. J. Quantum. Chem.* **1982**, *23*, 153. [CrossRef]

51. Dimitriev, V.F.; Rumer, Y.B. $O(2,1)$ Algebra and the Hydrogen atom. *Theor. Math. Phys.* **1970**, *5*, 1146. [CrossRef]

52. Martínez-Yépez, R.P.; Salas-Brito, A.L. An $su(1,1)$ algebraic method for the Hydrogen atom. *J. Phys. A* **2005**, *38*, 8579. [CrossRef]

53. Tsai, W.-Y. Third-order Stark effect: An operator approach. *Phys. Rev. A.* **1974**, *9*, 1081. [CrossRef]

54. Romero-Lópes, A.; Leal-Ferreira, P. Algebraic treatment of the Stark effect for Hydrogen. *Il Nuovo C* **1971**, *3*, 23. [CrossRef]

55. Greiner, W. *Quantum Mechanics: An Introduction*; Springer Science & Business Media: Berlin, Germany, 2011.

56. Davydov, A.S. *Quantum Mechanics*, 2nd ed.; Pergamon Press: New York, NY, USA, 1965.

57. Gilmore, R. *Catastrophe Theory for Scientists and Engineers*; Wiley: New York, NY, USA, 1981.

58. Castaños, O.; Calixto, M.; Pérez-Bernal, F.;Romera, E. Identifying the order of a quantum phase transition by means of Wehrl entropy in phase space. *Phys. Rev. E* **2015**, *92*, 052106.
59. Lemus, R. An algebraic approach to calculate Franck–Condon factors. *J. Math. Chem.* **2020**, *58*, 29. [CrossRef]

Publisher's Note: MDPI stays neutral with regard to jurisdictional claims in published maps and institutional affiliations.

 *symmetry*

Review

Dynamical Symmetries of the H Atom, One of the Most Important Tools of Modern Physics: SO(4) to SO(4,2), Background, Theory, and Use in Calculating Radiative Shifts

G. Jordan Maclay

Quantum Fields LLC, St. Charles, IL 60174, USA; jordanmaclay@quantumfields.com

Received: 18 June 2020; Accepted: 31 July 2020; Published: 7 August 2020

Abstract: Understanding the hydrogen atom has been at the heart of modern physics. Exploring the symmetry of the most fundamental two body system has led to advances in atomic physics, quantum mechanics, quantum electrodynamics, and elementary particle physics. In this pedagogic review, we present an integrated treatment of the symmetries of the Schrodinger hydrogen atom, including the classical atom, the SO(4) degeneracy group, the non-invariance group or spectrum generating group SO(4,1), and the expanded group SO(4,2). After giving a brief history of these discoveries, most of which took place from 1935–1975, we focus on the physics of the hydrogen atom, providing a background discussion of the symmetries, providing explicit expressions for all of the manifestly Hermitian generators in terms of position and momenta operators in a Cartesian space, explaining the action of the generators on the basis states, and giving a unified treatment of the bound and continuum states in terms of eigenfunctions that have the same quantum numbers as the ordinary bound states. We present some new results from SO(4,2) group theory that are useful in a practical application, the computation of the first order Lamb shift in the hydrogen atom. By using SO(4,2) methods, we are able to obtain a generating function for the radiative shift for all levels. Students, non-experts, and the new generation of scientists may find the clearer, integrated presentation of the symmetries of the hydrogen atom helpful and illuminating. Experts will find new perspectives, even some surprises.

Keywords: symmetry; hydrogen atom; group theory; SO(4); SO(4,2); dynamical symmetry; non-invariance group; spectrum generating algebra; Runge-Lenz; Lamb shift

---

1. Introduction

1.1. Objective of This Paper

This pedagogic review is focused on the symmetries of the Schrodinger nonrelativistic hydrogen atom exclusively to give it the attention that we believe it deserves. The fundamental results of the early work are known and do not need to be derived again. However, having this knowledge permits us to use the modern language of group theory to do a clearer, more focused presentation, and to use arguments from physics to develop the proper forms for the generators, rather than dealing with detailed, mathematical derivations to prove results we know are correct.

There are numerous articles about the symmetry of the Schrodinger hydrogen atom, particularly the SO(4) group of the degenerate energy eigenstates, including discussions from classical perspectives. The spectrum generating group SO(4,1) and the non-invariance group SO(4,2) have been discussed, but, in many fewer articles, often in appendices, with different bases for the representations. For example, numerous papers employ Schrodinger wave functions in parabolic coordinates, not with the familiar nlm quantum numbers, often with the emphasis on the details of the mathematical structures, not the physics, and with the emphasis on the potential role of the symmetry in elementary particle physics. Generators may be expressed in complex and unfamiliar terms, for example, in terms of the raising and lowering operators for quantum numbers characteristic of parabolic coordinates. Other approaches involve regularizing Schrodinger's equation by, for example, multiplying by r. The approach results in generators that are not always Hermitian or manifestly Hermitian, and the need for nonstandard inner products. Indeed, most of the seminal articles do not have the words "hydrogen atom" in the title but are focused on the prize down the road, understanding what was called at the time "the elementary particle zoo".

Yet, these foundational articles and books taken together present the information we have about the symmetry of the H atom, of which experts in the field are aware. On the other hand, to the non-expert, the student, or a researcher new to the field, it does not appear that the relevant information is in a form that is conveniently accessible. Since the hydrogen atom is the most fundamental physical system with an interaction, whose exploration and understanding has led to much of the progress in atomic physics, quantum physics, and quantum electrodynamics, we believe that a comprehensive treatment is warranted and, since most of the relevant papers were published four decades ago, is timely. Many younger physicists may not be acquainted with these results.

Unlike in a number of the foundational papers, here the operators are all Hermitian, and given in terms of the canonical position and momentum variables in the simplest forms. The transformations they generate are clearly explained, and we provide brief explanations for the group theory used in the derivations.

In most papers, a separate treatment for bound and scattering states in needed. In contrast we are able to clarify and simplify the exposition since we use a set of basis states which are eigenfunctions of the inverse of the coupling constant [1], that include both the bound states and the scattering states in a uniform way, and that employ the usual Cartesian position and momenta, with the usual inner product, with the exact same quantum numbers nlm as the ordinary bound states; a separate treatment for bound and scattering states is not required. In addition, we have two equivalent varieties of this uniform basis, one that is more suitable for momentum space calculations and one more suitable for configuration space calculations. This advantage again allows us to simplify the exposition.

We focus on the utility of group theoretic methods using our representation and derive expressions for the unitary transformation of group elements and some new results that allow for us to readily compute the first order radiative shift (Lamb shift) of a spinless electron, which accounts for about 95% of the total shift. This approach allows for us to obtain a generating function for the shifts for all energy levels. For comparison, we derive an expression for the Bethe log.

In summary, we present a unified treatment of the symmetries of the Schrodinger hydrogen atom, from the classical atom to SO(4,2) that focuses on the physics of the hydrogen atom, that gives explicit expressions for all the manifestly Hermitian generators in terms of position and momenta operators in a Cartesian space, that explains the action of the generators on the basis states, which evaluates the Casimir operators characterizing the group representations, and that gives a unified treatment of the bound and continuum states in terms of wave functions that have the same quantum numbers as the ordinary bound states. We give an example of the use of SO(4,2) in a practical application, the computation of the first order radiative shift in the hydrogen atom.

Hopefully, students and non-experts and the new generation of scientists will find this review helpful and illuminating, perhaps motivating some to use these methods in various new contexts. Senior researchers will find new perspectives, even some surprises and encouragements.

1.2. Outline of This Paper

In the remainder of Section 1 we give a brief historical account of the role of symmetry in quantum mechanics and of the work done in order to explore the symmetries of the Schrodinger hydrogen atom.

In Section 2, we provide some general background observations regarding symmetry groups and non-invariance groups and discuss the degeneracy groups for the Schrodinger, Dirac, and Klein-Gordon equations. We also introduce the uncommon $(Z\alpha)^{-1}$ eigenstates that allow us to treat the bound and scattering states in a uniform way, using the usual quantum numbers. In Section 3, the classical equations of motion of the nonrelativistic hydrogen atom in configuration and momentum space are derived from symmetry considerations. The physical meaning of the symmetry transformations and the structure of the degeneracy group SO(4) is discussed. In Section 4, we discuss the symmetries using the language of quantum mechanics. In order to display the symmetries in quantum mechanics in the most elegant and uniform way we use a basis of eigenstates of the inverse of the coupling constant $(Z\alpha)^{-1}$. In Section 5 we discuss these wave functions in momentum and configuration space, how they transform and their classical limit for Rydberg states.

In Section 6, we discuss the noninvariance or spectrum generating group of the hydrogen atom SO(4,l) and relate it to the conformal group in momentum space. In Section 7, the enlarged spectrum generating group SO(4,2) is introduced, with a discussion of the physical meaning of the generators. All of the physical states together form a basis for a unitary irreducible representation of these noninvariance groups. We derive manifestly Hermitian expressions in terms of the momentum and position canonical variables for the generators of the group transformations and obtain the values possible for the Casimir operators. We discuss the important subgroups of SO(4,2).

In Section 8, we use the group theory of SO(4,2) to determine the radiative shifts in energy levels due to the interaction of a spinless electron with its own radiation field, or equivalently with the quantum vacuum. In the nonrelativistic or dipole approximation the level shift contains a matrix element of a rotation operator of an O(1,2) subgroup of the group SO(4,2). We can sum this over all states, obtaining the character of the representation, yielding a single integral that is a generating function for the radiative shift for any level in the nonrelativistic or dipole approximation. A brief conclusion follows.

1.3. Brief History of Symmetry in Quantum Mechanics and Its Role in Understanding the Schrodinger Hydrogen Atom

The hydrogen atom is the fundamental two-body system and perhaps the most important tool of atomic physics and the continual challenge is to continually improve our understanding of the hydrogen atom and to calculate its properties to the highest accuracy possible. The current QED theory is the most precise of any physical theory [2]:

> The study of the hydrogen atom has been at the heart of the development of modern physics...theoretical calculations reach precision up to the 12th decimal place...high resolution laser spectroscopy experiments...reach to the 15th decimal place for the 1S–2S transition...The Rydberg constant is known to six parts in 10^{12} [2,3]. Today, the precision is so great that measurement of the energy levels in the H atom has been used to determine the radius of the proton.

Continual progress in understanding the properties of the hydrogen atom has been central to progress in quantum physics [4]. Understanding the atomic spectra of the hydrogen atom drove the discovery of quantum mechanics in the 1920's. The measurement of the Lamb shift in 1947 and its explanation by Bethe in terms of atom's interaction with the quantum vacuum fluctuations ushered in a revolution in quantum electrodynamics [5–7]. Exploring the symmetries of the hydrogen atom has been an essential part of this progress. Symmetry is a concept that has played a broader role in physics in general; for example, in understanding the dynamics of the planets, atomic, and molecular spectra, and the masses of elementary particles.

When applied to an isolated system, Newton's equations of motion imply the conservation of momentum, angular momentum and energy. But the significance of these conservation laws was not really understood until 1911 when Emily Nother established the connection between symmetry and conservation laws [8]. Rotational invariance in a system results in the conservation of angular momentum; translational invariance in space results in conservation of momentum; and translational invariance in time results in the conservation of energy. We will discuss Nother's Theorem in more detail in Section 2.

Another critical ingredient of knowledge, on which Nother based her proof, was the idea of an infinitesimal transformation, such as a infinitesimal rotation generated by the angular momentum operators in quantum mechanics. These ideas of infinitesimal transformations originated with the Norwegian mathematician Sophus Lie who was studying differential equations in the latter half of the nineteenth century. He studied the collection of infinitesimal transformations that would leave a differential equation invariant [9]. In 1918, German physicist and mathematician Hermann Weyl, in his classic book with the translated title "The Theory of Groups and Quantum Mechanics", would refer to this collection of differential generators leaving an operator invariant as a linear algebra, ushering in a little of the terminology of modern group theory [10]. Still this was a very early stage in understanding the role of symmetry in the language of quantum theory. When he introduced the new idea of a commutator on page 264, he put the word "commutator" in quotes. In the preface Weyl made a prescient observation: ".. the essence of the new Heisenberg-Schrodinger-Dirac quantum mechanics is to be found in the fact that there is associated with each physical system a set of quantities, constituting a non-commutative algebra in the technical mathematical sense, the elements of which are the physical quantities themselves".

A few years later Eugene Wigner published in German, "Group Theory and Its Application to the Quantum Mechanics of Atomic Spectra [11]". One might ask why was this classic not translated into English until 1959. In the preface to the English edition, Prof. Wigner recalled: "When the first edition was published in 1931, there was a great reluctance among physicists toward accepting group theoretical arguments and the group theoretical point of view. It pleases the author that this reluctance has virtually vanished.." It was the application of group theory in particle physics in the early sixties, such as SU(3) and chiral symmetry, which reinvigorated interest in Wigner's book and the field in general. In the 1940's, Wigner and Bargmann developed the representation theory of the Poincare group that later provided an infrastructure for the development of relativistic quantum mechanics [11,12].

The progress in understanding the symmetries of the hydrogen atom, in particular, has some parallels to the history of symmetry in general: there were some decades of interest but after the 1930's interest waned for about three decades in both fields, until stimulated by the work on symmetry in particle physics.

Probably, the first major advance in understanding the role of symmetry in the classical treatment of the Kepler problem after Newton's discovery of universal gravitation, elliptical orbits, and Kepler's Laws, was made two centuries ago by Laplace when he discovered the existence of three new constants of the motion in addition to the components of the angular momentum [13]. These additional conserved quantities are the components of a vector which determines the direction of the perihelion of the motion (point closest to the focus) and whose magnitude is the eccentricity of the orbit. The Laplace vector was later rediscovered by Jacobi and has since been rediscovered numerous times under different names. Today it is generally referred to as the Runge-Lenz vector. However, the significance of this conserved quantity was not well understood until the nineteen thirties.

In 1924, Pauli made the next major step forward in understanding the role of symmetry in the hydrogen atom [14]. He used the conserved Runge-Lenz vector **A** and the conserved angular momentum vector **L** to solve for the energy spectrum of the hydrogen atom by purely algebraic means, a beautiful result, yet he did not explicitly identify that **L** and **A** formed the symmetry group SO(4) corresponding to the degeneracy. At this time, the degree of degeneracy in the hydrogen energy levels was believed to be n^2 for a state with principal quantum number n, clearly greater than the degeneracy due to rotational symmetry which is $(2l + 1)$. The n^2 degeneracy arises from the possible values of the angular momentum $l = 0, 1, 2, \ldots n - 1$, and the $2l + 1$ values of the angular momentum along the azimuthal axis $m = -l, -l + 1, \ldots 0, 1, 2, l + 1$. The additional degeneracy was referred to as "accidental degeneracy [15]."

Six years after Pauli's paper, Hulthen used the new Heisenberg matrix mechanics to simplify the derivation of the energy eigenvalues of Pauli by showing that the sum of the squares $\mathbf{L}^2 + \mathbf{A}^2$ could be used to express the Hamiltonian and so could be used to find the energy eigenvalues [16]. In a one sentence footnote in this three page paper, Hulthen gives probably the most important information in his paper: Prof. Otto Klein, who had collaborated for years with Sophus Lie, had noticed that the two conserved vectors formed the generators of the Lorentz group, which we can describe as rotations in four dimensions, the fourth dimension being time. This is the non-compact group SO(3,1), the special orthogonal group in four dimension whose transformations leave the magnitude $g_{\mu\nu}z^{\mu}z^{\nu} = -t^2 + x^2 + y^2 + z^2$ unchanged [17]. Klein's perceptive observation triggered the introduction of group theory to understanding the hydrogen atom.

About a decade later, in 1935, the Russian physicist Vladimir Fock published a major article in Zeitshrift fur Physik, the journal in which all the key articles about the hydrogen atom cited were published [18]. He transformed Schrodinger's equation for a given energy eigenvalue from configuration space to momentum space, and did a stereographic projection onto a unit sphere, and showed that the bound state momentum space wave functions were spherical harmonics in four dimensions. He stated that this showed that rotations in four dimensions corresponded to the symmetry of the degenerate bound state energy levels in momentum space, realizing the group SO(4), the group of special orthogonal transformations which leaves the norm of a four-vector $U_0^2 + U_1^2 + U_2^2 + U_3^2$ constant. By counting the number of four-dimensional spherical harmonics Y_{nlm} in momentum space ($m = -l, -l+1 ... 0, 1, ... l$, where the angular momentum l can equal $l = n - 1, n - 2, ... 0$), he determined that the degree of degeneracy for the energy level characterized by the principal quantum number n was n^2. It is interesting that Fock did not cite the work by Pauli, implying the four dimensional rotational symmetry in configuration space. Fock also presented some ideas about using this symmetry in calculating form factors for atoms.

A year later, the German-American mathematician and physicist Valentine Bargmann showed that for bound states (E < 0) Pauli's conserved operators, the angular momentum $\mathbf{L}$ and the Runge–Lenz vector $\mathbf{A}$, obeyed the commutation rules of the SO(4) [12]. His use of commutators was so early in the field of quantum mechanics, that Bargmann explained the square bracket notation he used for a commutator in a footnote [19]. He gave differential expression for the operators, adapting the approach of Lie generators in the calculation of the commutators. He linked solutions to Schrodinger's equation in parabolic coordinates to the existence of the conserved Runge–Lenz vector and was thereby able to establish the relationship of Fock's results to the algebraic representation of SO(4) for bound states implied by Fock and Pauli [12]. He also pointed out that the scattering states (E > 0) could provide a representation of the group SO(3,1). In a note at the end of the paper, Bargmann, who was at the University in Zurich, thanked Pauli for pointing out the paper of Hulthen and the observation by Klein that the Lie algebra of $\mathbf{L}$ and $\mathbf{A}$ was the same as the infinitesimal Lorentz group, which is how he referred to a Lie algebra. Bargmann's work was a milestone demonstrating the relationship of symmetry to conserved quantities and it clearly showed that to fully understand a physical system one needed to go beyond the usual ideas of geometrical symmetry. This work was the birth, in 1936, without much fanfare, of the idea of dynamical symmetry.

Little attention was paid to these developments until the 1960's, when interest arose primarily because of the applications of group theory in particle physics, particularly modeling the mass spectra of hadrons. Particle physicists were faced with the challenge of achieving a quantitative description of hadron properties, particularly the mass spectra and form factors, in terms of quark models. Since little was know about quark dynamics they turned to group-theoretical arguments, exploring groups like SU(3), chiral U(3)xU(3), U(6)xU(6), etc. The success of the eight-fold way of SU(3) (special unitary group in three dimensions) of American physicist Murray Gell-Mann in 1962 brought attention to the use of symmetry considerations and group theory as tools for exploring systems in which one was unsure of the exact dynamics [20].

In 1964, three decades after Fock's work, American physicist Julian Schwinger published a paper using SO(4) symmetry to construct a Green function for the Coulomb potential, which he noted was based on a class he taught at Harvard in 1949 [21]. The publication was a response to the then current emphasis on group theory and symmetry, which led to, as Israeli physicist Yuval Ne'eman described it, "'the great-leap-forward' in particle physics during the years 1961–1966 [22]". Some of the principal researchers leading this effort were Ne'eman [23], Gell-Mann [20,24,25], Israeli physicist Y. Dothan [26], Japanese physicist Yochiri Nambu [27], and English-American Freeman Dyson [28]. Advantage was taken of the mathematical infrastructures of group theory developed years earlier [10–12,29,30].

Interest was particularly strong in systems with wave equations with an infinite number of components, which characterize non-compact groups. In about 1965, this interest in particle physics gave birth to the identification of SO(4,1) and SO(4,2) as Spectrum Generating Algebras that might serve as models for hadronic masses. The hydrogen atom was seen as a model to explore the infinite dimensional representations of non-compact groups. The first mention of SO(4,1) was by Barut, Budini, and Fronsdal [31], where the H atom was presented as an illustration of a system characterized by non-compact representation, and so comprising an infinite number of states. The first mention of a six dimensional symmetry, referred to as the "non-compact group O(6)", appears to be by the Russian physicists I. Malin and V. Man'ko of the Moscow Physico-technical Institute [32]. In a careful three page paper, they showed that all of the bound states of the H atom energy spectrum in Fock coordinates provided a representation of this group, and they calculated the Casimir operators for their symmetric tensor representation in parabolic coordinates.

Very shortly thereafter, Turkish-American theoretical physicist Asim Barut and his student at University of Colorado, German theoretical physicist Hagen Kleinert, showed that including the dipole operator *er* as a generator led to the expansion of SO(4,1) to SO(4,2), and that all the bound states of the H atom formed a representation of SO(4,2) [33]. This allowed them to calculate dipole transition matrix elements algebraically. They give a position representation of the generators based on the use of parabolic coordinates. The generators of the transformations are given in terms of the raising and lowering operators for the quantum numbers for solutions to the H atom in parabolic coordinates. The dilation operator is used to go from one SO(4) subspace with one energy to a SO(4) subspace with different energy and it has a rather complicated form. They also used SO(4,2) symmetry to compute form factors [34].

The papers of the Polish-American physicist Myron Bander and French physicist Claude Itzakson published in 1966, when both were working at SLAC (Stanford Linear Accelerator in California) provide the first mathematically rigorous and "succinct" review of the O(4) symmetry of the H atom and provide an introduction to SO(4,1) [35,36], which is referred to as a spectrum generating algebra SGA, meaning that it includes generators that take the basis states from one energy level to another. They use two approaches in their mathematical analysis, the first is referred to as "the infinitesimal method," based on the two symmetry operators, **L** and **A** and the O(4) group they form, and the other, referred to as the "global method", first done by Fock, converts the Schrodinger equation to an integral equation with a manifest four dimensional symmetry in momentum space. They establish the equivalence of the two approaches by appealing to the solutions of the H atom in parabolic coordinates, and demonstrate that the symmetry operators in the momentum space correspond to the symmetry operators in the configuration space. As they note, the stereographic projection depends on the energy, so the statements for a SO(4) subgroup are valid only in a subspace of constant energy. They then explore the expansion of the SO(4) group to include scale changes so the energy can be changed, transforming between states of different principal quantum number, which correspond to different subspaces of SO(4). To insure that this expansion results in a group, they include other transformations, which results in the the generators forming the conformal group O(4,1). Their mathematical analysis introducing SO(4,1) is based on the projection of a p dimensional space (4 in the case of interest) on a parabaloid in p + 1 dimensions (5 dimensions). In their derivation they treat bound states in their first paper [35] and scattering states in the second paper [36].

As we have indicated the interest in the SO(4,2) symmetry of the Schrodinger equation was driven by a program focused on developing equations for composite systems that had infinite multiplets of energy solutions and ultimately could lead to equations that could be used to predict masses of elementary particles, perhaps using other than four dimensions [27,35–40]. In 1969 Jordan and Pratt showed that one could add spin to the generators A and L, and still form a SO(4) degeneracy group. By defining $J = \frac{1}{2}(L + A) + S$, they showed one could obtain a representation of O(4,1) for any spin s [41].

In their review of the symmetry properties of the hydrogen atom, Bander and Itzakson emphasize this purpose for exploring the group theory of the hydrogen atom [35]:

> The construction of unitary representations of non-compact groups that have the property that the irreducible representations of their maximal subgroup appear at most with multiplicity one is of certain interest for physical applications. The method of construction used here in the Coulomb potential case can be extended to various other cases. The geometrical emphasis may help to visualize things and provide a global form of the transformations.

Special attention was also given to solutions for the hydrogen atom from the two body Bethe-Salpeter equation for a proton and electron interacting by a Coulomb potential, since the symmetry was that of a relativistic non-compact group [27,40,42,43].

Finally in 1969, five years after it was published, Schwinger's form of the Coulomb Green's function based on the SO(4) symmetry was used to calculate the Lamb shift by Michael Lieber, one of Schwinger's students at Harvard [21,44]. A year later, Robert Huff, a student of Christian Fronsdal at UCLA, focused on the use of the results from SO(4,2) group theory to compute the Lamb shift [45]. He converted the conventional expression for the Lamb shift into a matrix element containing generators of SO(4,2), and was able to perform rotations and scale changes to simplify and evaluate the matrix elements. After clever mathematical manipulation, he obtained an expression for the Bethe log in terms of a rapidly terminating series for the level shifts. He provided an appendix with a brief discussion of the fundamental of SO(4,2) representations for the H atom, showing the expressions for the three generators needed to express the Schrodinger equation.

In the next few years, the researchers published a few mathematically oriented papers [41,46–50], a short book [51] dealing with the symmetries of the Coulomb problem, and a paper by Barut presenting a SO(4,2) formulation of symmetry breaking in relativistic Kepler problems, with a 1 page summary of the application of SO(4,2) for the non-relativistic hydrogen atom [34,52]. Bednar published a paper applying group theory to a variety of modified Coulomb potentials, which included some matrix elements of SO(4,2) using hydrogen atom basis states with quantum numbers nlm [53]. There also was interest in application of the symmetry methods and dynamical groups in molecular chemistry [54] and atomic spectroscopy [55].

In the 1970's, researchers focused on developing methods of group theory and on understanding dynamical symmetries in diverse systems [56–58]. A book on group theory and its applications appeared in 1971 [59]. Barut and his collaborators published a series of papers dealing with the hydrogen atom as a relativistic elementary particle, leading to an infinite component wave equation and mass formula [60–63].

Papers on the classical Kepler problem, the Runge–Lenz vector, and SO(4) for the hydrogen atom have continued to appear over the years, from 1959 to today. Many were published in the 1970's [64–70] and some since 1980, including [71–75]. Papers dealing with SO(4,2) are much less frequent. In 1986 Barut, A. Bohm, and Ne'eman published a book on dynamical symmetries that included some material on the hydrogen atom [76]. In 1986, Greiner and Muller published the second edition of Quantum Mechanics Symmetries, which had six pages on the Hydrogen atom, covering only the SO(4) symmetry [77]. The 2005 book by Gilmore on Lie algebras has 4 pages of homework problems on the H atom to duplicate results in early papers [78]. The last papers I am aware of that used SO(4,2) were applications in molecular physics [79,80] and more general in scope [81]. Carl Wulfman published a book on dynamical symmetries in 2011, which provides a helpful discussion of dynamical symmetries for the hydrogen atom [82]. He regularizes the Schrodinger equation, essentially multiplying by r, obtaining Sturmian wave functions in parabolic coordinates. This approach allows for him to treat bound and scattering states for SO(4,2) at one time, but requires redefining the inner product, and leads to a non-Hermitian position operator.

1.4. The Dirac Hydrogen Atom

We have focused our discussion on the symmetries of the non-relativistic hydrogen atom described by the Schrodinger equation. Quantum mechanics also describes the hydrogen atom in terms of the relativistic Dirac equation, which we will only discuss briefly in this paper.

The gradual understanding of the dynamical symmetry of the Dirac atom parallels that of the Schrodinger atom, but it has received much less attention, probably because the system has less relevance for particle physics and for other applications. It was known that the rotational symmetry was present and that the equation predicted that the energy depended on the principal quantum number and the quantum number for the total angular momentum j, but not the spin s or orbital angular momentum l separately. This remarkable fact meant that, in some sense, angular momentum contributed the same to the total energy no matter whether it was intrinsic or orbital in origin. This degeneracy is lifted if we include the radiative interactions which leads to the Lamb shift.

In order to understand the symmetry group for the Dirac equation consider that for a given total angular momentum quantum number $J > 0$ there are two degenerate levels for each energy level of the Dirac hydrogen atom: one level has $l = J + 1/2$ and the other has $l = J - 1/2$. Since the l values differ by unity, the two levels have opposite parity. Dirac described a generalized parity operator K, which was conserved. For an operator Λ to transform one degenerate state into the other, it follows that the operator has to commute with J and have parity -1. This means it has to anticommute with K, and so it is a conserved pseudoscalar operator.

The parity $(-1)^{l+J-1/2}$ is conserved in time, so the states are parity eigenstates. Using the two symmetry operators Λ and K, one can build a SU(2) algebra. If we include the O(3) symmetry due to the conservation of angular momentum, we obtain the full symmetry group $SU(2)xO(3)$ which is isomorphic to $SO(4)$ for the degeneracy of the Dirac hydrogen atom.

In 1950, M.Johnson and B. Lippman discovered the operator Λ [83]. Further work was done on understanding Λ by Biedenharn [84]. The Johnson–Lippman operator has been rediscovered and reviewed several times over the decades [85–87]. It has been interpreted in the non-relativistic limit as the projection of the Runge–Lenz vector onto the spin angular momentum [87–89]. The SO(4) group can be expanded to include all states, and then the spectrum generating group is SO(4,1) or SO(4,2) depending on the assumptions regarding relativistic properties and the charges present [33,38]. We will not discuss the symmetries of the Dirac H atom further.

2. Background

2.1. The Relationship between Symmetry and Conserved Quantities

The nature of the relationship between symmetry, degeneracy, and conserved operators is implicit in the equation

$$[H, S] = 0 \tag{1}$$

where H is the Hamiltonian of our system, S is a Hermitian operator, and the brackets signify a commutator if we are discussing a quantum mechanical system, or i times a Poisson bracket if we are discussing a classical system. If S is viewed as the generator of a transformation on H, then Equation (1) says the transformation leaves H unchanged. Therefore, we say S is a symmetry operator of H and leaves the energy invariant. The fact that a non- trivial S exists means that there is a degeneracy. To show this, consider the action of the commutator on an energy eigenstate $|E\rangle$:

$$[H, S]|E\rangle = 0 \tag{2}$$

or

$$H(S|E\rangle) = E(S|E\rangle). \tag{3}$$

If S is nontrivial then $S|E\rangle$ is a different state from $|E\rangle$ but has the same energy eigenvalue. If we label all such degenerate states by

$$|E, m\rangle, m = 1, ..., N \tag{4}$$

then clearly $S|E, m\rangle$ is a linear combination of degenerate states:

$$S|E, m\rangle = S_{ms}|E, s\rangle. \tag{5}$$

S_{ms} is a matrix representation of S in the subspace of degenerate states. In a classical Kepler system S generates an orbit deformation that leaves H invariant. The existence of a nontrivial S therefore implies degeneracy, in which multiple states have the same energy eigenvalue. We can show that the complete set of symmetry operators for H forms a Lie algebra by applying Jacobi's identity to our set of Hermitian operators S_i :

$$[H, S_i] = 0, \quad i = 1, \dots, L. \tag{6}$$

$$[S_j, [H, S_i]] + [S_i, [S_j, H]] + [H, [S_i, S_j]] = 0 \tag{7}$$

so

$$[H, [S_i, S_j]] = 0. \tag{8}$$

Therefore, the commutator of S_i and S_j is a symmetry operator of H. Either the commutator is a linear combination of all the symmetry operators $S_i, i = 1, ..., L$:

$$[S_i, S_j] = C_{ij}^k S_k \tag{9}$$

or the commutator defines a new symmetry operator which we label S_{L+1}. We repeat this procedure until the Lie algebra closes as in Equation (9).

By exponentiation, we assume that we can locally associate a group of unitary transformations

$$e^{iS_i a^i} \tag{10}$$

for real a^i with our Lie algebra and so conclude that a group of transformations exists under which the Hamiltonian is invariant [90]. We call this the symmetry or degeneracy group of H. Our energy eigenstates states form a realization of this group.

It is possible to form scalar operators, called Casimir operators, from the generators of the group that commute with all the generators of the group, and, therefore, have numerical values. The values of the Casimir operators characterize the particular representation of the group. For example, for the rotation group in three dimensions, the generators are $L = (L_1, L_2, L_3)$ and the quantity $L^2 = L(L+1)$ commutes with all of the generators. L can have any positive integer value for a particular representation. The Casimir operator for O(3) is L^2. The number of Casimir operators that characterize a group is called the rank of the group. O(3) is rank 1 and SO(4,2) is rank 3.

Now let us consider Equation (1) in a different way. If we view H as the generator of translations in time, then we recall that the total time derivative of an operator S_i is

$$\frac{dS_i}{dt} = \frac{i}{\hbar} [H, S_i] + \frac{\partial S_i}{\partial t} \tag{11}$$

where the commutator and the partial derivative give the implicit and explicit time dependence respectively. Provided that the symmetry operators have no explicit time dependence $\left(\frac{\partial S_i}{\partial t} = 0\right)$, then Equation (11) implies that Equation (1) means that the symmetry operators S are conserved in time and $\frac{dS_i}{dt} = 0$. Conversely, we can say that conserved Hermitian operators with no explicit time dependence are symmetry operators of H. This very important relationship between conserved Hermitian operators and symmetry was first discovered by German mathematician Emmy Nother in 1917, and is called Nother's Theorem [8,91–95].

2.2. Non-Invariance Groups and Spectrum Generating Group

As we have discussed, the symmetry algebra contains conserved generators S_i that transform one energy eigenstate into a linear combination of eigenstates all with the same energy. In order to illustrate with hydrogen atom eigenstates:

$$S_i|nlm\rangle = \sum_{l',m'} S_{nlm}^{nl'm'}|nl'm'\rangle \tag{12}$$

where $|nlm\rangle$ refers to a state with energy E_n, angular momentum $l(l+1)$ and $l_z = m$.

A non-invariance algebra contains generators D_i that can be used to transform one energy eigenstate $|nlm\rangle$ into a linear combination of other eigenstates, with the same or a different energy, different angular momentum l, and different azimuthal angular momentum m:

$$D|nlm\rangle = \sum_{n',l',m'} D_{nlm}^{n'l'm'}|n'l'm'\rangle. \tag{13}$$

Because the set of energy eigenstates is complete, the action of the most general operator would be identical to that shown to Equation (13). Therefore this requirement alone is not sufficient to select the generators needed.

The goal is to expand the degeneracy group with its generators S_i into a larger group, so that some or all of the eigenstates form a representation of the larger group with the degeneracy group as a subgroup. Thus, we need to add generators G_i, such that the combined set of generators

$$\{S_i, G_j; \text{ for all } i, j\} \equiv \{D_k; \text{ for all } k\}$$

forms an algebra that closes

$$[D_i, D_j] = i\epsilon_{ijk}D_k. \tag{14}$$

This is the Lie algebra for the expanded group. To illustrate with a specific example, consider the O(4) degeneracy group with six generators [96]. One can expand the group to O(5) or O(4,1) which has ten generators by adding a four-vector of generators. One component might be a scalar and the other three a three-vector. The question then is can some or all of the energy eigenstates provide a representation of O(5)? If so, then this would be considered a non-invariance group. The group might be expanded further in order to obtain generators of a certain type or to include all states in the representation. For the H atom the generators D_i can transform between different energy eigenvalues meaning between eigenfunctions with different principal quantum numbers.

Another way to view the expansion of the Lie algebra of the symmetry group is to consider additional generators D_i that are constants in time [97] but do not commute with the Hamiltonian so

$$\frac{dD_i}{dt} = 0 = \frac{i}{\hbar}[H, \; D_i] + \frac{\partial D_i}{\partial t}.$$

If we make the additional assumption that the time dependence of the generators is harmonic

$$\frac{\partial^2 D_i(t)}{\partial t^2} = \omega_{in}D_n(t).$$

then generators D_i, and the first and second partial derivatives with respect to time could close under commutation, forming an algebra. This approach does not tell us what generators to add, but, as we demonstrate in Section 7.5, it does reflect the behavior of the generators that have been added to form the spectrum generating group in the case of the hydrogen atom.

We may look for the largest set of generators D_i, which can transform the set of solutions into itself in an irreducible fashion (meaning no more generators than necessary). These generators form the Lie algebra for the non-invariance or spectrum generating algebra [32,98]. If the generators for the spectrum generating algebra can be exponentiated, then we have a group of transformations for the spectrum generating group. The corresponding wave functions form the basis for a single irreducible representation of this group. This group generates transformations among all the solutions for all energy eigenvalues and it is called the Spectrum Generating Group [99]. For the H atom, SO(4,1) is a spectrum generating group or non-invariance group, which can be reduced to contain one separate SO(4) subgroup for each value of n.

To get a representation of SO(4,1), we need an infinite number of states, which we have for the H atom. This group has been expanded by adding a five vector to form SO(4,2) because the additional generators can be used to express the Hamiltonian and the dipole transition operator. The group SO(p,q) is the group of orthogonal transformations that preserve the quantity $X = x_1^2 + x_2^2 + ... + x_p^2 - ... - x_{p+q}^2$, which may be viewed as the norm or a p+q-dimensional vector in a space that has a metric with p plus signs and q minus signs. The letters SO stand for special orthogonal, meaning the orthogonal transformations have determinant equal to +1.

In terms of group theory, there is a significant difference between a group like SO(4) and SO(4,1). SO(4) and SO(3) are both compact groups, while SO(4,1) and SO(4,2) are non-compact groups. A continuous group G is compact if each function f(g), continuous for all elements g of the group G, is bounded. The rotation group in three dimensions O(3), which conserves the quantity $r^2 = x_1^2 + x_2^2 + x_3^2$, is an example of a compact group.

For a non-compact group, consider the Lorentz group O(3,1) of transformations to a coordinate system moving with a velocity v. The transformations preserve the quantity $r^2 - c^2t^2$. The matrix elements of the Lorentz transformations are proportional to $1/\sqrt{1 - \beta^2}$, where $\beta = v/c$, and are not bounded as $\beta \to 1$. Therefore, r and ct may increase without bound, while the difference of the squares remains constant. Unitary representations of non-compact groups are infinite dimensional, for example, the representation of the non-invariance group SO(4,1) has an infinite number of states. Unitary representations of compact groups can be finite dimensional, for example, our representation of SO(4) for an energy level E_n has dimension n^2.

In the nineteen sixties and later, the spectrum generating group was of special interest in particle physics, because it was believed it could provide guidance where the precise particle dynamics were not known. The hydrogen atom provided a physical system as a model. Because the application was in particle physics, there was less interest in exploring representations in terms of the dynamical variables for position and momentum.

The expansion of the group from SO(4,1) to SO(4,2) was motivated by the fact that the additional generators could be used to write Schrodinger's equation entirely in terms of the generators, and to express the dipole transition operator. This allowed for algebraic techniques and group theoretical methods to be used to obtain solutions, calculate matrix elements, and other quantities [33,38].

2.3. Basic Idea of Eigenstates of $(Z\alpha)^{-1}$

We briefly introduce the idea behind these states, since they are unfamiliar [1]. The full derivation is given in Section 4. Schrodinger's equation in momentum space for bound states can be written as

$$\left[p^2 + a^2 - \frac{2mZ\alpha}{r} \right] |a\rangle = 0.$$

where $a^2 = -2mE > 0$ and $Z\alpha$ is the coupling constant, which we will now view as a parameter. This equation has well behaved solutions for certain discrete eigenvalues of the energy or a^2, namely

$$a_n^2 = -2mE_n$$

where $E_n = -\frac{1}{2}\frac{m(Z\alpha)^2}{n^2}$. We can write the eigenvalue condition equivalently as

$$\left(\frac{a_n}{mZ\alpha}\right) = \frac{1}{n}.$$

This last equation shows that solutions exist for certain values a_n of the RMS momentum a. To introduce eigenstates of $(Z\alpha)^{-1}$ we simply take a different view of this last equation and say that instead of quantizing a and obtaining a_n, we imagine that we quantize $(Z\alpha)^{-1}$, let a remain unchanged, obtaining

$$\frac{a}{m(Z\alpha)_n} = \frac{1}{n}.$$

Accordingly, now we can interpret Schrodinger's equation as an eigenvalue equation that has solutions for certain values of $(Z\alpha)^{-1}$ namely

$$(Z\alpha)_n^{-1} = \frac{m}{an}.$$

We have the same equation but can view the eigenvalues differently but equivalently. Instead of quantizing a we quantize $(Z\alpha)^{-1}$.

This roughly conveys the basic idea of eigenstates of the inverse of $(Z\alpha)$, but this simplified version does at all reveal the advantages of our reformulation because we have left the Hamiltonian unchanged. In Section 4, we transform Schrodinger's equation to an eigenvalue equation in a, so that the kernel is bounded, which means that there are no states with E > 0, no scattering states, and all states have the usual quantum numbers. Other important advantages to this approach will also be discussed.

2.4. Degeneracy Groups for Schrodinger, Dirac and Klein-Gordon Equations

The degeneracy groups for the bound states described by the different equations of the hydrogen atom are summarized in Table 1. The degeneracy (column 2) is due to the presence of conserved operators which are also symmetry operators (column 3), forming a degeneracy symmetry group (column 4). For example, The symmetry operators for the degeneracy group in the Schrodinger hydrogen atom are the angular momentum **L** and the Runge–Lenz vector **A**. In Section 3, it will be shown that together these are the generators for the direct product SO(3)xSO(3) which is isomorphic to SO(4). Column 5 gives the particular representations present. These numbers are the allowed values of the Casimir operators for the group and they determine the degree of degeneracy (last column) and the corresponding allowed values of the quantum numbers for the degenerate states.

The Casimir operators, which are made from generators of the group, have to commute with all of the members of the group, and the only way this can happen is if they are actually constants for the representation. The generators are formed from the dynamical variables of the H atom, so the Casimir operators are invariants under the group composed of the generators, and their allowed numerical values reflect the underlying physics of the system and determine the appropriate representations of the group [9,82,100]. For example, L^2 is the Casimir operator for the group O(3) and can have the values $l(l+1)$. The relationship between Casimir operators and group representations is true for all irreducible group representations, including the SO(4) degeneracy group, as well as the spectrum generating group SO(4,2) [43,53].

For the Schrodinger equation, there are n^2 states $|nlm\rangle$ that form a representation of the degeneracy group SO(4) formed by L and A. These states correspond to the principal quantum number n, the n

different values of the angular momentum quantum number l, and $2l + 1$ different values of the z component of the angular momentum $l_z = m$.

For the Dirac equation, the $2(2J + 1)$ dimensional degeneracy group for bound states is realized by the total angular momentum operator **J**, the generalized parity operator K, and the Johnson–Lippman operator Λ, which together form the Lie algebra for SO(4).

For the fully relativistic Klein–Gordon equation, only the symmetry from rotational symmetry survives, leading to the degeneracy group O(3). If the V^2 term, the four-potential term squared is dropped in a semi-relativistic approximation as we describe in Section 4.3, then the equation can be rewritten in the same form as the non-relativistic Schrodinger equation, so a Runge–Lenz vector can be defined and the degeneracy group is again SO(4).

Table 1. In the table **L** = **rxp** is the orbital angular momentum; **A** is the Runge-Lenz vector; **J** = **L** + $\sigma/2$ is the total angular momentum; K is the generalized parity operator; Λ is the conserved pseudoscalar operator.

Degeneracy Groups for Bound States in a Coulomb Potential					
Equation	**Degeneracy**	**Conserved Quantities**	**Degeneracy Group**	**Representation**	**Dimension**
Schrodinger	E indep. of l, l_z	**A, L**	$SO(4)$	$(\frac{n-1}{2}, \frac{n-1}{2})$	n^2
Klein-Gordon	E indep. of l_z	**L**	$O(3)$	Casimir op. is $l(l+1)$	$2l+1$
Klein-Gordon without V^2 term	E indep. of l, l_z	**A, L**	$SO(4)$	$(\frac{n-1}{2}, \frac{n-1}{2})$	n^2
Dirac	E depends on J, n only	Λ, K, J	$SO(4)$	$(1/2, J)$	$2(2J+1)$

3. Classical Theory of the H Atom

In order to discuss orbital motion and the continuous deformation or orbits we give this discussion in terms of classical mechanics, but much of it is valid in terms of the Heisenberg representation of quantum mechanics if the Poisson brackets are converted to commutators, as will be discussed in Section 4.

For a charged particle in a Coulomb potential, there are two classical conserved vectors: the angular momentum **L**, which is perpendicular to the plane of the orbit, and the Runge–Lenz vector **A**, which goes from the focus corresponding to the center of mass and force along the semi-major axis to the perihelion (closest point) of the elliptical orbit. The conservation of **A** is related to the fact that non-relativistically the orbits do not precess. The Hamiltonian of our bound state classical system with an energy E < 0 is [17]

$$H = \frac{p^2}{2m} - \frac{Z\alpha}{r} = E \tag{15}$$

where m=mass of the electron, r is the location of the electron, p is its momentum, α is the fine structure constant, and E is the total non-relativistic energy.

The Runge–Lenz vector is

$$\mathbf{A} = \frac{1}{\sqrt{-2mH}}(\mathbf{p} \times \mathbf{L} - mZ\alpha\frac{\mathbf{r}}{r}) \tag{16}$$

where **L** is the angular momentum. From Hamilton's equation, $H = E$, so

$$\mathbf{A} = \frac{\mathbf{p} \times \mathbf{L}}{a} - \frac{mZ\alpha}{a}\frac{\mathbf{r}}{r} \tag{17}$$

where a is defined by

$$a = \sqrt{-2mE}. \tag{18}$$

From the virial theorem, the average momentum $\langle p^2 \rangle = -2mE$ so a is the root mean square momentum. We are discussing bound states so E < 0. It is straightforward to verify that **A** is conserved in time:

$$[\mathbf{A}, H] = \frac{d\mathbf{A}}{dt} = 0. \tag{19}$$

From the definition of **A** and the definition of angular momentum

$$\mathbf{L} = \mathbf{r} \times \mathbf{p} \tag{20}$$

if follows that **A** is orthogonal to the angular momentum vector

$$\mathbf{A} \cdot \mathbf{L} = 0. \tag{21}$$

Using the fact that **A** and **L** are conserved, we can easily obtain equations for the orbits in configuration and momentum space and the eccentricity, and other quantities, all usually derived by directly solving the equations of motion.

We will show that A and L are the generators of the group O(4). If we introduce the linear combinations $N = \frac{1}{2}(L + A)$ and $M = \frac{1}{2}(L - A)$, we find that N and M commute reducing the nonsimple group O(4) to SU(2)x SU(2), which we will discuss in Section 4.2 in the language of quantum mechanics.

3.1. Orbit in Configuration Space

In order to obtain the equation of the orbit one computes

$$\mathbf{r} \cdot \mathbf{A} = rA\cos\phi_r = -r\frac{mZ\alpha}{a} + \mathbf{r} \cdot \mathbf{p} \times \mathbf{L}. \tag{22}$$

Noting that $\mathbf{r} \cdot \mathbf{p} \times 1 = L^2$, we can solve for r

$$r = \frac{L^2/mZ\alpha}{(a/mZ\alpha)A\,\cos\phi_r + 1}. \tag{23}$$

This is the equation of an ellipse with eccentricity $e = (a/mZ\alpha)A$ and a focus at the origin (Figure 1). To find e we calculate $\mathbf{A} \cdot \mathbf{A}$ using the identity $\mathbf{p} \times \mathbf{L} \cdot \mathbf{p} \times \mathbf{L} = p^2 L^2$ and obtain

$$A^2 = \frac{p^2 L^2}{a^2} - \frac{2mZ\alpha}{a^2}\frac{L^2}{r} + \left(\frac{mZ\alpha}{a}\right)^2. \tag{24}$$

Substituting E for the Hamiltonian Equation (15) gives the usual result

$$\frac{a}{mZ\alpha}A = e = \sqrt{\frac{2EL^2}{m(Z\alpha)^2} + 1}. \tag{25}$$

The length r_c of the semi-major axis is the average of the radii at the turning points

$$r_c = \frac{r_1 + r_2}{2}. \tag{26}$$

Using the orbit equation we find

$$r_c = \frac{L^2}{mZ\alpha} \frac{1}{1 - e^2} \tag{27}$$

or

$$r_c = -\frac{Z\alpha}{2E} = \frac{mZ\alpha}{a^2}. \tag{28}$$

The energy depends only on the length of the semi-major axis r_c, not on the eccentricity. This important result is a consequence of the symmetry of the problem. It is convenient to parameterize the eccentricity in terms of the angle v (see Figure 1) where

$$e = \sin v \tag{29}$$

From this definition and from Equations (25), (27), and (28) follow the useful results

$$L = r_c a \cos v, \quad A = r_c a \sin v \tag{30}$$

which immediately imply

$$L^2 + A^2 = (r_c a)^2 = \left(\frac{mZ\alpha}{a}\right)^2. \tag{31}$$

This equation is the classical analogue of an important quantum mechanical result first obtained by Pauli and Hulthen allowing us to determine the energy levels from symmetry properties alone [14,16]. From Figure 1, it is apparent that this equation is a statement of Pythagoras's theorem for right triangles.

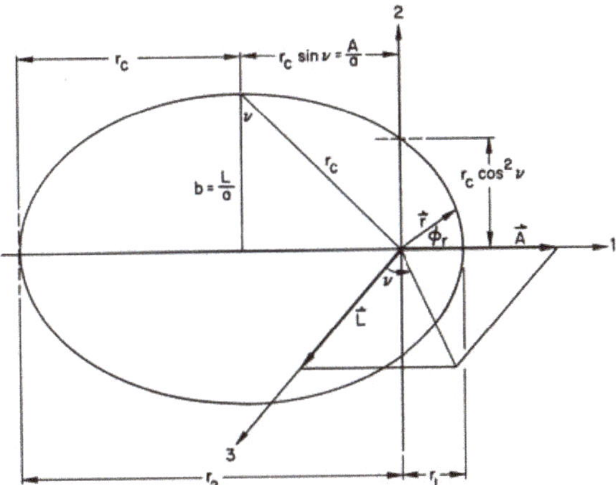

Figure 1. Classical Kepler orbit in configuration space. The orbit is in the 1–2 plane (plane of the paper). One focus, where the proton charge is located, is the origin. The semi-minor axis is $b = r_c \sin v$. The semi-major axis is r_c.

The energy equation (Equation (15)) and the orbit equation (Equation (23)) respectively may be rewritten in terms of a, r, and v:

$$\frac{r_c}{r} = \frac{p^2 + a^2}{2a^2} \tag{32}$$

$$r = \frac{r_c \cos^2 v}{1 + \sin v \cos \phi_r}. \tag{33}$$

3.2. The Period

To obtain the period, we use the geometrical definition of the eccentricity

$$e = \sqrt{1 - (b/r_c)^2} \tag{34}$$

where b is the semi-minor axis. Using $e = \sin \nu$ we find

$$b = r_c \cos \nu \tag{35}$$

so from Equation (30), we obtain

$$L = ab. \tag{36}$$

From classical mechanics, we know the magnitude of the angular momentum is equal to twice the mass times the area swept out by the radius vector per unit time. The area of the ellipse is $\pi b r_c$. It the period of the classical motion is T, then $L = 2m\pi b r_c / T$. Therefore, the classical period is

$$T = 2\pi \frac{m r_c}{a} = 2\pi \sqrt{\frac{m(Z\alpha)^2}{-8E^3}}. \tag{37}$$

and the classical frequency $\omega_{cl} = 2\pi / T$ is

$$\omega_{cl} = \frac{a}{m r_c}. \tag{38}$$

3.3. Group Structure SO(4)

The generators of our symmetry operations form the closed Poisson bracket algebra of $O(4)$:

$$[L_i, L_j] = i\epsilon_{ijk} L_k \ , \quad [L_i, A_j] = i\epsilon_{ijk} A_k \ , \quad [A_i, A_j] = i\epsilon_{ijk} L_k \ . \tag{39}$$

The brackets mean i times the Poisson bracket, which is the classical limit of a commutator. The first bracket says that the angular momentum generates rotations and forms a closed Lie algebra corresponding to O(3). The second bracket says that the Runge-Lenz vector transforms as a vector under rotations generated by the angular momentum. The last commutator says that the multiple transformations generated by the Runge-Lenz vector are equivalent to a rotation. Taken together the commutators form the Lie algebra of O(4). The connected symmetry group for the classical bound state Kepler problem is obtained by exponentiating our algebra giving the symmetry group SO(4). The scattering states with $E > 0$ form an infinite dimensional representation of the non-compact group SO(3,1).

We now want to determine the nature of the transformations generated by A_i and L_i. Clearly, $\mathbf{L} \cdot \delta \boldsymbol{\omega}$ generates a rotation of the elliptical orbit about the axis $\delta \boldsymbol{\omega}$ by an amount $\delta \omega$. To investigate the transformations generated by $\mathbf{A} \cdot \delta \boldsymbol{\nu}$ we assume a particular orientation of the orbit, namely that it is in the 1–2 (or x-y) plane and that $\mathbf{A}$ is along the 1-axis (see Figure 1). The more general problem is obtained by a rotation generated by L_i. For an example, we choose a transformation with $\delta \boldsymbol{\nu}$ pointing along the 2-axis, so that $\mathbf{A} \cdot \delta \boldsymbol{\nu} = A_2 \delta \nu$. The change in $\mathbf{A}$ is defined by $\delta \mathbf{A}$, where

$$\delta \mathbf{A} = i[\mathbf{A} \cdot \delta \boldsymbol{\nu}, \mathbf{A}]. \tag{40}$$

From the Poisson bracket relations, we find for this particular case:

$$\delta A_1 = L_3 \delta \nu \ , \quad \delta A_2 = 0 \ , \quad \delta A_3 = -L_1 \delta \nu \tag{41}$$

For our orbit, $L_1 = 0$ so $\delta A_3 = 0$. We perform a similar computation to find δL. We find we can characterize the transformation by

$$
\begin{aligned}
\delta A_1 &= L_3 \delta v \quad &\text{or} \quad &\delta A = L \delta v \\
\delta L_3 &= -A_1 \delta v \quad &\text{or} \quad &\delta L = -A \delta v \\
\delta e &= \sqrt{1 - e^2}\, \delta v \quad &\text{or} \quad &\delta(\sin v) = \cos v \, \delta v
\end{aligned}
\tag{42}
$$

Recalling $e = \sin v$ and Equation (30), we see that these transformations are equivalent to the substitution

$$
v \longrightarrow v + \delta v.
\tag{43}
$$

In other words, the eccentricity of the orbit, and therefore A and L are all changed in such a way that the energy, a and r_c (length of the semi-major axis) remain constant. In our example, both L and A are changed in length but not direction, so the plane and orientation of the orbit are unchanged. The general transformation $A \cdot \delta\mu$ will also rotate the plane of the orbit or the semi-major axis.

Figure 2 shows a set of orbits in configurations space with different values of the eccentricity $e = \sin v$, but the same total energy and the same semimajor axis r_c, which is the bold hypotenuse. The bold vertical and horizontal legs are A/a and L/a and they are related to the hypotenuse r_c by Pythagoras's theorem. The generator $A_2 v$ produces a deformation of the circular orbit into the various elliptical orbits shown. This classical degeneracy corresponds to the quantum mechanical degeneracy in energy levels that occurs for different eigenvalues of the angular momentum with a fixed principal quantum number.

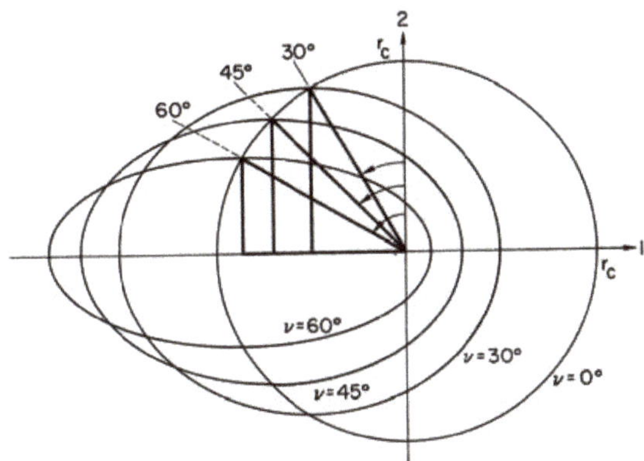

Figure 2. Kepler bound state orbits in configuration space for a fixed energy and different values of the eccentricity $e = \sin v$. The bold hypotenuse is the semi-major axis r_c, which makes an angle v with the vertical 2-axis.

We can visualize all possible elliptical orbits for a fixed total energy or semi-major axis through a simple device. It is possible to produce an elliptical orbit with eccentricity $\sin v$ as the shadow of a circle of radius r_c which is rotated an amount v about an axis perpendicular to the illuminating light. With a complete rotation of the circle, we will see all possible classical elliptical orbits corresponding to a given total energy. In quantum mechanics only certain angles of rotation would be possible corresponding to the quantized values of L. As the circle is rotated, we must imagine that the force center shifts as the sine of the angle of rotation, so that it always remains at the focus [101].

3.4. The Classical Hydrogen Atom in Momentum Space

We can derive the equation for the classical orbit in momentum space of a particle bound in a Coulomb potential using the conserved operators L and A. For convenience, we assume that we have rotated our axes, so that L lies along the 3-axis and A the 1-axis, as shown in Figure 1. We compute

$$p \cdot A = p_1 A = \frac{-mZ\alpha}{a} p \cdot \frac{r}{r} \equiv \frac{-mZ\alpha}{a} p_r \tag{44}$$

and we employ Equations (28) and (30) to show $A = \frac{mZ\alpha}{a} \sin v$, to obtain [102]

$$p_r = -\sin v p_1 \tag{45}$$

which we substitute into the identity

$$p_r^2 + \frac{L^2}{r^2} = p^2 = p_1^2 + p_2^2 \tag{46}$$

Using Equations (30) and (32) we find

$$1 = \left(\frac{2ap_1}{p^2 + a^2} \right)^2 + \left(\frac{2ap_2}{p^2 + a^2} \right)^2 \frac{1}{\cos^2 v} \tag{47}$$

and

$$p^2 - a^2 = 2ap_2 \tan v \tag{48}$$

which may also be written as

$$p_1^2 + (p_2 - a \tan v)^2 = \frac{a^2}{\cos^2 v}. \tag{49}$$

From Equation (49), we see the orbit in momentum space is a circle of radius $a/\cos v$ with its center displaced from the origin a distance $a \tan v$ along the 2-axis. Figure 3 shows the momentum space orbit that corresponds to the configuration space orbit in Figure 1.

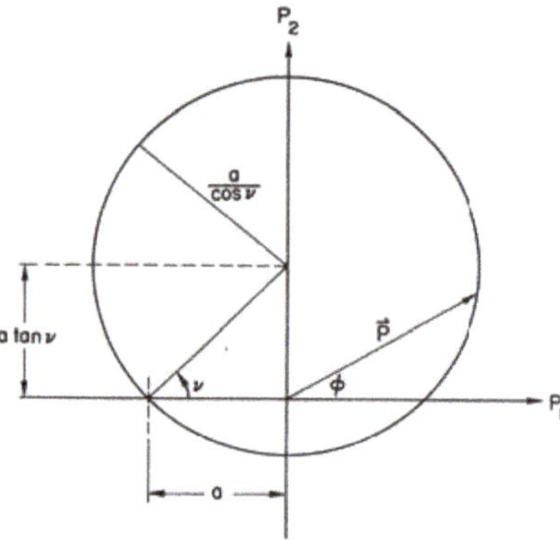

Figure 3. Kepler orbit in momentum space of radius $a/\cos v$, with its center at $p_2 = a \tan v$, corresponding to the orbit in configuration space shown in Figure 1. A circular orbit in configuration space corresponds to a circular orbit in momentum space centered on the origin with radius a.

As an alternative method of showing the momentum space orbits are circular, we can compute [103]

$$\left(p - a \frac{L \times A}{L^2} \right)^2 = C^2. \tag{50}$$

Using the lemma

$$p \times A = -\frac{L}{2a}(p^2 - a^2),\tag{51}$$

the fact that $L \cdot A = 0$, and Equation (30), we find $C = \frac{a}{\cos v}$. The orbit is a circle of radius $\frac{a}{\cos v}$ whose center lies at $a\frac{L \times A}{L^2}$, in agreement with the previous result.

We now consider what the generators A_i and L_i do to the orbit in momentum space. Clearly, L generates a rotation of the axes. For an A_i transformation, consider the same situation that we considered in our discussion of the configuration space orbit (see Figures 1 and 3). Because the generator $A_2 \delta v$ changes v to $v + \delta v$, we conclude that in momentum space this shifts the center of the orbit along the 2-axis and changes the radius of the circle. However, the distance a from the 2-axis to the intersection of the orbit with the 1-axis remains unchanged. Figure 4 shows a set of momentum space orbits for a fixed energy that correspond to the set of orbits in configuration space shown in Figure 2.

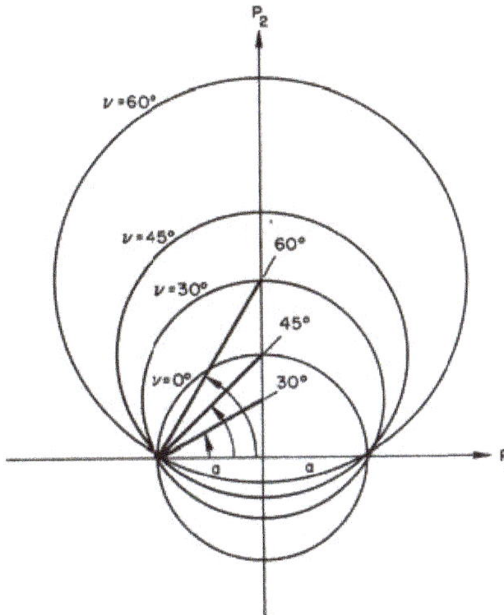

Figure 4. Kepler orbits in momentum space for a fixed energy and RMS momentum a with different values of the eccentricity $e = \sin v$ corresponding to the orbits in configuration space shown in Figure 2.

3.5. Four-Dimensional Stereographic Projection in Momentum Space

It is interesting that in classical mechanics the bound state orbits in a Coulomb potential are simpler in momentum space than in configuration space. In quantum mechanics the momentum space wave functions become simply four-dimensional spherical harmonics if one normalizes the momentum p by dividing by the RMS momentum $a = \sqrt{-2mE}$ and performs a stereographic projection onto a unit hyper-sphere in a four-dimensional space [18,35]. We will do the analogous projection procedure for the classical orbits. The three-dimensional momentum space hyperplane passes through the center of the four-dimensional hypersphere, as shown in Figure 5. The unit vector in the fourth direction is $\hat{n} = (1,0,0,0)$. The unit vector $\hat{U}$ goes from the center of the sphere to the surface of the hypersphere, where it is intersected by the line connecting the vector p/a to the north pole of the sphere.

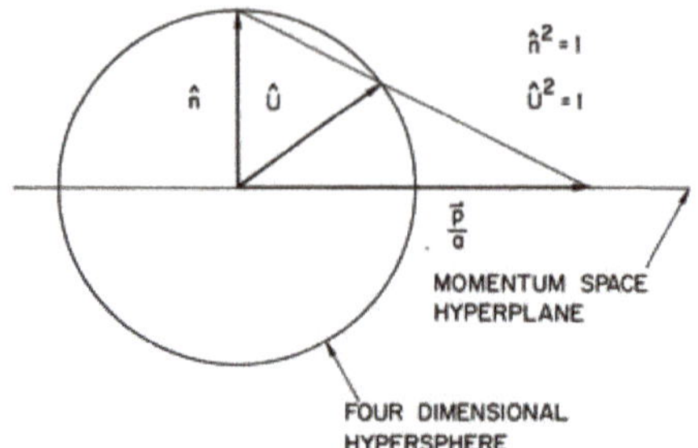

Figure 5. Stereographic projection in momentum space for a fixed energy, mapping p/a into $\hat{U}$. The unit vector in the four direction is $\hat{n}$ and $\hat{n} \cdot \hat{U} = \cos \Theta_4$.

We find

$$U_i = \frac{2ap_i}{p^2 + a^2} \quad i = 1, 2, 3. \quad U_4 = \frac{p^2 - a^2}{p^2 + a^2} \tag{52}$$

or inverting,

$$p_i = \frac{aU_i}{1 - U_4} \quad p^2 = a^2 \frac{1 + U_4}{1 - U_4}. \tag{53}$$

Momentum space vectors for which $p/a < 1$ are mapped onto the lower hyperhemisphere. The advantage of this projection over one in which the hypersphere is tangent to the hyperplane is that we may have $|\hat{n}| = |\hat{U}| = 1$. At times it is convenient to describe $\hat{U}$ in terms of spherical polar coordinates in four dimensions. Because $\hat{U}$ is a unit vector we define

$$\begin{aligned}
U_4 &= \cos \theta_4 \\
U_3 &= \sin \theta_4 \cos \theta \\
U_2 &= \sin \theta_4 \sin \theta \sin \phi \\
U_1 &= \sin \theta_4 \sin \theta \cos \phi
\end{aligned} \tag{54}$$

where θ and ϕ are the usual coordinates in three dimensions. By comparison to Equation (52), we have

$$\theta_4 = 2 \cot^{-1} \frac{p}{a} \quad \theta = \cos^{-1} \frac{p_3}{p} \quad \phi = \tan^{-1} \frac{p_2}{p_1} \tag{55}$$

3.6. Orbit in U space

We want to find the trajectory of the particle on the surface of the hypersphere corresponding to the Kepler orbits in configuration space or the displaced circles in momentum space. We assume we have rotated the axes in configuration space so that L is along the 3-axis and A is along the 1-axis, as shown in Figure 1. The equation for the orbit in three-dimensional momentum space is given by Equations (48) or (50). Dividing Equation (48) by $p^2 + a^2$ immediately gives a parametric equation for the projected orbit in U space:

$$U_4 = U_2 \tan \nu. \tag{56}$$

Because the orbit is in the 1–2 plane in configurations space, $U_3 = 0$. The orbit lies in a hyperplane perpendicular to the 2–4 plane that goes through the origin and makes an angle $\pi/2 - \nu$ with the 4-axis, as shown in Figure 6 [104]. The orbit is the intersection of this plane with the hypersphere and it is therefore a great circle. To derive the exact equation for the projected orbit we express p in Cartesian components p_1 and p_2 in Equation (48) and substitute Equation (52) obtaining

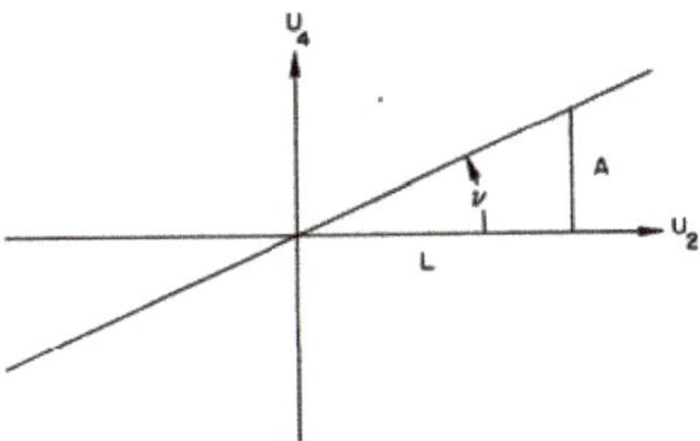

Figure 6. Showing the hyperplane containing the orbit making an angle v with the U_2 plane. Notice $\tan v = A/L$ as required by Equation (30).

$$U_l^2 + U_2^2 - 2\tan v\, U_2(l - U_4) = (1 - U_4)^2. \tag{57}$$

To interpret this equation, we consider it in a rotated coordinate system. If we perform a rotation by an amount δv about the 1–3 plane ($A_2\delta v$ is the generator of this rotation), the equations of transformation may be written [105]

$$\begin{aligned} U_2 &= U_2' \cos \delta v + U_4' \sin \delta v, \quad U_3 = U_3' \\ U_4 &= U_4' \cos \delta v - U_2' \sin \delta v, \quad U_1 = U_1'. \end{aligned} \tag{58}$$

This transformation is equivalent to making the substitution $v \longrightarrow v + \delta v$ in the equations relating to the orbit. For example, Equation (56) becomes

$$U_4' = U_2' \tan (v + \delta v). \tag{59}$$

We choose $\delta v = -v$, which means the orbital plane becomes $U_4' = 0$. Writing Equation (57) in terms of the primed coordinates, we find

$$U_1'^2 + U_2'^2 = 1 \tag{60}$$

which in the original system is

$$U_1^2 + (U_2 \cos v + U_4 \sin v)^2 = 1. \tag{61}$$

This is the equation of a great hypercircle $(v, 0)$ centered at the origin and lying in a hyperplane making an angle $\pi/2 - v$ with the 4-axis and an angle $\pi/2 - 0$ with the 3-axis. If L did not lie along the 3-axis, but, for example, was in the 1–3 plane, at an angle Θ from the 3-axis, then Equation (61) would be modified by the substitution

$$U_1 \longrightarrow U_1 \cos\Theta + U_3 \sin \Theta \tag{62}$$

which follows, since U_i transforms as a three-vector. The corresponding great circle (v, Θ) lies in a hyperplane making an angle $\pi/2 - v$ with the 4-axis and $\pi/2 - \Theta$ with the 3-axis.

The motion of the orbiting particle corresponds to a dot moving along the great circle $(v, 0$ or $\Theta)$ with a period T given by the classical period Equation (37). The velocity in configuration space can be expressed in terms of U_4 by using its definition in terms of p^2 Equation (53) or in terms of θ_4 Equation (55). The particle is moving at maximum velocity when θ_4 is a minimum, which occurs at the perihelion when $\theta_4 = \pi/2 - v$:

$$max(\frac{p}{a}) = \sqrt{\frac{1+e}{1-e}} = \sqrt{\frac{r_2}{r_1}} \tag{63}$$

and at a minimum velocity when $\theta_4 = \pi/2 + v$:

$$min\left(\frac{p}{a}\right) = \sqrt{\frac{1-e}{1+e}}. \tag{64}$$

These values of θ_4 correspond to turning points, at which r and p have extreme values. This is apparent when we use Equation (32) for the total energy to show

$$U_4 = \frac{r_c - r}{r_c}. \tag{65}$$

When $r > r_c$ then $p^2 < a^2$, so the particle is moving more slowly than the RMS velocity. Applying the virial theorem to any orbit we find $\langle p^2 \rangle = a^2$ so as expected a is the RMS momentum and $\langle \frac{1}{1-U_4} \rangle = 1 = \langle r_c/r \rangle$.

Figure 7 is a picture of a simple device illustrating the stereographic projection of the orbit in p/a-space onto the four-dimensional hypersphere in U-space. We assume that the orbit is in the 1–2 plane and that A lies along the 1-axis, so $p_3 = 0$, $U_3 = 0$. Because of this trivial dependence on p_3, we have omitted the 3-axis. The vertical pin or rod represents the unit vector $\hat{n}$ lying along the 4-axis. The circumference of the larger circle perpendicular to the 4-axis represents the orbit in p/a-space. One can see that it is displaced from the origin along the 2-axis. Centered at the origin, we must imagine a hypersphere of unit radius $\hat{U}^2 = 1$. The stereographic projection $\hat{U}$ of the vector p/a is obtained by placing the string coming from the top of n directly at the head of the vector p/a. The intersection of the string with the unit hypersphere defines $\hat{U}$. As the string is moved along the orbit in p/a-space, it intersects the hypersphere along a great circle shown by the circumference of the unit circle making an angle v with the 1–2 plane. We can see for example, that at the closest approach, θ_4 is a minimum and U4 is a maximum, $\hat{U} \cdot A = U_1$ is a minimum, and p/a is a maximum.

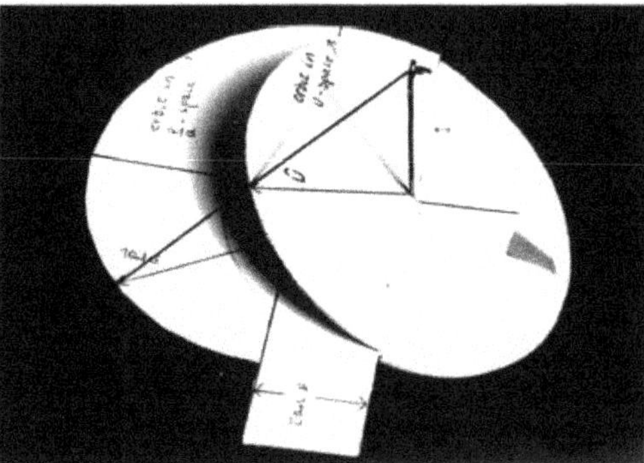

Figure 7. Model illustrating the stereographic projection from the 1–2 plane to a 4-d hypersphere. The pin represents the unit vector $\hat{n}$ along the 4-axis, normal to the 1–2 plane.

3.7. Classical Time Dependence of Orbital Motion

We can determine the time dependence of the orbital motion by integrating the expression for the angular momentum $L = mr^2 d\phi_r/dt$

$$\int dt = \int \frac{mr^2}{L} d\phi_r \tag{66}$$

where r is given by the orbit equation Equation (33) and we are assuming the orbit is in the 1–2 plane. After integrating, we can use the equations relating the momentum space and configuration space variables to obtain the time dependence in p-space and U-space. We obtain

$$\frac{1}{r_c^2}\frac{1}{\cos v}\frac{L}{m}\int_0^t dt = \cos^3 v \int_0^{\phi_r(t)} \frac{d\phi_r}{(1+\sin v\cos\phi_r)^2}. \tag{67}$$

The left-hand side of this equation is equal $\omega_{cl}t$, where ω_{cl} is the classical frequency. This follows by substituting Equations (30) and (38)

$$L = ar_c\cos v \qquad \omega_{cl} = \frac{a}{mr_c}. \tag{68}$$

The integral on the right side gives [106]

$$\omega_{cl}t = -\frac{\sin v\cos v\sin\phi_r}{1+\sin v\cos\phi_r} - \tan^{-1}\frac{\cos v\sin\phi_r}{\sin v+\cos\phi_r} \tag{69}$$

which may be simplified as

$$\omega_{cl}t = -\frac{A}{L}\frac{y}{r_c} - \tan^{-1}\frac{A}{L}\frac{y}{r_c-r} \tag{70}$$

where $y = r\sin\phi_r$.

The relationship between the angle $\phi \equiv \phi_p$ in momentum space and ϕ_r in configuration space follows by either differentiating the orbit equation Equation (33) with respect to time and using $L = mr^2\dot\phi_r$ or by simultaneously solving the configuration space orbit, the momentum space orbit equation Equation (48) and the energy equation Equation (32). We find

$$\sin\phi_r = -p\cos\phi\frac{\cos v}{a} \qquad \cos\phi_r = p\sin\phi\frac{\cos v}{a} - \sin v. \tag{71}$$

From these equations, the definitions of the U_i, Equation (52), and the orbit and energy equations, it follows that for the classical orbit in the 1–2 plane

$$\begin{aligned}
U_1 &= -\frac{r\sin\phi_r}{r_c\cos v}\\
U_2 &= \frac{r}{r_c}\frac{\sin v+\cos\phi_r}{\cos v} = U_4\cot v\\
U_3 &= 0\\
U_4 &= \frac{r_c-r}{r_c}.
\end{aligned} \tag{72}$$

Using these results in Equation (69) gives

$$\omega_{cl}t = U_1\sin v + \tan^{-1}\left(\frac{U_2}{U_1\cos v}\right) \tag{73}$$

which gives the time dependence in U space, and it agrees with the results of [97,107]. We could do rotations to generalize this result [102]. We can also rewrite the inverse tangent as $\cos^{-1}U_1$ using

$$U_1^2 + \frac{U_2^2}{\cos^2 v} = 1. \tag{74}$$

If we consider the last two equation for circular orbits with $e = \sin v = 0$ we obtain $\tan^{-1}(U_2/U_1) = \phi(t) = \omega_{cl}t = \phi_r(t) + \pi/2$, and our familiar circle $U_1^2 + U_2^2 = 1$.

Remark on Harmonic Oscillator

We can find a conserved Runge–Lenz vector for the non-relativistic hydrogen atom because the elliptical orbit does not precess, as it does for the relativistic atom. The only central force laws that yield classical elliptical orbits that do not precess are the inverse Kepler force and the linear harmonic oscillator force [108]. Thus, it seems reasonable that one could construct a constant vector similar to A for the oscillator, although the force center for the atom is at a focus and for the oscillator it is at the center of the ellipse. However, it is not readily possible [109]. Instead, one can construct a constant Hermitian second rank tensor T_{ij}:

$$T_{ij} = \frac{1}{m\omega_0} p_i p_j + m\omega_0 x_i x_j. \tag{75}$$

This constant tensor is analogous to the moment of inertia tensor for rigid body motion. The eigenvectors of the tensor will be constant vectors along the principal axes for the particular orbit being considered. The existence of the conserved tensor leads to the U(3) symmetry algebra of the oscillator. The generators are $\lambda_{ij}^a T_{ij}$, where the λ's are are the usual $U(3)$ matrices [22]. The spectrum generating algebra is SU(3,1).

In another approach, the Schrodinger equation for the hydrogen atom has been transformed into an equation for a four dimensional harmonic oscillator or two two dimensional harmonic oscillators. This approach which fits well with parabolic coordinates was used especially in the 1980's to analyze the group structure of the atom and relate it to SU(3) [110–119]. We will not discuss this approach further.

4. The Hydrogenlike Atom in Quantum Mechanics; Eigenstates of the Inverse of the Coupling Constant

In this section we switch from classical dynamics to quantum mechanics and discuss the group structure and exploit it to determine the bound state energy spectrum directly, as Pauli and followers did almost a century ago [14,16]. In Section 4.3 we introduce a new set of basis states for the hydrogenlike atom, eigenstates of the coupling constant. Using these states allows us to display the symmetries in the most convenient manner and to treat bound and scattering states uniformly.

4.1. The Degeneracy Group SO(4)

The quantum mechanical Hamiltonian is

$$H = \frac{p^2}{2m} - \frac{Z\alpha}{r} = E. \tag{76}$$

The classical expression for the Runge–Lenz vector needs to be symmetrized to insure the corresponding quantum mechanical operator A is Hermitian:

$$A = \frac{1}{\sqrt{-2mH}} \left(\frac{p \times L - L \times p}{2} - mZ\alpha\frac{r}{r} \right). \tag{77}$$

We may verify that A and $L = r \times p$ both commute with the Hamiltonian H. The commutation relations of L_i and A_i are the same as the corresponding classical Poisson bracket relations for bound states:

$$[L_i, L_j] = i\epsilon_{ijk}L_k \,, \quad [L_i, A_j] = i\epsilon_{ijk}A_k \,, \quad [A_i, A_j] = i\epsilon_{ijk}L_k \tag{78}$$

and form the algebra of O(4) [35]. We can write the commutation relations in a single equation which makes the 0(4) symmetry explicit. If we define

$$S_{ij} = \epsilon_{ijk}L_k \qquad S_{i4} = A_i \tag{79}$$

then

$$[S_{ab}, S_{cd}] = i(\delta_{ac}S_{bd} + \delta_{bd}S_{ac} - \delta_{ad}S_{bc} - \delta_{bc}S_{ad}) \quad a, b = 1, 2, 3, 4. \tag{80}$$

The Kronecker delta function δ_{ab} acts like a metric tensor.

4.2. Derivation of the Energy Levels

We can obtain the energy levels by determining which representations of the group SO(4) are realized by the degenerate eigenstates of the hydrogenlike atom [12,14,35]. The representations of SO(4) may be characterized by the numerical values of the two Casimir operators for SO(4):

$$C_1 = L \cdot A \quad C_2 = L^2 + A^2 \tag{81}$$

Once we know the value of C_2, then the eigenvalues of H follow from the quantum mechanical form of Equation (31), namely

$$L^2 + A^2 + 1 = \frac{(mZ\alpha)^2}{-2mH}. \tag{82}$$

In order to determine the possible values of C_2, we factor the 0(4) algebra into two disjoint SU(2) algebras [120], each of which has the same commutation relations as the ordinary angular momentum operators,

$$N = \tfrac{1}{2}(L + A) \quad M = \tfrac{1}{2}(L - A). \tag{83}$$

The commutation relations are

$$[M_i, N_j] = 0 \quad [M_i, M_j] = i\epsilon_{ijk}M_k \quad [N_i, N_j] = i\epsilon_{ijk}N_k. \tag{84}$$

In analogy with the results for the ordinary angular momentum operators, the Casimir operators are

$$\begin{aligned} M^2 &= j_1(j_2 + 1), & j_1 &= 0, \tfrac{1}{2}, 1, \ldots \\ N^2 &= j_2(j_2 + 1), & j_2 &= 0, \tfrac{1}{2}, 1, \ldots \end{aligned} \tag{85}$$

The numbers j_1 and j_2, which may have half-integral values for SU(2) but not O(3), define the (j_1, j_2) representation of SO(4). From the definitions of A and L in terms of the canonical variables, it follows that $C_1 = L \cdot A = 0$ which means $j_1 = j_2 = j$, as in the classical case. For our representations, we find

$$M^2 = N^2 = \tfrac{1}{4}(L^2 + A^2) = j(j+1), j = 0, \tfrac{1}{2}, 1, .. \tag{86}$$

and therefore

$$L^2 + A^2 + 1 = (2j + 1)^2. \tag{87}$$

Substituting this result into Equation (82) gives the usual formula for the bound state energy levels of the hydrogen atom:

$$H' = -\frac{m(Z\alpha)^2}{2n^2} = E_n \tag{88}$$

where the principal quantum number n = 2j+1 = 1,2,.. and the prime on H signifies an eigenvalue of the operator H.

Within a subspace of energy E_n, the Runge–Lenz vector is

$$A = \frac{1}{a_n}\left(\frac{p \times L - L \times p}{2} - mZ\alpha\frac{r}{r}\right). \tag{89}$$

where $a_n = \sqrt{-2mE_n} = \frac{mZ\alpha}{n}$.

Our considerations of the Casimir operators have shown that the hydrogen atom provides completely symmetrical tensor representations of SO(4), namely, $(j, j) = (\frac{n-1}{2}, \frac{n-1}{2})$, $n = 1, 2, \ldots$ The dimensionality is n^2, corresponding to the n^2 degenerate states. The appearance of only symmetrical tensor representations $(j_1 = j_2)$ may be traced to $L \cdot A$ vanishing, which is a consequence of the structure of L and A in terms of the dynamical variables for the hydrogenlike atom. For systems other than the hydrogenlike atom, it is not generally possible to find the expression for the energy levels in terms of all the different quantum numbers alone. It worked here, since we could express the Hamiltonian as a function of the Casimir operators that contained all quantum numbers explicitly.

There are a variety of possible basis states. We could choose basis states for the SO(4) representation that reflect the SU(2) decomposition, namely eigenstates of M^2, N^2, M_3 and N_3 [120]. Another possibility is to have a basis with eigenstates of the Casimir operator C_2, and A_3, and M_3. This choice fits well with the use of parabolic coordinates [60]. A more physically understandable choice is to choose the common basis states $|nlm\rangle$ that are eigenstates of C_2, L^2, and L_3. For this set of basis states, we have

$$\sqrt{(L^2 + A^2 + 1)}|nlm\rangle = n|nlm\rangle \qquad L^2|nlm\rangle = l(l+1)|nlm\rangle \qquad L_3|nlm\rangle = m|nlm\rangle \tag{90}$$

We can define raising and lowering operators for m:

$$L_\pm = L_1 \pm iL_2 \tag{91}$$

which obeys the commutation relations

$$[L^2, L_\pm] = 0] \qquad [L_3, L_\pm] = \pm L_\pm. \tag{92}$$

Therefore, we can use $L_\pm$ to change the value of m for the basis states

$$L_\pm|nlm\rangle = \sqrt{(l(l+1) - m(m \pm 1)}|nl\ m \pm 1\rangle \tag{93}$$

for $l \geq 1$. We can also use the generators A to change the angular momentum. A general SO(4) transformation can be expressed as a rotation induced by L, followed by a rotation induced by A_3, followed by another rotation generated by L [121]. Our interest is primarily in changing the angular momentum l, which is most directly done while using A_3, which commutes with L_3 and C_2, and so only changes l:

$$A_3|nlm\rangle = \left(\frac{(n^2 - (l+1)^2)((l+1)^2 - m^2)}{4(l+1)^2 - 1}\right)^{\frac{1}{2}} |n\ l+1\ m\rangle + \left(\frac{(n^2 - l^2)(l^2 - m^2)}{4l^2 - 1}\right)^{\frac{1}{2}} |n\ l-1\ m\rangle. \tag{94}$$

for $l \geq 1$.

4.3. Relativistic and Semi-Relativistic Spinless Particles in the Coulomb Potential and Klein–Gordon Equation

The Klein–Gordon equation

$$\left(p^2 - (\tilde{E} - V)^2 + m^2\right)\tilde{\psi} = 0$$

where $\tilde{E}$ is the relativistic total energy, may be solved exactly for a Coulomb potential, $V = -(Z\alpha)/r$ [122]. The energy levels depend on a principal quantum number and on the magnitude of the angular momentum but not its direction. The only degeneracy present is associated with the O(3) symmetry of the Hamilton. For a relativistic scalar particle, there is no degeneracy to be lifted by a Lamb shift.

If we neglect the V^2 term the resulting equation can be written in the form

$$\left(\frac{p^2}{2\tilde{E}} + V - \frac{\tilde{E}^2 - m^2}{2\tilde{E}}\right)\tilde{\psi} = 0$$

This is exactly the same as the nonrelativistic Schrodinger equation with the substitutions

$$m \to \tilde{E} \qquad E \to \frac{\tilde{E}^2 - m^2}{2\tilde{E}}.$$

Thus, we regain the O(4) symmetry of the nonrelativistic hydrogen atom, and can define two conserved vectors, as indicated in Table 1. It is possible to take the "square root" of this approximate Klein-Gordon equation (in the same sense that the Dirac equation is the square root of the Klein–Gordon equation) and get an approximate Dirac equation whose energy eigenvalues are independent of the orbital angular momentum [123].

4.4. Eigenstates of the Inverse Coupling Constant $(Z\alpha)^{-1}$

Solutions to Schrodinger's equation for a particle of energy $E = -\frac{a^2}{2m}$ in a Coulomb potential

$$\left[p^2 + a^2 - \frac{2mZ\alpha}{r} \right] |a\rangle = 0 \tag{95}$$

may be found for certain critical values of the energy $E_n = -\frac{a_n^2}{2m}$ where $a_n = \frac{mZ\alpha}{n}$. The corresponding eigenstates of the Hamiltonian are $|nlm\rangle$ which satisfy Equation (95) with a replaced by a_n. In addition to the bound states, because there is no upper bound on p^2 in the Hamiltonian, we also have the continuum of scattering states that have $E > 0$.

Because the quantity that must have discrete values for a solution to exist is actually $\frac{a}{mZ\alpha}$, as noted in Section 2.3, we might ask if eigensolutions to Equation (95) exist for certain critical values of $Z\alpha$ while keeping a and the energy fixed [1]. To investigate such solutions it is convenient to algebraically transform Equation (95):

$$\left[\frac{1}{\sqrt{\rho(a)}} \left(\frac{p^2 + a^2}{a} \right) \frac{1}{\sqrt{\rho(a)}} - \frac{1}{\sqrt{\rho(a)}} \left(\frac{2mZ\alpha}{ar} \right) \frac{1}{\sqrt{\rho(a)}} \right] \sqrt{\rho(a)} |a\rangle = 0$$

where

$$\rho(a) = \frac{p^2 + a^2}{2a^2}. \tag{96}$$

Because $\rho(a)$ commutes with p^2, we obtain the eigenvalue equation

$$\left[\left(\frac{a}{mZ\alpha} \right) - K(a) \right] |a\rangle = 0 \tag{97}$$

where the totally symmetric and real kernel is

$$K(a) = \sqrt{\frac{2a^2}{p^2 + a^2}} \frac{1}{ar} \sqrt{\frac{2a^2}{p^2 + a^2}} \tag{98}$$

and

$$|a\rangle = (\rho(a))^{\frac{1}{2}} |a\rangle. \tag{99}$$

As before, solutions to this transformed equation may found for the eigenvalues

$$K'(a) = \left(\frac{a}{mZ\alpha} \right)' = \frac{1}{n}. \tag{100}$$

If we hold $Z\alpha$ constant and let a vary, we obtain the usual spectrum $a_n = \sqrt{-2mE_n} = \frac{mZ\alpha}{n}$. For $a = a_n$, Equation (99) reduces to the equation for the eigenstates

$$\sqrt{\rho(a_n)} |nlm\rangle = |nlm; a_n\rangle. \tag{101}$$

Alternatively, if we hold a constant, then $Z\alpha$ has the spectrum

$$(Z\alpha)_n = \frac{na}{m} \tag{102}$$

with the corresponding eigenstates of $(Z\alpha)^{-1}$ being

$$|nlm;a). \tag{103}$$

The relationship between the usual energy eigenstates of the energy $|nlm\rangle$ and the eigenstates $|nlm;a)$ of $(Z\alpha)^{-1}$ is

$$|nlm\rangle = \frac{1}{\sqrt{\rho(a_n)}}|nlm;a_n), \tag{104}$$

which requires that both sets of states have the same quantum numbers. Note that the magnitude $(a|K(a)|a)$ is proportional to $\langle 1/ar \rangle_n = (1/a_n)(1/n^2 a_0)$ where a_0 is the Bohr radius for the ground state, and so is positive and bounded. The kernel K(a) is real and symmetric in p and r and manifestly Hermitean. Because the kernel K in Equation (98) is bounded, definite, and Hermitian with respect to the eigenstates $|nlm;a_n)$ [124] the set of normalized eigenstates

$$|nlm;a) \quad n = 0,1,2....; \quad l = 0,1,..n-2,n-1; \quad m = -l,-l+1,...l-1,l. \tag{105}$$

where

$$\left(\frac{1}{n} - K(a)\right) |nlm;a) = 0 \tag{106}$$

is a complete orthonormal basis for the hydrogenlike atom:

$$(nlm;a|n'l'm';a) = \delta_{nn'}\delta_{ll'}\delta_{mm'} \tag{107}$$

$$\sum_{nlm} |nlm;a)(nlm;a| = 1. \tag{108}$$

There are several important points to notice with regard to these eigenstates of the inverse of the coupling constant:

(1) **Because of the boundedness of K, there is no continuum portion in the eigenvalue spectrum of $(Z\alpha)^{-1}$, the eigenvalues are discrete.** Because K is a positive definite Hermitian operator, all eigenvalues are positive, real numbers. This feature leads to a unified treatment of all states of the hydrogenlike atom as opposed to the treatment in terms of energy eigenstates in which we must consider separately the bound states and the continuum of scattering states.

(2) It follows from Equation (104) that **the quantum numbers, multiplicities, and degeneracies of these states $|nlm;a)$ are precisely the same as those of the usual bound energy eigenstates.** For example, there are n^2 eigenstates of $(Z\alpha)^{-1}$ that have the principal quantum number equal to n or $(Z\alpha)$ equal to $\frac{na}{m}$.

(3) **A single value of the RMS momentum a or the energy $E = \frac{-a^2}{2m}$ applies to all the states in our complete basis**, as opposed to the usual energy eigenstates where each nondegenerate state has a different value of a. We have made this explicit by including a in the notation for the states: $|nlm;a)$. Sometimes we will write the states as $|nlm)$, provided the value of a has been specified. This behavior in which a single value of a applies to all states will prove to be very useful. In essence, it permits us to generalize from statements applicable in a subspace of Hilbert space with energy E_n or energy parameter a_n to the entire Hilbert space.

(4) **By a suitable scale change or dilation, we may give the quantity a any positive value we desire.** This is affected by the unitary operator

$$D(\lambda) = e^{i\frac{1}{2}(\mathbf{p}\cdot\mathbf{r}+\mathbf{r}\cdot\mathbf{p})\lambda} \tag{109}$$

which transforms the canonical variables

$$D(\lambda)pD^{-1}(\lambda) = e^{-\lambda}p \quad D(\lambda)rD^{-1}(\lambda) = e^{\lambda}r \tag{110}$$

and the kernel K(a)

$$D(\lambda)K(a)D^{-1}(\lambda) = K(ae^{\lambda}). \tag{111}$$

and the eigenvalue equation

$$\left(\frac{1}{n} - K(ae^{\lambda})\right)D(\lambda)|nlm;a\rangle = 0. \tag{112}$$

Therefore the states transform as

$$D(\lambda)|nlm;a\rangle = |nlm;ae^{\lambda}\rangle. \tag{113}$$

These transformed states form a new basis corresponding to the new value $e^{\lambda}a$ of the RMS momentum.

The relationship between the energy eigenstates and the $(Z\alpha)^{-1}$ eigenstates can be written while using the dilation operator:

$$|nlm\rangle = \frac{1}{\sqrt{\rho(a_n)}}D(\lambda_n)|n\ell m;a\rangle \quad \text{where } e^{\lambda_n} = \frac{a_n}{a}. \tag{114}$$

The usual energy eigenstates $|nlm\rangle$ are obtained from the eigenstates of $(Z\alpha)^{-1}$ by first performing a scale change to ensure that the energy parameter a has the value a_n and then multiplying by a factor $\rho^{-1/2}$. The need for the scale change is apparent from dimensional considerations: from the $(Z\alpha)^{-1}$ eigenvalue equation, we see that the eigenfunctions are functions of p/a or ar while from the energy eigenvalue equation the eigenfunctions are functions of p/a_n and $a_n r$. The factor $\rho^{-1/2}$ was required in order to convert Schrodinger's equation to one involving a bounded Hermitean operator.

Using the eigenstates of $(Z\alpha)^{-1}$ as our basis allows for us to analyze the mathematical and physical structure of the hydrogenlike atom in the easiest and clearest way.

4.5. Another Set of Eigenstates of $(Z\alpha)^{-1}$

We may transform Schrodinger's equation Equation (95) to an eigenvalue equation for $(Z\alpha)^{-1}$ that differs from Equation (96) by similar methods:

$$\left(\frac{1}{n} - K_1(a)\right)|nlm;a\rangle = 0 \tag{115}$$

where

$$K_1(a) = \frac{1}{\sqrt{ar}}\frac{2a^2}{p^2+a^2}\frac{1}{\sqrt{ar}} \quad \rho(a) = n/ar \quad \sqrt{\rho(a_n)}|nlm\rangle = |nlm;a_n\rangle \tag{116}$$

This kernel, like K(a), is a bounded, positive definite, Hermitian operator so the eigenstates form a complete basis. The relationship of these basis states to the energy eigenstates is the same as that of the previously discussed eigenstates of $(Z\alpha)^{-1}$ Equation (114) but with $\rho(a) = n/ar$. The n insures that the two sets of eigenstates have consistent normalization, which may be checked by means of the virial theorem. The n cancels out when similarity transforming from the basis of energy eigenstates to the basis of $(Z\alpha)^{-1}$. Note that, classically, both kernels equal $1/ar_c$.

The first set of basis states of $(Z\alpha)^{-1}$ with $\rho(a) = \frac{p^2+a^2}{2a^2}$ is more convenient to use when working in momentum space and the second set with $\rho(a) = n/ar$ is more convenient in configuration space.

Other researchers have used other approaches to secure a bounded kernel for the Schrodinger hydrogen atom, for example, by multiplying the equation from the left by r to regularize it [82].

However, the methods used have not symmetrized the kernels to make them Hermitian, nor are all the generators of the corresponding groups Hermitian, and they have to redefine the inner product [38,82].

4.6. Transformation of A and L to the New Basis States

We must transform A as given in Equation (89) and $L = r \times p$ when we change our basis states from eigenstates of the energy to eigenstates of the inverse coupling constant. The correct transformation may be derived by requiring that the transformed generators produce the same linear combination of new states as the original generators produced of the old states. Thus, because

$$A|nlm\rangle = \sum_{l',m'} |nl'm'\rangle A^{lm}_{l'm'} \tag{117}$$

where the coefficients $A^{lm}_{l'm'}$ are the matrix elements of A, we require that the transformed generator a satisfies the equation

$$a|nlm) = \sum_{l'm'} |nl'm') A^{lm}_{l'm'}. \tag{118}$$

In other words, since the Runge–Lenz vector A is a symmetry operator of the original energy eigenstates, a will be a symmetry operator of the new states with precisely the same properties and matrix elements. Because A is Hermitian, a is Hermitian.

To obtain a differential expression for a acting on the new states, we need to transform the generator using Equation (114):

$$a = D^{-1}(\lambda_n) \left(\sqrt{\rho(a_n)} A \frac{1}{\sqrt{\rho(a_n)}} \right) D(\lambda_n). \tag{119}$$

The effect of the scale change on the quantity in large parenthesis is to replace a_n everywhere by a. By explicit calculation, we find

$$a = \frac{1}{2a} \left(\frac{rp^2 + p^2 r}{2} - r \cdot pp - pp \cdot r \right) - \frac{ar}{2} \tag{120}$$

for $\rho(a) = (p^2 + a^2)/2a^2$. And we obtain

$$a = \frac{1}{2a} \left(\frac{rp^2 + p^2 r}{2} - r \cdot pp - pp \cdot r - \frac{r}{4r^2} \right) - \frac{ar}{2} \tag{121}$$

for $\rho(a) = n/ar$.

Both of these expressions for a are manifestly Hermitian. In addition, since there is no dependence on the principal quantum number these expressions are valid in the entire Hilbert space, and not just in a subspace spanned by the degenerate states, as was the case when we used the energy eigenstates as a basis (Equation (89)).

The angular momentum operator is invariant under scale changes and it commutes with scalar operators. Therefore L is invariant under the similarity transformation $\frac{1}{\sqrt{\rho(a_n)}} D(\lambda_n)$ and the expression for the angular momentum operator with respect to the eigenstates of $(Z\alpha)^{-1}$ is the same as the expression with respect to the eigenstates of the energy.

4.7. The $\langle U'|$ Representation

The U' coordinates provide the natural representation for the investigation of the symmetries of the hydrogenlike atom in quantum mechanics, as in classical mechanics [125]. Therefore, we briefly consider the relevant features of this representation and, in particular, its relationship to the momentum representation. The eigenstate $< U'|$ of U_b, $b = 1,2,3,4$, is defined by

$$\langle U'| \, U_b = U_b' \, \langle U'| \tag{122}$$

These states are complete on the unit hypersphere in four dimensions:

$$\int |U'\rangle\langle U'| \, d^3\Omega' = 1 \tag{123}$$

where Ω refers to the angles (θ_4, θ, ϕ) defined in Equation (54). The U variables are defined in terms of the momentum variables and the quantity a in Equation (52). Therefore, the momentum and the U operators commute

$$[p_i, U_b] = 0 \tag{124}$$

and the state $\langle U'|$ is proportional to a momentum eigenstate $\langle p'|$:

$$\langle U'| = \langle p'| \, \sqrt{J(p)} \tag{125}$$

where the momentum eigenstate is defined by $\langle p'|\boldsymbol{p} = \boldsymbol{p}'\langle p|$ and

$$\int d^3p' |p'\rangle\langle p'| = 1. \tag{126}$$

The function $J(p')$ may be determined by equating the completeness conditions and substituting Equation (125):

$$1 = \int d^3p' |p'\rangle\langle p'| = \int d^3\Omega' |U'\rangle\langle U'| = \int d^3\Omega' J(p') |p'\rangle\langle p'| \tag{127}$$

which leads to the identification of the differential quantities

$$d^3p' = d^3\Omega' J(p') \tag{128}$$

demonstrating that $J(p')$ is the Jacobian of the transformation from the p- to the U- space. Noting that on the unit sphere

$$U_4^2 = 1 - U_i U_i \tag{129}$$

we can compute the Jacobian

$$J(p') = \left[\frac{p'^2 + a^2}{2a}\right]^3 . \tag{130}$$

Therefore from Equation (125) we have the important result

$$\langle U'| = \langle p'| \left[\frac{p^2 + a^2}{2a}\right]^{3/2} . \tag{131}$$

We can use this result to compute the action of $\boldsymbol{r}$ on $\langle U'|$ in terms of the differential operators. Using the equation

$$\langle p'|\boldsymbol{r} = i\boldsymbol{\nabla}_{p'}\langle p'| \tag{132}$$

we obtain

$$\langle U'|\boldsymbol{r} = \left(i\boldsymbol{\nabla}_{p'} - \frac{3i\boldsymbol{p}'}{p'^2 + a^2}\right)\langle U'|. \tag{133}$$

Action of a and L on $\langle U'|$

Using Equation (133) for the action of r on $\langle U'|$ and using the expression Equation (120) for a since $\langle U'|$ is proportional to $\langle p'|$, we immediately find that when acting on $\langle U'|$, a has the differential representation

$$a' = \frac{i}{2a}\left((p'^2 - a^2)\nabla_{p'} - 2p'p' \cdot \nabla_{p'} \right) \tag{134}$$

where

$$\langle U'|a = a'\langle U'|. \tag{135}$$

We can also write a' in terms of the U' variables by using the relationship Equation (52) between the p and U variables:

$$a' = U_4' i\nabla_{U'} - U' i\partial/\partial_4' \tag{136}$$

where the spatial part of the four vector U' is $U = (U_1, U_2, U_3)$ and U_4 is the fourth component. This is the differential representation of a rotation operator mixing the spatial and the fourth components of U_a'. When acting on the state $\langle U'|$, clearly $e^{ia\cdot v}$ generates a four-dimensional rotation that produces a new eigenstate $\langle U''|$. To explicitly derive the form of the finite transformation, we compute

$$[a_j', U_j'] = iU_4'\delta_{ij} \qquad [a_j', U_4'] = -iU_i'. \tag{137}$$

For a finite transformation $a^{ia\cdot nv}$ with $n^2 = 1$, we have

$$\begin{aligned} U'' &= e^{ia'\cdot nv}U'e^{-ia'\cdot nv} \\ &= U' - nn\cdot U' + nn\cdot U'\cos v - nU_4'\sin v \end{aligned} \tag{138}$$

and

$$\begin{aligned} U_4'' &= e^{ia'\cdot nv}U_4'e^{-ia'\cdot nv} \\ &= U_4'\cos v + n\cdot U'\sin v \end{aligned} \tag{139}$$

These equations of transformation are like those for a Lorentz transformation of a four-vector (r, it). We can illustrate the equations for e^{ia_2v} (cf Equation (58)), which mixes the two and four components of U':

$$\begin{array}{ll} U_1'' = U_1' & U_3'' = U_3' \\ U_2'' = U_2'\cos v - U_4'\sin v & U_4'' = U_2'\sin v + U_4'\cos v \end{array} \tag{140}$$

When L acts on $\langle U'|$, it has the differential representation

$$L' = U' \times i\nabla_{U'} \tag{141}$$

This result follows directly, since U_i' equals p_i' times a factor that is a scalar under rotations in three dimensions. When $e^{iL\cdot\omega}$ acts on $\langle U'|$ it produces a new state $\langle U''|$, where the spatial components of U' have been rotated to produce U''.

In summary, we see that U' is a four-vector under rotations generated by a' and L'. Therefore the states $\langle U'|$ provide a vector representation of the group of rotations in four dimensions SO(4), with the generators a and L.

5. Wave Functions for the Hydrogenlike Atom

In this section, we analyze the wave functions of the hydrogenlike atom, working primarily in the $\langle U'|$ representation and using eigenstates of the inverse of the coupling constant $(Z\alpha)^{-1}$ for the basis states. In this representation the wave functions are spherical harmonics in four dimensions. We derive the relationship of the usual energy eigenfunctions in momentum space to the spherical harmonics and discuss the classical limits in momentum and configuration space.

5.1. Transformation Properties of the Wave Functions under the Symmetry Operations

We can show that the wave functions $Y_{nlm}(U')$ in the $\langle U'|$ representation with respect to the eigenstates of $(Z\alpha)^{-1}$

$$Y_{nlm}(U') \equiv \langle U'|nlm\rangle \tag{142}$$

transform as four-dimensional spherical harmonics under the four-dimensional rotations generated by the Runge-Lenz vector a and the angular momentum L. We note that the quantity a is implicit in both the bra and the ket in Equation (142). For our basis states we employ the set of $(Z\alpha)^{-1}$ eigenstates $|nlm\rangle$ of the inverse coupling constant that are convenient for momentum space calculations ($\rho = \frac{p^2+a^2}{2a^2}$). We choose these states rather than those convenient for configuration space calculations, because the $\langle U'|$ eigenstates are proportional to the $\langle p'|$ eigenstates.

If we transform our system by the unitary operator $e^{i\theta}$ where $\theta = L \cdot \omega + a \cdot v$, then the wave function in the new system is

$$Y'_{nlm}(U') = \langle U'|e^{i\theta}|nlm\rangle. \tag{143}$$

There are two ways in which we may interpret this transformation, corresponding to what have been called the active and the passive interpretations. In the passive interpretation we let $e^{i\theta}$ act on the coordinate eigenstate $\langle U'|$. As we have seen in Section 3.6, this produces a new eigenstate $\langle U''|$, where the four-vector U'' is obtained by a four-dimensional rotation of U' (Equations (138) and (139)). Thus, we have

$$Y'_{nlm}(U') = \langle U''|nlm\rangle = Y_{nlm}(U''). \tag{144}$$

In the active interpretation, we let $e^{i\theta}$ act on the basis state $|nlm\rangle$. Because L and a are symmetry operators of the system, transforming degenerate states into each other, it follows that $e^{i\theta}|nlm\rangle$ must be a linear combination of states with principal quantum number equal to n. Therefore, we have

$$Y'_{nlm}(U') = \sum_{l'm'}\langle U'|R^{nlm}_{nl'm'}|nl'm'\rangle \;=\; \sum_{l'm'}R^{nlm}_{nl'm'}Y_{nl'm'}(U'). \tag{145}$$

The wave functions for degenerate states with a given n transform irreducibly among themselves under the four-dimensional rotations, forming a basis for an irreducible representation of SO(4) of dimensions n^2. Equating the results of the two different interpretations gives

$$Y_{nlm}(U'') = \sum_{l'm'}R^{nlm}_{nl'm'}Y_{nl'm'}(U'). \tag{146}$$

The transformation properties Equation (146) of Y_{nlm} are precisely analogous to those of the three-dimensional spherical harmonic functions. It follows that the Y_{nlm} are four-dimensional spherical harmonics [35,124].

5.2. Differential Equation for the Four Dimensional Spherical Harmonics $Y_{nlm}(U')$

The differential equation for the $Y_{nlm}(U')$ may be obtained from the equation

$$(L'^2 + a'^2)Y_{nlm}(U') = (n^2 - 1)Y_{nlm}(U') \tag{147}$$

which follows from $C_2 = n^2 - 1$ and the definition of C_2, Equation (81). Substituting in the differential expressions Equations (134) and (141) for a' and L' we find that $L'^2 + a'^2$ equals $\nabla^2_{U'} - (U' \cdot \nabla_{U'})^2$, which is the angular part of the Laplacian operator in four dimensions (cf in three dimensions, $L^2/r^2 = p^2 - p_r^2$). Thus, Equation (147) is the differential equation for four-dimensional spherical harmonics with the degree of homogeneity equal to $n - 1$, which means n^2 such functions exist, in agreement with the know degree of degeneracy.

5.3. Energy Eigenfunctions in Momentum Space

We want to determine the relationship between the usual energy eigenfunctions in momentum space $\psi_{nlm}(p') \equiv \langle p'|nlm \rangle$ (with $a = a_n$) and the four-dimensional spherical harmonic eigenfunctions $Y_{nlm}(U'; a) = \langle U'|nlm; a \rangle$.

We choose the RMS momentum a to have the value a_n. If we use the expression Equation (131) for $\langle U'|$ in terms of $\langle p'|$

$$\langle U'| = \langle p'| \left(\frac{p^2 + a_n^2}{2a_n} \right)^{3/2} \tag{148}$$

and the expression Equation (104) for the eigenstates of $(Z\alpha)^{-1}$ in terms of the energy eigenstates

$$|nlm; a_n) = \sqrt{\frac{p^2 + a_n^2}{2a_n^2}}|nlm\rangle \tag{149}$$

we find the desired result

$$Y_{nlm}(U'; a_n) = \left(\frac{p^2 + a_n^2}{2a_n} \right)^2 \frac{1}{\sqrt{a_n}}\psi_{nlm}(p'). \tag{150}$$

The usual method of deriving this relationship between the wave function in momentum space and the corresponding spherical harmonics in four dimensions involves transforming the Schrodinger wave equation to an integral equation in momentum space [18,35]. As in the classical case, to do this we first replace p by p/a and perform a stereographic projection from the hyperplane corresponding to the three- dimensional momentum space to a unit hypersphere in a four-dimensional space. The resulting integral equation manifests a four-dimensional invariance. When the wave functions are normalized as in Equation (150), the solutions are spherical harmonics in four dimensions. As another alternative to this procedure, we can Fourier transform the configuration space wave functions directly [126].

5.4. Explicit Form for the Spherical Harmonics

The spherical harmonics in four dimensions can be expressed as [127]:

$$Y_{nlm}(\Omega) = N_1(n, l)(\sin \theta_4)^l C_{n-1-l}^{l+1}(\cos \theta_4) \cdot \left[N_2(l, m)(\sin \theta)^m C_{l-m}^{m+1/2}(\cos \theta) \frac{e^{im\phi}}{\sqrt{2\pi}} \right]. \tag{151}$$

The factor in brackets is equal to $Y_l^m(\theta, \phi)$, the usual spherical harmonic in three-dimensions [127]. The Gegenbauer polynomials C_n^λ of degree n and order λ are defined in terms of a generating function:

$$\frac{1}{(1 - 2tx + t^2)^\lambda} = \sum_{n=0}^{\infty} t^n C_n^\lambda. \tag{152}$$

$N_1(n, l)$ and $N_2(l, m)$ are chosen to normalize the Y_{nlm} on the surface of the unit sphere:

$$\int |Y_{nlm}(\Omega)|^2 d^3\Omega_U = 1 \tag{153}$$

where $d^3\Omega_U = \sin^2 \theta_4 \sin \theta d\theta d\phi$. We find

$$N_1(n, l) = \sqrt{\frac{2^{2l+1}}{\pi} \frac{n(n - l - 1)!(l!)^2}{(n + l)!}} \tag{154}$$

$$N_2(l, M) = \sqrt{\frac{2^{2m}}{\pi}(l + \frac{1}{2})\frac{(l - m)!}{(l + m)!}[\Gamma(m + \frac{1}{2})]^2} \; . \tag{155}$$

In the next section, we discuss the asymptotic behavior of Y_{nlm} for large quantum numbers and compare it to the classical results of Section 3.

5.5. Wave Functions in the Classical Limit

5.5.1. Rydberg Atoms

Advances in quantum optics, such as the development of ultra short laser pulses, microwave spectroscopy, and atom inteferometery, have opened new possibilities for experiments with atoms and Rydberg states, meaning hydrogenlike atoms in states with very large principal quantum numbers and correspondingly large diameter electron orbits. The pulsed electromagnetic fields can be used to modify the behavior of the orbital electrons. Semi-classical electron wave packets in hydrogenlike atoms were first generated in 1988 by ultrashort laser pulses, and today are often generated by unipolar teraherz pulses [128–130]. Over the last few decades, there has been interest in the classical limit of the hydrogenlike atom for n very large, Rydberg states, for a number of reasons [131]: 1. Rydberg states are at the border between bound states and the continuum, and any process which leads to excited bound states, ions or free electrons usually leads to the production of Rydberg states. This includes, for example, photo-ionization or the application of microwave fields. The very large cross section for scattering is unique. 2. Rydberg states can be used to model atoms with a higher atomic number that have an excited valence electron that orbits beyond the core. 3. In Rydberg states, the application of electric and magnetic fields breaks the symmetry of the atom and allows for the study of different phenomena, including the transition from classical to quantum chaos [132]. 4. Rydberg atoms can be used to study coherent transient excitation and relaxation, for example, the response to short laser pulses creating coherent quantum wave packets that behave like a classical particle.

The square of the wave function for a given quantum state gives a probability distribution for the electron that is independent of time. If we want to describe the movement of an electron in a semiclassical state, with a large radius, going around the nucleus with a classical time dependence, then we need to form a wave packet. The wave packet is built as a superposition of many wave functions with a band of principal quantum numbers.

A variety of theoretical methods have been used to derive expressions for the hydrogen atom wave functions and wave packets for highly excited states. There is general agreement on the wave functions for large n, and that the wave functions display the expected classical behavior, elliptical orbits in configurations space, and great circles in the four dimensional momentum space [133–136].

Researchers have proposed a variety of wave packets to describe Rydberg states. There are general similarities in the wave packets that describe electrons going in circular or elliptical orbits with a classical time dependence for some characteristic number of orbits, and it is maintained that the quantum mechanical wave packets provide results that agree with the classical results [129,130,133–139]. Most of the approaches exploit the SO(4) or SO(4,2) symmetry of the hydrogen atom which is used to rotate a circular orbital to an elliptical orbit. The starting orbital is often taken as a coherent state, which is usually considered a classical like state with minimum uncertainty. The most familiar example of a coherent state is for a one dimensional harmonic oscillator characterized by creation and annihilation operators $a^{\dagger}$ and a. The coherent state $|\alpha\rangle$ is a superposition of energy eigenstates that is an eigenstate of a where $a|\alpha\rangle = \alpha|\alpha\rangle$ for a complex α. This coherent state will execute harmonic motion like a classical particle [140]. To obtain a coherent state for the hydrogenlike atom, eigenstates of the operator than lowers the principal quantum number n (which will be discussed in Section 7.4) have been used [141], as well as lowering operators based on the equivalence of the four dimensional harmonic oscillator representation of the hydrogen atom [134,135,142].

In either case, this coherent eigenstate is characterized by a complex eigenvalue, which needs to be specified. Several constraints have been used to obtain the classical wave packet that presumably obeys Kepler's Laws, such as requiring that the orbit lie in a plane so $\langle z \rangle = 0$ for the orbital, or that $\langle r - r_{classical} \rangle$ be a minimum, or that some minimum uncertainty relationship is obtained. In addition, there are issues regarding the approximations used, in particular, those that relate to time. For times characteristic of the classical hydrogen atom, the wavepackets act like a classical system. For longer times, the wavepacket spreads in the azimuthal direction and after some number of classical revolutions of order 10 to 100 the spread is 2π, so the electron is uniformly spread over the entire orbit. The spread arises because the component wave functions forming the wavepacket have different momenta. In two derivations, still longer times were considered, and recoherence was predicted to occur after about $n/3$ (where n is the approximate principal quantum number) revolutions, although there is some difference in the predicted amount of recoherence [131,136]. Because of the conservation of L and A, the spread of the wave packets is inhibited, except in the azimuthal direction.

This is a system with very interesting physics. For example, one can view the transition from a well defined wave packet representing the electron to a 2π spread and back again as a dynamical illustration of the wave-particle duality.

Brown took a different approach to develop a wavepacket for a circular orbit [139]. He first developed the asymptotic wave function for large n and then optimized the coefficients in a Gaussian superposition to minimize the spread in ϕ, obtaining a predicted characteristic decoherence time of about 10 minutes, considerable longer than any other predicted decoherence time.

Other authors have explored the problem from the perspective of classical physics and the correspondence principle [133,137,143,144]. The results from the different methods are similar with the basic conclusion that the wave functions are peaked on the corresponding classical Kepler trajectories: "atomic elliptic states sew the wave flesh on the classical bones" [129].

With the variety of experimental methods used to generate Rydberg states, a variety of Rydberg wave packets are created, and it is not clear which theoretical model, if any, is preferred [131]. We take a very simple approach to forming a wavepacket and simply use a Gaussian weight for the different frequency components. This does not give an intentionally optimized wave packet, but it is a much simpler approach and the result has all of the expected classical behavior that is very similar to that obtained from much more complicated derivations. We start with a circular orbit and then do a SO(4) rotation to secure an elliptical orbit. We show that it has the classical period of rotation.

5.5.2. Wave Functions in the Semi-Classical Limit

We need to derive the semiclassical limit of the wave functions that correspond to circular orbits in configuration space. For this case, $\sin \nu$, which we interpret as the expectation value of the eccentricity, vanishes. We derive expressions for the wave functions in momentum space and then form a wavepacket. To obtain corresponding expressions for elliptical orbits, we perform a rotation by $e^{i\alpha \cdot \nu}$, which does not alter the energy but changes the eccentricity and the angular momentum.

Case 1: Circular orbits, $\sin \nu = e = 0$

We derive the asymptotic form of Y_{nlm} for large quantum numbers, where for simplicity we choose the quantum numbers $n - 1 = l = m$ corresponding to a circular orbit in the 1–2 plane. From Equation (151) we see we encounter Gegenbauer polynomials of the form C_0^λ, which, by Equation (152), are unity. For a very large l, $\sin^l \theta$ will have a very strong peak at $\theta = \pi/2$, so we make the expansion [103]

$$\sin \theta = \sin \left(\frac{\pi}{2} + \left(\theta - \frac{\pi}{2} \right) \right) = 1 - \frac{1}{2} \left(\theta - \frac{\pi}{2} \right)^2 + \cdots \quad \approx e^{-(1/2)(\theta - \pi/2)^2} \tag{156}$$

to obtain

$$\sin^l \theta \approx e^{-(1/2)l(\theta - \pi/2)^2}. \tag{157}$$

The asymptotic forms for N_1 and N_2 can be computed using the properties of Γ functions:

$$\Gamma(2z) = \left(\frac{1}{\sqrt{2\pi}} \right) 2^{2z - \frac{1}{2}} \Gamma(z) \Gamma \left(z - \frac{1}{2} \right)$$

$$\lim_{z \to \infty} \Gamma(az + b) \simeq \sqrt{2\pi} e^{-az} (az)^{az + b - \frac{1}{2}}. \tag{158}$$

We finally obtain [145]

$$Y_{n,n-1,n-1}(\Omega) = \sqrt{\frac{n}{2\pi^2}}\, e^{-\frac{1}{2}n\left(\theta_4 - \frac{\pi}{2}\right)^2} \cdot e^{-\frac{1}{2}n\left(\theta - \frac{\pi}{2}\right)^2} e^{i(n-1)\phi}, \tag{159}$$

which gives the probability density

$$\left|Y_{n,n-1,n-1}(\Omega)\right|^2 = \frac{n}{(2\pi^2)}\, e^{-n\left(\theta_4 - \frac{\pi}{2}\right)^2} \cdot e^{-n\left(\theta - \frac{\pi}{2}\right)^2}. \tag{160}$$

We have Gaussian probability distributions in θ_4 and θ about the value $\pi/2$. The distributions are quite narrow with widths $\Delta\theta_4 \approx \Delta\theta \approx 1/\sqrt{n}$ and the spherical harmonic essentially describes a circle $(\theta_4 = \theta = \pi/2)$ on the unit sphere in the 1–2 plane. As n becomes very large, $U_4 = \cos\theta_4 \approx (r - r_c)/r$ (Equations (54) and (72)) and $U_3 = \sin\theta_4 \cos\theta$, which is proportional to p_3, both go to zero as $1/\sqrt{n}$. The distribution approaches the great circle $U_1^2 + U_2^2 = 1$ that we found in Section 3.6 for a classical particle moving in a circular orbit in the 1–2 plane in configuration space. Note that this state is a stationary state with a constant probability density. To get the classical time dependence we need to form a wavepacket.

Forming a Wavepacket

We form a time dependent wavepacket for circular orbits by superposing circular energy eigenstates:

$$\chi(\Omega, t) = \sum_n e^{itE_n} Y_{n,n-1,n-1} A_{n-N} \tag{161}$$

where A_{n-N} is an amplitude peaked about $n = N \gg 1$. For $n \gg 1$ we expand E_n about E_N:

$$E_n = E_N + \left.\frac{\partial E}{\partial n}\right|_N s + \left.\frac{\partial^2 E}{\partial n^2}\right|_N s^2 + \dots \tag{162}$$

where $s = n - N$. From the equation for the energy levels, $E = -m(Z\alpha)^2/(2n^2)$ we compute

$$\left.\frac{\partial E_n}{\partial n}\right|_N = \frac{m(Z\alpha)^2}{N^3} = \sqrt{\frac{-8E_N^3}{m(Z\alpha)^2}}. \tag{163}$$

In agreement with the Bohr Correspondence Principle, the right-hand side of this equation is just the classical frequency ω_{cl} as given in Equation (38). For the second order derivative, we have

$$\left.\frac{\partial^2 E}{\partial n^2}\right|_N = -\frac{3}{N}\omega_{cl} \equiv \beta \tag{164}$$

which gives

$$\chi(\Omega, t) = e^{-itE_N} e^{i\phi(N-1)} \sum_{s=-N+1}^{\infty} e^{-i\left(\omega_{cl}st - \left(\frac{\beta}{2}\right)s^2 t - s\phi\right)} \cdot A_s \left|Y_{N+s,N+s-1,N+s-1}\right|. \tag{165}$$

We choose a Gaussian form for A_s

$$A_s = \frac{1}{\sqrt{2\pi N}} e^{-s^2/(2N)}. \tag{166}$$

Brown used $A_s = Ce^{-s^2 3\omega_{cl}t/N}$ which minimizes the diffusion in ϕ at time t [139]. Since $\left|Y_{N+s,N+s-1,N+s-1}\right|$ varies slowly with s for $N \gg 1$ we can take it outside the summation in Equation (165). We now replace the sum by an integral over s. Because A_s is peaked about N

we can integrate from $s = -\infty$ to $s = +\infty$. We perform the integral by completing the square in the usual way. The final result for the probability amplitude for a circular orbital wave packet is

$$|x(\Omega, t)|^2 = |Y_{N,N-1,N-1}|^2 \left(1 + \beta^2 t^2 N^2\right)^{-1}$$
$$\cdot \exp\left[-(\phi - \omega_{cl}t)^2 \frac{N}{1 + (\beta t N)^2}\right].$$
(167)

This represents a Gaussian distribution in ϕ that is centered about the classical value $\phi = \omega_{cl}t$, meaning that the wavepacket is traveling in the classical trajectory with the classical time dependence. The width of the ϕ distribution is

$$\Delta\phi = (N)^{(-1/2)}(1 + \beta^2 t^2 N^2)^{1/2} = (N)^{(-1/2)}(1 + 9\omega_{cl}^2 t^2)^{1/2}.$$
(168)

The distribution in ϕ at $t = 0$ is very narrow, proportional to $1/\sqrt{N}$, but after several orbits $\Delta\phi$ is increasing linearly with time.

The distributions in θ_4 and θ are Gaussian and centered about $\pi/2$ in each case as for the circular wave function (cf. Equation (160)) with widths equal to $(N)^{-1/2}$. The spreading of these distributions in time is inhibited because of the conservation of angular momentum and energy. The detailed behavior of the widths depend on our use of the Gaussian distribution. Other distributions will give different widths, although the general behavior is expected to be similar.

As a numerical example, consider a hydrogen atom that is in the semiclassical region when the orbital diameter is about 1 cm. The corresponding principal quantum number is about 10^4, the mean velocity is about 2.2×10^4 cm/s and the period about 1.5×10^{-4} s. After about 34 revolutions or 5×10^{-3} s, the spread in ϕ is about 2π, meaning the electron is spread uniformly about the entire circular orbit. This characteristic spreading time can be compared to 1.6×10^{-3} s for a fully optimized wave packets formed from coherent SO(4,2) states [136,146]. In order to make predictions about significantly longer times, we would need to retain more terms in the expansion Equation (162) of E_n.

Case 2: Elliptical orbits $\sin v = e \neq 0$

We can obtain the classical limit of the wave function for elliptical orbits by first writing our asymptotic form Equation (159) for $Y_{n,n-1,n-1}$ in terms of the U variables instead of the angular variables by using definitions Equation (52), and setting $a = a_n$. Retaining only the lowest order terms in $(\theta_4 - \pi/2)$ and $(\theta - \pi/2)$, we find

$$Y_{n,n-1,n-1}(U) = \left(\frac{n}{2\pi^2}\right)^{\frac{1}{2}} e^{i(n-1)\tan^{-1}\left(\frac{U_2}{U_1}\right)} \cdot e^{-\frac{1}{2}n(U_4)^2} e^{-\frac{1}{2}n(U_3)^2}$$
(169)

For large n, this represents a circular orbit in the 1–2 plane. We now perform a rotation by A_2v. which will change the eccentricity to $\sin v$, and change the angular momentum, but will not change the energy or the orbital plane. Using Equation (140) to express the old coordinates in terms of the new coordinates, we find to lowest order

$$Y'_{n,n-1,n-1}(U) = \left(\frac{n}{2\pi^2}\right)^{1/2} e^{i(n-1)\tan^{-1}\left(\frac{U_2}{U_1\cos v}\right)} \cdot e^{-\frac{1}{2}n\{U_2 \sin v - U_4 \cos v\}^2} \cdot e^{-\frac{1}{2}n(U_3)^2}.$$
(170)

In Section 3.6, we found that the vanishing of the term in braces $0 = U_2 \sin v - U_4 \cos v$ specifies the classical great hypercircle orbit (Equation (56)) corresponding to an ellipse in configuration space with eccentricity $e = \sin v$ and lying in the 1–2 plane. The probability density $|Y'_{n,n-1,n-1}(U)|^2$ vanishes except within a hypertorus with a narrow cross section of radius approximately $\frac{1}{\sqrt{n}}$, which is centered about the classical distribution. Because the width $\frac{1}{\sqrt{n}}$ of the distribution is constant in U space, it will not be constant when projected onto p space.

In terms of the original momentum space variables, the asymptotic spherical harmonic is

$$Y_{n,n-1,n-1}(\boldsymbol{p}) = \left(\frac{n}{2\pi^2}\right)^{\frac{1}{2}} e^{i(n-1)\tan^{-1}\left(\frac{p_2}{p_1\cos v}\right)}$$

$$\cdot \exp\left\{-\left(\frac{n}{2}\right)\left[p_1^2 + (p_2 - a\tan v)^2 - \frac{a^2}{\cos^2 v}\right]^2 \left(\frac{\cos v}{p^2+a^2}\right)^2\right\} \tag{171}$$

$$\cdot \exp\left\{-\left(\frac{n}{2}\right)\left(\frac{2p_3 a}{p^2+a^2}\right)\right\}^2 \bigg|_{a\,=\,a_n}$$

The expression in brackets corresponds to the momentum space classical orbit equation we found previously (Equation (49)). As we expect, p_3 is Gaussian about zero since the classical orbit is in the 1–2 plane. We can simplify the expressions for the widths by observing that to lowest order we can use Equation (48), which implies $p^2 + a^2 = 2a^2 + 2ap_2\tan v$ in the exponentials. The widths of both distributions therefore increase linearly with p_2. We also note that, since classically there exists a one-to-one correspondence between each point of the trajectory in momentum space and each point in configuration space, we may interpret the widths of the distributions using Equation (32) $\frac{p^2+a^2}{a^2} = \frac{2r_c}{r}$. Accordingly, the widths increase as the momentum increases or as the distance to the force center decreases (Figure 8).

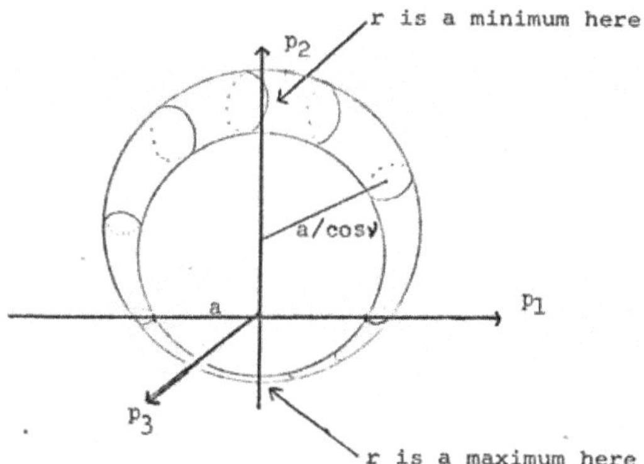

Figure 8. Wave function probability distribution $|Y'_{n,n-1,n-1}(\boldsymbol{p})|^2$ in momentum space for large n, showing the variation in the width of the momentum distribution about the classical circular orbit. The center of the distribution is at $p_2 = a\tan v$. The classical orbit is in the 1–2 plane.

Forming a Wave packet for Elliptical Motion

We may form a time dependent wavepacket superposing the wave functions of Equation (170). Care must be taken to include the first order dependence (through a_n) of $\tan^{-1}(U_2/U_1\cos v)$ on the principal quantum number when integrating over the Gaussian weight function. The result for the probability density is the same as before (Equation (167)), except $|Y'_{N,N-1,N-1}|^2$ (given in Equation (170)) replaces $|Y_{N,N-1,N-1}|^2$ and

$$\omega_{cl}t = \tan^{-1}\left(\frac{U_2}{U_1\cos v}\right) + \sin v\,(U_1) \tag{172}$$

where replaces $\omega_{cl}t = \phi$. The result is exactly the same as the classical time dependence Equation (73). The spreading of the wave packet will be controlled by the same factor as for the circular wave packet.

Remark on the Semiclassical Limit in Configuration Space

The time dependent quantum mechanical probability density follows the classical trajectory in momentum space meaning that the probability is greatest at the classical location of the particle in momentum space. Because the configuration space wave function is the Fourier transform of the momentum space wave function, the classical limit must also be obtained in configuration space. That this limit is obtained is made explicit by observing that the momentum space probability density is large when

$$(U_2 \sin v - U_4 \cos v)^2 \approx 0. \tag{173}$$

However from Section 3 we can show that

$$U_2 \sin v - U_4 \cos v = \cos v \left(\frac{r - r_{classical}}{r_{classical}} \right) \tag{174}$$

where $r_{classical}$ is given by the classical orbit Equation (33). Accordingly, we see that the configurations space probability will be large when

$$\left[\frac{r - r_{classical}}{r_{classical}} \right]^2 \approx 0. \tag{175}$$

5.6. Quantized Semiclassical Orbits

It is convenient at times to have a semiclassical model for the orbitals of the hydrogenlike atom. Historically this was first done by Pauling and Wilson [147]. We can obtain a model by interpreting the classical formulae for the geometrical properties of the orbits as corresponding to the expectation values of the appropriate quantum mechanical expressions. Thus, when the energy $E = -a^2/2m$ appears in a classical formula, we employ the expression for a for the quantized energy levels $a = \frac{1}{nr_0}$ where $r_0 = (mZ\alpha)^{-1}$, which is the radius 0.53 Angstrom of the ground state. Similarly if L^2 appears in a classical formula, we substitute $l(l+1)$, where l is quantized $l = 0, 1, 2..n-2, n-1$; and m, the component of L along the 3-axis is quantized: $m = -l, -l+1, -l+2, ..l$.

Orbits in Configuration Space

Recalling Equation (28) $r_c = \frac{mZ\alpha}{a^2}$, we see $ar_c = n$, which gives a semimajor axis of length $r_c = n^2 r_0$, where r_0 is the radius for the circular orbit of the ground state and r_c is for a circular orbit for a state with principal quantum number n. The equations for the magnitude of L and A are

$$L = r_c a \cos v = n \cos v = \sqrt{l(l+1)} \tag{176}$$

$$A = r_c a \sin v = n \sin v = \sqrt{n^2 - l(l+1)} \tag{177}$$

This gives an eccentricity $\sin v$ equal to

$$e = \sin v = \sqrt{1 - \frac{l(l+1)}{n^2}} \tag{178}$$

and a semi-major axis equal to

$$b = r_c \sin v = n\sqrt{l(l+1)}. \tag{179}$$

Note that the expression for e is limited in its meaning. For an s state, it always gives $e = 1$, and for states with $l = n - 1$ it give $e = \sqrt{1/n}$, not the classically expected 0 for a circular orbit.

Orbits in U-Space

The corresponding great hypercircle orbits (v, Θ) in U-space are described by giving the quantized angle v, between the three-dimensional hyperplane of the orbit and the 4-axis, and the quantized angle Θ, between the hyperplane of the orbit and the 3-axis:

$$\cos v = \sqrt{\frac{l(l+1)}{n^2}} \tag{180}$$

$$\cos \Theta = \sqrt{\frac{m^2}{l(l+1)}}. \tag{181}$$

Note the similarity in these two equations, suggesting that m relates to l the same way that l relates to n, which suggests a generalization of the usual vector model of the atom which only describes the precession of L about the z axis.

The results for orbits in configuration and momentum space illustrate some interesting features:

1. The equation $ar = n$ illustrates that the characteristic dimensions of an orbit in configuration space and the corresponding orbit in momentum space are inversely proportional, as expected, since they are related by a Fourier transform, consistent with the Heisenberg Uncertainty Principle.

2. If $l = 0$, then no classical state exists. The orbit in configuration space degenerates into a line passing through the origin while the corresponding circular orbit in momentum space attains an infinite radius and an infinite displacement from the origin. Although this seems peculiar from the pure classical viewpoint, quantum mechanically it follows, since for S states there is a nonvanishing probability of finding the electron within the nucleus.

In order to interpret these statements about quantized semiclassical elliptical orbits we observe that for quantum mechanical state of the hydrogenlike atom with definite n, l, m, the probability density is (1) independent of ϕ_r or ϕ_p and (2) it does not confine the electron to some orbital plane. Because the quantum mechanical distribution for such a state specifies no preferred direction in the 1–2 plane, we must imagine this distribution as corresponding in some way to an average over all possible orientations of the semiclassical elliptical orbit. This interpretation is supported by the fact that the region within which the quantum mechanical radial distribution function differs largely from zero is included between the values of r corresponding to the semiclassical turning points $r_c(1 \pm \sin v)$.

5.7. Four-Dimensional Vector Model of the Atom

In configuration space or momentum space,, the angle between the classical plane of the orbit and the 3-axis is Θ, which is usually interpreted in terms of the vector model of the atom in which we imagine L to be a vector of magnitude $\sqrt{l(l+1)}$ precessing about the 3-axis, with m as the component along the 3-axis. This precession may be linked to the ϕ_r independence of the probability and the absence of an orbital plane as mentioned at the end of the preceding section. The precession constitutes a classical mechanism which yields the desired average over all possible orientations of the semiclassical elliptical orbit. Because the angle Θ is restricted to have only certain discrete values one can say that there is a quantization of space.

The expression for $\cos v = \sqrt{l(l+1)/n^2}$ is quite analogous to that for Θ and so suggests a generalization of the vector model of the atom to four dimensions. The projection of the four-dimensional vector model onto the physical three-dimensional subspace must give the usual vector model. We can achieve this by imagining that a four-dimensional vector of length n, where n is the principal quantum number, is precessing in such a way that its three and four components are constants, while the one and two components vary periodically. The projection onto the 1–2–3 hyperplane is a vector of constant magnitude $\sqrt{l(l+1)}$ precessing about the 3-axis. The component along the 3-axis is m. The component along the 4-axis is $A = \sqrt{n^2 - l(l+1)}$ the magnitude of the vector A. The vectors L and A are perpendicular to each other. Thus, the precessing n vector makes a constant angle Θ with the 3-axis and a constant angle $\pi/2 - v$ with the 4-axis. Because both angles are restricted to certain values, we may say that we have a quantization of four-dimensional space.

6. The Spectrum Generating Group SO(4,1) for the Hydrogenlike Atom

We consider the Schrodinger hydrogen atom and its unitary "noninvariance" or spectrum generating operators $e^{iD_i\beta_i}$ where D_i is a generator and β_i is a real parameter, using eigenstates of $(Z\alpha)^{-1}$ for our basis of our representation. These operators transform an eigenstate of the kernel K with a definite value of the coupling constant (or principal quantum number) into a linear combination of eigenstates with different values of the coupling constant (or different principal quantum numbers), and different l and m. Unlike the invariance generators L and A, the noninvariance generators clearly do not generally commute with the kernel K, $[D_i, K] \neq 0$, so they change the principal quantum number.

The set of all invariance and noninvariance operators forms a group with which we may generate all eigenstates in our complete set from a given eigenstate. We show that this group, called the Spectrum Generating Group of the hydrogenlike atom, is SO(4,1), the group of orthogonal transformations in a 5-dimensional space with a metric $g_{AB} = (-1, 1, 1, 1, 1)$, where $A, B = 0, 1, 2, 3, 4$. The complete set of eigenstates of $(Z\alpha)^{-1}$ for the hydrogenlike atom forms a unitary, irreducible, infinite-dimensional representation of SO(4,1) which, we shall find, can be decomposed into an infinite sum of irreducible representations of SO(4), each corresponding to the degeneracy group for a particular principal quantum number. A unitary representation means all generators are unitary operators. An irreducible representation does not contain lower dimensional representations of the same group. In Section 6.3, we discuss the isomorphism between the spectrum generating group SO(4,1) and the group of conformal transformations in momentum space. An isomorphism means the groups have the same structure and can be mapped into each other.

6.1. Motivation for Introducing the Spectrum Generating Group Group SO(4,1)

We have examined the group structure for the degenerate eigenstates of $(Z\alpha)^{-1}$ for the Schrodinger hydrogenlike atom: the n^2 degenerate states form an irreducible representation of SO(4). The next question we might ask is: Do all or some of the states with different principal quantum numbers form an irreducible representation of some larger group which is reducible into SO(4) subgroups? If such a group exists then it clearly is not an invariance group of the kernel K (Equation (106)). If we want our noninvariance group to include just some of the states then it will be a compact group, since unitary representations of compact groups can be finite dimensional. If we want to include all states then it will be a noncompact group since there are an infinite number of eigenstates of $(Z\alpha)^{-1}$ and all unitary representations of noncompact groups are infinite dimensional [9].

We can find a compact noninvariance group for the first N levels of the coupling constant, $n = 1, 2, ...N$. The dimensionality of our representation is

$$\sum_{n=1}^{N} n^2 = \frac{N(N+1)(2N+1)}{6}. \tag{182}$$

Mathematical analysis of the group SO(5) shows that this is the dimensionality of the irreducible symmetrical tensor representation of SO(5), given by the tensor with five upper indices $T^{abc\cdots}$, where $a, b, .. = 1, 2, 3, 4$ or 5 [11]. Reducing this representation of SO(5) into its SO(4) components gives

$$(\text{symm} \cdot \text{tensor } N)_{SO(5)} = (0,0) \oplus \left(\frac{1}{2}, \frac{1}{2}\right) \oplus \ldots \oplus \left(\frac{N-1}{2}, \frac{N-1}{2}\right)$$

$$= (\text{symm} \cdot \text{tensor } n = 1)_{SO(4)} \oplus$$
$$(\text{symm} \cdot \text{tensor } n = 2)_{SO(4)} \oplus \tag{183}$$
$$\ldots \oplus \ (\text{symm} \cdot \text{tensor } n = N)_{SO(4)}$$

which is precisely the structure of the first N levels of a hydrogenlike atom. If we want to include all levels then we guess that the appropriate noncompact group is SO(4,1), whose maximal compact subgroup is SO(4). Thus, we conjecture that all states form a representation of SO(4,1).

Consider the Lie algebra of O(4,1) and the general structure of its generators in terms of the canonical variables. The algebra of O(n) has $\frac{n(n-1)}{2}$ generators so to extend the algebra of O(n) to O(n+1) takes n generators, which can be taken as the components of a n-vector. To extend the Lie algebra from O(4) to O(5) or O(4,1) we can choose the additional generators G_a to be components of a four-vector G under O(4):

$$[S_{ab}, G_c] = i(G_b\delta_{ac} - G_a\delta_{bc}) \quad a,b,c = 1,2,3,4. \tag{184}$$

If we apply Jacobi's identity to S_{ab}, G_a, and G_b and use Equation (184) we find

$$[S_{ab}, [G_a, G_b]] = 0. \tag{185}$$

We require that the Lie algebra closes, so $[G_a, G_b]$ must be a linear combination of the generators, clearly proportional to S_{ab} and we choose the normalization, such that

$$[G_a, G_b] = -iS_{ab}. \tag{186}$$

If we define

$$G_4 = S_{40} = S; \quad G_i = S_{i0} = B_i \tag{187}$$

and recall Equation (79)

$$L_i = e_{ijk}S_{jk} \quad A_i = S_{i4}$$

then the additional commutation relations that realize SO(4,1) may be written in terms of $L, A, B,$ and S:

$$\begin{aligned}
[L_i, B_j] &= i\epsilon_{ijk}B_k & [L_i, S] &= 0 \\
[S, A_j] &= iB_j & [S, B_j] &= iA_j \\
[A_j, B_k] &= i\delta_{jk}S & [B_i, B_j] &= -i\epsilon_{ijk}L_k.
\end{aligned} \tag{188}$$

The top two commutators show that **B** transforms as a three-vector under O(3) rotations and that S is a scalar under rotations. Alternatively, we can write the commutation relations in terms of the generators S_{AB}, $A, B = 0, 1, 2, 3, 4$:

$$[S_{AB}, S_{CD}] = i(g_{AC}S_{BD} + g_{BD}S_{AC} - g_{AD}S_{BC} - g_{BC}S_{AD}) \tag{189}$$

where $g_{00} = -1, g_{aa} = 1$.

The commutators above follow directly from the mathematical theory of SO(4,1), but the theory does not tell us what these generators represent, just their commutation properties. We now investigate the general features of the representations of SO(4,1) provided by the hydrogenlike atom and how to represent the generators in terms of the canonical variables.

6.2. Casimir Operators

The two Casimir operators of SO(4,1) are [120]

$$Q_2 = -\frac{1}{2}S_{AB}S^{AB} = S^2 + B^2 - A^2 - L^2 \tag{190}$$

and

$$Q_4 = -w_Aw^A = (SL - A x B)^2 - \frac{1}{4}[L \cdot (A + B) - (A + B) \cdot L]^2 \tag{191}$$

where $w_A = \frac{1}{8}\epsilon_{ABCDE}S^{BC}S^{DE}$.

For SO(4), we recall that for the SO(4) representations the structure of the generators in terms of the canonical variables led to the vanishing of one Casimir operator $C_1 = L \cdot A$ and, consequently, the appearance of only symmetrical tensor representations. We will find Q_4 vanishes for analogous reasons.

If B is a pseudovector, it is proportional to L, which is the only independent pseudovector that can be constructed from the dynamical variables. The coefficient of proportionality, a scalar, X need not commute with H:

$$B = XL \quad [X, L] = 0 \quad [X, H] \neq 0 \tag{192}$$

Because $[B_i, B_j] = -ie_{ijk} L_k$ it follows that $X^2 = -1$ and B would therefore be a constant multiple of L and not an independent generator. Thus B must be a vector and expressible as

$$B = fr + hp \tag{193}$$

where f and h are scalar functions of r, p^2, and $r \cdot p$. Accordingly we find

$$B \cdot L = L \cdot B = 0 \tag{194}$$

Further, since B is a vector and A is a vector, $A \times B$ is a pseudovector and therefore is proportional to L :

$$A \times B = YL \quad , \quad [Y, L] = 0 \tag{195}$$

For this equation to be consistent with the SO(4,1) commutation relations we find $Y = S$ and, therefore

$$A \times B = SL. \tag{196}$$

It follows from substituting $L \cdot A = 0$ and Equations (194) and (196) into Equation (191) that for the SO(4,1) representations realized by the hydrogenlike atom

$$Q_4 = 0. \tag{197}$$

As with the SO(4) symmetry, the dynamics of the hydrogen atom require that only certain representations of SO(4,1) appear. From the mathematical theory of irreducible infinite dimensional unitary representations of SO(4,1), we have the following results:

$$\text{Class I:} \quad Q_4 = 0; \quad Q_2 \text{ real, } > 0$$
$$SU(2) \times SU(2) \text{ content:}$$
$$(Q)^I = (0,0) \oplus \left(\frac{1}{2}, \frac{1}{2}\right) \oplus (1,1) \oplus \cdots \tag{198}$$

$$\text{Class II: } Q_4 = 0, Q_2 = -(s-1)(s+2), s = \text{ integer } > 0$$
$$SU(2) \times SU(2) \text{ content:}$$
$$(Q)^{II} = \left(\frac{s}{2}, \frac{s}{2}\right) \oplus \left(\frac{s+1}{2}, \frac{s+1}{2}\right) \oplus \cdots \tag{199}$$

The class I representations are realized by the complete set of eigenstates of $(Z\alpha)^{-1}$ for the hydrogenlike atom. Note, however, that we have an infinite number of such class I representations since Q_2 may have any positive real value. We shall find that for $Q_2 = 2$ we may extend our group from SO(4,1) to SO(4,2). The class II representations are realized by the eigenstates of $(Z\alpha)^{-1}$ with principal quantum numbers from $n = s + 1$ to n becomes infinite. The first s levels could, if we desire, be described by SO(5).

In this section, we have analyzed the group structure and the representations using the complete set of eigenstates of $(Z\alpha)^{-1}$ for our basis. We might ask: what if we used energy eigenstates instead as a basis for the representations? From Section 4.3 we know that the quantum numbers and multiplicities of the $(Z\alpha)^{-1}$ eigenstates are precisely the same as those of the bound energy eigenstates. Thus, with the energy eigenstates as our basis, we would reach the same conclusions about the group structure as before but we would be including only the bound states in our representations and we would be ignoring all scattering states.

6.3. Relationship of the Dynamical Group SO(4,1) to the Conformal Group in Momentum Space

We can give a more complete analysis of the hydrogenlike atom in terms of SO(4,1) by considering the relationship between the four-dimensional rotations of the four-vector U'_a, with $a = 1,2,3,4$, which we discussed in Section 4.6, and the group of conformal transformations in momentum space. Conformal transformations preserve the angles between directed curves, but not necessarily lengths. The rotations generated by the Runge-Lenz vector a and the angular momentum L leave the scalar product $U_a V^a$ of four-vectors invariant and, therefore, are conformal transformations. The stereographic projection we employed is also a conformal transformation. Since the product of two conformal transformations is itself a conformal transformation, we must conclude that a generates a conformal transformation of the momentum three-vector p.

We must introduce two additional operators that correspond to the operators B and S introduced in Section 6.1 in order to express the most general conformal transformation. By employing the isomorphism between the generators L, a, B, and S of SO(4,1) and the generators of conformal transformations in momentum space we can immediately obtain expressions for the additional generators B and S in terms of the canonical variables, which is our objective. We need these additional generators to complete our SO(4,1) group for the hydrogen atom.

To derive the isomorphism we use the most convenient representation, namely that based on eigenstates of $(Z\alpha)^{-1}$ convenient for momentum space calculations ($\rho = \frac{(p^2+a^2)}{2a^2}$). Once established, the isomorphism becomes a group theoretical statement and it is independent of the particular representation.

The Conformal Group in Momentum Space

An arbitrary infinitesimal conformal transformation in momentum three-space may be written as

$$\delta p_j = \delta a_j + \delta \omega_{jk} p_k + \delta \rho p_j + \left(p^2 \delta c_j - 2 p_j \boldsymbol{p} \cdot \delta \boldsymbol{c} \right) \tag{200}$$

where $\delta \omega_{jk} = -\delta \omega_{kj}$.

The terms in δp_j arise as follows:

$$
\begin{aligned}
&\delta a_j \text{ translation generated by } \boldsymbol{R} \cdot \delta \boldsymbol{a} \\
&\delta \omega_{jk} \text{ rotation generated by } \boldsymbol{J} \cdot \delta \boldsymbol{\omega}, J_{ij} = \epsilon_{ijk} J_k \\
&\delta \rho \text{ dilation generated by } D \delta \rho \\
&\delta c_j \text{ special conformal transformation generated by } \boldsymbol{K} \cdot \delta \boldsymbol{c}
\end{aligned}
\tag{201}
$$

This is a ten parameter group with the generators $(\boldsymbol{R}, \boldsymbol{J}, D, \boldsymbol{K})$ which obey the following commutation relations:

$$
\begin{aligned}
&\left[D, R_j \right] \cdot = i R_j && \left[D, J_i \right] = 0 \\
&\left[D, K_j \right] = -i K_j && \left[R_i, J_k \right] = i \epsilon_{ijk} R_k \\
&\left[K_n, R_m \right] = 2i \epsilon_{nmr} J_r - 2i \delta_{mn} D && \left[J_i, J_k \right] = i \epsilon_{ikm} J_m \\
&\left[R_i, R_j \right] = 0 && \left[K_i, J_k \right] = i \epsilon_{ikm} K_m \\
&\left[K_j, K_j \right] = 0
\end{aligned}
\tag{202}
$$

There is an isomorphism between the algebra of the generators of conformal transformations and the dynamical noninvariance algebra of SO(4,1) of the hydrogen atom. Because J_i is the generator of spatial rotations we make the association $L_i = J_i$. Comparing the differential change in p_i from a transformation generated by $\boldsymbol{a} \cdot \delta\boldsymbol{v}$ (in the representation with $\rho = \frac{(p^2+A^2)}{2a^2}$)

$$\delta p_i = i\,[\boldsymbol{a} \cdot \delta\boldsymbol{v}, p_i]$$
$$= -\frac{1}{2a}\left[\left(p^2 - a^2\right)\delta v_i - 2\boldsymbol{p} \cdot \delta\boldsymbol{v}\,p_i\right] \tag{203}$$

to the differential change in p_i from a conformal transformation leads to the association

$$a_i = \frac{1}{2}\left(\frac{K_i}{a} - aR_i\right). \tag{204}$$

To confirm the identification we can use the commutation relations of the conformal group to show that the O(4) algebra of $\boldsymbol{L}$ and $\boldsymbol{a}$ corresponds precisely to that of $\boldsymbol{J}$ and $\frac{1}{2}(\frac{\boldsymbol{K}}{a} - a\boldsymbol{R})$. This result alone suggests that our SO(4) degeneracy group should be considered as a subgroup of the larger group SO(4,1). It suggests introducing the operators

$$\boldsymbol{B} = \frac{1}{2}\left(\frac{\boldsymbol{K}}{a} + a\boldsymbol{R}\right) \qquad S = D. \tag{205}$$

The commutation relations of S and $\boldsymbol{B}$ which follow from Equations (205) and the commutation relations Equation (202) are identical to the commutation relations given for S and $\boldsymbol{B}$ in Section 6.1. Thus, by considering the $\boldsymbol{a}$ and $\boldsymbol{L}$ transformations in momentum space as conformal transformations, we were led to introduce the generators $\boldsymbol{B}$ and S and obtain the dynamical algebra SO(4,1). Further, we are led to the expressions for these generators in terms of the canonical variables.

By comparing the expression for $\boldsymbol{a}$ in terms of the conformal generators with our known expressions for $\boldsymbol{a}$, Equation (120) or Equation (121), we obtain expressions for K_i and R_i in terms of the canonical variables. If we use the eigenstates convenient for configuration space calculations ($\rho = n/ar$) we make the identifications

$$\boldsymbol{K} = \tfrac{1}{2}(rp^2 + p^2r) - \boldsymbol{r} \cdot \boldsymbol{p}\boldsymbol{p} - \boldsymbol{p}\boldsymbol{p} \cdot \boldsymbol{r} - r\tfrac{1}{4r^2}$$
$$\boldsymbol{R} = \boldsymbol{r}. \tag{206}$$

Substituting these results in the equation for $\boldsymbol{B}$ we find

$$\boldsymbol{B} = \frac{1}{2a}\left(\frac{p^2r + rp^2}{2} - \boldsymbol{r} \cdot \boldsymbol{p}\boldsymbol{p} - \boldsymbol{p}\boldsymbol{p} \cdot \boldsymbol{r} - \frac{r}{4r^2}\right) + \frac{ar}{2} \tag{207}$$

which is a manifestly Hermitean operator valid throughout Hilbert space. From Equations (204) and (205) we note that

$$\boldsymbol{B} - \boldsymbol{a} = a\boldsymbol{r}. \tag{208}$$

To compute D, we substitute the expressions for $\boldsymbol{K}$ and $\boldsymbol{R}$ into the commutation relation

$$D = \frac{i}{2}[K_i, R_i]$$

obtaining the result

$$S = \frac{1}{2}(\boldsymbol{p} \cdot \boldsymbol{r} + \boldsymbol{r} \cdot \boldsymbol{p}) = D \tag{209}$$

which is identical to the generator of the scale change transformation $D(\lambda)$ defined in Equation (109) in Section 4.3.

The significance of the generator $D = S$ of the scale change in terms of SO(4,1) is apparent if we compute

$$e^{i\lambda D} \left(\frac{K}{a} \pm a R \right) e^{-i\lambda D} = \frac{K}{a'} \pm a' R \tag{210}$$

where $a' = e^{\lambda} a$.

The unitary transformation $e^{i\lambda D}$ may be viewed as generating an inner automorphism of SO(4,1) which is an equivalent representation of SO(4) that is characterized by a different value of the quantity a or the energy. In other words, under the scale change $e^{i\lambda D}$, the basis states for our representation of SO(4,1), $|nlm; a\rangle$, transform to a new set, $|nlm; e^{\lambda} a\rangle$ in agreement with our discussion in Section 4.3.

Because the algebra of our generators closes, we may also view $e^{i\lambda D}$ as transforming a given generator into a linear combination of the generators. With the definitions of a and B (Equation (204) and (205)), we can easily show that Equation (210), with the upper sign, may also be written

$$e^{i\lambda D} B e^{-i\lambda D} = B \cosh \lambda + a \sinh \lambda. \tag{211}$$

7. The Group SO(4,2)

7.1. Motivation for Introducing SO(4,2)

We would like to express Schrodinger's equation as an algebraic equation in the generators of some group [37,38]. As we are unable to do this with our SO(4,1) generators S_{AB} we again expand the group. To guide us, we recall that to expand SO(3) to SO(4) we added a three-vector of generators A, and to expand SO(4) to SO(4,1), we added a four-vector of generators (S, B). In both cases, this type of expansion produced a set of generators convenient for the study of the hydrogenlike atom. We guess that the appropriate expansion of SO(4,1) is obtained by adding a five-vector (under SO(4,1)) of generators Γ_A to obtain SO(4,2) [37,38]. We can provide additional motivation for this choice by considering Schrodinger's equation. The generators in terms of which we want to express this equation must be scalars under L_i rotations. Additionally we know $S = S_{40}$ (Equation (187)) generates scale changes of Schrodinger's equation. The fact that S_{40} mixes the zero and four components of a five-vector suggests that Schrodinger's equation may be expressed in terms of the components Γ_0 and Γ_4 which are scalars under L_i, of the five vector Γ_A. Since Γ_A is a five-vector under SO(4,1), it must satisfy the equation

$$[S_{AB}, \Gamma_C] = i \left(\Gamma_B g_{AC} - \Gamma_A g_{BC} \right). \tag{212}$$

The spatial components of Γ_A which are $(\Gamma_1, \Gamma_2, \Gamma_3) = \Gamma$ transform as a vector under rotations generated by L.

To construct the Lie algebra of SO(4,2) we require that the set of operators $\{\Gamma_A, S_{AB}; A, B = 0, 1, 2, 3, 4\}$ must close under the operations of commutation. By applying Jacobi's identity to Γ_A, Γ_B, and S_{AB}, and requiring that Γ_A and Γ_B do not commute, we find

$$[S_{AB}, [\Gamma_A, \Gamma_B]] = 0 \quad A, B = 0, 1, 2, 3, 4.$$

Because we require that our Lie algebra closes, the commutator of Γ_A and Γ_B must be proportional to S_{AB}. We normalize Γ, so

$$[\Gamma_A, \Gamma_B] = -i S_{AB} \quad A, B = 0, 1, 2, 3, 4. \tag{213}$$

If we define

$$S_{A5} = \Gamma_A = -S_{5A} \quad A = 0, 1, 2, 3, 4. \tag{214}$$

and recall

$$A_i = S_{i4} \quad B_i = S_{io} \quad L_i = e_{ijk} S_{jk} \quad S = S_{40}$$

then we may unite all the commutations relations of Γ_A and S_{AB} in the single equation :

$$[S_{AB}, S_{CD}] = i(g_{AC}S_{BD} + g_{BD}S_{AC} - g_{AD}S_{BC} - g_{BC}S_{AD}) \tag{215}$$

where $\mathcal{A}, \mathcal{B}, .. = 0, 1, 2, 3, 4, 5$ and $g_{00} = g_{55} = -1; g_{aa} = 1, a = 1, 2, 3, 4$.

These are the commutation relation for the Lie algebra of SO(4,2). In terms of A, B, L, S and Γ_A the additional commutation relations for the noncommuting generators are [53]:

$$
\begin{aligned}
&[B_i, \Gamma_j] = i\Gamma_0 \delta_{ij} \quad && [\Gamma_i, \Gamma_j] = -i\epsilon_{ijk}L_k \\
&[A_i, \Gamma_j] = i\Gamma_4 \delta_{ij} \quad && [\Gamma_i, \Gamma_0] = -iB_i \\
&[L_i, \Gamma_j] = i\epsilon_{ijk}\Gamma_k \quad && [\Gamma_i, \Gamma_4] = -iA_i \\
& && [\Gamma_4, \Gamma_0] = -iS \\
&[\,B_i, \Gamma_0] = i\Gamma_i \quad && [A_i, \Gamma_4] = -i\Gamma_i \\
&[\,S, \Gamma_0\,] = i\Gamma_4 \quad && [\,S, \Gamma_4] = i\Gamma_0
\end{aligned}
\tag{216}
$$

7.2. Casimir Operators

The Lie algebra of SO(4,2) is rank three, so it has three Casimir operators W_2, W_3, and W_4 [43]:

$$W_2 = -\frac{1}{2}S_{AB}S^{AB} = Q_2 + \Gamma_A \Gamma^A \tag{217}$$

where Q_2 is the nonvanishing SO(4,1) Casimir operator Equation (190) and

$$W_3 = \epsilon^{ABCDEF}S_{AB}S_{CD}S_{EF} \tag{218}$$

$$W_4 = S_{AB}S^{BC}S_{CD}S^{DA}. \tag{219}$$

Computation of W_3

We can show that $W_3 = 0$ from dynamical considerations similar to those used in the discussion of SO(4,1) Casimir operators. The only terms that can be included in W_3 are scalars that are formed from products of three generators with different indices

$$
\begin{aligned}
&B \cdot A \times \Gamma, \quad A \cdot \Gamma \times B, \quad \Gamma \cdot B \times A \\
&\Gamma_4 L \cdot B, \quad \Gamma_0 L \cdot A, \quad S\Gamma \cdot L
\end{aligned}
\tag{220}
$$

It is interesting that these terms are actually all pseudoscalars. Terms like $B \cdot A \times L$ are simply not possible because of the structure of W_3. We know that $\Gamma = (\Gamma_1, \Gamma_2, \Gamma_3)$ must not be pseudovector, otherwise it would be proportional to L. Since it is a vector, it must equal a linear combination of r and p. Therefore, we conclude

$$\Gamma \cdot L = L \cdot \Gamma = 0. \tag{221}$$

Because Γ and B are both vectors and L is the only pseudovector we have $\Gamma \times B = \Lambda L$. In order to determine the scalar Λ we evaluate the commutators

$$[B_k, (\Gamma \times B)_k], \quad \text{and} \quad [\Gamma_k, (\Gamma \times B)_k] \tag{222}$$

and find

$$\Gamma \times B = \Gamma_0 L = -B \times L. \tag{223}$$

The analogous equations for A and Γ, and for A and B, are

$$
\begin{aligned}
\Gamma \times A &= -\Gamma_4 L = -A \times \Gamma \\
A \times B &= \quad SL \quad = -B \times A.
\end{aligned}
\tag{224}
$$

From Equations (223) and (224), we see that because of the dynamical structure of the generators each of the quantities in the first line of Equation (220) is proportional to the quantity directly below in the second line. We also have shown that (Equations (194) and (221))

$$L \cdot B = L \cdot A = \Gamma \cdot L = 0. \tag{225}$$

Accordingly, each scalar in our list vanishes and

$$W_3 = 0. \tag{226}$$

Computation of W_2

In order to compute W_2 we need to evaluate

$$\Gamma^2 \equiv \Gamma_A \Gamma^A = \Gamma_4^2 + \Gamma_i \Gamma^i - \Gamma_0^2. \tag{227}$$

From the structure of W_2 as shown in Equation (217), we see that Γ^2 must be a number since W_2 and Q_2 are both Casimir operators and therefore equal numbers for a particular representation. Accordingly, we have

$$[\Gamma^2, \Gamma_A] = 0. \tag{228}$$

From this equation, we can deduce a lemma allowing us to easily evaluate W_2 and W_4 in terms of the number Γ^2. Using Equation (227) and the definition of S_{AB} Equation (213) we find

$$\Gamma^A S_{AB} + S_{AB} \Gamma^A = 0.$$

Contracting Equation (212) with g_{AC} gives

$$S_{AB} \Gamma^A - \Gamma^A S_{AB} = 4i\Gamma_B.$$

Consequently, it must follow that

$$S_{AB} \Gamma^A = 2i\Gamma_B = -\Gamma^A S_{AB}. \tag{229}$$

We are now able to evaluate the quantity

$$S_{AB} S^B_{\ C} = iS_{AB}[\Gamma^B, \Gamma_C] = i(S_{AB}\Gamma^B \Gamma_C - S_{AB}\Gamma_C \Gamma^B).$$

Using Equation (212) for the commutator of S_{AB} with Γ_C and Equation (229) for the contraction $S_{AB}\Gamma^B$ we prove the lemma

$$S_{AB} S^B_{\ C} = 2iS_{CA} - \Gamma_A \Gamma_C + \Gamma^2 g_{AC}. \tag{230}$$

The value of the SO(4,1) Casimir operator $Q_2 = \frac{1}{2}g^{AC}S_{AB}S^B_{\ C}$ follows directly from the lemma:

$$Q_2 = 2\Gamma^2. \tag{231}$$

Accordingly, we have from Equation (217)

$$W_2 = 3\Gamma^2. \tag{232}$$

Computation of W_4

The Casimir operator W_4 can be written as

$$W_4 = S_{AB}S^{BC}S_{CD}S^{DA} + S_{AB}S^{B5}S_{5D}S^{DA} + S_{5B}S^{BC}S_{CD}S^{D5} + S_{5B}S^{B5}S_{5D}S^{D5}. \tag{233}$$

where $\mathcal{B}, \mathcal{D} = 0, 1, 2, 3, 4, 5$ and $A, C = 0, 1, 2, 3, 4$.

In order to evaluate W_4 in terms of Γ^2 we compute $S_{AB}S^{BC}$. Recalling $\Gamma_A = S_{A5}$ we see

$$S_{AB}S^{BC} = \Gamma_A \Gamma^C + S_{AB}S^{BC}. \tag{234}$$

Substituting the lemma Equation (230), we find

$$S_{AB}S^{BC} = 2iS_A^C - \Gamma^2 g_A^C. \tag{235}$$

From Equation (229), it follows that

$$S_{5B}S^{BC} = 2i\Gamma^C. \tag{236}$$

Substituting Equations (231), (232), (235), and (236) into Equation (233) for W_4 we find

$$W_4 = 6(\Gamma^2)^2 - 24\Gamma^2. \tag{237}$$

The fact that the nonvanishing Casimir operators (Q_2, W_2, and W_4) for SO(4,1) and SO(4,2) are given in terms of Γ^2 implies that the representation of SO(4,2) determines the particular representation of SO(4,1) appropriate to the hydrogenlike atom. In turn the value of Γ^2 is determined by the structure of the Γs in terms of the canonical variables. In Section 7.4, we derive these structures and find that

$$\Gamma^2 = 1.$$

Therefore, the quadratic SO(4,1) Casimir operator Q_2 has the value

$$Q_2 = 2$$

and the SO(4,2) Casimir operators have the values:

$$W_2 = 3 \quad W_3 = 0 \quad W_4 = -18.$$

The researchers that have published different representations of SO(4,2) based on the hydrogen atom that give their Casimir operators all have $W_2 = 3$ (or its equivalent) and $W_3 = 0$ [34,79,82], however, two authors have representations with $W_4 = 0$ [34,82] and one [79] has $W_4 = -12$, as compared to our value of -18.

From the mathematical theory of representations, it follows that our representations of SO(4,1) and SO(4,2) are both unitary and irreducible. This means there is no subset of basis vectors that transform among themselves as either SO(4,1) or as SO(4,2).

7.3. Some Group Theoretical Results

In this section, we derive the transformation properties of the generators of SO(4,2) and then a novel contraction formula that will prove useful for situations in which we want to employ perturbation theory, for example, in our calculation of the radiative shift for the hydrogen atom in Section 8. We will work primarily with the SO(4,2) generators expressed as the combination of the SO(4,1) generators S_{AB} and the five-vector Γ, with $g_{AB} = (-1,1,1,1,1)$ where $A, B = 0,1,2,3,4$.

Transformation Properties of the Generators

We can evaluate quantities like

$$^{AB}\Gamma_B(\theta) \equiv e^{iS_{AB}\theta}\Gamma_B e^{-iS_{AB}\theta} \quad \text{no sum over A or B} \tag{238}$$

by expanding the exponentials in an infinite series and then using the commutation relations Equations (212) and (213) of the generators S_{AB} and Γ_A, Γ_B repeatedly. However, it is easier to solve the differential equations satisfied by $^{AB}\Gamma_B$ and to use the appropriate boundary conditions. Differentiating Equation (238) and using the commutation relations, we obtain the equations

$$\frac{d}{d\theta}\,^{AB}\Gamma_B = -g_{BB}\,^{AB}\Gamma_A \qquad \frac{d^2}{d\theta^2}\,^{AB}\Gamma_B = -g_{AA}g_{BB}\,^{AB}\Gamma_B \tag{239}$$

which have the solution

$$^{AB}\Gamma_B = \Gamma_B \cos\sqrt{g_{AA}g_{BB}}\,\theta + \frac{g_{BB}}{\sqrt{g_{AA}g_{BB}}}\Gamma_A \sin\sqrt{g_{AA}g_{BB}}\,\theta. \tag{240}$$

Using a similar procedure we find

$$e^{i\Gamma_A\theta}S_{AB}e^{-i\Gamma_A\theta} = S_{AB}\cosh\sqrt{g_{AA}}\,\theta + \sqrt{g_{AA}}\Gamma_B \sinh\sqrt{g_{AA}}\,\theta \tag{241}$$

$$e^{i\Gamma_A\theta}\Gamma_B e^{-i\Gamma_A\theta} = \Gamma_B \cosh\sqrt{g_{AA}}\,\theta + \frac{1}{\sqrt{g_{AA}}}S_{AB}\sinh\sqrt{g_{AA}}\,\theta \tag{242}$$

where no summation over A or B is implied.

These formulae, Equation (240)–(242), give the SO(4,2) transformation properties of the SO(4,2) generators.

<u>The Contraction Formula</u>

If we multiply Equation (241) from the right by $e^{i\Gamma_A\theta}$ and then contract from the left with Γ_B, we obtain

$$\sum_B \Gamma^B e^{i\Gamma_A\theta}\Gamma_B = \left[(1 - g_{AA}\Gamma_A^2)\cosh\sqrt{g_{AA}}\,\theta + \frac{2i\Gamma_A}{\sqrt{g_{AA}}}\sinh\sqrt{g_{AA}}\,\theta\right]e^{i\Gamma_A\theta} + g_{AA}\Gamma_A^2 e^{i\Gamma_A\theta} \tag{243}$$

where we have used $\Gamma^2 = 1$ and Equation (229). Expanding the hyperbolic functions in terms of exponentials and collecting terms gives

$$\sum_B \Gamma^B e^{i\Gamma_A\theta}\Gamma_B = \frac{1}{2}\left(1 + \frac{i\Gamma_A}{\sqrt{g_{AA}}}\right)^2 e^{i(\Gamma_A - i\sqrt{g_{AA}})\theta} + \frac{1}{2}\left(1 - \frac{i\Gamma_A}{\sqrt{g_{AA}}}\right)^2 e^{i(\Gamma_A + i\sqrt{g_{AA}})\theta} + g_{AA}\Gamma_A^2 e^{i\Gamma_A\theta}. \tag{244}$$

A Fourier decomposition of a function Γ_A may be written

$$f(\Gamma_A) = \frac{1}{2\pi}\int d\theta\, h(\theta)e^{i\Gamma_A\theta}. \tag{245}$$

Consequently we have

$$\sum_B \Gamma^B f(\Gamma_A)\Gamma_B = \frac{1}{2}\left(1 + \frac{i\Gamma_A}{\sqrt{g_{AA}}}\right)^2 f(\Gamma_A - i\sqrt{g_{AA}}) + \frac{1}{2}\left(1 - \frac{i\Gamma_A}{\sqrt{g_{AA}}}\right)^2 f(\Gamma_A + i\sqrt{g_{AA}}) + g_{AA}\Gamma_A^2 f(\Gamma_A). \tag{246}$$

By performing a suitable rotation we can generalize this formula from functions of Γ_A to functions of $\Gamma_A n^A$ where $n_A n^A = \pm 1$. For $n^2 = -1$ we start with $\Gamma_A = \Gamma_0$ and rotate to obtain a very general result

$$\sum_B \Gamma_B f(n\Gamma)\Gamma^B = \frac{1}{2}(n\Gamma + 1)^2 f(n\Gamma + 1) + \frac{1}{2}(n\Gamma - 1)^2 f(n\Gamma - 1) - (n\Gamma)^2 f(n\Gamma). \tag{247}$$

We will have occasion to apply this formula for the special case

$$f(n\Gamma) = \frac{1}{\Gamma n - \nu}. \tag{248}$$

Using the representation

$$\frac{1}{\Gamma n - v} = \int_0^\infty ds\, e^{vs} e^{-\Gamma n s} \tag{249}$$

we obtain the result

$$\Gamma_A \frac{1}{\Gamma n - v} \Gamma^A = -2v \int_0^\infty ds\, e^{vs} \frac{d}{ds}\left(\sinh^2 \frac{s}{2}\, e^{-\Gamma n s}\right) \tag{250}$$

which is in a form convenient for perturbation calculations.

Derivation of the Γ_A in terms of the Canonical Variables

For our basis states, we shall use eigenstates of $(Z\alpha)^{-1}$ convenient for configuration space calculations ($\rho = na/r$). We choose these states rather than those convenient for momentum space calculations because they lead to simpler expressions for the Γ_A in terms of the canonical variables, although the expression for a is slightly more complicated. Thus, our states obey the equation

$$\left[\frac{1}{K_1(a)} - n\right] |nlm) = 0 \tag{251}$$

where $K_1(a)$ is given by Equation (116). We know that K_1^{-1} must commute with the generators of the SO(4) symmetry group $(a)_i = S_{i4}$ and $S_{ij} = \epsilon_{ijk}L_k$. This suggests that we choose

$$\Gamma_0 = [K_1(a)]^{-1} = \sqrt{ar}\frac{p^2 + a^2}{2a^2}\sqrt{ar} = \frac{1}{2}\left(\frac{\sqrt{r}p^2\sqrt{r}}{a} + ar\right) \tag{252}$$

so that

$$(\Gamma_0 - n)\,|nlm) = 0. \tag{253}$$

This last equation is the Schrodinger equation expressed in our language of SO(4,2): our states $|nlm)$ are eigenstates of Γ_0 with eigenvalue n.

To find Γ_4, we calculate $\Gamma_4 = -i[S, \Gamma_0]$, using Equation (209) for S,

$$\Gamma_4 = \sqrt{ar}\frac{p^2 - a^2}{2a^2}\sqrt{ar} = \frac{1}{2}\left(\frac{\sqrt{r}p^2\sqrt{r}}{a} - ar\right). \tag{254}$$

Sometimes it is convenient to use the linear combinations

$$\Gamma_0 - \Gamma_4 = ar \qquad \Gamma_0 + \Gamma_4 = \frac{\sqrt{r}p^2\sqrt{r}}{a} \tag{255}$$

which can be used to express the dipole transition operator [33]. We can find Γ_i from Equation (216), $\Gamma_i = -i[B_i, \Gamma_0]$

$$\Gamma_i = \sqrt{r}p_i\sqrt{r} \tag{256}$$

which we might have guessed initially since $[rp_i, rp_j] \sim L_k$. Every component of Γ_A is Hermitean, consequently the generators S_{AB} given by the commutators Equation (213) are also Hermitian. We may explicitly verify that these expressions for Γ_A lead to a consistent representation of all generators in the SO (4,2) Lie algebra.

Under a scale change generated by S, Γ_i is invariant and Γ_4 and Γ_0 transform in the same manner as a and B (Equation (210)): they retain their form but a is transformed into $e^\lambda a$:

$$e^{i\lambda S}\left\{\begin{array}{c}\Gamma_0 \\ \Gamma_4\end{array}\right\} e^{-i\lambda S} = \frac{1}{2}\left(\frac{\sqrt{r}p^2\sqrt{r}}{e^\lambda a} \pm e^\lambda ar\right) \tag{257}$$

The scale change generates an inner automorphism of SO(4,2) characterized by a different value of the parameter a.

7.4. Subgroups of SO(4,2)

The two most significant subgroups are [53]:

1. L_i, a_i or S_{jk}, S_{i4}, forming an SO(4) subgroup. These generators commute with Γ_0 and therefore constitute the degeneracy group for states of energy $-a^2/(2m)$ and fixed principal quantum number n (or fixed coupling constant na/m). The Casimir operator for this subgroup is

$$a^2 + L^2 = n^2 - 1 = \Gamma_0^2 - 1. \tag{258}$$

We discussed this subgroup in Section 4.2 in terms of L and A and the states $|nlm\rangle$. The same results are obtained with the generators L and a with the states $|nlm\rangle$. For example, we have the raising and lowering operators for m and l (Equations (93) and (94)). With the definition

$$L_\pm = L_1 \pm iL_2 \tag{259}$$

it follows that

$$[L_3, L_\pm] = \pm L_\pm \tag{260}$$

which gives

$$L_\pm|nlm\rangle = \sqrt{(l(l+1) - m(m\pm1)}|nl\,m\pm1\rangle \qquad L_3|nlm\rangle = m|nlm\rangle. \tag{261}$$

for $l \geq 1$. In analogy to $L_\pm$ one can define

$$a_\pm = a_1 \pm ia_2 \tag{262}$$

which obey the relations

$$[a_3, a_\pm] = \pm L_3 \qquad [L_3, a_\pm] = \pm a_\pm \tag{263}$$

and

$$\begin{aligned}
a_\pm|nlm\rangle = \mp &\left(\frac{(n^2 - (l+1)^2)(l+2\pm m)(l+1\pm m)}{4(l+1)^2 - 1}\right)^{\frac{1}{2}} |n\,l+1\,m\pm1\rangle \\
\pm &\left(\frac{(n^2 - l^2)(l\mp m)(l-1\mp m)}{4l^2 - 1}\right)^{\frac{1}{2}} |n\,l-1\,m\pm1\rangle
\end{aligned} \tag{264}$$

for $l \geq 1$. The action of $a_\pm$ is not directly analogous to that of $L_\pm$, because we are using $|nlm\rangle$ as basis states. If we used $|na_3l_3 = m\rangle$ as basis states, the action would be similar. An operator that only changes the angular momentum is a_3

$$a_3|nlm\rangle = \left(\frac{(n^2 - (l+1)^2)((l+1)^2 - m^2)}{4(l+1)^2 - 1}\right)^{\frac{1}{2}} |n\,l+1\,m\rangle + \left(\frac{(n^2 - l^2)(l^2 - m^2)}{4l^2 - 1}\right)^{\frac{1}{2}} |n\,l-1\,m\rangle. \tag{265}$$

for $l \geq 1$. Since a_3 commutes with L_3 and Γ_0, it does not change n or m.

2. $\Gamma_4, S = S_{40}, \Gamma_0$, forming a SO(2,1) subgroup. These operators commute with L but not with Γ_0, hence then can change n but not L or m. The Casimir operator for this subgroup is

$$\Gamma_0^2 - \Gamma_4^2 - S^2 = L^2 = l(l+1). \tag{266}$$

We can define the operators [53]

$$j_1 = \Gamma_4 \qquad j_2 = S \qquad j_3 = \Gamma_0 \tag{267}$$

with commutators

$$[j_1, j_2] = -ij_3 \qquad [j_2, j_3] = ij_1 \qquad [j_3, j_1] = ij_2 \tag{268}$$

We can define the raising and lowering operators

$$j_\pm = j_1 \pm ij_2 = \Gamma_4 \pm iS \tag{269}$$

which obey the commutation relations

$$[j_\pm, j_3] = \mp j_\pm. \tag{270}$$

We find (in analogy to Equation (261))

$$\Gamma_0|nlm) = n|nlm) \quad (\Gamma_4 \pm iS)|nlm) = \sqrt{n(n \pm 1) - l(l+1)}|n \pm 1 \; lm) \tag{271}$$

We can express the action of $\Gamma_0 - \Gamma_4 = ar$ on our states

$$ar|nlm) = \tfrac{1}{2}\left((n)(n-l) - l(l+1)\right)^{\frac{1}{2}}|n-1 \; lm) \; + n|n \; lm) \; + \tfrac{1}{2}\left((n)(n+l) - l(l+1)\right)^{\frac{1}{2}}|n+1 \; lm) \tag{272}$$

As previously mentioned, the operator S generates scale changes as shown in Equation (257), where the value of a is changed. We can also express the action of S equivalently as transforming Γ_0 into Γ_4

$$e^{iS\lambda}\Gamma_0 e^{-iS\lambda} = \Gamma_0 \cosh \lambda - \Gamma_4 \sinh \lambda \quad e^{iS\lambda}\Gamma_4 e^{-iS\lambda} = \Gamma_4 \cosh \lambda - \Gamma_0 \sinh \lambda. \tag{273}$$

7.5. Time Dependence of SO(4,2) Generators

For a generator to be a constant it must commute with the Hamiltonian as discussed in Section 2.1. Because the SO(4,2) group is the non-invariance or spectrum generating group, the additional generators do not all commute with the Hamiltonian and may have a harmonic time dependence as discussed in Section 2.2. It is notable that as far as we know only one paper considers the time dependence of the generators of non-invariance groups in general and one considers SO(4,2) specifically [97,136]. Our results certainly clarify and make explicit the time dependence, and show that it is just a particular aspect of the SO(4,2) transformations. In our representation with basis states $|nlm; a)$, the Hamiltonian, which is the generator of translations in time, has been transformed into Γ_0 and the Schrodinger energy eigenvalue equation has become $\Gamma_0|nlm) = n|nlm)$. Accordingly, all of the generators that commute with Γ_0 are constants of the motion, which includes a, L. The other operators, B, Γ, S, Γ_4 have a time dependence given by Equations (241) and (242), for example

$$S(t) = e^{iHt}S(0)E^{-iHt} = e^{i\Gamma_0 t}Se^{-i\Gamma_0 t} = S \cos t + \Gamma_4 \sin t. \tag{274}$$

$$\Gamma_4(t) = e^{iHt}\Gamma_4(0)E^{-iHt} = e^{i\Gamma_0 t}\Gamma_4 e^{-i\Gamma_0 t} = \Gamma_4 \cos t - S \sin t. \tag{275}$$

Consequently, terms like $j_\pm$ have a simple exponential time dependence

$$j_\pm(t) = j_\pm(0)e^{\pm it}. \tag{276}$$

Similarly $\Gamma \pm iB$ has an exponential time dependence.

7.6. Expressing the Schrodinger Equation in Terms of the Generators of SO(4,2)

We can write the Schrodinger equation for the energy eigenstate $E_n = -a_n^2/2m$ of a particle in a Coulomb potential in terms of the SO(4,2) generators, which are expressed in terms of the energy $-a^2/2m$, by making a scale change. From Section 4.3, Equation (114), the relationship between the Schrodinger energy eigenstate $|nlm\rangle$ and the eigenstate of $(Z\alpha)^{-1}$ is:

$$|nlm; a) = e^{-iS\lambda_n}\sqrt{\rho(a_n)}|nlm\rangle \tag{277}$$

where

$$e^{\lambda_n} = \frac{a_n}{a} \qquad \rho(a_n) = \frac{n}{a_n r}. \tag{278}$$

Substituting Equation (277) into the eigenvalue equation Equation (253) for $|nlm; a\rangle$ and employing the transformation Equation (273), we find the usual Schrodinger equation can be expressed in SO(4,2) terms as

$$(\Gamma n - n)\sqrt{\rho(a_n)}|nlm\rangle = 0 \tag{279}$$

where

$$\Gamma n \equiv \Gamma_A n^A = \Gamma_0 n^0 + \Gamma_i n^i + \Gamma_4 n^4 \tag{280}$$

$$n^0 = \cosh \lambda_n = \frac{a^2 + a_n^2}{2 a a_n}, \quad n^i = 0, \quad n^4 = -\sinh \lambda_n = \frac{a^2 - a_n^2}{2 a a_n} \tag{281}$$

and $n_A n^A = n_4^2 - n_0^2 = -1$.

Equation (279) expresses Schrodinger's equation for an ordinary energy eigenstate $|nlm\rangle$ with energy $E_N = -a_n^2/2m$ in the language of SO(4,2). It shows the relationship between these energy eigenstates and the basis states of $(Z\alpha)^{-1}$ used for the SO(4,2) representation,

8. SO(4,2) Calculation of the Radiative Shift for the Schrodinger Hydrogen Atom

In the 1930's, it was generally believed that the Dirac equation predicted the energy levels of the hydrogen atom with excellent accuracy, but there were some questions about the prediction that the energy levels for a given principal quantum number and given total angular momentum were independent of the orbital angular momentum. To finally resolve this issue, in 1947, Willis Lamb and his student Robert Retherford at Columbia University in New York City employed rf spectroscopy and exploited the metastability of the hydrogen $2s_{1/2}$ level in a beautiful experiment and determined that the $2s_{1/2}$ and $2p_{1/2}$ levels were not degenerate and that the energy difference between the levels was about 1050 MHz, or 1 part in 10^6 of the $2s_{1/2}$ level [5,148]. Shortly thereafter Hans Bethe [6] published a ground breaking nonrelativistic quantum theoretical calculation of the shift assuming it was due to the interaction of the electron with the ground state electromagnetic field of the quantum vacuum field. This radiative shift accounted for about 96% of the measured shift. The insight that one needed to include the interaction of the atom with the vacuum fluctuations and how one could actually do it ushered in the modern world of quantum electrodynamics [7]. Here, we compute in the non-relativistic dipole approximation and to first order in the radiation field, as did Bethe, the radiative shift, but we use group theoretical methods based on the SO(4,2) symmetry of the non-relativistic hydrogen atom as developed in this paper. Bethe's calculation required the numerical sum over intermediate states to obtain the average value of the energy of the states contributing to the shift. In our calculation, we do not use intermediate states, and we derive an integral equivalent to Bethe's log, and more generally derive the shift for all levels in terms of a double integral.

An expression for the radiative shift Δ_{NL} for energy level E_N of a hydrogen atom in a state $|NL\rangle$ can be readily obtained using second order perturbation theory (to first order in α the radiation field) [6,149–151]

$$\Delta_{NL} = \frac{2\alpha c}{3\pi m^2} \sum_n^s \int_0^{\omega_c} d\omega \frac{(E_n - E_N)\langle NL|p_i|n\rangle\langle n|p_i|NL\rangle}{E_n - E_N + \omega - i\epsilon}, \tag{282}$$

where ω_c is a cutoff frequency for the integration that we will take as $\omega_c = m$.

This expression, which is the same as Bethe's, has been derived by inserting a complete set of states $|n\rangle\langle n|$, a step that we eliminate with our group theoretical approach:

$$\Delta_{NL} = \frac{2\alpha}{3\pi m^2} \int_0^{\omega_c} d\omega \langle NL|p_i \frac{H - E}{H - (E - \omega) - i\epsilon} p_i|NL\rangle \tag{283}$$

If we add and subtract ω from the numerator, we find the real part of the shift is

$$Re\Delta_{NL} = \frac{2\alpha}{3\pi m^2} Re \int_0^{\omega_c} d\omega [\langle NL|p^2|NL\rangle - \omega\Omega_{NL}] \tag{284}$$

where

$$\Omega_{NL} = \langle NL|p_i \frac{1}{H - E_N + \omega - i\epsilon} p_i|NL\rangle \tag{285}$$

and

$$H = \frac{p^2}{2m} - \frac{Z\alpha}{r}. \tag{286}$$

The imaginary part of the shift gives the width of the level [7].

The matrix element Ω_{NL} can be converted to a matrix element of a function of the generators Γ_A taken between eigenstates $|nlm\rangle$ of $(Z\alpha)^{-1}$. To do this we insert factors of $1 = \sqrt{r}\frac{1}{\sqrt{r}}$ and use the definitions of the Γ_A in terms of the canonical variables, Equations (254)–(256). Letting the parameter a take the value a_N, we obtain the result

$$\Omega_{NL} = \frac{m\nu}{N^2}(NL|\Gamma_i \frac{1}{\Gamma n(\xi) - \nu}\Gamma_i|NL) \tag{287}$$

where

$$n^0(\xi) = \frac{2+\xi}{2\sqrt{1+\xi}} = \cosh\phi \qquad n^i = 0 \qquad n^4(\xi) = -\frac{\xi}{2\sqrt{1+\xi}} = -\sinh\phi \tag{288}$$

and

$$\xi = \frac{\omega}{|E_N|} \qquad \nu = \frac{N}{\sqrt{1+\xi}} = Ne^{-\phi}. \tag{289}$$

From the definitions we see $\phi = \frac{1}{2}ln(1 + \xi) > 0$ and $n_A(\xi)n^A(\xi) = -1$. The quantity

$$\nu = \frac{mZ\alpha}{\sqrt{-2m(E_N - \omega)}}$$

may be considered the effective principal quantum number for a state of energy $E_N - \omega$. The contraction over i in Ω_{NL} may be evaluated using the group theoretical formula Equation (250):

$$\begin{aligned}
\Omega_{NL} = &-2\frac{m\nu^2}{N^2}\int_0^\infty ds e^{\nu s}\frac{d}{ds}\left(\sinh^2 \frac{s}{2}M_{NL}(s)\right) \\
&- m\frac{\nu}{N^2}(NL|\Gamma_4 \frac{1}{\Gamma n(\xi) - \nu}\Gamma_4|NL) + m\frac{\nu}{N^2}(NL|\Gamma_0 \frac{1}{\Gamma n(\xi) - \nu}\Gamma_0|NL)
\end{aligned} \tag{290}$$

where

$$M_{NL}(s) = (NL|e^{-\Gamma n(\xi)s}|NL). \tag{291}$$

In order to evaluate the last-two terms in Ω_{NL}, we can express the action of Γ_4 on our states in terms of $\Gamma n(\xi) - \nu$. Substituting the equation

$$\Gamma_0|NL) = N|NL) \tag{292}$$

into the expression for $\Gamma n(\xi) - \nu$, with $n(\xi)$ given by Equation (288), gives

$$\Gamma_4 = N - \left(\frac{1}{\sinh\phi}\right)(\Gamma n(\xi) - \nu) \tag{293}$$

when acting on the state $|NL\rangle$. If we substitute Equation (293) into the expression for the $Re\Delta E_{NL}$ Equation (284) and simplify using the virial theorem

$$(NL|p^2|NL) = a_N^2$$

we find that the term in p^2 exactly cancels the last two terms in Ω_{NL}, yielding the result

$$Re\Delta E_{NL} = \frac{4m\alpha(Z\alpha)^4}{3\pi N^4} \int_0^{\phi_c} d\phi \sinh \phi e^\phi \int_0^\infty ds\, e^{vs} \frac{d}{ds} \left(\sinh^2 \frac{s}{2} M_{NL}(s) \right) \tag{294}$$

where

$$\phi_c = \frac{1}{2}ln\left(1 + \frac{\omega_c}{|E_N|}\right) = \frac{1}{2}ln\left(1 + \frac{2N^2}{(Z\alpha)^2}\right) \tag{295}$$

and $\omega_c = m$.

Comparison to the Bethe Logarithm

The first order non-relativistic radiative shift is commonly given in terms of the Bethe logarithm $\gamma(N, L)$, which is interpreted as the average over all states, including scattering states, of $ln\frac{|E_n - E_N|}{\frac{1}{2}m(Z\alpha)^2}$ [149] :

$$\begin{aligned}
&\gamma(N, L) \sum_n^S (E_n - E_N) \langle N0 |p_i| n\rangle \langle n |p_i| N0\rangle \\
&= \sum_n^S (E_n - E_N) \langle NL |p_i| n\rangle \langle n |p_i| NL\rangle \ln \frac{|E_n - E_N|}{\frac{1}{2}m(Z\alpha)^2}
\end{aligned} \cdot \tag{296}$$

We use the dipole sum rule [150]

$$2\sum_n^s (E_n - E_N) \langle N |p_i| n\rangle \langle n |p_i| N\rangle = -\left\langle N \left| \nabla^2 V \right| N \right\rangle \tag{297}$$

and apply it for the Coulomb potential $\nabla^2 V(r) = 4\pi Z\alpha\delta(r)$. The use of the Bethe log allowed Bethe to take the logarithmic expression obtained from the frequency integration outside the summation over the states, and replace it with the average value. Only the S states contribute to the expectation value in Equation (297), giving, from Equation (282), an expression for the shift

$$Re\,\Delta E_{NL} = \left[\frac{4m}{3\pi}\alpha(Z\alpha)^4\right] \frac{1}{N^3} \left\{ \delta_{L0} \ln \frac{2}{(Z\alpha)^2} - \gamma(N, L) \right\}. \tag{298}$$

Comparing the shift in terms of M_{NL} Equation (294) to the shift in terms of $\gamma(N, L)$ we find that the Bethe log is

$$\gamma(N, L) = \int_0^{\phi_c} d\phi \sinh \phi\, e^\phi \int_0^\infty ds\, e^{vs} \frac{d}{ds} \left(\sinh^2 \frac{s}{2} M_{NL}(s) \right) - \delta_{L0}ln\frac{1}{(Z\alpha)^2} \tag{299}$$

8.1. Generating Function for the Shifts

We can derive a generating function for the shifts for any eigenstate characterized by N and L if we multiply Equation (291) by $N^4 e^{\beta N}$ and sum over all $N, N \geq L + 1$. To simplify the right side of the resulting equation, we use the fact that the O(2,1) algebra of Γ_0, Γ_4, and S closes. We can compute the sum on the right hand side:

$$\sum_{N=L+1}^\infty e^{-\beta N} M_{NL} = \sum_{N=L+1}^\infty (NL|e^{-j\cdot\psi}|NL). \tag{300}$$

where

$$e^{-j\cdot\psi} \equiv e^{-\beta\Gamma_0}e^{-s\Gamma n(\xi)}. \tag{301}$$

We perform a j transformation, such that

$$e^{-j\cdot\psi} \rightarrow e^{-j_3\psi} = e^{-\Gamma_0\psi} \tag{302}$$

so

$$\sum_{N=L+1}^{\infty} e^{-\beta N} M_{NL} = \sum_{N=L+1}^{\infty} (NL|e^{-j_3\psi}|NL) = \sum_{N=L+1}^{\infty} e^{-N\psi} \tag{303}$$

$$= \frac{e^{-\psi(L+1)}}{1-e^{-\psi}}. \tag{304}$$

In order to find a particular M_{NL}, we must expand the right hand side of the equation in powers of $e^{-\beta}$ and equate the coefficients to those on the left hand side. First, we need an equation for $e^{-\psi}$. This can be obtained using the isomorphism between j and the Pauli σ matrices:

$$(\Gamma_4, S, \Gamma_0) \rightarrow (j_1, j_2, j_3) \rightarrow (\tfrac{i}{2}\sigma_1, \tfrac{i}{2}\sigma_2, \tfrac{1}{2}\sigma_3) \tag{305}$$

Using the formula

$$e^{\frac{i}{2}sn\cdot\sigma} = \cos\frac{s}{2} + in\cdot\sigma\sin\frac{s}{2} \tag{306}$$

where $|n| = 1$, we find

$$\cosh\frac{\psi}{2} = \cosh\frac{\beta}{2}\cosh\frac{s}{2} + \sinh\frac{\beta}{2}\sinh\frac{s}{2}\cosh\phi. \tag{307}$$

We can rewrite this equation in a form easier for expansion

$$e^{+\frac{1}{2}\psi} = de^{\frac{1}{2}\beta} + be^{-\frac{1}{2}\beta} - e^{-\frac{1}{2}\psi} \tag{308}$$

where

$$\begin{aligned} d &= \cosh\tfrac{s}{2} + \sinh\tfrac{s}{2}\cosh\phi \\ b &= \cosh\tfrac{s}{2} - \sinh\tfrac{s}{2}\cosh\phi \end{aligned} \tag{309}$$

Let β become very large and iterate the equation for $e^{-\frac{1}{2}\psi}$ to obtain the result

$$e^{-\psi} = Ae^{-\beta}\left[1 + A_1 e^{-\beta} + A_2 e^{-2\beta} + \ldots\right] \tag{310}$$

where

$$A = A_0 = \frac{1}{d^2}$$

$$A_1 = -\left(\frac{2}{d}\right)\left(b - d^{-1}\right)$$

$$A_2 = 3d^{-2}\left(b - d^{-1}\right)^2 - 2^{-2}\left(b - d^{-1}\right)$$

$$\vdots$$

$$\tag{311}$$

Note $b - d^{-1} = -d^{-1}\sinh^2\tfrac{s}{2}\sinh^2\phi$.

8.2. The Shift between Degenerate Levels

Expressions for the energy shift between degenerate levels with the same value of N may be obtained directly from the generating function using Equations (294) and (304). We find

$$\sum_{N=L+1}^{\infty} e^{-\beta N} N^4 \, \mathrm{Re}\, \Delta E_{NL} - \sum_{N=L'+1}^{\infty} e^{-\beta N} N^4 \, \mathrm{Re}\, \Delta E_{NL'} =$$

$$\frac{4m\alpha(Z\alpha)^4}{3\pi} \int_0^{\phi_c} d\phi e^\phi \sinh\phi \int_0^\infty ds e^{vs} \frac{d}{ds}\left(\sinh^2\frac{s}{2}\frac{e^{-\psi(L+1)} - e^{-\psi(L'+1)}}{1 - e^{-\psi}}\right). \tag{312}$$

For an example, consider $L = 1, L' = 0$. For the shifts between levels we obtain

$$\sum_{N=2}^\infty e^{-\beta N} N^4 \, \text{Re}\left(\Delta E_{NO} - \Delta E_{N1}\right) + \text{Re} \, \Delta E_{10} \, e^{-\beta} =$$

$$\frac{4m\alpha(z\alpha)^4}{3\pi} \int_0^{\phi_c} d\phi \, e^\phi \sinh\phi \int_0^\infty ds e^{vs} \frac{d}{ds}\left(\sinh^2\frac{s}{2}e^{-\psi}\right) \tag{313}$$

Substituting Equation (310) for $e^{-\psi}$, using the coefficient $A A_{N-1}$ of $e^{-N\beta}$, gives

$$Re(\Delta E_{N0} - \Delta E_{N1}) = \frac{4m\alpha(Z\alpha)^4}{3\pi N^4} \int_0^{\phi_c} d\phi e^\phi \sinh\phi \int_0^\infty ds e^{vs} \frac{d}{ds}\left(\sinh^2\frac{s}{2} A A_{N-1}\right). \tag{314}$$

where A and A_{N-1} are given in Equation (311) in terms of the variables of integration s and ϕ.

General Expression for M_{NL}

Once we have a general expression for M_{NL}, we can use Equation (294) to calculate the shift for any level E_{NL}. We can obtain expressions for the values of M_{NL} by letting β become large, expanding the denominator in Equation (304) and equating coefficients of powers of $e^{-\beta}$. For large β, we have large ψ. We have

$$\frac{e^{-\psi(L+1)}}{1 - e^{-\psi}} = \sum_{m=1}^\infty e^{-\psi(m+L)}$$

and for large β it follows from Equation (310) that

$$\sum_{N=L+1}^\infty e^{-\beta N} M_{NL} = \sum_{m=1}^\infty \left[e^{-\beta} A \left(1 + A_1 e^{-\beta} + \dots\right)\right]^{m+L}. \tag{315}$$

Using the multinomial theorem [124], the right side of the equation becomes

$$\sum_{m=1}^\infty A^{m+L} \sum_{r,s,t,\dots} \frac{(m+L)!}{r!s!t!\dots} A_1{}^s A_2^t \dots e^{-\beta(m+L+s+2t+\dots)}. \tag{316}$$

where $r + s + t + \dots = m + L$.

To obtain the expression for M_{NL}, we note N is the coefficient of β so $N = m + L + s + 2t + \dots = r + 2s + 3t + \dots$ Accordingly we find

$$M_{NL} = \sum_{r,s,t,\dots} A^{(r+s+t+\dots)} \frac{(r+s+t+\dots)!}{r!s!t!\dots} A_1^s A_2^t \dots \tag{317}$$

where $r + 2s + 3t + \dots = N$ and $r + s + t + \dots > L$.

By applying this formula, we obtain the results:

$N = 1$:

$$M_{10} = A \tag{318}$$

$N = 2$:

$$M_{20} = A^2 + A A_1$$
$$M_{21} = A^2 \tag{319}$$

Shifts for N = 1 and N = 2

To illustrate these results, we can calculate the shift for a given energy level using Equation (294). For $N = 1$, we note from Equation (318) that $M_{10} = A$, and from Equation (311) that $A = 1/d^2$. We find that the real part of the radiative shift for the $1S$ ground state is

$$Re\Delta E_{10} = \frac{4m\alpha(Z\alpha)^4}{3\pi} \int_0^{\phi_c} d\phi e^{\phi} \sinh\phi \int_0^{\infty} ds e^{se^{-\phi}} \frac{d}{ds} \frac{1}{\left(\coth\frac{s}{2} + \cosh\phi\right)^2} \tag{320}$$

For the shift between two states Equation (314) can be used. For the N = 2 Lamb shift between 2S-2P states, the radiative shift to first order in α is

$$Re(\Delta E_{20} - \Delta E_{21}) = \frac{m\alpha(Z\alpha)^4}{6\pi} \int_0^{\phi_c} d\phi e^{\phi} \sinh^3\phi \int_0^{\infty} ds e^{2se^{-\phi}} \frac{d}{ds} \frac{1}{\left(\coth\frac{s}{2} + \cosh\phi\right)^4} \tag{321}$$

The s integral can be computed in terms of a Jacobi function of the second kind [127].

As one check on our group theoretical methods, we can compare our matrix elements $(10|e^{iS\phi}|n0)$ with those of Huff [45]. To go from Equation (301) to Equation (302), we did a rotation $R(\phi) = e^{i\phi S}$ generated by S that transformed Γn into Γ_0. For $N, L = 1, 0$ we have

$$M_{10} = (10|e^{-\Gamma ns}|10) = (10|R(\phi)e^{-\Gamma_0 s}R^{-1}(\phi)|10) = \frac{1}{(\cosh\frac{s}{2} + \sinh\frac{s}{2}\cosh\phi)^2} \tag{322}$$

Expanding the hyperbolic functions, we get

$$M_{10} = \frac{4e^{-s}}{(1+\cosh\phi)^2} \left[1 - e^{-s}\tanh^2\frac{s}{2}\right]^{-2}$$

$$= \frac{4}{(1+\cosh\phi)^2} \sum_{n=1}^{\infty} ne^{-ns} \left(\tanh^2\frac{\phi}{2}\right)^{n-1}. \tag{323}$$

We can also compute M_{10} by inserting a complete set of states and using $\Gamma_0|n0) = n|n0)$ in Equation (322). Because the generator S is a scalar, only states with $L = 0, m = 0$ can contribute:

$$M_{10} = \sum_{nlm} e^{-ns}|n(10|R(\phi)|n0)|^2. \tag{324}$$

Comparing this to Equation (323), we make the identification

$$|(10|R(\phi)|n0)|^2 = \frac{4n}{(1+\cosh\phi)^2} \left(\tanh^2\frac{\phi}{2}\right)^{n-1}. \tag{325}$$

Huff computes this matrix element by analytically continuing the known $O(3)$ matrix element of $e^{iJ_y\phi}$ obtaining

$$|\langle 10|R(\phi)|n0\rangle|^2 = \frac{4n}{\cosh^2\phi - 1} \left(\tanh^2\frac{\phi}{2}\right)^{n} \cdot [{}_2F_1(0, -1; n; \frac{1}{2}(1 - \cosh\phi))]^2. \tag{326}$$

By algebraic manipulation and using ${}_2F_1 = 1$ for the arguments here, we see that this result agrees with our much more simply expressed result from group theory.

9. Conclusions and Future Research

Measuring and explaining the properties of the hydrogen atom has been central to the development of modern physics over the last century. One of the most useful and profound ways to understand its properties is through its symmetries, which we have explored in this paper, beginning with the symmetry of the Hamiltonian, which reflects the symmetry of the degenerate levels, then the larger non-invariance and spectrum-generating groups, which include all of the states. The successes in using symmetry to explore the hydrogen atom led to use of symmetry to understand and model other physical systems, particularly elementary particles.

The hydrogen atom will doubtless continue to be one of testing grounds for fundamental physics. Researchers are exploring the relationship between the hydrogen atom and quantum information [152], the effect of non-commuting canonical variables $[x_i, x_j] \neq 0$ on energy levels [153–155], muonic hydrogen spectra [156], and new physics using Rydberg states [157–162]. The ultra high precision of the measurement of the energy levels has led to new understanding of low Z two body systems, including muonium, positronium, and tritium [151]. As mentioned in the introduction, measurements of levels shifts are currently being used to measure the radius of the proton [2]. We can expect that further investigations of the hydrogen atom and hydrogenlike atoms will continue to reveal new vistas of physics and that symmetry considerations will play an important part.

Funding: This research received no external funding.

Acknowledgments: I would especially like to thank Lowell S. Brown very much for the time and energy he has spent on my education as a physicist. I also thank Peter Milonni for his encouragement and his helpful comments and many insightful discussions and MDPI for the invitation to be the Guest Editor for this special issue of Symmetry on Symmetries in Quantum Mechanics.

Conflicts of Interest: The author declares no conflict of interest.

References and Notes

1. Brown, L. Bound on Screening Corrections in Beta Decay. *Phys. Rev.* **1964**, *135*, B314. [CrossRef]
2. Beyer, A. The Rydberg constant and proton size from atomic hydrogen. *Science* **2017**, *358*, 79–85. [CrossRef]
3. Mohr, P.D.B.N.; Taylor, B.N. CODATA recommended values of the fundamental physical constants: 2014. *Rev. Mod. Phys.* **2016**, *88*, 035009. [CrossRef]
4. Rigden, J. *Hydrogen, The Essential Element*; Harvard University Press: Cambridge, MA, USA, 2002.
5. Lamb, W.; Retherford, R. Fine Structure of the Hydrogen Atom by a Microwave Method. *Phys. Rev.* **1947**, *72*, 241. [CrossRef]
6. Bethe, H. The Electromagnetic Shift of Energy Levels. *Phys. Rev.* **1947**, *72*, 339. [CrossRef]
7. Maclay, J. History and Some Aspects of the Lamb Shift. *Physics* **2020**, *2*, 8. [CrossRef]
8. Noether, E. Invariante Variationsprobleme. In *Nachrichten von der Gesellschaft der Wissenschaften zu Göttingen*; Akademie der Wissenschaften zu Göttingen: Göttingen, Germany, 1918; pp. 235–257.
9. Hamermesh, M. *Group Theory*; Adddison-Wesley Publishing Co.: Reading, MA, USA, 1962.
10. Weyl, H. *The Theory of Groups and Quantum Mechanics*, 2nd ed.; Dover Reprint; Dover Publications: New York, NY, USA, 1928.
11. Wigner, E. *Group Theory and Its Application to the Quantum Mechanics of Atomic Spectra*; Academic Press: New York, NY, USA, 1959.
12. Bargmann, V. Zur Theorie des Wasserstffatoms. *Z. Phys.* **1936**, *99*, 576. [CrossRef]
13. Laplace, P. *A Treatise of Celestial Mechanics*; Forgotten Books: Dublin, Ireland, 1827.
14. Pauli, W. Uber das Wasserstoffspektrum vom Standpunkt der neuen Quantummechanik. *Z. Phys.* **1926**, *36*, 336–363. [CrossRef]
15. McIntosh, H. On Accidental Degeneracy in Classical and Quantum Mechanics. *Am. J. Phys.* **1959**, *27*, 620–625. [CrossRef]
16. Hulthen, E. Über die quantenmechanische Herleitung der Balmerterme. *Z. Phys.* **1933**, *86*, 21–23. [CrossRef]
17. We employ natural Gaussian units so $\hbar = 1$, $c = 1$, and $\alpha = (e^2/\hbar c) \approx 1/137$. The notation for indices and vectors is $\mu, \nu, .. = 0, 1, 2, 3$; $i, j, . = 1, 2, 3$; $p_\mu p^\mu = -p_0^2 + \boldsymbol{p}^2$, $\boldsymbol{p} = (p_1, p_2, p_3)$, $g_{\mu\nu} = (-1, 1, 1, 1)$.
18. Fock, V. Zur Theorie des Wasserstoffatoms. *Z. Phys.* **1935**, *98*, 145–154. [CrossRef]
19. Dirac, P. *Quantum Mechanics*, 1st ed.; Oxford University Press: Oxford, UK, 1930.
20. Gell-Mann, M. Symmetries of Baryons and Mesons. *Phys. Rev.* **1962**, *125*, 1067. [CrossRef]
21. Schwinger, J. Coulomb's Green's Function. *J. Math. Phys.* **1964**, *5*, 1606–1608. [CrossRef]
22. Ne'eman, Y. *Algebraic Theory of Particle Physics*; Benjamin: New York, NY, USA, 1967.
23. Ne'eman, Y. Derivation of strong interactions from a gauge invariance. *Nucl. Phys.* **1961**, *26*, 222–229. [CrossRef]
24. Gell-Mann, M. A schematic model of baryons and mesons. *Phys. Lett.* **1964**, *8*, 214–215. [CrossRef]
25. Gell-Mann, M.; Ne'eman, Y. *The Eightfold Way*; Benjamin: New York, NY, USA, 1964.

26. Dothan, Y.; Gell-Man, M.; Ne'eman, Y. Series of Hadron Energy Levels as Representations of Non-Compact Groups. *Phys. Letters* **1965**, *17*, 148. [CrossRef]
27. Nambu, Y. Infinite-Component Wave Equations with Hydrogenlike Mass Spectra. *Phys. Rev.* **1967**, *160*, 1171. [CrossRef]
28. Dyson, F. *Symmetry Groups in Nuclear and Particle Physics*; Benjamin: New York, NY, USA, 1966.
29. Thomas, L. On the Unitary Representations of the Group of de Sitter Space. *Ann. Math.* **1941**, *42*, 113–126. [CrossRef]
30. Harish-Chandra, Representations of Semisimple Lie Groups II. *Trans. Am. Math. Soc.* **1954**, *76*, 26. [CrossRef]
31. Barut, A.; Budini, P.; Fronsdal, C. Two examples of covariant theories with internal symmetries involving spin. *Proc. Roy. Soc.* **1966**, *A291*, 106–112.
32. Malkin, I.; Man'ko, V. Symmetry of the Hydrogen Atom. *Sov. Phys. Jetp Lett.* **1966**, *2*, 146.
33. Barut, A.; Kleinert, K. Transition Probabilities of the Hydrogen Atom from Noncompact Dynamical Groups. *Phy. Rev.* **1967**, *156*, 1541. [CrossRef]
34. Barut, A.; Kleinert, H. Transition Form Factors in the H Atom. *Phys. Rev.* **1967**, *160*, 1149. [CrossRef]
35. Bander, M.; Itzykson, C. Group Theory and the Hydrogen Atom (I). *Rev. Mod. Phys.* **1966**, *38*, 330. [CrossRef]
36. Bander, M.; Itzykson, C. Group Theory and the Hydrogen Atom (II). *Rev. Mod. Phys.* **1966**, *38*, 346. [CrossRef]
37. Fronsdal, C. Infinite Multiplets and Local Fields. *Phys. Rev.* **1967**, *156*, 1653. [CrossRef]
38. Fronsdal, C. Infinite Multiplets and the Hydrogen Atom. *Phys. Rev.* **1967**, *156*, 1665. [CrossRef]
39. Barut, A.; Fronsdal, C. On Non-Compact Groups. II Representations of the 2+1 Lorentz Group. *Proc. R. Soc.* **1965**, *A287*, 532–548.
40. Fronsdal, C. Relativistic Lagrangian Field Theory for Composite Systems. *Phys. Rev.* **1968**, *171*, 1811. [CrossRef]
41. Pratt, R.; Jordan, T. Coulomb Group Theory for and Spin. *Phys. Rev.* **1969**, *188*, 2534.
42. Fronsdal, C. Relativistic and Realistic Classical Mechanics of Two Interacting Point Particles. *Phys. Rev. D* **1971**, *4*, 1689. [CrossRef]
43. Kyriakopoulos, R. Dynamical Groups and the Bethe-Salpeter Equation. *Phys. Rev.* **1968**, *174*, 1846. [CrossRef]
44. Lieber, M. O(4) Symmetry of the Hydrogen Atom and the Lamb Shift. *Phys. Rev.* **1968**, *174*, 2037. [CrossRef]
45. Huff, R. Simplified Calculation of Lamb Shift Using Algebraic Techniques. *Phys. Rev.* **1969**, *186*, 1367. [CrossRef]
46. Musto, R. Generators of SO(4,1) for the Quantum Mechanical Hydrogen Atom. *Phys. Rev.* **1966**, *148*, 1274. [CrossRef]
47. Barut, A.; Bornzin, G. SO(4,2)-Formulation of the Symmetry Breaking in Relativistic Kepler Problems with of without Magnetic Charge. *J. Math. Phys.* **1971**, *12*, 841–846. [CrossRef]
48. Barut, A.; Kleinert, H. Current Operators and Majorana Equation for the Hydrogen Atom from Dynamical Groups. *Phys. Rev.* **1967**, *157*, 1180. [CrossRef]
49. Mack, G.; Todorov, I. Irreducibility of the Ladder representations when restricted to the Poincare Subgroup. *J. Math Phys.* **1969**, *10*, 2078–2085. [CrossRef]
50. Decoster, A. Realization of the Symmetry Groups of the Nonrelativistic Hydrogen Atom. *Nuovo Cimento* **1970**, *68A*, 105–117. [CrossRef]
51. Englefield, M. *Group Theory and the Coulomb Problem*; Wiley-Interscience: New York, NY, USA, 1972.
52. Barut, A. *Dynamical Groups*; University of Canterbury Press: Christchurch, New Zealand, 1972.
53. Bednar, M. Algebraic Treatment of Quantum-Mechanical Models with Modified Coulomb Potentials. *Ann. Phys.* **1973**, *75*, 305–331. [CrossRef]
54. Wulfman, C.; Takahata, Y. Noninvariance Groups in Molecular Quantum Mechanics. *J. Chem. Phys.* **1967**, *47*, 488–498. [CrossRef]
55. Wybourne, B. Symmetry Principles in Atomic Spectroscopy. *J. Phys.* **1970**, *31*, C4-33. [CrossRef]
56. Mariwalla, K. Dynamical Symmetries in Mechanics. *Phys. Rep.* **1975**, *20*, 287–362. [CrossRef]
57. Akyildiz, Y. On the dynamical symmetries of the Kepler problem. *J. Math. Phys.* **1980**, *21*, 665–670. [CrossRef]
58. Fronsdal, C.; Huff, R. Two-Body Problem in Quantum Field Theory. *Phys. Rev. D* **1971**, *3*, 933. [CrossRef]
59. Loebl, E. (Ed.) *Group Theory and Its Applications*; Academic Press: New York, NY, USA, 1971.
60. Barut, A.; Rasmussen, W. The hydrogen atom as a relativistic elementary particle I. The wave equation and mass formulae. *J. Phys.* **1973**, *B6*, 1695. [CrossRef]

61. Barut, A.; Rasmussen, W. The hydrogen atom as a relativistic elementary particle II. Relativistic scattering problems and photo-effect. *J. Phys.* **1973**, *B6*, 1713. [CrossRef]
62. Barut, A.; Bornzin, G. Unification of the external conformal symmetry group and the internal conformal dynamical group. *J. Math. Phys.* **1974**, *15*, 1000–1006. [CrossRef]
63. Barut, A.; Schneider, C.; Wilson, R. Quantum theory of infinite component fields. *J. Math. Phys.* **1979**, *20*, 2244–2256. [CrossRef]
64. Shibuya, T.; Wulfman, C. The Kepler Problem in Two-Dimensional Momentum Space. *Am. J. Phys.* **1965**, *33*, 570–574. [CrossRef]
65. Dahl, J. Physical Interpretation of the Runge-Lenz Vector. *Phys. Let.* **1968**, *27A*, 62–63. [CrossRef]
66. Collas, P. Algebraic Solution of the Kepler Problem Using the Runge-Lenz Vector. *Am. J. Phys.* **1970**, *38*, 253–255. [CrossRef]
67. Rodgers, H. Symmetry transformations of the classical Kepler problem. *J. Math. Phys.* **1973**, *14*, 1125–1129. [CrossRef]
68. Majumdar, S.; Basu, D. O(3,1) symmetry of the hydrogen atom. *J. Phys. Math. Nuc. Gen.* **1974**, *7*, 787. [CrossRef]
69. Stickforth, J. The classical Kepler problem in momentum space. *Am. J. Phys.* **1978**, *46*, 74–75. [CrossRef]
70. Ligon, T.; Schaaf, M. On the Global Symmetry of the Classical Kepler Problem. *Rep. Math. Phys.* **1976**, *9*, 281–300. [CrossRef]
71. Lakshmanan, M.; Hasegawa, H. On the canonical equivalence of the Kepler problem in coordinate and momentum space. *J. Phys. Math. Gen.* **1984**, *17*, L889. [CrossRef]
72. O'Connell, R.; Jagannathan, K. Illustrating dynamical symmetries in classical mechanics: The Laplace-Runge-Lenz vector revisited. *Am. J. Phys.* **2003**, *71*, 243–246. [CrossRef]
73. Valent, G. The hydrogen atom in electric and magnetic fields: Pauli's 1926 article. *Am. J. Phys.* **2003**, *71*, 171–175. [CrossRef]
74. Morehead, J. Visualizing the extra symmetry of the Kepler problem. *Am. J. Phys.* **2005**, *73*, 234–239. [CrossRef]
75. Huntington, L.; Nooijen, M. An SO(4) invariant Hamiltonian and the two-body bound state. I: Coulomb interaction between two spinless particles. *Int. J. Quant. Chem.* **2009**, *109*, 2885–2896. [CrossRef]
76. Barut, A.; Bohm, A.; Neeman, Y. *Dynamical Groups and Spectrum Generating Algebras*; World Scientific: Singapore, 1986.
77. Greiner, W.; Muller, B. *Quantum Mechanics, Symmetries*; Springer: Berlin, Germany, 1989.
78. Gilmore, R. *Lie Groups, Lie Algegras and Some of Their Applications*; Dover Books on Mathmatics; Dover: Mineola, NY, USA, 2005.
79. Kibler, M. On the use of the group SO(4,2) in atomic and molecular physics. *Mol. Phys.* **2004**, *102*, 1221–1229. [CrossRef]
80. Hammond, I.; Chu, S. Irregular wavefunction behavior in dimagnetic Rydberg atoms:a dynamical SO(4,2) group study. *Chem. Phys. Let.* **1991**, *182*, 63. [CrossRef]
81. Lev, F. Symmetries in Foundation of Quantum Theory and Mathematics. *Symmetry* **2020**, *12*, 409. [CrossRef]
82. Wulfman, C. *Dynamical Symmetry*; World Scientific Publishing: Singapore, 2011.
83. Johnson, M.; Lippmann, B.A. Relativistic Kepler problem. *Phys. Rev.* **1950**, *78*, 329.
84. Biedenharn, L. Remarks on the relativistic Kepler problem. *Phys. Rev.* **1962**, *126*, 845–851. [CrossRef]
85. Lanik, J. The Reformulations of the Klein-Gordon and Dirac Equations for the Hydrogen Atom to Algebraic Forms. *Czech. J. Phys.* **1969**, *B19*, 1540–1548. [CrossRef]
86. Stahlhofen, A. Algebraic solutions of relativistic Coulomb problems. *Helv. Phys. Acta* **1997**, *70*, 1141.
87. Chen, J.; Deng, D.; Hu, M. SO(4) symmetry in the relativistic hydrogen atom. *Phys. Rev. A* **2008**, *77*, 034102. [CrossRef]
88. Khachidze, T.; Khelashvili, A. The hidden symmetry of the Coulomb problem in relativistic quantum mechanics: From Pauli to Dirac. *Am. J. Phys.* **2006**, *74*, 628–632. [CrossRef]
89. Zhang, F.; Fu, B.; Chen, J. Dynamical symmetry of Dirac hydrogen atom with spin symmetry and its connection to Ginocchio's oscillator. *Phys. Rev. A* **2008**, *78*, 040101(R). [CrossRef]
90. Heine, V. *Group theory in Quantum Mechanics*; Dover Publications: New York NY, USA, 1993.
91. Noether, E.; Mort, T. Invariant Variation Problems. *Transp. Theory Stat. Phys.* **1971**, *1*, 186–207. [CrossRef]
92. Neuenschwander, D.E. *Emmy Noether's Wonderful Theorem*; Johns Hopkins University Press: Baltimore, MD, USA, 2010.

93. Hanca, J.; Tulejab, S.; Hancova, M. Symmetries and conservation laws: Consequences of Noether's theorem. *Am. J. Phys.* **2004**, *72*, 428–435. [CrossRef]

94. Byers, N.E. Noether's Discovery of the Deep Connection Between Symmetries and Conservation Laws. *arXiv* **1998**, arXiv:physics/9807044.

95. The daughter of a mathematician, she wanted to be a mathematician, but since women were not allowed to take classes at the University of Erlingen, she audited courses. She did so well in the exams, that she received a degree and was allowed to enroll in the university and received a PhD in 1907. She remained at the university, unpaid, in an unofficial status, for 8 years. Then she went to the University at Gottengen, where she worked for 8 years with no pay or status before being appointed as Lecturer with a modest salary. She was invited in 1915 by Felix Klein and David Hilbert, two of the most famous mathematicians in the world at the time, to work with them and address issues in Einstein's theory of General Relativity about energy conservation. She discovered Nother's First Theorem (and a second theorem also). She remained there until 1933 when she, as a Jew, lost her job. At Einstein's suggestion, she went to Bryn Mawr College in Pennsylvania. She died from ovarian cysts two years later.

96. A rotation in 4 dimension can be represented by an antisymmetric 4×4 matrix which has $3 + 2 + 1 = 6$ independent non-diagonal elements corresponding to 6 generators. Similarly a rotation in 5 dimensions has 10 independent elements or 10 generators.

97. Dothan, Y. Finite-Dimensional Spectrum-Generating Algebras. *Phys. Rev.* **1970**, *D2*, 2944. [CrossRef]

98. Mukanda, N.; Sudarshan, E. Characteristic Noninvariance Groups of Dynamical Systems. *Phys. Rev. Lett.* **1965**, *15*, 1041. [CrossRef]

99. Kyriakopoulos. E. Algebraic Equations for Bethe-Salpeter and Coulomb Green's Functions. *J. Math. Phys.* **1972**, *13*, 1729–1735. [CrossRef]

100. Lipkin, H. *Lie Groups for Pedestrians*; Dover Publications: New York, NY, USA, 2001.

101. Were it not for this displacement of the force center, the observation that a rotated circle projects onto a plane as an ellipse would manifest the four-dimensional symmetry of the hydrogenlike atom directly in configuration space. The elliptical orbits could be viewed as projections of a rotated hypercircle onto a three-dimensional hyperplane. These considerations can be applied with some modification to the three-dimensional harmonic oscillator for which the force center and the center of the ellipse coincide.

102. This equation and any other equation written in this specific coordinate system can be generalized to an arbitrary coordinate system by noting that the Cartesian unit vectors may be written in a manner that is independent of the coordinate system: $i = \frac{A}{A}$, $j = \frac{L \times A}{LA}$, $k = \frac{L}{L}$.

103. Brown, L. University of Washington, Seattle, WA, Unpublished lecture notes, 1972. Unpublished lecture notes.

104. We define the angle between a three-dimensional hyper-plane and a line as $\pi/2$ minus the angle between the line and the normal to the hyperplane.

105. It is desirable to first show that A (and of course L) generate rotations of the hypersphere or $\hat{U}$. However, as we prefer to do the necessary calculations in terms of commutators rather than Poisson brackets, we defer these considerations to Section 4. There we show that the generator L_i rotates $\hat{U}$ about the $i - 4$ plane; the generator A_1 rotates $\hat{U}$ about the 2–3 plane, etc., thereby changing the orbit with respect to the 4-axis and changing the eccentricity.

106. Bois, G. *Tables of Indefinite Integrals*; Dover Publications: New York, NY, USA, 1961; p. 123.

107. Using Equation (44) and [102], Equation (73) may be written as $cos^{-1}(U \cdot A/A) = p \cdot r/ar_c + \omega_{cl}t$. This agrees with the time dependent function $\phi = p \cdot r/ar_c - \omega_{cl}t$ Equation (70) defined in [97].

108. Brown, L. Forces giving no orbit precession. *Am. J. Phys.* **1978**, *46*, 930–931. [CrossRef]

109. Bacry, H. *Lectures in Theoretical Physics*; Brittin, W.E., Barut, A.O., Guenin, M., Eds.; Gordon and Breach: New York, NY, USA, 1967.

110. Barut, A. Dynamics of a Broken SU_N Symmetry for the Oscillator. *Phys. Rev.* **1965**, *139*, B1433. [CrossRef]

111. Boiteux, M. The Three-Dimensional Hydrogen Atom as a Restricted Four-Dimensional Harmonic Oscillator. *Physica* **1972**, *65*, 381–395. [CrossRef]

112. Hughes, J. The harmonic oscillator:values of the SU(3) invariants. *J. Phys. A Math. Gen.* **1973**, *6*, 453. [CrossRef]

113. Chen, A. Hydrogen atom as a four-dimensional oscillator. *Phys. Rev. A* **1980**, *22*, 333. [CrossRef]

114. Chen, A. Homomorphism between SO(4,2) and SU(2,2). *Phys. Rev. A* **1981**, *23*, 1653. [CrossRef]

115. Kibler, M.; Negadi, T. Connection between the hydrogen atom and the harmonic oscillator: The zero-energy case. *Phys. Rev. A* **1984**, *29*, 2891. [CrossRef]

116. Chen, A.; Kibler, M. Connection between the hydrogen atom and the four-dimensional oscillator. *Phys. Rev. A* **1985**, *31*, 3960. [CrossRef]

117. Gerry, C. Coherent states and the Kepler-Coulomb problem. *Phys. Rev. A* **1986**, *33*, 6. [CrossRef]

118. Chen, A. Coulomb–Kepler problem and the harmonic oscillator. *Am. J. Phys.* **1987**, *55*, 250–252. [CrossRef]

119. Van der Meer, J. The Kepler system as a reduced 4D oscillator. *J. Geom. Phys.* **2015**, *92*, 181–193. [CrossRef]

120. Bacry, H. The de Sitter Group $L_{4,1}$ and the Bound States of the Hydrogen Atom. *Nuovo Cimento* **1966**, *41*, 222–234. [CrossRef]

121. Biedenharn, L. Wigner Coefficients for the R_4 Group and Some Applications. *J. Math. Phys.* **1961**, *2*, 433–441. [CrossRef]

122. Shiff, L. *Quantum Mechanics*; McGraw Hill: New York, NY, USA, 1955.

123. Biedenharn, L.; Swamy, N. Remarks on the Relativistic Kepler Problem. II. Approximate Dirac-Coulomb Hamiltonian Possessing Two Vector Invariants. *Phys. Rev.* **1964**, *133*, B1353. [CrossRef]

124. Morse, P.; Feshbach, H. *Methods of Theoretical Physics, Vol. 1*; McGraw-Hill: New York, NY, USA, 1953.

125. The primes indicates eigenvalues of operators, and unprimed quantities indicate abstract operators. The quantity x' means the four-vector $(t', \vec{r}')$.

126. Morse, P.; Feshbach, H. *Methods of Theoretical Physics, Vol. 2*; McGraw-Hill: New York, NY, USA, 1953.

127. Erdeli, A. (Ed.) *Higher Transcendental Functions, Bateman Manuscript Project*; McGraw-Hill Book Co.: New York, NY, USA, 1953.

128. Makowski, A.; Pepłowski, P. Zero-energy wave packets that follow classical orbits. *Phys. Rev. A* **2012**, *86*, 042117. [CrossRef]

129. Bellomo, P.; Stroud, C., Jr. Classical evolution of quantum elliptical orbits. *Phys. Rev. A* **1999**, *59*, 2139. [CrossRef]

130. Berry, M.; Mount, K.E. Semiclassical approximations in wave mechanics. *Rep. Prog. Phys.* **1972**, *35*, 315. [CrossRef]

131. Eberly, J.; Stroud, C. Chapters 14 (Rydberg Atoms) and Chapter 73 (Coherent Transients). In *Springer Handbook of Atomic, Molecular, and Optical Physics*; Drake, G., Ed.; Springer Science and Business Media: New York, NY, USA, 2006.

132. Lakshmanan, M.; Ganesan, K. Rydberg atoms and molecules-Testing grounds for quantum manifestations of chaos. *Curr. Sci.* **1995**, *68*, 38–44.

133. Kay, K. Exact Wave Functions for the Coulomb Problem from Classical Orbits. *Phys. Rev.* **1999**, *25*, 5190. [CrossRef]

134. Lena, C.; Deland, D.; Gay, J. Wave functions of Atomic Elliptic States. *Europhys. Lett.* **1991**, *15*, 697. [CrossRef]

135. Bhaumik, D.; Dutta-Roy, B.; Ghosh, G. Classical limit of the hydrogen atom. *J. Phys. A Math. Gen.* **1986**, *19*, 1355. [CrossRef]

136. McAnally, D.; Bracken, A. Quasiclassical states of the Coulomb system and SO(4, 2). *J. Phys. A Math. Gen.* **1990**, *23*, 2027. [CrossRef]

137. Pitak, A.; Mostowski, J. Classical limit of position and matrix elements for Rydberg atoms. *Eur. J. Phys.* **2018**, *39*, 025402. [CrossRef]

138. Nauenberg, M. Quantum wavepackets on Kepler elliptical orbits. *Phys. Rev. A* **1989**, *40l*, 1133. [CrossRef] [PubMed]

139. Brown, L.S. Classical limit of the hydrogen atom. *Am. J. Phys.* **1973**, *41*, 525–530. [CrossRef]

140. Leonhardt, U. *Measuring the Quantum State of Light*; Cambridge University Press: Cambridge, UK, 1997.

141. Barut, A.; Girardello, L. New "coherent" states associated with non-compact groups. *Commun. Math. Phys.* **1971**, *21*, 41–55. [CrossRef]

142. Satyanarayana, M. Squeezed coherent states of the hydrogen atom. *J. Phys. A Math. Gen.* **1986**, *19*, 1973. [CrossRef]

143. Liu, Q.; Hu, B. The hydrogen atom's quantum-to-classical correspondence in Heisenberg's correspondence principle. *J. Phys. A Math. Gen.* **2001**, *34*, 5713. [CrossRef]

144. Zverev, V.; Rubinstein, B. Dynamical symmetries and well-localized hydrogenic wave packets. *Proc. Inst. Math. Nas Ukr.* **2004**, *50*, 1018.

145. The wave function in momentum space $\psi(p)$ is obtained by multiplying Y_{nlm} by the normalizing factor $\frac{(a_n)^{3/2}}{(1-U_4)^2}$, cf Equation (150).

146. Nandi, S.; Shastry, C. Classical limit of the two-dimensional and three-dimensional hydrogen atom. *J. Phys. A Math. Gen.* **1989**, *22*, 1005. [CrossRef]

147. Pauling, L.; Wilson, E.B. *Introduction to Quantum Mechanics*; McGraw-Hill: New York, NY, USA, 1935; p. 36.

148. Lamb, W.; Retherford, R. Fine Structure of the H Atom, Part I. *Phys. Rev.* **1950**, *79*, 549. [CrossRef]

149. Bethe, H.; Salpeter, E. *The Quantum Mechanics of One and Two Electron Atoms*; Springer: Berlin, Germany, 1957.

150. Milonni, P. *The Quantum Vacuum*; Academic Press: San Diego, CA, USA, 1994.

151. Eides, M.; Grotch, H.; Shelyuto, V. *Theory of Light Hydrogenic Bound States, Springer Tracts in Modern Physics 222*; Springer: Berlin, Germany, 2007.

152. Rau, A.; Alber, G. Shared symmetries of the hydrogen atom and the two-bit system. *J. Phys. B At. Mol. Opt.* **2017**, *50*, 242001. [CrossRef]

153. Castro, P.; Kullock, R. Physics of the $SO_p(4)$ Hydrogen Atom. *Theo. Math. Phys.* **2015**, *185*, 1678. [CrossRef]

154. Alavi, A.; Rezaei, N. Dirac equation, hydrogen atom spectrum and the Lamb shift in dynamical non-commutative spaces. *Pramana-J. Phys.* **2017**, *88*, 5. [CrossRef]

155. Gnatenko, K.P.; Krynytskyi, Y.S.; Tkachuk, V.M. Perturbation of the ns levels of the hydrogen atom in rotationally invariant noncommutative space. *Mod. Phys. Lett.* **2015**, *30*, 8. [CrossRef]

156. Haghighat, M.; Khorsandi, M. Hydrogen and muonic hydrogen atomic spectra in non-commutative space-time. *Eur. Phys. J.* **2015**, *75*, 1. [CrossRef]

157. Praxmeyer, L. Hydrogen atom in phase space: The Wigner representation. *J. Phys. A Math. Gen.* **2006**, *39*, 14143. [CrossRef]

158. Jones, M.; Potvliege, M.R.; Spannowsky, M. Probing new physics using Rydberg states of atomic hydrogen. *Phys. Rev. Res.* **2020**, *2*, 013244. [CrossRef]

159. Jentschura, U.; Mohr, P. Calculation of hydrogenic Bethe logarithms for Rydberg States. *Phys. Rev. A* **2002**, *72*, 012110. [CrossRef]

160. Jentschura, U.; LeBigot, E.; Evers, J.; Mohr, P.; Keitel, C. Relativistic and radiative shifts for Rydberg states. *J. Phys. B At. Mol. Opt. Phys.* **2005**, *38*, S97. [CrossRef]

161. Jentschura, U.; Mohr, P.; Tan, J. Fundamental constants and tests of theory in Rydberg states of one-electron ions. *J. Phys. B At. Mol. Opt. Phys.* **2010**, *43*, 074002. [CrossRef]

162. Cantu, S.H.; Venkatramani, A.V.; Xu, W. Repulsive photons in a quantum nonlinear medium. *Nat. Phys.* **2020**. [CrossRef]

Article

Symmetries in Teleportation Assisted by N-Channels under Indefinite Causal Order and Post-Measurement

Carlos Cardoso-Isidoro [†] and Francisco Delgado [*,†]

School of Engineering and Sciences, Tecnologico de Monterrey, Atizapan 52926, Mexico; A01750267@itesm.mx
* Correspondence: fdelgado@tec.mx; Tel.: +52-55-5864-5670
† These authors contributed equally to this work.

Received: 10 October 2020; Accepted: 16 November 2020; Published: 20 November 2020

Abstract: Quantum teleportation has had notorious advances in the last decade, being successfully deployed in the experimental domain. In other terrains, the understanding of indefinite causal order has demonstrated a valuable enhancement in quantum communication to correct channel imperfections. In this work, we address the symmetries underlying imperfect teleportation when it is assisted by indefinite causal order to correct the use of noisy entangled resources. In the strategy being presented, indefinite causal order introduces a control state to address the causal ordering. Then, by using post-selection, it fulfills the teleportation enhancement to recover the teleported state by constructive interference. By analysing primarily sequential teleportation under definite causal order, we perform a comparison basis for notable outcomes derived from indefinite causal order. After, the analysis is conducted by increasing the number of teleportation processes, thus suggesting additional alternatives to exploit the most valuable outcomes in the process by adding weak measurement as a complementary strategy. Finally, we discuss the current affordability for an experimental implementation.

Keywords: teleportation; indefinite causal order; weak measurement; quantum algorithm

---

1. Introduction

Quantum communication has always looked for improvements and new outstanding approaches. Particularly, it has been shown that certain enhancements in information transmission can be reached through the superposition of quantum communication channels. That enhancement has shown that the interference of causal orders using sequential extreme imperfect depolarizing channels surprisingly produces a transparent quantum channel due to constructive superposition in the components of the state being transmitted [1]. Since that discovery, a growing interest in indefinite causal order has emerged boosting a deep study of this topic. Experimental implementations have been proposed in order to find, to understand, and to control their advantages [2].

1.1. Background of Indefinite Causal Order in Communication

In quantum communication with extremely noisy channels, only limited information can be transmitted. If we continue applying such quantum channels sequentially, no information becomes transmitted, obtaining the so-called depolarizing quantum channel. Despite, it has been shown that when such channels are applied in a superposition of causal orders, we can still transmit information, and notably, the quality of information transmitted becomes improved while more channels are applied under this scheme. Concretely for the case of two quantum channels, some works considering controllable strengths of depolarization have shown that combining a superposition of causal orders, it is still possible to transmit information (instead of worsening it as it obviously happens for the simpler sequential case) [1,3]. The success of the causal orders superposition has been experimentally verified for two channels transmitting information [4].

Following such a trend in communication, it has been found the possibility to extrapolate the increasing number of causal orders superposed (with more than two channels) by developing a combinatoric approach to the problem [5,6]. As a matter of fact, it has been shown that the amount of information transmitted, in comparison with the two-channel scenario, increases for the three-channels scenario [5]. Therefore, it has been concluded that the amount of classical information transmitted becomes higher if the number of causal orders increases.

Some notorious approaches regarding the indefiniteness of causal orders have been explored, exhibiting the capability to transmit information in a more efficient way. It highlights the importance to extend this approach on teleportation, as a genuine communication process [7,8].

1.2. Approaches to Teleportation under Causal Order Schemes

Information can be transmitted from one party to another as a quantum state if it is prepared in combination with an Einstein–Podolsky–Rosen state [9]. Such a quantum communication process is called quantum teleportation. It plays an important role related to quantum information and quantum communication. Teleportation algorithm for one single qubit is performed using one entangled Bell state and one channel for classical communication in order to achieve it [10]. Symmetries in the conformation of such quantum entangled state automatically transfer a state into another party if post-measurement is applied. The same algorithm has also been useful to teleport states of larger systems if they are composed of two-level systems [11]. The teleportation algorithm has been widely studied and new approaches have been discovered, as well as variants on the algorithm in order to make it either more efficient in terms of the quantum resources used [12] or more adaptive to some specific quantum systems [13–15]. Additionally, several successful tests have been experimentally performed in order to prove the feasibility of teleportation when the distance increases [16–18]. Tests with larger multidimensional states rather than qubits have been performed successfully [19]. Recently, a new approach has shown that the assistance of indefinite causal order in teleportation improves its performance when imperfect entangled resources become involved [7], which is equivalent to a quantum noisy communication channel.

Teleportation assisted by indefinite causal order and measurement has been introduced in [7] by pointing out that teleportation is a quantum channel itself (here, entanglement distribution is assumed to be performed through a transparent communication channel). The last proposal has been criticized in [20] arguing the entanglement distribution in teleportation is a critical aspect not being considered there (due to the large distances and communication issues involved). Instead, as in [7], the most recent work [20] interestingly has analysed the use of indefinite causal order in the form of a quantum switch for the entanglement distribution process as a part of the teleportation algorithm, thus making an analysis to quantify the performance gained by such a switch. Nevertheless, nowadays teleportation has been achieved through kilometers in the free space or through optical fiber, with still high fidelities [21] without considerable deformation in the entangled resource other than that the introduced in its imperfect generation. Thus, we believe both approaches are still valuable in the quest of understanding creative ways to implement indefinite causal order in teleportation. Both approaches show interesting features in the quantification of indefinite causal order issues applied to teleportation.

In [7], the quantum teleportation uses imperfect singlets showing that despite those noisy singlets make impossible a faithful teleportation, there is still a stochastic possibility of teleporting perfectly the state by applying indefinite causal order as the superposition of two teleportation channels. Such teleportation process has been conducted considering two identical teleportation channels with the same imperfect entangled resources, but in a superposition of causal orders through an evenly quantum control system. Finally, the outcome is measured on a specific basis in order to improve the fidelity of the teleportation process in the best possible way by recovering the symmetrical composition of the teleported state. Following this analysis and considering the same two imperfect channels but with an arbitrary initialized quantum control system, it has been also found the possibility to get again the highest possible transmission by post-selecting the appropriate outputs under alternative scenarios [22]: a proper selection of the post-measurement state on the control system, thus extending the interesting outcomes obtained in [7]. In addition, it has been shown that for the less noisy cases, the effect becomes still limited [7].

In teleportation, the traditional algorithm [9] is entirely represented as a quantum channel T in Figure 1a. In order to carry out the teleportation, it is necessary an entangled resource shown as $|\chi\rangle$. When this resource is the Bell state $|\beta_{00}\rangle = \frac{1}{\sqrt{2}}(|00\rangle + |11\rangle)$, a perfect teleportation is then achieved, but if such state is imperfect (it can be generally expressed as a mixture of all Bell states), teleportation process does not work properly. In Figure 1b, an alternative (but still equivalent) circuit is presented assuming that Bell states measurements could be performed. In such a case, no gates are required, due to teleportation is just reached due to the non-locality of the entangled resource $|\beta_{00}\rangle$ (or imperfectly, $|\chi\rangle$). This fact will be useful at the end of the article for a tentative experimental proposal.

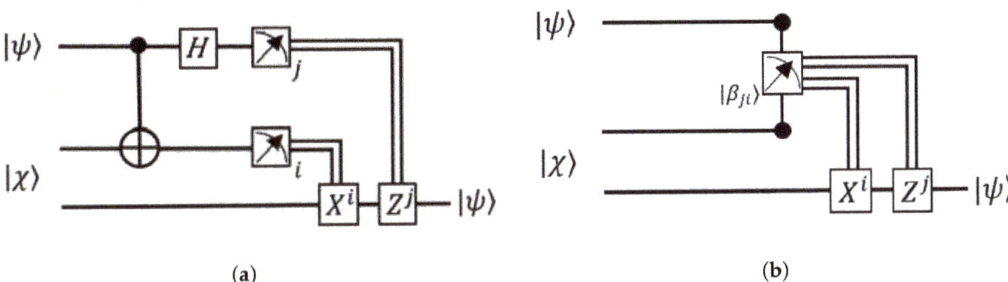

(a)   (b)

Figure 1. (**a**) Traditional teleportation circuit T where $|\psi\rangle = \alpha|0\rangle + \beta|1\rangle$ and ideally $|\chi\rangle$ is the Bell state $|\beta_{00}\rangle = \frac{1}{\sqrt{2}}(|00\rangle + |11\rangle)$. Measurements refer to one single qubit measurement and the double line to classical communication channels. (**b**) Modified teleportation circuit considering a Bell states measurement (which are generated by enclosing the gates on (**a**) within the measurement gadget).

Still, applying a sequence of two imperfect teleportation channels, the outcome worsens. In [7], it has been shown that for the worst deformed case of $|\chi\rangle$, the fidelity of single teleportation goes down. However, if two teleportation channels are used in an indefinite causal order with the superposition ruled by a quantum control system, surprisingly the previous worst-case arises with fidelity equal to 1. The analysis has been extended in [22] considering a wider kind of measurements required in the original approach. In this sense, the use of indefinite causal order improves the teleportation process. Thus, it is possible to correct this lack of fidelity working with the worst entangled state by applying indefinite causal order, together, with some appropriate selection in the control used and in the measurement performed, making possible to reach perfect teleportation.

In the current work, we deal with an extended version of the algorithm presented in [7,22] by using several sequential channels in order to benchmark the outcomes obtained by increasing the number of channels [5]. Section 2 develops the case of sequential channels in a definite causal order as a comparison basis. Section 3 develops the same situation but considering an indefinite causal order superposition using N channels. Section 4 uses the last formalism with more than two teleportation channels under indefinite causal order widening the spectrum of analysis. Section 5 revisits the problem but implementing additionally weak measurement proposing an improved procedure. Finally, Section 6 discusses the affordability of a possible experimental implementation for two teleportation channels under indefinite causal order using the current experimental developments. The last section gives the conclusions and future work to extend our findings.

2. Teleportation Algorithm as a Quantum Channel and N-Redundant Teleportation Problem

2.1. Quantum Teleportation as a Quantum Channel

Traditional quantum teleportation algorithm developed originally in [10] has become a central procedure in quantum information theory. This process uses an entangled resource in the form of the Bell state $|\beta_{00}\rangle = \frac{1}{\sqrt{2}}(|00\rangle + |11\rangle)$. Experimentally, such an entangled state becomes difficult to create and to sustain. For this reason, it could arrive imperfect to the process. Thus, considering a general variation of this resource in the form of the general state $|\chi\rangle = \sum_{i=0}^{3} \sqrt{p_i} |\beta_i\rangle$, where $|\beta_i\rangle$ is a short notation for the Bell basis $|\beta_0\rangle = |\beta_{00}\rangle$, $|\beta_1\rangle = |\beta_{01}\rangle$, $|\beta_2\rangle = |\beta_{11}\rangle$ and $|\beta_3\rangle = |\beta_{10}\rangle$. The traditional teleportation algorithm running under this resource (instead the perfect case with $p_0 = 1$ and $p_1 = p_2 = p_3 = 0$) becomes a quantum channel whose output expression in terms of Kraus operators is given by [23]:

$$\Lambda[\rho] = \sum_{i=0}^{3} p_i \widetilde{\sigma}_i \rho \widetilde{\sigma}_i^{\dagger} = \sum_{i=0}^{3} p_i \sigma_i \rho \sigma_i \tag{1}$$

with $\widetilde{\sigma}_i = \sigma_i$ if $i = 0, 1, 3$ and $\widetilde{\sigma}_2 = i\sigma_2$. $\rho = |\psi\rangle \langle\psi|$ is the state to teleportate (in the current work we will restrict the analysis to pure state cases, despite our outcomes can be extended to mixed states [8]). This formula, regarding teleportation algorithm as a communication channel will be discussed at the end of the article in terms of possible and current available experimental developments for its implementation. It means Kraus operators are $K_i = \sqrt{p_i}\sigma_i$. In the terms stated before, we are interested to assess the corresponding fidelity of the process as function of the p_i values under several schemes. It has the form of a Pauli channel [24] and it has been recently studied to characterize its properties under indefinite causal order and measurement [8] exhibiting notable properties and symmetries of communication enhancement as function of the parameters p_i. In the current approach, the set $\{p_i | i = 0, 1, 2, 3\}$ plays an additional role because it is associated with the quantum resource $|\chi\rangle$.

In the current article, we will use the fidelity to measure the channel performance:

$$\mathcal{F}(\rho, \Lambda[\rho]) = \left[\mathrm{Tr} \left(\sqrt{\sqrt{\rho}\Lambda[\rho]\sqrt{\rho}} \right) \right]^2, \tag{2}$$

because we will restrict to the case when ρ is a pure state $\rho = |\psi\rangle\langle\psi|$, then $\sqrt{\rho} = \rho$. Those facts still give the easier formula: $\mathcal{F}(\rho, \Lambda[\rho]) = \langle\psi|\Lambda[\rho]|\psi\rangle = \mathrm{Tr}(\rho\Lambda[\rho])$. Then, in the following, we will express the fidelity briefly as $\mathcal{F}_\Lambda \equiv \mathcal{F}(\rho, \Lambda[\rho])$.

2.2. N-Redundant Quantum Teleportation

In this section, we will study the effect on the fidelity of imperfect teleportation as it was previously depicted. For such reason, we first consider a set of identical and redundant N teleportation channels in a definite causal order as a composition of the depicted channel in (1). In addition, we consider for the sake of simplicity that each channel is identical to others in the redundant application:

$$(\bigcirc_N \Lambda)[\rho] \equiv \Lambda[\Lambda[\ldots \Lambda[\rho]\ldots]] = \sum_{i_1,\ldots,i_n=0}^{3} p_{i_1} \cdots p_{i_n} \sigma_{i_N} \cdots \sigma_{i_1} \rho \sigma_{i_1} \cdots \sigma_{i_N}. \tag{3}$$

If $p_1 = p_2 = p_3 \equiv p$, with $0 \le p \le \frac{1}{3}$ for simplicity (to avoid the increasing parameters involved), we have gotten the expressions for the corresponding fidelity $\mathcal{F}_{\bigcirc_N \Lambda} \equiv \mathrm{Tr}(\rho(\bigcirc_N \Lambda)[\rho])$ for the first five cases of redundant sequential applications of teleportation (assuming ρ is a pure state), getting:

$$\mathcal{F}_{\bigcirc_1 \Lambda} = 1 - 2p \tag{4}$$

$$\mathcal{F}_{\bigcirc_2 \Lambda} = 1 - 4p + 8p^2 \tag{5}$$

$$\mathcal{F}_{\bigcirc_3 \Lambda} = 1 - 6p + 24p^2 - 32p^3 \tag{6}$$

$$\mathcal{F}_{\bigcirc_4 \Lambda} = 1 - 8p + 48p^2 - 128p^3 + 128p^4 \tag{7}$$

$$\mathcal{F}_{\bigcirc_5 \Lambda} = 1 - 10p + 80p^2 - 320p^3 + 640p^4 - 512p^5. \tag{8}$$

Interestingly, those outcomes are independent from the state to teleport (a consequence from the symmetric simplification $p_1 = p_2 = p_3 = p$ and the algebraic properties of Pauli operators). Such cases can be computationally developed to get last outcomes (and other for larger cases). Figure 2 exhibits the behavior of such applications as function of p. The gray zone sets the middle point $\mathcal{F}_{\bigcirc_1 \Lambda} = \frac{2}{3}$ of fidelity $\mathcal{F}_{\bigcirc_1 \Lambda} \in [\frac{1}{3}, 1]$ for the case $N = 1$ as a reference (as it was remarked in [7]). The single case $N = 1$ sets the expected outcome about the effect of p on $\mathcal{F}_{\bigcirc_1 \Lambda}$ giving the worst value for $p = \frac{1}{3}$. For $N > 1$, the outcome becomes as it could be expected, each application of a new teleportation worsens the output state teleported. Despite this, there are certain recoveries for $p = \frac{1}{3}$, useful only for the lowest values of N. A convergent value $\mathcal{F}_{N \to \infty} = \frac{1}{2}$ appears (it corresponds to the behavior of total depolarization for the channel, $\rho_{\text{out}} \equiv (\bigcirc_N \Lambda)[\rho] = \frac{\sigma_0}{2}$). The cases $p = \frac{1}{4}$ coincide for all N because for $N = 1$ the total depolarized state $\frac{\sigma_0}{2}$ is obtained, then any further application of the teleportation cannot worsen the outcome.

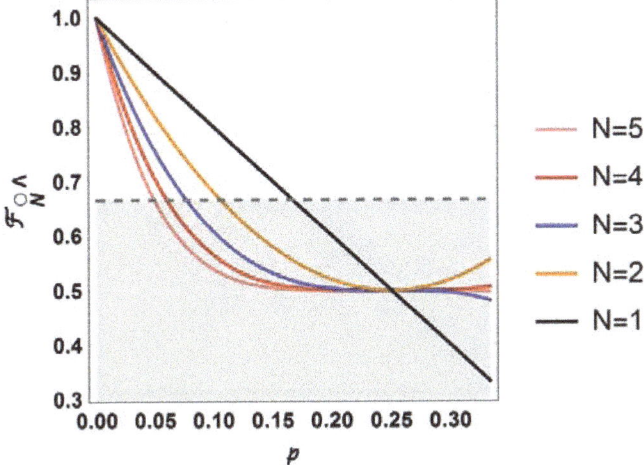

Figure 2. Sequential fidelity as function of the number N of channels being applied, and p is the deformation strength in $|\chi\rangle$.

3. Quantum Teleportation Assisted by Indefinite Causal Order with N Channels

In this section, we will consider a generalization of some variants of the process under indefinite causal order as they are presented in [7,22] by considering N channels in a superposition of causal orders. By applying N channels in a superposition of causal orders, we could have $N!$ combinations with different orders. Thus, we will need a control state with such number of dimensions ($|0\rangle$ sets for the normal sequential order of gates $T_1, T_2, ..., T_N$) to rule the application of each causal order:

$$\rho_c = \left(\sum_{i=0}^{N!-1} \sqrt{q_i} |i\rangle_c \right) \left(\sum_{j=0}^{N!-1} \sqrt{q_j} \langle j|_c \right) = \sum_{i,j=0}^{N!-1} \sqrt{q_i q_j} |i\rangle_c \langle j|. \tag{9}$$

For a definite causal order of teleportation channels $T_{i_1}, T_{i_2}, ..., T_{i_N}$ given by the element $\pi_k \in \Sigma_N$ in the symmetric group of permutations Σ_N from the ordered case, it has the effect:

$$\pi_k = \begin{pmatrix} T_{i_1} & T_{i_2} & \cdots & T_{i_N} \\ T_{i_{j_1}} & T_{i_{j_2}} & \cdots & T_{i_{j_N}} \end{pmatrix} \to \pi_k(K_{i_1} K_{i_2} \cdots K_{i_N}) = K_{i_{j_1}} K_{i_{j_2}} \cdots K_{i_{j_N}}, \tag{10}$$

and symbolically corresponding to the control state $|k\rangle_c$. Then, the corresponding Kraus operators $W_{i_1,i_2,\dots,i_N}$ are:

$$W_{i_1,i_2,\dots,i_N} = \sum_{k=0}^{N!-1} \pi_k (K_{i_1} K_{i_2} \dots K_{i_N}) \otimes |k\rangle_c \langle k|, \tag{11}$$

where in the following, we will drop the tensor product symbol $\otimes$ in the sake of simplicity. Thus, the output for N-channels in superposition is given by:

$$
\begin{aligned}
\Lambda^N [\rho \otimes \rho_c] &= \sum_{i_1,i_2,\dots,i_N} W_{i_1,i_2,\dots,i_N} \rho \otimes \rho_c \left(W_{i_1,i_2,\dots,i_N} \right)^\dagger \tag{12} \\[2mm]
&= \sum_{i_1,i_2,\dots,i_N} \left(\sum_k \pi_k (K_{i_1} K_{i_2} \dots K_{i_N}) |k\rangle\langle k| \right) \rho \otimes \rho_c \left(\sum_{k'} \pi_{k'} (K_{i_1} K_{i_2} \dots K_{i_N}) |k'\rangle\langle k'| \right)^\dagger \\[2mm]
&= \sum_{i_1,i_2,\dots,i_N} p_{i_1} \cdots p_{i_N} \left(\sum_k \pi_k (\sigma_{i_1} \cdots \sigma_{i_N}) |k\rangle\langle k| \right) \rho \otimes \rho_c \left(\sum_{k'} \pi_{k'}^\dagger (\sigma_{i_1} \cdots \sigma_{i_N}) |k'\rangle\langle k'| \right) \tag{13} \\[2mm]
&= \sum_{\substack{i_1,i_2,\dots,i_N \\ k,k'}} p_{i_1} \cdots p_{i_N} \sqrt{q_k q_{k'}} |k\rangle\langle k'| \otimes \pi_k (\sigma_{i_1} \cdots \sigma_{i_N}) \rho \pi_{k'}^\dagger (\sigma_{i_1} \cdots \sigma_{i_N}).
\end{aligned}
$$

Still, we can use the last formula to reach a simpler expression using combinatorics and then the properties of Pauli operators. In fact, noting that the sum in (14) includes all different values given to each $i_1, i_2, \dots, i_N$, after they are permuted as distinguishable objects by π_k and $\pi_{k'}$, it can be transformed into:

$$\sum_{i_1=0}^{3} \sum_{i_2=0}^{3} \dots \sum_{i_N=0}^{3} \longrightarrow \sum_{t_1=0}^{N} \sum_{t_2=0}^{N-t_1} \sum_{t_3=0}^{N-t_1-t_2} \sum_{p=1}^{N'}, \tag{14}$$

where t_j is the number of scripts in $i_1, i_2, \dots, i_N$ equal to $j = 0, 1, 2, 3$ ($t_0 = N - t_1 - t_2 - t_3$). Sum over p runs on the distinguishable arrangements obtained with a fix number t_j of operators σ_j departing from $\sigma_0^{t_0} \sigma_1^{t_1} \sigma_2^{t_2} \sigma_3^{t_3}$ by means of a certain permutation $\pi_{k_p^{t_1,t_2,t_3}}$. Then, the permutations among identical operators in each one of the four types $\sigma_0, \sigma_1, \sigma_2, \sigma_3$ are indistinguishable. There, $N' = \frac{N!}{t_0! t_1! t_2! t_3!}$. In such case, Formula (14) can be written as:

$$
\begin{aligned}
\Lambda^N [\rho \otimes \rho_c] &= \sum_k \sum_{k'} \sqrt{q_k q_{k'}} |k\rangle\langle k'| \sum_{t_1=0}^{N} \sum_{t_2=0}^{N-t_1} \sum_{t_3=0}^{N-t_1-t_2} \prod_{j=0}^{3} p_j^{t_j} \otimes \tag{15} \\[2mm]
&\sum_{p=1}^{N'} \pi_k \left(\pi_{k_p^{t_1,t_2,t_3}} \left(\sigma_0^{t_0} \sigma_1^{t_1} \sigma_2^{t_2} \sigma_3^{t_3} \right) \right) \rho \left(\pi_{k'} \left(\pi_{k_p^{t_1,t_2,t_3}} \left(\sigma_0^{t_0} \sigma_1^{t_1} \sigma_2^{t_2} \sigma_3^{t_3} \right) \right) \right)^\dagger,
\end{aligned}
$$

providing an easier formula for $\Lambda^N [\rho \otimes \rho_c]$ in terms of a definite number of sums and with the teleported state separated from the control state. From the properties of Pauli operators algebra, it is clear that both permutation terms besides ρ in (15) becomes equal until a sign. In addition, each one becomes in the set $\{\sigma_j | j = 0, 1, 2, 3\}$. Thus, (15) becomes a mixed state obtained as a linear combination of syndromes $\sigma_j \rho \sigma_j, j = 0, 1, 2, 3$ and normally entangled with the control state.

Following to [7], then we select an adequate basis to perform a measurement on the control state: $\mathcal{B} = \{|\psi_{M_i}\rangle | i = 1, 2, \dots, N!\}$. Such a measurement post-selects the original symmetry of the teleported state mixed with the control and the imperfect entangled state. In such a basis, we hope to find a privileged state $|\psi_m\rangle \in \mathcal{B}$ to stochastically maximize the fidelity with probability $\mathcal{P}_m$ (assuming ρ is a pure state). $\mathcal{P}_m$ sets the probability of success of the process. If the measurement of control does not conduct to $|\psi_m\rangle$, then other undesired teleportation outcome will be obtained. Then, if the desired outcome is not obtained, we disregard the output state. The fidelity and the success probability are:

$$\mathcal{F}_N = \frac{\mathrm{Tr}(\rho\langle\psi_m|\Lambda^N[\rho\otimes\rho_c]|\psi_m\rangle)}{\mathcal{P}_m} \tag{16}$$

$$\mathcal{P}_m = \mathrm{Tr}(\langle\psi_m|\Lambda^N[\rho\otimes\rho_c]|\psi_m\rangle). \tag{17}$$

The process is depicted by Figure 3, where $N!$ causal orders are considered to arrive to the pictorial representation of a complete superposition of causal orders on the right. Each causal order corresponds to one definite order in the application of channels T_i ruled by the control state ρ_c above it.

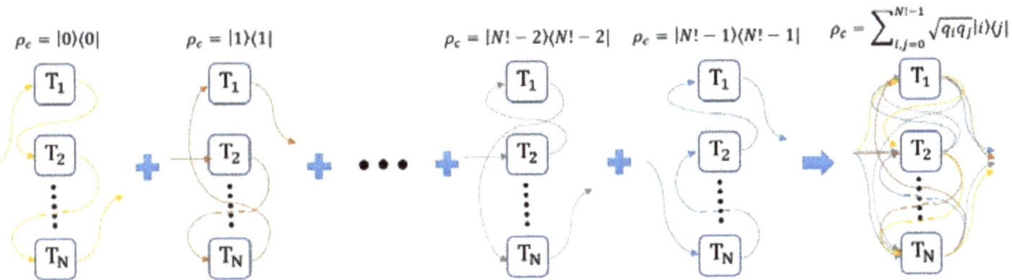

Figure 3. $N!$ causal order combinations for N identical teleportation channels $T_i, i = 1, 2, ..., N$ finally conforming a superposition of it. Each one is ruled by the control state above.

4. Analysis of Quantum Teleportation Assisted by the First Indefinite Causal Orders

In the following section, we deal with the analysis for the increasing number of teleportation channels after to remark some outcomes for the case $N = 2$ guiding the further analysis.

4.1. Teleportation with $N = 2$ Teleportation Channels in an Indefinite Causal Order Superposition

For the case $N = 2$, it has been obtained in [22] that (16) reduces to:

$$\mathcal{F}_2 = \frac{\sum_{i,j=0}^{3} p_i p_j\left(\left(\frac{1}{2}+(q_0-\frac{1}{2})\right)\cos\theta\right)\mathrm{Tr}(\rho\sigma_i\sigma_j\rho\sigma_j\sigma_i)+\sqrt{q_0 q_1}\sin\theta\cos\phi\,\mathrm{Tr}(\rho\sigma_i\sigma_j\rho\sigma_i\sigma_j)\right)}{\sum_{i,j=0}^{3} p_i p_j\left(\left(\frac{1}{2}+(q_0-\frac{1}{2})\right)\cos\theta\right)\mathrm{Tr}(\sigma_i\sigma_j\rho\sigma_j\sigma_i)+\sqrt{q_0 q_1}\sin\theta\cos\phi\,\mathrm{Tr}(\sigma_i\sigma_j\rho\sigma_i\sigma_j)\right)}, \tag{18}$$

then, a measurement on the control is made on the basis $\mathcal{B} = \{|\psi_m\rangle = \cos\frac{\theta}{2}|0\rangle + \sin\frac{\theta}{2}e^{i\phi}|1\rangle, |\psi_m^\perp\rangle = \sin\frac{\theta}{2}|0\rangle - \cos\frac{\theta}{2}e^{-i\phi}|1\rangle\}$, being $|\psi_m\rangle$ the supposed state maximizing $\mathcal{F}_2$. The corresponding probability to get that outcome becomes:

$$\mathcal{P}_m = \sum_{i,j=0}^{3} p_i p_j\left(\left(\frac{1}{2}+(q_0-\frac{1}{2})\right)\cos\theta\right)\mathrm{Tr}(\sigma_i\sigma_j\rho\sigma_j\sigma_i)+\sqrt{q_0 q_1}\sin\theta\cos\phi\,\mathrm{Tr}(\sigma_i\sigma_j\rho\sigma_i\sigma_j)\right). \tag{19}$$

Last formulas, (18) and (19) become reduced for pure states $\rho = |\psi\rangle\langle\psi|, |\psi\rangle = \alpha|0\rangle + \beta|1\rangle$ and $p_0 = 1 - 3p, p_1 = p_2 = p_3 = p$ by considering the identities:

$$\sum_{i,j=0}^{3} p_i p_j \mathrm{Tr}(\rho\sigma_i\sigma_j\rho\sigma_j\sigma_i) = 1 - 4p + 8p^2 \tag{20}$$

$$\sum_{i,j=0}^{3} p_i p_j \mathrm{Tr}(\sigma_i\sigma_j\rho\sigma_j\sigma_i) = 1 \tag{21}$$

$$\sum_{i,j=0}^{3} p_i p_j \mathrm{Tr}(\rho\sigma_i\sigma_j\rho\sigma_i\sigma_j) = (1 - 2p)^2 \tag{22}$$

$$\sum_{i,j=0}^{3} p_i p_j \text{Tr}(\sigma_i \sigma_j \rho \sigma_i \sigma_j) \;=\; 1 - 12p^2. \tag{23}$$

Note that the combination of the two first formulas gives the sequential case in (5). The other two terms correspond to the interference terms. First and third formulas can be demonstrated noting that:

$$\rho \;=\; \frac{1}{2}(\sigma_0 + \hat{n} \cdot \vec{\sigma}) \tag{24}$$

$$\text{with}: \quad \hat{n} = (|\alpha|^2 - |\beta|^2, \alpha\beta^* + \alpha^*\beta, i(\alpha\beta^* - \alpha^*\beta)),$$
$$\vec{\sigma} = (\sigma_1, \sigma_2, \sigma_3).$$

This fact is not exclusive of the case $N = 2$. Due to the Pauli operators algebra and the regarding they are traceless (while, $\text{Tr}(\sigma_0) = 2$), introducing (24) in (16) and (17), we note for $\mathcal{P}_m$ that only the terms containing σ_0 become different from zero. For $\mathcal{F}_N$, only the quadratic terms in σ_0 and $\hat{n} \cdot \vec{\sigma}$ become different from zero. For the terms quadratic in $\hat{n} \cdot \vec{\sigma}$, the additional condition $p_i = p_j \forall i \neq j (i, j \neq 0)$ is required in order to reduce the terms containing $\sigma_\alpha \sigma_\beta$ to the magnitude of $\hat{n}$, thus removing all reference of the teleported state.

In [7], it has been demonstrated that for $|\psi_m\rangle = |+\rangle$ the worst deformed state $|\chi\rangle$ with $p = \frac{1}{3}$ still lets a perfect teleportation with probability $\mathcal{P}_m = \frac{1}{3}$. In fact, Figure 4 summarizes the findings for the fidelity considering the two families of measurements with $|-\rangle$ (dashed orange lines) and $|+\rangle$ (dashed blue lines). The sequential case with $N = 2$ is reported as a continuous line black together with the single teleportation channel $N = 1$ (continuous red line). Dashed blue and orange lines go folded from $q_0 = 0, 1$ (two channels in definite causal order) nearest to the two sequential channels case in black to the outermost lines for $q_0 = \frac{1}{2}$ (the evenly distributed control state) reaching $\mathcal{F} = 1$ in $p = 0, \frac{1}{3}$ (blue for $|\psi_m\rangle = |+\rangle$) and $\mathcal{F} = \frac{1}{3}, \forall p$ (orange for $|\psi_m\rangle = |-\rangle$).

For the case $N = 2$, [22] has shown that for different values of $q_0 = \frac{1}{2}$, other measurements $|\psi_m\rangle = \cos\frac{\theta}{2}|0\rangle + \sin\frac{\theta}{2}e^{i\phi}|1\rangle$ are possible in order to achieve $\mathcal{F} = 1$ when $p = \frac{1}{3}$ giving $\phi = 0$ and θ distributed as in the Figure 5 as function of q_0. Thus, the best fidelities $\mathcal{F}_2$ depend entirely from p (see the color-scale besides in Figure 5) but the corresponding values of $\mathcal{P}_m$ go down far from $q_0 = \frac{1}{2}$ ($\theta = \frac{\pi}{2}$). The red dotted line is the threshold setting the minimum fidelity reached in the optimal case for $p = \frac{3-\sqrt{3}}{6}$, $\mathcal{F}_2 = \frac{1}{\sqrt{3}}$ [22]. Thus, we conclude that for $p = p_1 = p_2 = p_3$, the best state for the control is $q_0 = \frac{1}{2}$ in order to maximize $\mathcal{P}_m$, despite only for $p = \frac{1}{3}$ and $p \to 0$ it is possible to approach $\mathcal{F}_2 \to 1$. The last outstanding outcome for $p = \frac{1}{3}$ is a consequence of the two-folded interference introduced by the indefinite causal order together with the post-selection induced by the measurement which filters only constructive interference among the terms belonging to the original state.

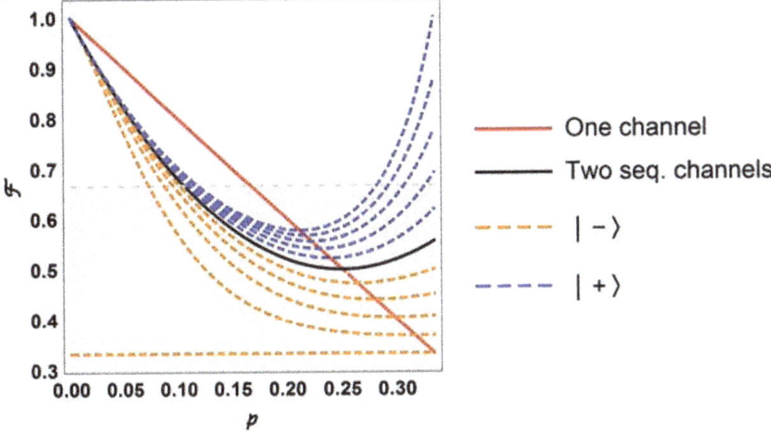

Figure 4. Fidelity for the case of two channels in indefinite causal order as function of p. The blue dashed upper line corresponds to $|\psi_m\rangle = |+\rangle$ and $q_0 = \frac{1}{2}$ reaching $\mathcal{F} = 1$ in $p = \frac{1}{3}$.

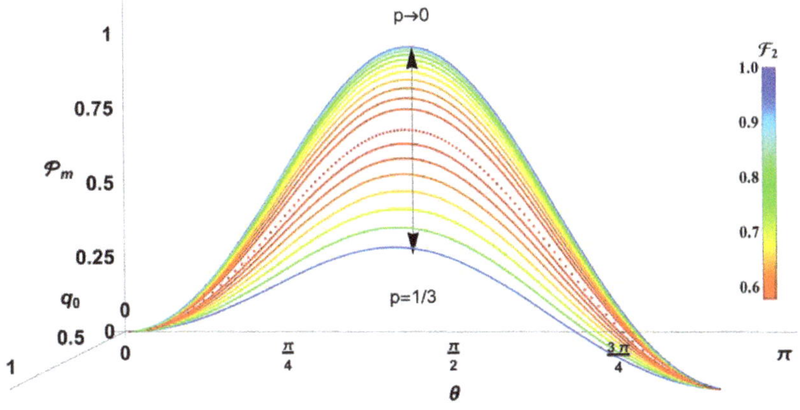

Figure 5. Condensed outcomes for the case $N = 2$. The respective probability $\mathcal{P}_m$ of measurements are included as function of q_0 and θ in $|\psi_m\rangle = \cos\frac{\theta}{2}|0\rangle + \sin\frac{\theta}{2}e^{i\phi}|1\rangle$ ($\phi = 0$ in the optimal measurement). Fidelity depends entirely from p, and $\mathcal{P}_m$ goes down while $p \to \frac{1}{3}$.

Fidelity (18) can be still analysed for independent values of p_1, p_2, p_3. Figure 6 shows a numerical analysis to search the best possible fidelity (achieved for certain teleported state) $\max_{|\psi_m\rangle, q_0}(\mathcal{F}_2)$ for all possible $|\psi_m\rangle$ and $0 \le q_0 \le 1$. The value of fidelity $\mathcal{F}_2$ is represented in color in agreement with the color-scale bar besides. Figure 6a shows a cut from the entire plot showing the inner core where fidelity goes down (three parts are symmetric). The higher values of fidelity on the faces of polyhedron suggest that better solutions can be reached for other cases with unequal values of $p_i, i = 1, 2, 3$, particularly for the frontal face $p_0 = 0$ completely colored in blue in Figure 6. The case $p_1 = p_2 = p_3 \equiv p$ falls in the central red dashed division crossing the clearer core reflecting the outcome in Figure 4, where not good values of $\mathcal{F}_2$ are inevitably obtained far from $p = 0$ and $p = \frac{1}{3}$. In addition, complementary information for such cases is given by $\mathcal{P}_m$ in Figure 6b, the probability to reach the corresponding higher fidelity in each process assisted by an intermediate optimal measurement on the control qubit. The plot depicts disperse outcomes barely around of $\mathcal{P}_m \approx 0.5$. Note that the computer process to obtain Figure 6a,b requires optimization on lots of parameters, thus requiring a considerable time of processing. The region (p_1, p_2, p_3) was divided in 10^7 points to perform such optimization. After, each point is reported as a colored sphere to fill the space in order to give a representation in color about the continuity of $\mathcal{F}_2$ and $\mathcal{P}_m$. Such an approach gives a certain impression of blurring in the figures, but they are reported with the best precision available under numerical processing. Particularly, Figure 6b is a collage of colored dots due to $\mathcal{P}_m$ is reported on an average basis, due the optimization was made on $\mathcal{F}_2$ on the left. By performing a numerical statistics of our outcomes for each $\mathcal{P}_m$, we get an approximation to its statistical distribution $\rho_{\mathcal{P}_m}$ included in the upper inset in Figure 6b. This distribution shows symmetric behavior around of $\mathcal{P}_m = 0.5$ as it could be expected for the numerical optimization.

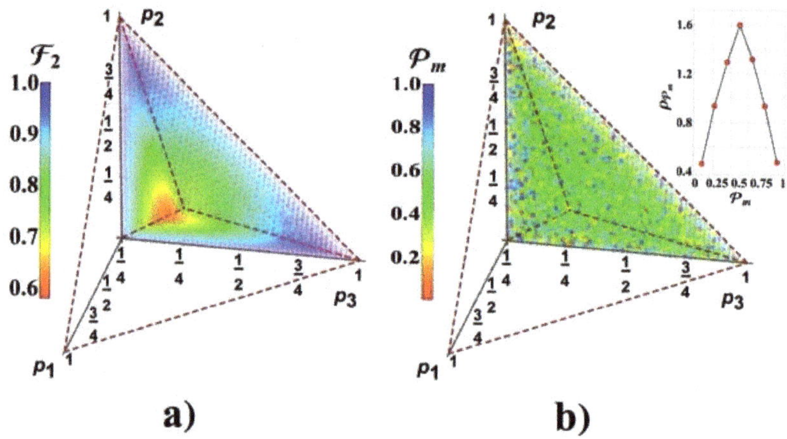

Figure 6. (**a**) Best fidelity $\mathcal{F}_2$ for the two-channels case as function of p_1, p_2, p_3. Each point inside the polyhedron corresponds to their acceptable values and it is coloured in agreement with its fidelity value (see the color-scale besides); the cut of polyhedron region exhibits the inner structure; (**b**) The corresponding values for measurement probabilities $\mathcal{P}_m$ denoting disperse values around 0.5. The upper inset confirms the statistical distribution $\rho_{\mathcal{P}_m}$ exhibiting symmetry around $\mathcal{P}_m = 0.5$.

4.2. Teleportation with an Increasing Number of Teleportation Channels in an iNdefinite Causal Order Superposition

Formula (15) exhibits the superposition of terms finally involving the states $\rho, \sigma_1 \rho \sigma_1, \sigma_2 \rho \sigma_2$ and $\sigma_3 \rho \sigma_3$ while they become entangled with the control state ρ_c. In the next sections, we deal with two cases of interest for the use of the teleportation algorithm under indefinite causal order.

4.2.1. Case $p_1 = p_2 = p_3 \equiv p$

First, we will address with the case $p = p_1 = p_2 = p_3$ widely used in the literature for simplicity. In [7], it has been suggested that for $|\psi_m\rangle$ having one of the following forms:

$$|\varphi_m^{\pm}\rangle \equiv \frac{1}{\sqrt{N!}} \sum_{i=0}^{N!-1} (\pm 1)^{\sigma(\pi_i)} |i\rangle. \tag{25}$$

The teleportation fidelity becomes optimal. There, σ is the signature of the parity of each order $|i\rangle$. By considering (15) together with (25) and the control state with $q_k = \frac{1}{N!} \forall k = 0, 1, ..., N! - 1$:

$$\langle \varphi_m^{\pm} | \Lambda^N [\rho \otimes \rho_c] | \varphi_m^{\pm} \rangle = \sum_k \sum_{k'} \frac{1}{N!^2} (\pm 1)^{\sigma(\pi_k)+\sigma(\pi_{k'})} \sum_{t_1=0}^{N} \sum_{t_2=0}^{N-t_1} \sum_{t_3=0}^{N-t_1-t_2} \prod_{j=0}^{3} p_j^{t_j} \cdot \tag{26}$$

$$\sum_{p=1}^{N'} \pi_k \left(\pi_{k_p^{t_1, t_2, t_3}} \left(\sigma_0^{t_0} \sigma_1^{t_1} \sigma_2^{t_2} \sigma_3^{t_3} \right) \right) \rho \left(\pi_{k'} \left(\pi_{k_p^{t_1, t_2, t_3}} \left(\sigma_0^{t_0} \sigma_1^{t_1} \sigma_2^{t_2} \sigma_3^{t_3} \right) \right) \right)^{\dagger}.$$

Then, we have developed the Formulas (14) and (16) with $|\psi_m\rangle = |\varphi_m^{\pm}\rangle$ in (25) to get both $\mathcal{F}_N$ and $\mathcal{P}_N$ for $N = 2, 3, 4$. Those formulas have been plotted (they are not reported here because their complexity, despite they are included in the Appendix A), the outcomes are shown in Figure 7 showing that a perfect fidelity $\mathcal{F}_N = 1$ for $p = \frac{1}{3}$ is achieved when $|\varphi_m^{\pm}\rangle$ meets with the same parity to N (p is indicated in the color-scale besides). Despite, for $p = \frac{1}{3}$ the success probabilities $\mathcal{P}_m$ decrease while N increases. For $|\varphi^-\rangle$ and $N = 4$, we get $\mathcal{P}_m = 0$, thus $\mathcal{F}_4$ becomes undefined in such a case. While $p \in [0, \frac{1}{6}]$ the best election is the single teleportation channel, for $p \in [\frac{1}{6}, \frac{1}{3}]$, the assistance of the causal order becomes an alternative to enhance the fidelity of teleportation, particularly with $N = 2$ channels.

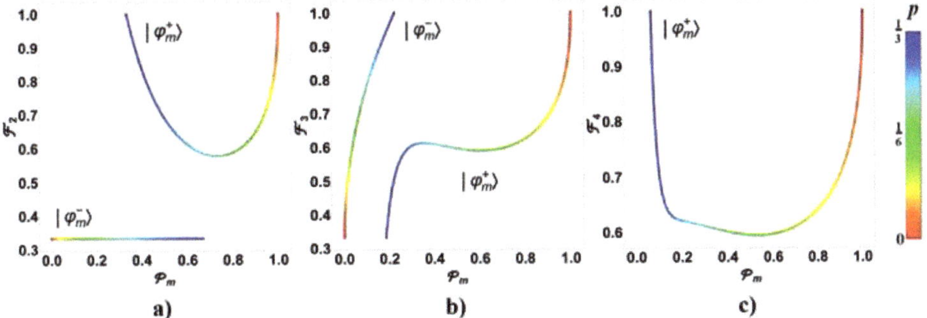

Figure 7. Probability $\mathcal{P}_m$ to obtain different values of fidelity $\mathcal{F}_N$ when the measurement states $|\varphi_+\rangle$ or $|\varphi_-\rangle$ are applied for cases (**a**) $N = 2$, (**b**) $N = 3$ and (**c**) $N = 4$. Color-scale bar depicts the respective value for p for $N = 2, 3, 4$.

Figure 8 again compares the fidelity $\mathcal{F}_N$ versus p for both measurements with the corresponding sequential case showing the alternated optimization of $\mathcal{F}_N$ as function of the parity of N and $|\varphi_m^{\pm}\rangle$. Despite, the outcomes in Figure 6 suggest analysing the behavior of $\mathcal{F}_N$ for independent values of p_1, p_2, p_3.

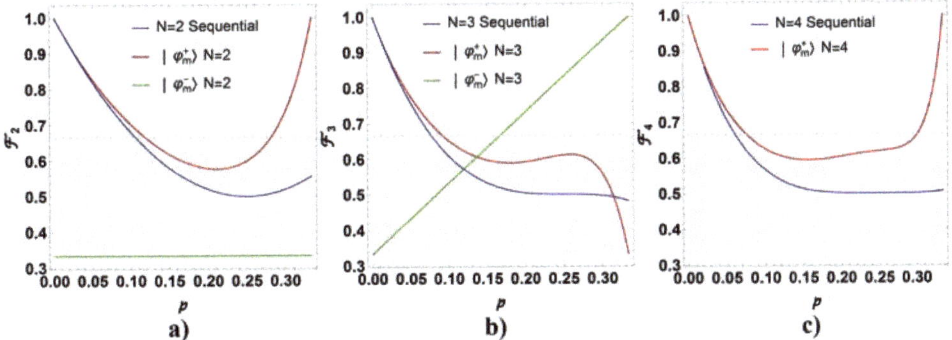

Figure 8. Comparison of fidelity obtained when the channels are applied sequentially (blue) and with indefinite causal order depending on the measurement state $|\varphi_m^{\pm}\rangle$ (red) and $|\varphi_m^-\rangle$ (green), for the cases (**a**) $N = 2$, (**b**) $N = 3$, and (**c**) $N = 4$ (in this last case, the fidelity becomes undefined for $|\varphi_m^-\rangle$).

4.2.2. Case $p_j \ll 1$, $j = 1, 2, 3$

In some practical cases, the expected values for the entangled resource $|\chi\rangle$ vary slightly from a perfect entangled state: $p_j \ll 1$ for $j = 1, 2, 3$. Thus, the outcome described through Formula (15) becomes in this case (developing to first order for $p_j, j = 1, 2, 3$ the factor $\prod_{j=0}^{3} p_j^{t_j}$ there):

$$\Lambda^N[\rho \otimes \rho_c] \approx \left[\left(1 - N \sum_{j=1}^{3} p_j \right) \rho + N \sum_{j=1}^{3} p_j \sigma_j \rho \sigma_j \right] \otimes \rho_c \equiv \rho_{out} \otimes \rho_c. \tag{27}$$

Note that under this approximation, ρ_c becomes unaltered and separated from the system state. Thus, the optimal way to teleport the state implies to measure the control state considering $|\psi_m\rangle = \sum_k \sqrt{q_k}|k\rangle$. In the following, we assume such an optimal measurement made on the control state.

For the particular case where $p_j = \frac{1}{4N}$ with $j = 1, 2, 3$, last formula can be written as:

$$\Lambda^N[\rho \otimes \rho_c] \approx \frac{1}{2}\sigma_0 \otimes \rho_c. \tag{28}$$

Obtaining the totally depolarized state $\frac{1}{2}\sigma_0$. Notice that it is only applicable for very large values of N (due to the assumption $p_j \ll 1, j = 1, 2, 3$). This aspect is advised in the Figure 7 where the fidelity drops more rapidly to $\frac{1}{2}$ when N grows around of $p = 0$.

In general, the probability and fidelity given in (27) will become respectively (developing to first order in $p_j, j = 1, 2, 3$):

$$\mathcal{P}_m \approx \text{Tr}[\rho_{out}] = 1 \tag{29}$$

$$\mathcal{F}_N \approx \frac{\text{Tr}[\rho \rho_{out}]}{\mathcal{P}_m} = 1 - N \sum_{j=1}^{3} p_j(1 - n_j^2) \equiv 1 - N p_{ts} \sum_{j=1}^{3} \alpha_j(1 - n_j^2) \equiv 1 - N p_{ts} \Delta_{\theta,\phi}^{\alpha_1,\alpha_2,\alpha_3}, \tag{30}$$

where ρ was written as in (24). We are introduced the reduced parameters $\alpha_j \in [0, 1]$ and the threshold probability $p_{ts} \ll 1$ to limit the validity of the current approximation ($p_j = p_{ts}\alpha_j \ll 1, j = 1, 2, 3$). We note in any case that the increasing of N worsens the fidelity. Note each term in the sum in (30) is non-negative, thus the fidelity becomes commonly reduced. Because only one of $n_j^2, j = 1, 2, 3$ could be one at the time, then it is necessary in addition that two p_j become zero to get $\mathcal{F}_N = 1$. Otherwise, $\mathcal{F}_N < 1$ with a notable decreasing if N is large. The outcome in (29) exhibits a combination of the three error-syndromes $\sigma_1 \rho \sigma_1, \sigma_2 \rho \sigma_2, \sigma_3 \rho \sigma_3$ reflected through the terms $\alpha_j(1 - n_j^2)$ as function of α_j. Thus, for each syndrome $\sigma_j \rho \sigma_j$ the best states being teleported are those closer to the eigenstates of σ_j, otherwise while several $\alpha_j \neq 0$ the teleportation capacity is widely reduced.

Considering $\rho = |\psi\rangle\langle\psi|$ with $|\psi\rangle = \cos\frac{\theta}{2}|0\rangle + \sin\frac{\theta}{2}e^{i\phi}|1\rangle$ on the Bloch sphere: $n_1 = \sin\theta\cos\phi, n_2 = \sin\theta\sin\phi, n_3 = \cos\theta$. Then, we analyze each syndrome and its impact on the fidelity through the quantity $\Delta_{\theta,\phi}^{\alpha_1,\alpha_2,\alpha_3}$. As lower it becomes, higher becomes $\mathcal{F}_N$. Figure 9a shows the simple behavior of $\Delta_{\theta,\phi}^{\alpha_1,\alpha_2,\alpha_3}$ for each state on the Bloch sphere under each syndrome: $p_1 = 1, p_2 = p_3 = 0$; $p_2 = 1, p_1 = p_3 = 0$; and $p_3 = 1, p_1 = p_2 = 0$ in such order. We have denoted as $|0_j\rangle$ and $|1_j\rangle$ to the eigenstates of $\sigma_j, j = 1, 2, 3$ (or $j = x, y, z$). Note the behavior commented in the previous paragraph.

Despite, the most interesting issue is centered in the fact that the entanglement resource $|\chi\rangle$ is normally unknown but with a tiny variation of $|\beta_0\rangle$ through the deformation parameters p_1, p_2, p_3. By calculating the average and the standard deviation of $\Delta_{\theta,\phi}^{\alpha_1,\alpha_2,\alpha_3}$ on the parameters $\alpha_1, \alpha_2, \alpha_3 \in [0, 1]$:

$$\mu_{\Delta_{\theta,\phi}^{\alpha_1,\alpha_2,\alpha_3}} = \int_0^1 \int_0^1 \int_0^1 \Delta_{\theta,\phi}^{\alpha_1,\alpha_2,\alpha_3} d\alpha_1 d\alpha_2 d\alpha_3 = 1 \rightarrow \mu_{\mathcal{F}_N} = 1 - N p_{ts} \tag{31}$$

$$\sigma_{\Delta_{\theta,\phi}^{\alpha_1,\alpha_2,\alpha_3}} = \sqrt{\mu_{(\Delta_{\theta,\phi}^{\alpha_1,\alpha_2,\alpha_3})^2} - (\mu_{\Delta_{\theta,\phi}^{\alpha_1,\alpha_2,\alpha_3}})^2}$$

$$= \frac{1}{8\sqrt{6}} \sqrt{53 + \sin^4(\theta)\cos(4\phi) + 4\cos(2\theta) + 7\cos(4\theta)} \in [\frac{1}{3}, \frac{1}{\sqrt{6}}] \tag{32}$$

$$\rightarrow \sigma_{\mathcal{F}_N} = N p_{ts} \sigma_{\Delta_{\theta,\phi}^{\alpha_1,\alpha_2,\alpha_3}}. \tag{33}$$

We note that the average value of fidelity $\mathcal{F}_N = 1 - N p_{ts}$ becomes independent from the state being teleported. While, the dispersion for $\Delta_{\theta,\phi}^{\alpha_1,\alpha_2,\alpha_3}$ on the values p_1, p_2, p_3 depends from the teleported state and it becomes lowest for the eigenstates of $\sigma_1, \sigma_2, \sigma_3$. In fact, the exact result for the case of $N = 1$ is precisely (30) with such value in (1): $\mathcal{F}_1 = 1 - \sum_{j=1}^{3} p_j (1 - n_j^2)$, thus the values in (33) are scaled from it by a factor N. The reason is easily noticed, the ρ_{out} in (27) obtained by linearization from (3) coincides with the sequential case (3) under linearization, so both cases exactly meet under the current limit. It implies that indefinite causal order procedure in teleportation becomes unpractical in this limit.

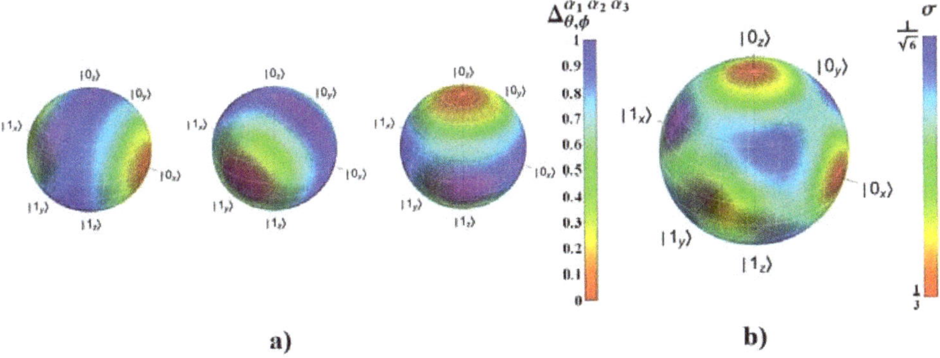

Figure 9. Bloch sphere showing under the assumption $p_j \ll 1$, $j = 1, 2, 3$ for each state: (a) $\Delta_{\theta,\phi}^{\alpha_1,\alpha_2,\alpha_3}$ in color obtained for each syndrome in (27), $\sigma_1 \rho \sigma_1, \sigma_2 \rho \sigma_2, \sigma_3 \rho \sigma_3$ respectively, and (b) the standard deviation $\sigma_{\Delta_{\theta,\phi}^{\alpha_1,\alpha_2,\alpha_3}}$ in (33). Red is the best fidelity in (a) and the lower dispersion in (b).

4.3. Notable Behavior on the Frontal Face of Parametric Region: Case $p_0 = 0$

The behavior of $\mathcal{F}_2$ on the frontal face ($p_0 = 0$) in Figure 6 can be now better advised in Figure 10. There, we have calculated numerically (for 10^5 states covering the frontal face), the best fidelity obtained using two teleportation channels under indefinite causal order by taking the optimal measurement on the control state together with the best state able to be teleported. Thus, it represents naively the best possible scenario.

In the last process, for each $|\chi\rangle$ on the frontal face, we have additionally taken a sample of 10^2 sets of values for $q_0 \in [0, 1]$ (the initialization value for the control state for $N = 2$), $\theta \in [0, \pi], \phi \in [0, 2\pi]$ for $|\psi_m\rangle$ and $\theta_0 \in [0, \pi], \phi_0 \in [0, 2\pi]$ for the teleported state $|\psi\rangle = \cos\frac{\theta_0}{2}|0\rangle + \sin\frac{\theta_0}{2}e^{i\phi_0}|1\rangle$. Each value is used as initial condition to find a local maximum for the fidelity $\mathcal{F}_2$. Then, those values are used to predict the global maximum of $\mathcal{F}_2$ for each point on the frontal face. Figure 10a shows the best fidelity on the face together with the statistical distribution of the fidelities on the frontal face in the upper image of Figure 10c, which suggests that $\mathcal{F}_2 = 1$ could be obtained on the face always (the little dispersion with lower values of $\mathcal{F}_2 \in [0.9, 1]$ are due to the numerical procedure followed). The same follows for $\mathcal{P}_m$ (Figure 10b,c lower) but denoting that such probabilities of success are centrally distributed around $\frac{1}{2}$ (note they are not the best probabilities because the process is centred on maximize $\mathcal{F}_2$). As in Figure 6, images in Figure 10 appear blurred due to the limited number of points considered because the time processing.

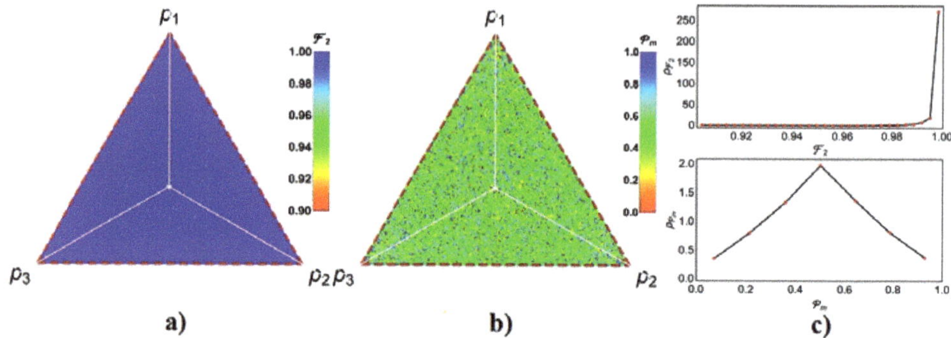

Figure 10. Optimal fidelity using two teleportation channels in indefinite causal order followed by an appropriate measurement $|\varphi_m\rangle$. (**a**) The best fidelity obtained for certain teleported state if optimal control measurement is obtained, (**b**) the probability $\mathcal{P}_m$ of success for the last process, and (**c**) the statistical distribution for $\mathcal{F}_2$ and $\mathcal{P}_m$.

Nevertheless, the last fact is in reality a blind strategy. A more critical view of Formulas (18) and (19) and referring to [22] which numerically suggests that $q_0 = \sin^2\frac{\theta}{2} = \frac{1}{2}(1 - \cos\theta), \phi = 0$ is related with the optimal case for the case $p = p_1 = p_2 = p_3 = \frac{1}{3}$. In fact, in such case, last formulas become reduced to:

$$\mathcal{F}_2 = \frac{\sum_{i,j=0}^{3} p_i p_j \left(\mathrm{Tr}(\rho\sigma_i\sigma_j\rho\sigma_j\sigma_i) + \mathrm{Tr}(\rho\sigma_i\sigma_j\rho\sigma_i\sigma_j)\right)}{\sum_{i,j=0}^{3} p_i p_j \left(\mathrm{Tr}(\sigma_i\sigma_j\rho\sigma_j\sigma_i) + \mathrm{Tr}(\sigma_i\sigma_j\rho\sigma_i\sigma_j)\right)} \tag{34}$$

$$\mathcal{P}_m = \frac{\sin^2\theta}{2} \sum_{i,j=0}^{3} p_i p_j \left(\mathrm{Tr}(\sigma_i\sigma_j\rho\sigma_j\sigma_i) + \mathrm{Tr}(\sigma_i\sigma_j\rho\sigma_i\sigma_j)\right). \tag{35}$$

Last formula explains the reason because the case $\theta = \frac{\pi}{2}$ is optimal for $\mathcal{P}_m$. Moreover, on the frontal face $p_0 = 0$ (then $i, j = 1, 2, 3$), then (34) and (35) clearly become (by splitting the cases $i = j$ from $i \neq j$, noting for the last case $\sigma_i\sigma_j = -\sigma_j\sigma_i$ and the fact that we are dealing with pure states):

$$\mathcal{F}_2 = 1 \tag{36}$$

$$\mathcal{P}_m = \sin^2\theta \sum_{i=1}^{3} p_i^2, \quad \text{with} : \sum_{i=1}^{3} p_i = 1. \tag{37}$$

Thus, the last conditions make the teleportation optimal not only for $p = p_1 = p_2 = p_3 = \frac{1}{3}$ but also for the entire cases on the frontal face, being independent from the teleported state. Nevertheless, the probability of success depends entirely from the values of p_i (considering only the best case $\theta = \frac{\pi}{2}$). Figure 11 shows the distribution of $\mathcal{P}_m$ on the frontal face (in some cases we will denote this probability by $\mathcal{P}_{m,N=2}^{\mathrm{ff},\{p_i\}}$ to state $\theta = \frac{\pi}{2}, p_0 = 0$ and p_i arbitrary but fulfilling $p_1 + p_2 + p_3 = 1$), which ranges on $[\frac{1}{3}, 1]$. In fact, the case $p = p_1 = p_2 = p_3 = \frac{1}{3}$ corresponds to the worst case for $\mathcal{P}_m$ in the center of the face. We have constructed the norm on the frontal face to report such distribution. The mean $\mu_{\mathcal{P}_m} = \frac{1}{2}$ and the standard deviation $\sigma_{\mathcal{P}_m} \approx 0.13$ were calculated using such distribution.

In order to solve the cases for $N > 2$ by including further teleportation channels under indefinite causal order, last analysis suggests for arbitrary N that the procurement of an analytical formula for (15) is in order at least for the case $p_0 = 0$, implying $t_0 = 0$:

$$\Lambda^N\left[\rho\otimes\rho_c\right] \;=\; \sum_k\sum_{k'}\sqrt{q_k q_{k'}}\,|k\rangle\langle k'|\sum_{t_1=0}^{N}\sum_{t_2=0}^{N-t_1}\prod_{j=1}^{3}p_j^{t_j}\otimes \tag{38}$$

$$\sum_{p=1}^{N'}\pi_k\left(\pi_{k_p^{t_1,t_2,t_3}}\left(\sigma_1^{t_1}\sigma_2^{t_2}\sigma_3^{t_3}\right)\right)\rho\left(\pi_{k'}\left(\pi_{k_p^{t_1,t_2,t_3}}\left(\sigma_1^{t_1}\sigma_2^{t_2}\sigma_3^{t_3}\right)\right)\right)^{\dagger}$$

$$=\;\sum_k\sum_{k'}\sqrt{q_k q_{k'}}\,|k\rangle\langle k'|\sum_{t_1=0}^{N}\sum_{t_2=0}^{N-t_1}\prod_{j=1}^{3}p_j^{t_j}\otimes\sum_{p=1}^{N'}\Sigma_{k_p}^{k}\Sigma_{k_p}^{k'}\left(\sigma_1^{t_1}\sigma_2^{t_2}\sigma_3^{t_3}\right)\rho\left(\sigma_1^{t_1}\sigma_2^{t_2}\sigma_3^{t_3}\right)^{\dagger} \tag{39}$$

and $t_3 = N - t_1 - t_2$. As it was previously mentioned, factors generated by π_k and $\pi_{k'}$ are equal until a sign. In addition, they always evolve to $\sigma_0, \sigma_1, \sigma_2$ or σ_3 (easily depending on the parity of t_1, t_2, t_3). Thus, those factors and their signs state the introduction of syndromes on ρ together with interference among them and the different orders. Such interference could be manipulated through the parameters q_k, p_j.

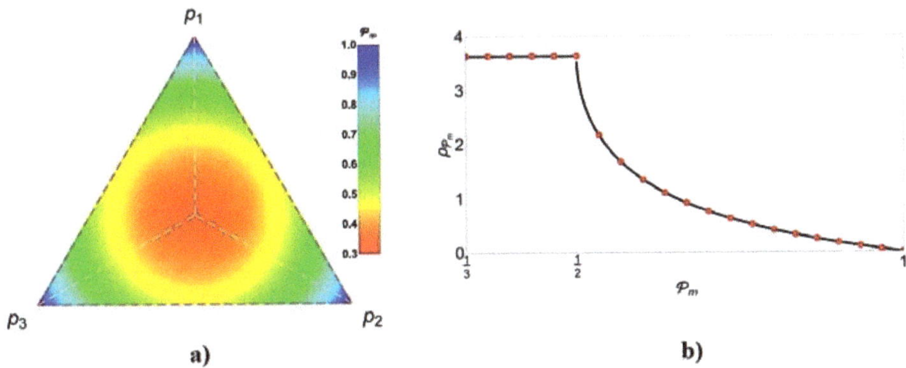

Figure 11. (a) Values of $\mathcal{P}_m$ on $p_0 = 0$ face , and (b) its corresponding statistical distribution $\rho_{\mathcal{P}_m}$ for two teleportation channels in indefinite causal order.

Even so, this formula is not easy to address in order to get a simpler closed result because the sign $\Sigma_{k_p}^{k}, \Sigma_{k_p}^{k'}$ introduced in the permutation with respect $\sigma_1^{t_1}\sigma_2^{t_2}\sigma_3^{t_3}$ cannot be advised easily (see a parallel analysis in [8]). Nevertheless, we can still to analyse computationally the cases for the lowest values of N (analytical cases addressed by computer aided methods due to the factorial increasing number of terms). Thus, formulas for $\mathcal{P}_{m,N}^{ff,\{p_i'\}}$ and $\mathcal{F}$ for N larger than two have been obtained using a computational treatment. The formulas obtained in the analysis are reported in Appendix B. As in our previous discussion for the case $p_1 = p_2 = p_3 = p$ in the Section 4.2.1, $\mathcal{F} = 1$ is obtained for all cases on the frontal face if the measurement in the indefinite causal order becomes $|\varphi_m^+\rangle$ for $N = 2, 4$ and $|\varphi_m^-\rangle$ for $N = 3$ independently of the teleported state. Again, it is a consequence of the order interference due to the indefinite causal order together the post-selection induced by the measurement. For complementary cases using other measurement outcomes, we get $\mathcal{F} \neq 1$ depending from p_1, p_2, p_3 or still undefined, and additionally depending from the teleported state (see Appendix B).

5. An Alternative Procedure Introducing Weak Measurement

In spite of the previous outcomes, we guess the indefinite causal order could not work properly at any point inside of region depicted in the Figure 6. Nevertheless, due to the outcomes in [7] for the case $p = p_1 = p_2 = p_3$ and those exhibited in the Figure 6, the teleportation process assisted by indefinite causal order (at least for two channels) becomes optimal on $p_0 = 0$ and $p_0 = 1$ (the origin and the frontal face in Figure 6a). Then, we propose an alternative strategy beginning with a weak measurement on the entangled resource.

5.1. General Case for $N = 2$ Assisted by a Weak Measurement

By considering first the following weak measurements on $|\chi\rangle$, we get the post-measurement states and their probabilities of occurrence:

$$P_0 = |\beta_0\rangle\langle\beta_0| \to |\chi_0\rangle = (P_0|\chi\rangle)_{\text{norm}} = |\beta_0\rangle, \quad \tilde{p}_0 = p_0 \tag{40}$$

$$P_1 = \mathbb{I} - P_0 \to |\chi_1\rangle = (P_1|\chi\rangle)_{\text{norm}} = \sum_{i=1}^{3} \sqrt{\frac{p_i}{\tilde{p}_1}}|\beta_i\rangle \equiv \sum_{i=1}^{3} \sqrt{p'_i}|\beta_i\rangle, \quad \tilde{p}_1 = \sum_{i=1}^{3} p_i \tag{41}$$

which projects the entangled state on one of the two states $|\chi_0\rangle$ or $|\chi_1\rangle$ with probabilities $\tilde{p}_0$ or $\tilde{p}_1$ respectively. Each state is located on the origin or otherwise on the frontal face of the region in Figure 6. Then, if $|\chi_0\rangle$ is obtained, the teleportation process can go as in the Figure 1, otherwise, if $|\chi_1\rangle$ is obtained, we can try the teleportation assisted by indefinite causal order (at this point the reader could note that clearly, we need two entangled resources). We will come back to discuss the weak measurement strategy widely in the last section).

The entire process is depicted in the Figure 12, a schematic diagram of the process as it was originally proposed by [7]. Given certain state to teleport, we use certain entangled resource $|\chi_a\rangle$. It goes through the weak measurement in (40) to get $|\chi_{a0}\rangle = |\beta_0\rangle$ with probability p_0. Then we perform a single teleportation. Instead, by obtaining $|\chi_{a1}\rangle$ with probability $1 - p_0$, then we prepare a second entangled resource $|\chi_b\rangle$ repeating with it the same procedure, if after of the weak measurement $|\chi_{b0}\rangle = |\beta_0\rangle$ is obtained with probability p_0, we disregard $|\chi_{a1}\rangle$ proceeding with a single teleportation using such state. Otherwise, if $|\chi_{b1}\rangle$ is obtained, we perform a two-channel teleportation assisted by indefinite causal order using the states previously obtained. There, the teleportation will become successful with probability $p'^2_1 + p'^2_2 + p'^2_3$, otherwise it becomes unsuccessful and we need disregard the process. Thus, the global probability of success is (there, $\mathcal{P}^{\text{ff}}_{m,N=2}$ corresponds to (37) with $\theta = \frac{\pi}{2}, \phi = 0$, renaming p_i as p'_i, with $p'_1 + p'_2 + p'_3 = 1$):

$$\begin{aligned}
\mathcal{P}_{\text{Tot}} &= p_0 + (1-p_0)p_0 + (1-p_0)^2 \mathcal{P}^{\text{ff},\{p'_i\}}_{m,N=2} \\
&= p_0 + (1-p_0)p_0 + (1-p_0)^2 \sum_{i=1}^{3} p'^2_i = 1 - 2(p_1 p_2 + p_2 p_3 + p_3 p_1)
\end{aligned} \tag{42}$$

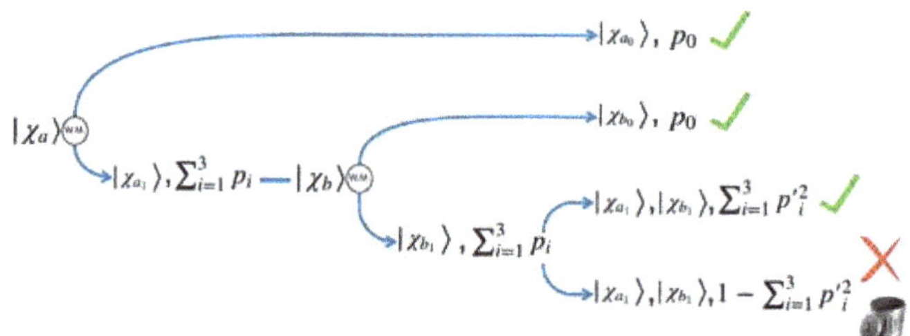

Figure 12. Schematic teleportation process assisted by weak measurement.

The last function has been represented in the plots of Figure 13. For each initial set (p_1, p_2, p_3) of the entangled resources (assumed identical), $\mathcal{P}_{\text{Tot}}$ is plotted in color in agreement with the bar besides in the Figure 13a. One-third of the plot has shown, due to its symmetry, to exhibit its inner structure. The corresponding statistical distribution is obtained numerically in the Figure 13b by uniformly sampling the space in the figure on the left. The mean value of $\mathcal{P}_{\text{Tot}}$ becomes 0.70 and their standard deviation 0.16.

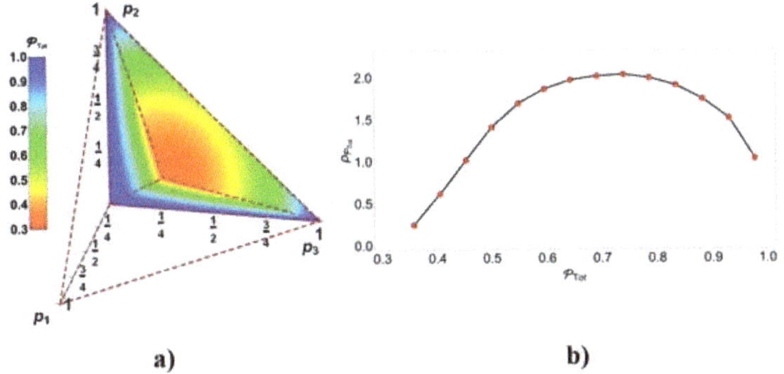

Figure 13. Distribution of $\mathcal{P}_{\text{Tot}}$: (**a**) as function of (p_1, p_2, p_3), and (**b**) as statistical distribution by itself obtained numerically from the data of (**a**).

5.2. Cases for $N \geq 2$ Assisted by a Weak Measurement

For $N \geq 2$, the procedure follows as in the previous section by introducing N imperfect entangled resources, $|\chi_i\rangle$ (assumed identical for simplicity) but in each step, we decide if after of the weak measurement, the state $|\chi_{j_0}\rangle = |\beta_0\rangle$ is used to perform a single teleportation or if we continue the process of weak measurement N times on identical entangled resources $|\chi_j\rangle$ to finally get $|\chi_{N1}\rangle = \sum_{i=1}^{3} p_i'|\beta_i\rangle$ as in the Figure 12. The corresponding situation is now depicted for the general case in the Figure 14. In this case, the global probability of success becomes:

$$\mathcal{P}_{\text{Tot}N} = \sum_{j=1}^{N} p_0(1-p_0)^{j-1} + (1-p_0)^N \mathcal{P}_{m,N}^{\text{ff},\{p_i'\}}. \tag{43}$$

Inserting the formulas for $\mathcal{P}_{m,N}^{\text{ff},\{p_i'\}}$ in Appendix B (specialized for the frontal face $p_0 = 0$ and changing p_i by p_i'). Then, we can get the outcomes for global probability $\mathcal{P}_{\text{Tot}N}$ for the last cases with $\mathcal{F} = 1$:

$$\mathcal{P}_{\text{Tot}2} = 1 - 2(p_1 p_2 + p_1 p_3 + p_2 p_3) \tag{44}$$

$$\mathcal{P}_{\text{Tot}3} = 1 - (p_1^3 + p_2^3 + p_3^3) - 3(p_1^2(p_2 + p_3) + p_2^2(p_1 + p_3) + p_3^2(p_1 + p_2)) \tag{45}$$

$$\mathcal{P}_{\text{Tot}4} = 1 - 4(p_1^3(p_2 + p_3) + p_2^3(p_1 + p_3) + p_3^3(p_1 + p_2))$$

$$-12 p_1 p_2 p_3(p_1 + p_2 + p_3) - \frac{16}{3}(p_1^2 p_3^2 + p_2^2 p_3^2 + p_1^2 p_2^2). \tag{46}$$

Now, we can visualize last outcomes for $\mathcal{P}_{\text{Tot}}$ in Figure 15. Again, all the entangled states used for the teleportation process are assumed to be identical by simplicity. Figures 15a–c depict the probability $\mathcal{P}_{\text{Tot}N}$ to reach $\mathcal{F} = 1$ in the entire process represented in color. Each color bar shows the entire range of values for such probabilities on the graphs. According to the color, the blue zone represents the region where $\mathcal{P}_{\text{Tot}} \to 1$, observing for the case $N = 4$ a larger blue area, suggesting still the goodness of increase the number of teleportation channels under indefinite causal order combined with post-measurement.

Figure 15d depicts a numerical analysis of statistical distribution for the cases $N = 2, 3, 4$. Note that for $N = 3$, all greater values for the probability occur almost evenly. For the case $N = 4$, it is observed a larger amount of success probabilities than failure probabilities compared with $N = 3$. Despite, $\mu_{\mathcal{P}_{\text{Tot}2}} \approx 0.702, \sigma_{\mathcal{P}_{\text{Tot}2}} \approx 0.158$ and $\mu_{\mathcal{P}_{\text{Tot}4}} \approx 0.667, \sigma_{\mathcal{P}_{\text{Tot}4}} \approx 0.249$ (because for $N = 2$ there are a large distribution for medium success probabilities). In any case, the most successful outcomes of teleportation appears for $N = 4$.

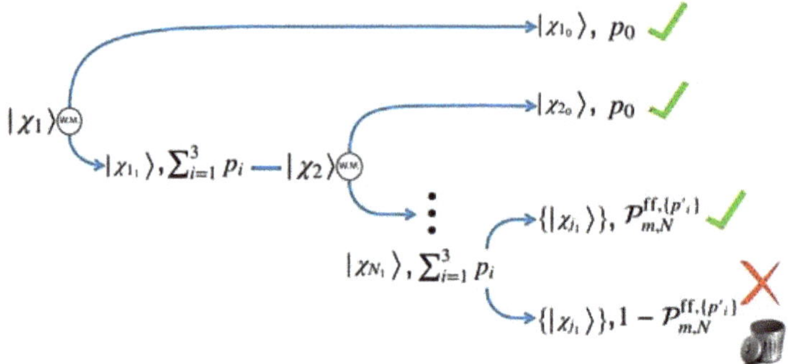

Figure 14. Schematic teleportation process assisted by indefinite causal order using N-teleportation channels and weak measurement.

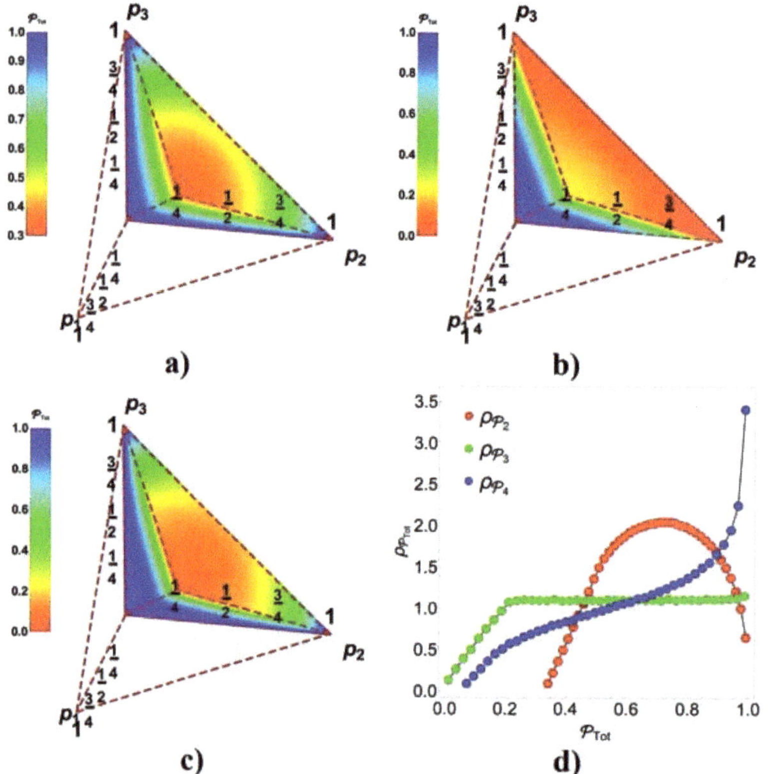

Figure 15. (**a**–**c**) values of $\mathcal{P}_{\text{Tot}}$ as function of (p_1, p_2, p_3), for N_2, N_3 and N_4 respectively. (**d**) Statistical distribution numerically obtained for $\mathcal{P}_{\text{Tot2}}$, $\mathcal{P}_{\text{Tot3}}$ and $\mathcal{P}_{\text{Tot4}}$.

6. Experimental Deployment of Teleportation with Indefinite Causal Order

In this section, we comment on some main experimental developments for a possible deployment of indefinite causal order in teleportation. We begin with the procedure to set the weak measurement used in Section 5.1. Afterwards, we set some elements and experimental developments to propose the implementation of the theoretical proposal before presented.

6.1. Implementation of Weak Measurement to Project $|\chi\rangle$

In Section 5.1, we stated the implementation of a weak measurement to project $|\chi\rangle$ conveniently onto $|\chi_0\rangle = |\beta_0\rangle$ or $|\chi_1\rangle = \sum_{i=1}^{3} p_i'|\beta_i\rangle$. Despite, in the experimental approach, there are certain differences due to the resources been used. In this section, we present how to afford the weak measurement stated in (40). We use an ancilla qubit $|0_a\rangle$ to do the measurement minimizing the impact on $|\chi\rangle$ as is desired. In this implementation, we will use as a central resource the Toffoli gate. In order to prepare the $|\chi\rangle$ stated properly for such measurement, we combine it with the ancilla. Then, we send the combined system into the circuit presented in Figure 16a. This circuit employs the Toffoli gate $\mathcal{T}_{1,2,a}$ on channels 1, 2 for $|\chi\rangle$ and a for $|0_a\rangle$ together with the $C^1 Not_2$ gate (already developed for ions [25,26] and photons [27]). In fact, it is well-known that Toffoli gate can be performed using $CNOT$ gates and single-qubit gates [14] o by means of the Sleator–Weinfurter construction [28], despite other more efficient developments are known for ions [29] and photons [30]. Some single-qubit gates as Hadamard ($\mathcal{H}$) and Not ($\mathcal{X}$) are also used. In the following development, we write $|\chi\rangle = \sum_{i=0}^{3} \sqrt{p_i^*}|\beta_i\rangle$ as the imperfect entangled resource (be aware that $*$ not means complex conjugation). Thus, all necessary quantum gates have been experimentally developed in our days at least in quantum optics.

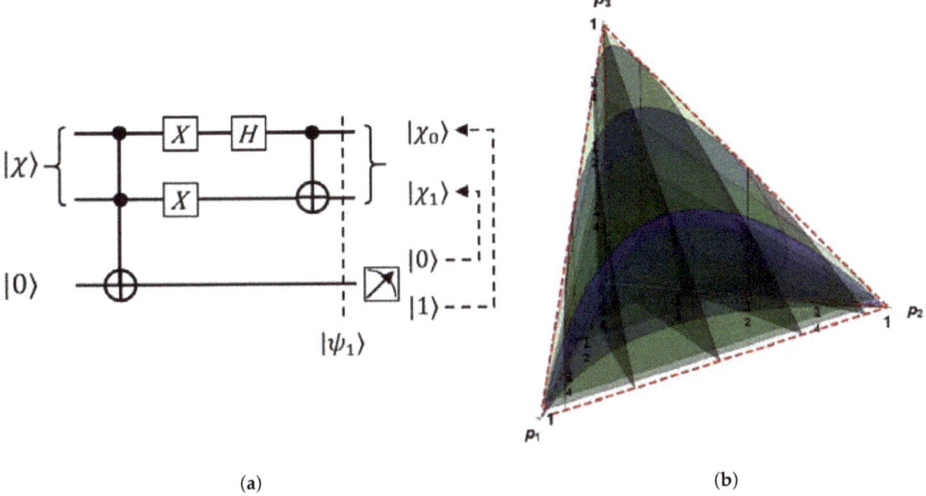

(a) (b)

Figure 16. (a) Quantum circuit generating the weak measurement on $|\chi\rangle$, and (b) contour plots for the map on the region (p_1, p_2, p_3) between those probabilities and (p_1^*, p_2^*, p_3^*).

A direct calculation shows that this circuit performs the following transformation on $|\psi_0\rangle = |\chi\rangle \otimes |0_a\rangle$ into:

$$|\psi_1\rangle \quad = \quad \sqrt{p_0}|\beta_0\rangle \otimes |1_a\rangle + \left(\sqrt{p_1}|\beta_1\rangle + \sqrt{p_2}|\beta_2\rangle + \sqrt{p_3}|\beta_3\rangle\right) \otimes |0_a\rangle \tag{47}$$

$$\text{with:} \quad \sqrt{2p_0} = \sqrt{p_0^*} - \sqrt{p_3^*}, \quad \sqrt{2p_1} = \sqrt{p_1^*} - \sqrt{p_2^*}, \tag{48}$$

$$\sqrt{2p_2} = \sqrt{p_0^*} + \sqrt{p_3^*}, \quad \sqrt{2p_3} = \sqrt{p_1^*} + \sqrt{p_2^*}$$

Just before of the projective measurement on the qubit a shown in the Figure 16a. Clearly, after measurement, two possible outcomes arise in the qubit a, $|1_a\rangle, |0_a\rangle$ while on qubits $1, 2$ the outcomes are $|\chi_0\rangle = |\beta_0\rangle$, $|\chi_1\rangle = \sum_{i=1}^{3} \sqrt{p_i'}|\beta_i\rangle$ respectively as in the Section 5.1, thus completing the weak measurement. The only difference with respect our previous development is that those coefficients are not the original $\{p_i^*\}$. Despite this, in the event that such coefficients are unknown, this fact is not important, the really outstanding outcome is that this procedure projects the state into the perfect Bell state to perform the teleportation $|\beta_0\rangle$ or otherwise on the frontal face if this resource is planned to be used under indefinite causal order and measurement (as it was previously depicted in the procedure of Section 5.1). Anyway, Figure 16b shows the contour plots of p_1^* (red), p_2^* (green) and p_3^* (blue) in the region (p_1, p_2, p_3) as a reference of the involved geometric transformations.

6.2. An Insight View about Teleportation Implementing Indefinite Causal Orders Experimentally with Light

Formula (1) regards the teleportation algorithm as a quantum communication channel. Despite this formula being a useful simplification for the theoretical analysis, it expresses the teleportation channel with the input and output through the same system, which is not precisely the real experimental situation. Then, as it was true for the original implementation of the original teleportation proposal [9] in [31], the deployment should be modified to have a correct approach to the theory. In this section we discuss an insight view into the experimental deployment together with indefinite causal order based on current techniques and experimental developments.

A possible implementation with light should to consider an initial state with at least three initial photons exhibiting each one at least a pair of quantum variables as polarization, frequency or spatial localization (**k**-vector state) among others (as in the original experimental teleportation proposal [31]): $|\psi_0\rangle = |v\rangle_1 \otimes |v\rangle_a \otimes |v\rangle_b$, using polarization in the vertical direction as instance. Those photons should then be converted into five photons by splitting the last two into entangled pairs using Spontaneous Parametric Down Conversion (SPDC) [32] as instance, while the first state is arbitrarily rotated by a quartz polarization rotator (QPR) [33] -to generate the state to teleportate-: $|\psi_1\rangle = (\alpha|v\rangle_1 + \beta|h\rangle_1) \otimes \frac{1}{\sqrt{2}}(|v\rangle_2|h\rangle_3 + |h\rangle_2|v\rangle_3) \otimes \frac{1}{\sqrt{2}}(|v\rangle_4|h\rangle_5 + |h\rangle_4|v\rangle_5)$. After, five photons should be sent together into two alternative directions (through a dichroic beamsplitter—a splitting wavelength dependent—instead a polarization beamsplitter) coincidentally, not independently (it means five photons will travel through corresponding paths labeled by p_A or p_B). This beamsplitter (BS) works as our control state superposing the two path states (the two causal orders further). Last effect should be solved based on the frequency of original photons which should be quantum generated to let a quantum splitting of all beams (or otherwise based on the previous generation of a GHZ state [34]). This necessary beamsplitter is still a cutting-edge technology. Such spatial quantization introduces an additional quantum variable thus converting the initial state into (removing the tensor product symbols for the sake of simplicity):

$$|\psi_2\rangle = \frac{1}{\sqrt{8}}\Big((\alpha|v\rangle_1 + \beta|h\rangle_1)|p_A\rangle_1(|v\rangle_2|h\rangle_3 + |h\rangle_2|v\rangle_3)|p_A\rangle_2|p_A\rangle_3(|v\rangle_4|h\rangle_5 + |h\rangle_4|v\rangle_5)|p_A\rangle_4|p_A\rangle_5$$

$$+ (\alpha|v\rangle_1 + \beta|h\rangle_1)|p_B\rangle_1(|v\rangle_2|h\rangle_3 + |h\rangle_2|v\rangle_3)|p_B\rangle_2|p_B\rangle_3(|v\rangle_4|h\rangle_5 + |h\rangle_4|v\rangle_5)|p_B\rangle_4|p_B\rangle_5 \Big) \tag{49}$$

If additionally we introduce certain optical distortion in the SPDC, we get imperfect entangled states then changing each $\frac{1}{\sqrt{2}}(|v\rangle_i|h\rangle_j + |h\rangle_i|v\rangle_j)$ by $|\chi\rangle_{ij}$. In the following, we will change v, h by $0, 1$ respectively for simplicity.

Note that teleportation is, in a certain sense, automatically generated due to the non-locality of the resource $|\beta_0\rangle$ (or imperfectly by $|\chi\rangle$), then collapsed on four adequate outcomes involving an additional correction as a function of those outcomes using classical communication (Figure 1a). In addition, for two sequential teleportation channels, the process can be achieved by post-measurement at the end of both processes. Nevertheless, the implementation of indefinite causal order in teleportation introduces additional challenges due to the connectivity of paths and measurements. In the process, it will be required the implementation of the $SWAP$ gate, which has already been experimentally performed in optics [35,36].

Thus, Figure 17 depicts a possible implementation for two teleportation processes assisted by indefinite causal order. The first photon goes to the QRP and then the five photons go through the coordinated BS. The proposed process can be then followed in the Figure 17 with paths labeled by p_A in green and p_B in red. For simplicity, teleportation processes are assumed to perform measurements on the Bell states basis as in Figure 1b, thus avoiding the use of H and $CNOT$ gates in the analysis. Due to the above construction (post-measurement and measurement assumed on the Bell basis), almost no gates are present in the process, just two $SWAP$ gates stating the causal connections. At the end of each path, a semi-transparent mirror should mix again the paths (but not the polarization) by pairs into the basis $|\pm\rangle_i = \frac{1}{2}(|p_A\rangle_i \pm |p_B\rangle_i)$ for each photon i, in order to erase the information of the path followed. We labeled each path (or the information being carried on it) by M_{ij}^k (in case that photon carries the information of one of the complementary systems not containing the output of teleportation) remarking the path type followed $k = A, B, +, -$; the final belonging teleportation process $i = 1, 2$; and the number of the sequential qubit to be measured there: $j = 1$ for the former input and $j = 2$ for the correspondent to the first qubit of the original entangled resource. Instead, the final outputs are labeled by S^k ($k = A, B, +, -$). By following the color, the reader should easily identify each path considering additionally the effect of the intermediate use of $SWAP$ gates which is discussed below.

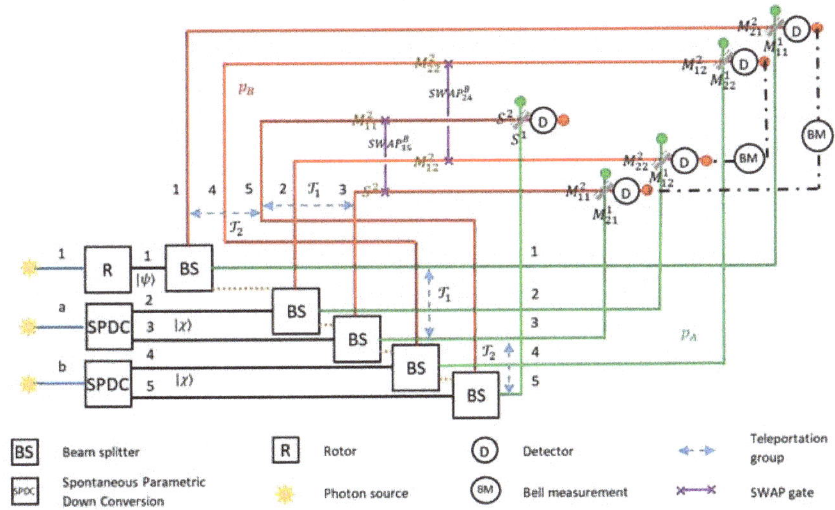

Figure 17. Diagram for implementation of teleportation with causal ordering as it is discussed in the text. Photons are split on two different set of paths to superpose the two causal orders of two sequential teleportation process.

By ignoring first the $SWAP$ gates in the Figure 17, we can realize that the circuit has not any effect. We have indicated each optical element described before. The dotted line connecting the BS's denotes the not independent functioning, all together should send the five photons on the green paths or on the red ones. States $|\psi\rangle$ and $|\chi\rangle$ are remarked on photons 1 and 2, 3, 4, 5 respectively. Each path (green or red) is labeled from 1 to 5 in agreement with the photon carried out. Blue arrow remarks the group of photons involved in each teleportation process $\mathcal{T}_1$ or $\mathcal{T}_2$ on each path (the first subscript in M_{ij}^k): $1,2,3$ and $3,4,5$ respectively for the green paths, and $1,4,5$ and $5,2,3$ respectively for the red ones. On each path, we reported the associated label for each system S^k or M_{ij}^k as it was depicted before. Note that brown labels correspond to the information being carried before of $SWAP$ gates, while black labels are the final states reported there at the end of the path but before of the recombining in the semi-transparent mirrors. The reason for the $SWAP$ gate between the paths 3 and 5 should be clear, we need to get the teleportation outputs on the same photon to generate the superposition of information. The $SWAP$ gate on the red paths 2 and 4 exchanges the information on those paths in order to generate the superposition at the end among path information M_{ij}^1 and $M_{i'j}^2$ with $i \neq i'$, $j = 1,2$ thus mixing both. Note that the set of states in M_{ij}^k are those to be measured in the teleportation process (here in the Bell basis by pairs) in order to correct the output states. The reader should advise this process does not reproduce exactly that depicted by (1) because such formula assumes the measurements are internal operations generating a mixed state coming from the corresponding projections and corrections. In this approach, we have the possibility to measure only four qubits instead of eight. Despite, we will note this procedure still reproduces some of the main previous features analyzed. At the end of the process, each semi-transparent mirror (diagonal in grey) mixes the information on the states $|\pm\rangle_i$ for each photon i on the red and green edges (with information $M_{ij}^\pm$ or $S^\pm$ respectively -red and green-, not represented in the Figure 17). On the red edges, a detector first decides if the photon exits through them (they are the projective measurement on $|\varphi_m^\pm\rangle$ states in our development). In addition, a Bell measurement is then performed on each pair $1,3$ and $2,4$ in order to inquire the information codified in the output S^+.

A direct but large calculation to expand (49) then applying the $SWAP$ gates and projecting on $|+\rangle_i, i = 1,...,5$ was performed. Finally, this output was written in terms of $|\beta_i\rangle_{1,3} \otimes |\beta_j\rangle_{2,4}, i,j = 0,...,3$ to ease the identification of final successful measurements. If $p_0 = 1$ or $p_0 = 0$, upon the measurement of $|\beta_i\rangle_{1,3} \otimes |\beta_j\rangle_{2,4}$ and then the application of $\sigma_i\sigma_j$ as correction, the output S^+ becomes $|\psi\rangle$ faithfully in the following cases:

- If $p_0 = 1$ for all $i,j = 0,...,3$ cases with a global successful probability of $\frac{1}{16}$.
- If $p_0 = 0$ for the cases $i = 0,...,3$ and $j = 2$ with a global successful probability of $\frac{(p_1 - p_2 + p_3)^2}{64}$.

This clearly resembles our main outcomes. For the second case, other measurement outcomes give imperfect teleportation thus rearranging the success probabilities with respect of those in our theoretical development. Thus, alternative experimental proposals should be developed to approach them into the ideal case considered in our theoretical results.

7. Conclusions

Quantum teleportation has an important role in quantum processing for the transmission of quantum information, nevertheless, there are possible issues on the entangled resource assisting the teleportation process mainly related to its maintenance and precise generation. It introduces imprecision in the teleported state. In this work, the implementation of indefinite causal order has been studied in order to propose an improved scheme to tackle such imprecision on the entangled state when it is combined with the measurement of the control assessing it.

The analysis for the redundant case where quantum channels are simply applied sequentially (assumed as identical) shows that the number N of channels applied, rapidly decreases the fidelity converging to the maximal depolarization of the teleported state thus obtaining $\mathcal{F}_{N \to \infty} = \frac{1}{2}$. By modifying the process under indefinite causal order for two or more teleportation channels as it was proposed by [7] and later discussed in [22], we advise advantages on the quantum fidelity of the teleported state for the first values N of sequential teleportation channels. From the outcomes, a categorization was performed to analyze the effects on the entangled state, thus obtaining a surprising enhancement for the most imperfect entangled resource, $p_0 = 0$ with the absence of the ideal entangled resource $|\beta_0\rangle$, and still for near regions of it with $p_0 \approx 0$ when N increases. Notably, in the first case, it is possible to obtain a perfect teleportation process with $\mathcal{F}_N = 1$. However, when N increases, the principal downside is the reduction of the probability of successful measurement $\mathcal{P}_m$, which decreases drastically as N increases.

In order to improve the global probability of success, we have proposed the combined use of weak measurement to first projecting the entangled resource to either $p_0 = 1$ with $p_1, p_2, p_3 = 0$ or $p_1 + p_2 + p_3 = 1$ with $p_0 = 0$, where the indefinite causal order generates the most notable enhancements. In such cases, $\mathcal{F} = 1$ is obtained always and $\mathcal{P}_m$ is improved. Those notable processes are possible as for pure as for mixed states [8]. A remarkable aspect is that for such a notable case the outcome is independent of the teleported state.

Finally, a more detailed process for the weak measurement (first barely discussed in the initial presentation) is after detailed and oriented to the practical implementation in terms of the current experimental developments for light and matter. The development of a Toffoli gate is advised as central in the implementation. In addition, an introductory analysis for a possible experimental implementation has been included for the teleportation process under indefinite causal order using two teleportation channels. Such an approach is still imperfect and not optimal. Despite this, it reproduces the main features found in our development. In the proposal, recent experiments and technological developments in optics become central, particularly the implementation of the $SWAP$ gate and the generation of $|GHZ\rangle$ states. A valuable aspect being noticed is the use of post-measurement in the teleportation process. Additional theoretical and experimental developments should still improve the vast possibilities of indefinite causal order in the teleportation research field.

Author Contributions: C.C.-I. performed the research related with the analysis of indefinite causal order using 2 channels setting the structure for the further research. Both authors performed the computer algorithms to reproduce the output states after of indefinite causal orders. F.D. performed the theoretical development of indefinite causal order using N channels. C.C.-I. contributed in the development of computer simulations and computer graphics in the entire manuscript. C.C.-I. perform the research about the feasibility of indefinite causal order using the current technologies. F.D. developed the experimental proposal to implement indefinite causal order in Section 6.2. All authors contributed evenly in the writing of the manuscript. All authors have read and agreed to the published version of the manuscript.

Funding: This research received no external funding.

Acknowledgments: F.D. acknowledges to Jesus Ramírez Joachín his taught and thoroughly discussions in combinatorics during 1982, without which some parts of this work will not be possible. Both authors acknowledge the economic support to publish this article to the School of Engineering and Science from Tecnologico de Monterrey. The support of CONACYT is also acknowledged.

Conflicts of Interest: The authors declare no conflict of interest.

Appendix A. Formulas for $\mathcal{P}_m$ and $\mathcal{F}$ for the Case $p_1 = p_2 = p_3 = p$

Formulas for the fidelity and the success probability as the number of channels in indefinite causal order increases when $p_1 = p_2 = p_3 = p$ have been obtained. For the case $N = 2$, when $|\psi_m\rangle = |\varphi_m^-\rangle$ the outcomes are:

$$\mathcal{F}_2 = \frac{1}{3}, \qquad \mathcal{P}_m = 6p^2 \tag{A1}$$

and for the case when $|\psi_m\rangle = |\varphi_m^+\rangle$, the expressions become:

$$\mathcal{F}_2 = \frac{6p^2 - 4p + 1}{1 - 6p^2}, \qquad \mathcal{P}_m = 1 - 6p^2. \tag{A2}$$

For the case $N = 3$, when $|\psi_m\rangle = |\varphi_m^-\rangle$ the outcomes are:

$$\mathcal{F}_3 = \frac{1}{3} + 2p, \qquad \mathcal{P}_m = 2p^2 \tag{A3}$$

and for the case when $|\psi_m\rangle = |\varphi_m^+\rangle$, we get:

$$\mathcal{F}_3 = \frac{-76p^3 + 54p^2 - 18p + 3}{96p^3 - 54p^2 + 3}, \qquad \mathcal{P}_m = 1 - 18p^2 + 32p^3. \tag{A4}$$

For the case $N = 4$, when $|\psi_m\rangle = |\varphi_m^-\rangle$ we get $\mathcal{P}_m = 0$, thus $\mathcal{F}_4$ becomes undefined in such case, while for the case when $|\psi_m\rangle = |\varphi_m^+\rangle$, we get the expressions:

$$\mathcal{F}_4 = \frac{360p^4 - 304p^3 + 108p^2 - 24p + 3}{-408p^4 + 384p^3 - 108p^2 + 3}, \qquad \mathcal{P}_m = 1 - 36p^2 + 128p^3 - 136p^4. \tag{A5}$$

Appendix B. Formulas for $P_{m,N}^{\mathrm{ff},\{p_i\}}$ and $\mathcal{F}$ for the Case $p_0 = 0$

In this section, formulas for $\mathcal{F}$ and $\mathcal{P}_{m,N}^{\mathrm{ff},\{p_i\}}$ when the entangled state has different values for p_1, p_2 and p_3 (note they are restricted to the frontal face $p_0 = 0$ of the parametric space) and the measurement of the control state is either $|\varphi_m^+\rangle$ or $|\varphi_m^-\rangle$. In those results, the angles θ and ϕ corresponds to the state being teleported ($|\psi\rangle = \cos\frac{\theta}{2}|0\rangle + \sin\frac{\theta}{2}e^{i\phi}|1\rangle$), thus meaning a dependence of those values on this state. For the case $N = 2$, with the privileged measurement state as $|\varphi_m^+\rangle$, the expressions become:

$$\mathcal{F}_2 = 1 \tag{A6}$$
$$\mathcal{P}_{m,N=2}^{\mathrm{ff},\{p_i\}} = p_1^2 + p_2^2 + p_3^2 \tag{A7}$$

and with the privileged state as $|\varphi_m^-\rangle$, the corresponding expressions are:

$$\mathcal{F}_2 = \frac{1}{2\mathcal{P}_{m,N=2}^{\mathrm{ff},\{p_i\}}}\Big(2p_1p_2(1 + \cos 2\theta) + p_3(p_1 + p_2)(1 - \cos 2\theta) \tag{A8}$$
$$+ 2p_3(p_2 - p_1)\sin^2\theta\cos 2\phi\Big)$$
$$\mathcal{P}_{m,N=2}^{\mathrm{ff},\{p_i\}} = 2(p_1p_2 + p_2p_3 + p_3p_1). \tag{A9}$$

For the case $N = 3$, with the privileged measurement state as $|\varphi_m^+\rangle$, the outcomes are:

$$\mathcal{F}_3 = \frac{1}{12\mathcal{P}_{m,N=3}^{\mathrm{ff},\{p_i\}}}\Big((3(p_1^3 + p_2^3 + 2p_3^3) + p_1(p_2^2 + p_3^2) + p_2(p_1^2 + p_3^2))(1 - \cos 2\theta) \tag{A10}$$
$$+ 2p_3(p_1^2 + p_2^2)(1 + \cos 2\theta) + 2(3(p_1^3 - p_2^3) + p_1(p_2^2 + p_3^2) - p_2(p_1^2 + p_3^2))\sin^2\theta\cos 2\phi\Big)$$
$$\mathcal{P}_{m,N=3}^{\mathrm{ff},\{p_i\}} = p_1^3 + p_2^3 + p_3^3 + \frac{1}{3}(p_1^2(p_2 + p_3) + p_2^2(p_1 + p_3) + p_3^2(p_1 + p_2)) \tag{A11}$$

while, with the privileged state as $|\varphi_m^-\rangle$, they become:

$$\mathcal{F}_3 = 1 \tag{A12}$$
$$\mathcal{P}_{m,N=3}^{\mathrm{ff},\{p_i'\}} = 6p_1p_2p_3. \tag{A13}$$

Finally, for the case $N = 4$, with the privileged measurement state as $|\varphi_m^+\rangle$, then:

$$\mathcal{F}_4 = 1 \tag{A14}$$
$$\mathcal{P}_{m,N=4}^{\mathrm{ff},\{p_i\}} = p_1^4 + p_2^4 + p_3^4 + \frac{2}{3}(p_1^2p_2^2 + p_1^2p_3^2 + p_2^2p_3^2) \tag{A15}$$

and if the privileged state is $|\varphi_m^-\rangle$, then we get $\mathcal{P}_{m,N=4}^{\mathrm{ff},\{p_i\}} = 0$, thus $\mathcal{F}$ gets undetermined.

References

1. Ebler, C.; Salek, S.; Chiribella, G. Enhanced Communication With the Assistance of Indefinite Causal Order. *Phys. Rev. Lett.* **2017**, *120*, 120502.

2. Chiribella, G.; Banik, M.; Bhattacharya, S.S.; Guha, T.; Alimuddin, M.; Roy, A.; Saha, S.; Agrawal, S.; Kar, G. Indefinite causal order enables perfect quantum communication with zero capacity channel. *arXiv* **2018**, arXiv:1810.10457v2.

3. Goswami, K.; Cao, Y.; Paz-Silva, G.A.; Romero, J.; White, A. Communicating via ignorance. *arXiv* **2019**, arXiv:1807.07383v3.

4. Procopio, L.M.; Moqanaki, A.; Araújo, M.; Costa, F.; Calafell, I.A.; Dowd, E.G.; Hamel, D.R.; Rozema, L.A.; Brukner, C.; Walther, P. Experimental superposition of orders of quantum gates. *Nat. Commun.* **2015**, *6*, 7913.

5. Procopio, L.M.; Delgado, F.; Enríquez, M.; Belabas, N.; Levenson, J.A. Communication enhancement through quantum coherent control of N channels in an indefinite causal-order scenario. *Entropy* **2019**, *21*, 1012.

6. Procopio, L.M.; Delgado, F.; Enríquez, M.; Belabas, N.; Levenson, J.A Sending classical information via three noisy channels in superposition of causal orders. *Phys. Rev. A* **2020**, *101*, 012346.

7. Mukhopadhyay, C.; Pati, A.K. Superposition of causal order enables perfect quantum teleportation with very noisy singlets. *J. Phys. Commun.* **2020**. [CrossRef]

8. Delgado, F.; Cardoso, F. Performance characterization of Pauli channels assisted by indefinite causal order and post-measurement. *Quantum Inf. Comput.* **2020**, *20*, 1261–1280.

9. Bennett, C.H.; Wiesner, S.J. Communication via One- and Two-Particle Operators on Einstein-Podolsky-Rosen States. *Phys. Rev. Lett.* **1992**, *69*, 2881-2884.

10. Bennett, C.H.; Brassard, G.; Crépeau, C.; Jozsa, R.; Peres, A.; Wootters, W.K.; Teleporting an Unknown Quantum State via Dual Classical and Einstein-Podolsky-Rosen Channels. *Phys. Rev. Lett.* **1993**, *70*, 1895.

11. Delgado, F. Teleportation based on control of anisotropic Ising interaction in three dimensions. *JPCS* **2015**, *624*, 012006.

12. Yan, F.; Yang, L. Economical teleportation of multiparticle quantum state. *Il Nuovo C. B* **2003**, *118*, 79.

13. Metcalf, B.J.; Spring, J.B.; Humphreys, P.C.; Thomas-Peter, N.; Barbieri, M.; Kolthammer, W.S.; Jin, X.-M.; Langford, N.K.; Kundys, F.; Gates, J.C.; et al. Quantum teleportation on a photonic chip. *Nat. Photonics* **2014**, *8*, 770.

14. Nielsen, M.A.; Knill, E.; Laflamme, R. Complete quantum teleportation using nuclear magnetic resonance. *Nature* **1998**, *396*, 52–55.

15. Delgado, F. Teleportation algorithm settled in a resonant cavity using non-local gates. In *Quantum Information and Measurement*; OSA Technical Digest: Rome, Italy, 2019.

16. Bouwmeester, D.; Pan, J.; Mattle, K.; Eibl, M.; Weinfurter, H.; Zeilinger, A. Experimental quantum teleportation. *Nature* **1997**, *390*, 575–579.

17. Ursin, R.; Jennewein, T.; Aspelmeyer, M.; Kaltenbaek, R.; Lindenthal, M.; Walther, P.; Zeilinger, A. Quantum teleportation across the Danube. *Nature* **2004**, *430*, 849.

18. Ma, X.S.; Herbst, T.; Scheidl, T.; Wang, D.; Kropatschek, S.; Naylor, W.; Wittmann, B.; Mech, A.; Kofler, J.; Anisimova, E.; et al. Quantum teleportation over 143 kilometres using active feed-forward. *Nature* **2012**, *489*, 269–273.

19. Luo, Y.H.; Zhong, H.S.; Erhard, M.; Wang, X.L.; Peng, L.C.; Krenn, M.; Jiang, X.; Li, L.; Liu, N.-L.; Lu, C.-Y.; et al. Quantum Teleportation in High Dimensions. *Phys. Rev. Lett.* **2019**, *123*, 070505.

20. Caleffi, M.; Cacciapuoti, A.S. Quantum Switch for the Quantum Internet: Noiseless Communications through Noisy Channels. *IEEE J. Sel. Areas Commun.* **2020**, *38*, 575.

21. Yin, J.; Ren, J.G.; Lu, H.; Cao, Y.; Yong, H.L.; Wu, Y.P.; Liu, C.; Liao, S.K.; Zhou, F.; Jiang, Y.; et al. Quantum teleportation and entanglement distribution over 100-kilometre free-space channels. *Nature* **2012**, *488*, 185–188.

22. Cardoso, C.; Delgado, F. Featuring causal order in teleportation of two quantum teleportation channels. *J. Phys. Conf. Ser.* **2020**, *1540*, 012024.

23. Bowen, G.; Bose, S. Teleportation as a Depolarizing Quantum Channel, Relative Entropy, and Classical Capacity. *Phys. Rev. Lett.* **2001**, *87*, 267901. [PubMed]

24. Petz, D. *Quantum Information Theory and Quantum Statistics*; Springer: Berlin/Heidelberg, Germany, 2008.

25. Schmidt-Kaler, F.; Häffner, H.; Riebe, M.; Gulde, S.; Lancaster, G.P.; Deuschle, T.; Becher, C.; Roos, C.F.; Eschner, J.; Blatt, R. Realization of the Cirac–Zoller controlled-NOT quantum gate. *Nature* **2003**, *422*, 408–411. [PubMed]

26. Maslov, D. Basic circuit compilation techniques for an ion-trap quantum machine. *New J. Phys.* **2017**, *19*, 023035.

27. Lopes, J.-H.; Soares, W.-C.; Bertúlio de Lima, B.; Caetano, D.P.; Askery, C. Linear optical CNOT gate with orbital angular momentum and polarization. *Quantum Inf. Process.* **2019**, *18*, 256.

28. Sleator, T.; Weinfurter, H. Elementary gates for quantum computation. *Phys. Rev. Lett.* **1995**, *74*, 4087.

29. Monz, T.; Kim, K.; Hänsel, W.; Riebe, M.; Villar, A.-S.; Schindler, P.; Chwalla, M.; Hennrich, M.; Blatt, R. Realization of the Quantum Toffoli Gate with Trapped Ions. *Phys. Rev. Lett.* **2009**, *102*, 040501.

30. Huang, H.L.; Bao, W.S.; Li, T.; Li, F.G.; Fu, X.Q.; Zhang, S.; Zhang, H.L.; Wang, X. Deterministic linear optical quantum Toffoli gate. *Phys. Lett. A* **2017**, *381*, 2673–2676.

31. Boschi, D.; Branca, S.; De Martini, F.; Hardy, L.; Popescu, S. Experimental Realization of Teleporting an Unknown Pure Quantum State via Dual Classical and Einstein-Podolsky-Rosen Channels. *Phys. Rev. Lett.* **1998**, *80*, 1121.

32. Boyd, R. *Nonlinear Optics*; Academic Press: Rochester, NY, USA, 2018.

33. Bass, M.; DeCusatis, C.; Enoch, J.; Lakshminarayanan, V.; Li, G.; MacDonald, C.; Mahajan, V.; Van Stryland, E. *Handbook of Optics*, 3rd ed.; McGraw-Hill Professiona: New York, NY, USA, 2009; Volume 1, pp. 3.46–3.61.

34. Erhard, M.; Malik, M.; Krenn, M.; Zeilinger, A. Experimental Greenberger–Horne–Zeilinger entanglement beyond qubits. *Nat. Photonics* **2018**, *12*, 759–764.

35. Zhu, M.; Ye, L. Implementation of swap gate and Fredkin gate using linear optical elements. *Int. J. Quantum Inf.* **2013**, *11*, 1350031.

36. Ono, T.; Okamoto, R.; Tanida1, M.; Hofmann, H.; Takeuchi, S. Implementation of a quantum controlled-SWAP gate with photonic circuits. *Sci. Rep.* **2017**, *7*, 45353. [PubMed]

Publisher's Note: MDPI stays neutral with regard to jurisdictional claims in published maps and institutional affiliations.

Article

Permutation Symmetry in Coherent Electrons Scattering by Disordered Media

Elena V. Orlenko [1,*] and Fedor E. Orlenko [2]

[1]   Higher School of Engineering Physics, Institute of Physics, Nanotechnology and Telecommunication, Peter the Great Saint Petersburg Polytechnic University, 195251 St. Petersburg, Russia

[2]   Scientific and Educational Center for Biophysical Research in the Field of Pharmaceuticals, Saint Petersburg State Chemical Pharmaceutical University (SPCPA), 197376 St. Petersburg, Russia; elena.orlenko@phmf.spbstu.ru

*   Correspondence: eorlenko@mail.ru; Tel.: +7-911-762-7228

Received: 27 October 2020; Accepted: 24 November 2020; Published: 28 November 2020

Abstract: A non-Anderson weak localization of an electron beam scattered from disordered matter is considered with respect to the principle of electron indistinguishability. A weak localization of electrons of a new type is essentially associated with inelastic processing. The origin of inelasticity is not essential. We take into account the identity principle for electron beam and electrons of the atom of the scatterer with an open shell. In spite of isotropic scattering by each individual scatterer, the electron exchange contribution has a hidden parameters effect on the resulting angular dependence of the scattering cross-section. In this case, the electrons of the open shell of an atomic scatterer can be in the s-state, that is, the atomic shell remains spherically symmetric. The methods of an invariant time-dependent exchange perturbation theory and a Green functions with exchange were applied. An additional angular dependence of the scattering cross-section appears during the coherent scattering process. It is shown exactly for the helium scatterer that the role of exchange effects in the case of a singlet is negligible, while for the triplet state, it is decisive, especially for those values of the energy of incident electrons when de Broglie's waves are commensurate with the atomic.

Keywords: identity principle; exchange contribution; new type weak localization; inelastic coherent scattering

---

1. Introduction

The phenomenon of weak localization of conduction electrons, which manifests itself in the enhancement of backscattering of classical waves in disordered media, has attracted scientific interest in recent decades [1–9]. Weak localization manifests itself mainly in an increase in the probability of elastic backscattering in a narrow range of solid angles, of the order of λ/l, where λ is the length of an electron or light wave, and l is the mean free path of electrons and photons. Coherent phenomena associated with the scattering of external particles (such as electrons or neutrons) with fixed excitation energies of a disordered medium were studied in [10–13], where weak localization was observed for electron beams with energies from 10 to 1000 electron volts. Neutron beams were also the subject of this work [14]. According to these works, coherent phenomena can be observed in the enhancement of particle backscattering during elastic interaction with a disordered medium, despite the relatively high energies of the particle beams. The influence of inelastic processes on the conductivity under conditions of weak localization has been studied in many works [15–18]. In these cases, two basic assumptions are usually made. First, multiple scattering is represented as forward multiple scattering and single large-angle scattering [19]. Second, the scattering by each individual diffuser is considered to be isotropic. It was shown that the role of inelastic processes at weak localization is secondary and negative, since inelastic collisions violate phase relations, thereby reducing the probability of coherent processes. However, there are cases when inelastic processes do not lead to the loss of phase memory by the system. It was shown for the first time in [20]. In this case, in addition to the usual weak localization of the Anderson type, there is a weak localization of electrons of a new type or a new weak localization, which is essentially associated with inelastic processing. Moreover, the origin of inelasticity is not essential, for example, it can be a plasmon, photon, phonon, or exciton. Moreover, quantum coherence can exist even if the electron is exposed to an incoherent electromagnetic field. In these cases, the situation is considered when, in the case of an inelastic interaction, the particle loses a fixed energy and enters the inelastic channel, having an energy different from the initial value in the incident beam. In addition to inelastic collision, the particle can still participate in at least one elastic process, after which it leaves the medium and can be registered. This effect demonstrates itself in the scattering processes of high-energy electron beams, where electrons velocities are relativistic. The investigation of the process of the resonant spontaneous bremsstrahlung of ultrarelativistic electrons in the fields of a nucleus and a weak quasimonochromatic electromagnetic wave was done in [21,22]. There is a coherent scattering process of photons with inelastic interaction with ultrarelativistics electrons. In this case, a characteristic angular dependence with a frequency shift appears for photons. The process has been studied in a special kinematic region, where stimulated processes with correlated emission and absorption of photons of the first and second waves predominate (the effect of parametric interference). It is interesting that the described effect of a new type of weak localization for photons does not depend on the type of inelasticity, the creation of an electron-positron, or something else. The amount of energy loss is important. In this interference kinematics, correspondence is established between the emission angle and the energy of the final electron.

There are two ways to implement this process, so it can start or end with an inelastic collision. The interference of electron waves associated with these additional processes turns out to be constructive [23]. This manifests itself in an increase in the scattering of electrons at an angle other than π, and this difference can be significant. There are certain differences in the localization of electrons for different mechanisms of inelastic scattering, but it turns out that the general features of this phenomenon prevail. The most striking difference between localization of a new type in the case of inelastic scattering from ordinary weak localization is the difference in the characteristic scattering angles. The scattering probability here is maximal in the range of scattering angles close to $\pi/2$, and the effect manifests itself in a much wider range of angles than in the case of traditional localization, in which the beam enhancement is observed in forward or backward scattering at an angle of π. As already mentioned, the main difference between conventional and new weak localization is the typical electron scattering angle. The angular distribution of particles and radiation in the case of ordinary weak localization in a disordered medium is usually described using the maximum cross (or so-called "fan") diagrams, which are used to calculate the electron radiation cross-section. Regular weak localization, in particular, can be described by a simple graphical method [24,25], which gives an idea of this phenomenon and explains why the angle pi is specific for regular weak localization. This method takes into account that an electron with momentum k is scattered through two complementary series of intermediate scattering states $k \to k_1 \to k_2 \to \ldots \to k_{n-1} \to k_n = -k$ and $k \to k\prime_1 = -k_{n-1} \to k\prime_2 = -k_{n-2} \to \ldots \to k\prime_{n-1} = -k_1 \to k_n = -k$ to the state $-k$. Momentum changes: $q_1, q_2, \ldots q_{n-1}, q_n$ for the first scattering chain and $q_n, q_{n-1}, \ldots q_2, q_1$ for the second. The amplitudes in the final state $-k$ are the same, and add up, and the waves in the forward and reverse directions are superimposed on each other constructively, reinforcing each other. This is due to the fact that the complementary scattering processes have the same changes in momentum, both in a straight line and in the opposite sequence. An explanation of why the coherent enhancement of electron scattering in the inelastic scattering channel occurs at angles other than π is proposed in [25]. A simple kinematic model is used to determine the basic properties of weak localization of electrons in the inelastic scattering channel. It easily reproduces the range of scattering angles characteristic of weak localization of electrons with energy loss. The results are consistent with the results based on the dynamic theory associated with the calculation of crossed and ladder diagrams. It is possible to trace the transition from a new type of weak localization to the usual weak localization with a decrease in energy losses. The new type of weak localization is consistent with regular weak localization if the energy losses are approximately equal to zero.

The range of angles in the elastic channel is of the order of λ/l, and in the inelastic channel is of the order $\gamma/\omega = (\lambda/l)(E/\hbar\omega)$, where γ is the electron collision frequency, E is its energy, and $\hbar\omega$ is the energy transferred to the medium. It is assumed that the energy of an electron incident on the medium is high enough to excite plasmons or atoms. However, this energy should not be too large so that the de Broglie wavelength of electrons was less but remained comparable to the distance between the centers of scattering, so that constructive interference of waves of scattered electrons inside condensed media could exist. The corresponding energy values for electron beams are in the range from hundreds of eV to keV. Moreover, the above statements remain valid both for the case of a small number of elastic collisions, and for a sufficiently large number; the main thing in this case is only one inelastic collision [23]. The main reason for the difficulty of fixing the indicated effect in a solid is that the very phenomenon of the new weak localization and all possible measurable parameters have been considered and calculated for an infinite three-dimensional medium, while the role of surface effects at a boundary of a condensed medium is large.

Taking into account the principle of indistinguishability of electrons in a beam and a medium, as was shown for a neutral atomic medium [26], but also takes place in a solid, this shifts the scattering enhancement to the parameters of an ordinary weak localization. As it is shown in [26], the scattering intensity includes the electron exchange terms varying directly as $\cos\chi$, where χ is a scattering angle. The expression for intensity contains two main terms, the first one is specific for the new weak localization, and the exchange interaction here contributes as an increasing pre-factor. The second term is entirely produced by electron exchange and is proportional to $\cos\chi$. Despite the obtained interesting exchange effect in a new type of weak localization for hydrogen scatterers, there was no systematic method allowing the principle of indistinguishability and permutation symmetry to be taken into account when describing scattering processes in the general case. The exchange effect considered in [24] is a good example of the importance of developing a general formalism that makes it possible to take into account the permutation symmetry of a many-electron system in the scattering problem in general and for weak localization, in particular.

In the presented work, we discuss how exchange affects the scattering of electrons in a disordered medium in a general case. We will take into account the identity of the incident electrons and electrons belonging to multielectron scatterer atoms. An invariant exchange perturbation theory method [27,28] is applied for the development of the cross-section for the weak localization scattering process while taking into account exchange effects. It is shown that, in spite of isotropic scattering by each individual scatterer, the number of electrons and their spin state in the open shell of the atom radically affect the resulting angular dependence of the scattering cross-section of incident electrons. In this case, the electrons themselves of the open shell of the scattering atom can be in the s-state, that is, the shell of the atom remains spherically symmetric. An additional angular dependence of the scattering cross-section appears, which is proportional to $\cos\chi$ and does not depend on the number and state of the spin of electrons with an open or closed shell. The main difference, depending on the number of electrons and their spin states, is the coefficient in front of $\cos\chi$.

2. Probability of Plasmon Emission with Atom Excitation and Electron Exchange

We consider the interaction of electrons with a medium, accompanying the excitation of atoms when electrons are scattered by them, by the same way as in [20] by using the interaction operator in the form of the sum of two contributions:

$$V = V_{at} + V_{pl}, \tag{1}$$

where the operator V_{at} describes electron interaction with atoms of medium, numbered by indexes l:

$$V_{at} = \sum_l V(r - R_l), \qquad (2)$$

and V_{pl} describes the interaction of the moving electron with the electric polarization field of the medium electrons, which arises under this action. Then, let the process of inelastic scattering of an electron be associated with the loss of energy on the plasmon:

$$V_{pl} = \int dr \hat{\rho}(\mathbf{r}) \hat{\phi}(r), \qquad (3)$$

where $\hat{\phi}(r)$ is the operator of the electric field potential due to plasma oscillation and $\hat{\rho}(r)$ is the charge density operator.

The initial state of the system corresponding to an incident electron and an atom of the medium (for example, a Zi-Ni alloy doped with Ca, Mg, or hydrogen or He) is described by the wave vector $|n, p\rangle$, antisymmetric taking into account the electronic permutations between the atom and the incident electron. Here, n is the set of quantum numbers of atoms in the initial state, and p is the momentum of the incident electron. Regarding the exchange perturbation theory method [27,28], the final state of the whole system can be described by the non-symmetric wave function $|m, p - Q) = |m\rangle \cdot |p - Q\rangle$, which is a simple product of the atomic function $|m\rangle$, antisymmetrized with internal electron permutations, and a free electron function $|p - Q\rangle$, where Q is the total momentum transferred to the medium. It should be underlined that the operator (2) describing electron interaction with atoms has a non-symmetric form regarding the interatomic electron permutations (the exchange of numbered atomic electrons with the free electron of incident):

$$\left[\hat{A} V_{at}\right] \neq 0, \qquad (4)$$

where $\hat{A}$ is the antisymmetrization operator [27,28]. Acting on a non-symmetric wave function, operator $\hat{A}$ performs antisymmetrization:

$$|n, p\rangle = \hat{A}|n, p) = \frac{1}{f_{np}} \sum_v (-1)^{g_v} |n, p)_{v'} \qquad (5)$$

where g_v is the parity of the $v - th$ permutation, $(n, p | n, p) = 1$, $f_{np} = \sum_v (-1)^{g_v} (n, p | n, p)_v$ are the normalization condition and a normalization factor, $|n, p) = |n\rangle \cdot |p\rangle \equiv |n, p)_0$ is the wave vector of the zeroth permutation with the initial arrangement of the electrons, and the subindex shows that the number of the permutation is $v = 0$. Here, we deal with the antisymmetric non-orthogonal basis, while $0 < (m, p\prime | n, p) < 1$, is the same as for $0 < \langle m, p\prime | n, p \rangle < 1$, but this set meet a completeness property:

$$\sum_{n,p} |n, p\rangle \frac{f_{np}}{N} (n, p| = \hat{1}, \qquad (6)$$

where N is the total number of electron permutations. The proof of completeness property (6) is described in detail in Appendix A. Generally speaking, both the "zero" Hamiltonian $\hat{H}_0$ describing a multicentre many-electron system without interatomic interaction, and the perturbation operator $\hat{V}$ describing this interatomic interaction are not invariant to the operation of antisymmetrization, taking into account the rearrangement of electrons between the two subsystems: $[\hat{A}\hat{H}_0] \neq 0, [\hat{A}V] \neq 0$. At the same time, the complete Hamiltonian of the system $\hat{H} = \hat{H}_0 + \hat{V}$ retains its invariance: $\left[\hat{A}\hat{H}\right] = 0$. This discrepancy means a serious problem related to the fact that the zero-order wave function, antisymmetrized with respect to center-to-center electron permutations, is not an eigenfunction of the non-invariant Hamiltonian $\hat{H}_0$ [27], and the corrections are obtained by applying an asymmetric operator of interactions containing non-physical contributions. In [27,28], a special symmetric form of the perturbation operator and zeroth Hamiltonian were developed. It allows corrections of the wave functions, properly antisymmetrised, and the energy corrections to be obtained by using the correct antisymmetric basis of the wavefunctions:

$$\hat{H}_0 = \left(\sum_{v=0}^{v=N} H_v^0 \Lambda^v \right) = \sum_{v=0}^{N} H_v^0 \sum_n \frac{|n,p\rangle_v \left(p,n|_v \right)}{f_{np}},$$

$$\hat{V}_{at} = \left(\sum_{v=0}^{v=N} V_v \Lambda^v \right) = \sum_{v=0}^{N} V_v \sum_n \frac{|n,p\rangle_v \left(p,n|_v \right)}{f_{np}},$$

where Λ^v denotes a projector onto an asymmetric state, corresponding to the $v - th$ permutation $\Lambda^v = \sum_n \frac{|n,p\rangle_v \left(p,n|_v \right)}{f_n^N}$, and $\Lambda^v |n,p\rangle = (-1)^{g_v} |n,p\rangle$, and N, as mentioned before, is the total number of permutations. In this case, the Hamiltonian $\hat{H}_0$ describing a system without the interaction of two subsystems has an eigenvector, which has an antisymmetric form taking into account the electronic permutations between these subsystems:

$$\hat{H}_0 |n,p\rangle = \left(E_n + p^2/2m \right) |n,p\rangle. \tag{7}$$

In our case, the second term of the perturbation operator (1) V_{pl} has a symmetric form with respect to the permutations of electrons. It is important to emphasize that after obtaining all the corrections to the energy and the wave vector in all orders of the perturbation theory in the general formalism, all matrix elements containing the complex symmetric form (6) of the perturbation operator can be analytically reduced to a simple form. This is a symmetry-adapted form that includes the usual asymmetric perturbation operator corresponding to the initial zero permutation of electrons. Such a form is more convenient for practical applications. We transform all matrix elements in the following way:

$$(k,p'|\hat{V}|n,p\rangle = (k,p'|\sum_{v=0}^{N} V_v \Lambda^v |n,p\rangle = (k,p'|\sum_{v=0}^{N} \frac{(-1)^{g_v}}{f_{np}} V_v |n,p\rangle_v =$$

$$\sum_{v=0}^{N} \frac{(-1)^{g_v}}{f_{kp}} (k,p'|_v V_{p=0}|n,p\rangle_{v=0} = \frac{f_{np}}{f_{kp}} k,p'|V_{v=0}|n,p\rangle. \tag{8}$$

The probability of a transition of the system in unit time from one state to another with transfer of momentum Q from the incident electron to the medium and with transition of the medium from state n to state m has the form obtained in [27,28], taking into account electron exchange

$$w_{mn,Q} = \frac{2\pi}{\hbar} \delta\left(E_n - E_m + E_p - E_{p-Q} \right) |\langle m, p - Q|\hat{T}|n,p\rangle|^2, \tag{9}$$

where operator $\hat{T}$ is the operator of transition on the energy surface [27–29], the general operator equation for which with taking into account electron permutations between subsystems is:

$$\hat{T} = V^{\mathbb{N}} + V^{\mathbb{N}} (E_i - H_0 + i\eta)^{-1} \left(\frac{\hat{f}}{N} \right)^{-1} \hat{T},$$

$$V^{\mathbb{N}} = V\left(\frac{\hat{f}}{N} \right), \underline{\quad} \hat{f}|n,p\rangle = f_{np}|n,p\rangle, \tag{10}$$

where η is the relaxation frequency in the general case. We rewrite it as a series:

$$\hat{T} = V^{\mathbb{N}} + V^{\mathbb{N}} (E_i - H_0 + i\eta)^{-1} \left(\frac{\hat{f}}{N} \right)^{-1} V^{\mathbb{N}} +$$

$$V^{\mathbb{N}} \left(E_i - H_{p=0}^0 + i\eta \right)^{-1} \left(\frac{\hat{f}}{N} \right)^{-1} V^{\mathbb{N}} \left(E_i - H_{p=0}^0 + i\eta \right)^{-1} \left(\frac{\hat{f}}{N} \right)^{-1} V^{\mathbb{N}} + \dots \tag{11}$$

In the matrix element $\langle m, p - Q | \hat{T} | n, p \rangle$ in Equation (9), the *bra*-vector has an antisymmetric form with respect to electronic permutations, and the *ket*-vector has a simple non-symmetric form. Then, a transition amplitude will be:

$$
\begin{aligned}
\langle m, p - Q | \hat{T} | n, p \rangle &= \langle m, p - Q | (V_{at} + V_{pl})^{N} | n, p \rangle + \langle m, p - Q | (V_{at} + V_{pl})^{N} \hat{G}_0^N (V_{at} + V_{pl})^{N} | n, p \rangle + \\
&\langle m, p - Q | (V_{at} + V_{pl})^{N} \hat{G}_0^N (V_{at} + V_{pl})^{N} \hat{G}_0^N (V_{at} + V_{pl})^{N} | n, p \rangle + \ldots = \langle m, p - Q | (V_{pl})^{N} \hat{G}_a^N (V_{at})^{N} | n, p \rangle + \\
&\langle m, p - Q | (V_{at})^{N} \hat{G}_a^N (V_{pl})^{N} | n, p \rangle + \langle m, p - Q | (V_{at})^{N} \hat{G}_a^N (V_{at})^{N} | n, p \rangle \ldots,
\end{aligned} \tag{12}
$$

where we denote the normalized (indexed N) resolvent operator $\hat{G}_0^N$ (or the Green's function for the coordinate representation) and the operator equation taking into account an antisymmetric nonorthogonal basis:

$$
\begin{aligned}
\hat{G}_0^N &= (E_i - H_0 + i\eta)^{-1} \left(\frac{f}{N} \right)^{-1} \\
\hat{G}_a^N &= \hat{G}_0^N + \hat{G}_0^N V_{at}^N \hat{G}_a^N.
\end{aligned} \tag{13}
$$

Since the plasmon excitation is small, we can consider this process in the Born approximation. We do not fix our attention on the features of plasmon excitation in different situations, as it was done in detail in [22,30], where the excitation and propagation of bulk and surface plasma waves by incident electrons were analyzed. The motion of electrons was considered both in vacuum when approaching the surface of the metal, and inside the metal, the boundary of which elastically and specularly reflected the internal nonequilibrium electrons. The effect of electron boundary scattering parameters on the structure of bulk and surface plasmon resonances was analyzed in [31]. The probability of transition radiation of bulk plasmon by an electron moving in vacuum was examined.

We leave only the first two terms in Equation (12), where the scattering process begins or ends with plasmon excitation, and omit all other terms where the inelastic process with a plasmon occurs between the processes of elastic scattering by atoms. This is because we are not interested in small-angle scattering by the medium as a whole. The mentioned two terms we shall write out in the form:

$$
\begin{aligned}
\langle m, p - Q | \hat{T} | \Psi_p^+ \rangle &= \Big\{ \langle m, p - Q | (V_{pl}) | m, p - Q + q \rangle \big[\langle m, p - Q + q | \delta_{Qq} | n, p \rangle + \langle m, p - Q + q | \hat{G}_a (V_{at})^{N} | n, p \rangle \big] + \\
&\big[\langle m, p - Q | (V_{at})^{N} \hat{G}_a | n, p - q \rangle + \langle m, p - Q | \delta_{Qq} | n, p - q \rangle \big] \langle n, p - q | V_{pl} | n, p \rangle \Big\} = \\
&\langle m, p - q | n, p - q \rangle \langle n, p - q | V_{pl} | n, p \rangle + \langle m, p - q | V_{pl} | m, p \rangle \langle m, p | n, p \rangle + \\
&\Big\{ \langle m, p - Q | (V_{pl}) | m, p - Q + q \rangle \langle m, p - Q + q | \hat{G}_a (V_{at})^{N} | n, p \rangle + \langle m, p - Q | (V_{at})^{N} \hat{G}_a | n, p - q \rangle \langle n, p - q | V_{pl} | n, p \rangle \Big\},
\end{aligned} \tag{14}
$$

where the first two matrix elements describe the process of plasmon excitation without atomic scattering. However, the exchange coefficients:

$$
S_{mnp-q} = \langle m, p - q | n, p - q \rangle \text{ and } S_{mnp} = \langle m, p | n, p \rangle, \tag{15}
$$

mean the exchange density due to the entanglement of electronic states. The third term of Equation (14) corresponds to a process that begins with scattering by atoms and ends with excitation of plasmons. The fourth term corresponds to the process in which the events occur in the opposite order. Since we are not interested in small-angle scattering by the medium as a whole, as mentioned earlier, we can omit the first two terms.

Dividing the result of Equation (9) by the flux density of incident particles ($j_p = \hbar k_p / m_e$, where k_p is the wavevector of the relative movement of the incident electrons and m_e is their mass), we obtain an expression for the cross-section of scattering events and reactions. Then, we multiply this expression for the cross-section by the number of final states in the volume per unit energy interval for scattering along the unit vector n_f into a solid angle element $d\Omega$, $d\rho(E_f) = \dfrac{m k_f}{(2\pi)^3 \hbar^2} d\Omega$, because the final state is within the continuous spectrum. Then, for the differential cross-section summed over the final states of the medium, it has the form:

$$d\sigma_{fi} = \frac{m_e^2 k_f}{(2\pi\hbar^2)^2 k_p}\left|\langle m, p - Q|\hat{T}|\Psi_{n,p}^+\rangle\right|^2 d\Omega d(\hbar\omega) \rightarrow$$

$$\left(\frac{m_e}{2\hbar^2}\right)^{3/2}\frac{(p^2 - 2m_e\hbar\omega)^{1/2}}{2\pi^2 p}\ \Sigma_q\left|\langle m, p - Q|\hat{T}|\Psi_{n,p}^+\rangle\right|^2 d\Omega_Q d\omega. \tag{16}$$

For electrons, we take into account the disorder of the medium, and for the plasmon electric field, we assume that the medium is completely homogeneous. Therefore, we can assume that the matrix elements in Equation (14) are approximately equal to each other:

$$\langle n, p - q|V_{pl}|n, p\rangle \approx \langle m, p - q|V_{pl}|m, p\rangle \approx \langle m, p - Q|\left(V_{pl}\right)|m, p - Q + q\rangle$$

and the exchange coefficients S_{mnp-q} and S_{mnp} have the same order and are proportional to $\dfrac{1}{[1+(p/\hbar)^2 a_B^2]^2} = \dfrac{1}{\left[1+(a_B/l_p)^2\right]^2}$, then we can rewrite Equation (16) in the form:

$$d\sigma_{fi} = \left(\frac{m_e}{2\hbar^2}\right)^{3/2}\frac{(p^2 - 2m_e\hbar\omega)^{1/2}}{2\pi^2 p}\times$$

$$\Sigma_q\left|\langle m, p - Q|\left(V_{pl}\right)|m, p - Q + q\rangle\langle m, p - Q + q|\hat{G}_a(V_{at})^N|n, p\rangle + \langle m, p - Q|(V_{at})^N\hat{G}_a|n, p - q\rangle\langle n, p - q|V_{pl}|n, p\rangle\right|^2 d\Omega_Q d\omega, \tag{17}$$

which is illustrated in Figure 1 by the set of so-called fan diagrams determining the effect of new angle dependence of the effective cross-section.

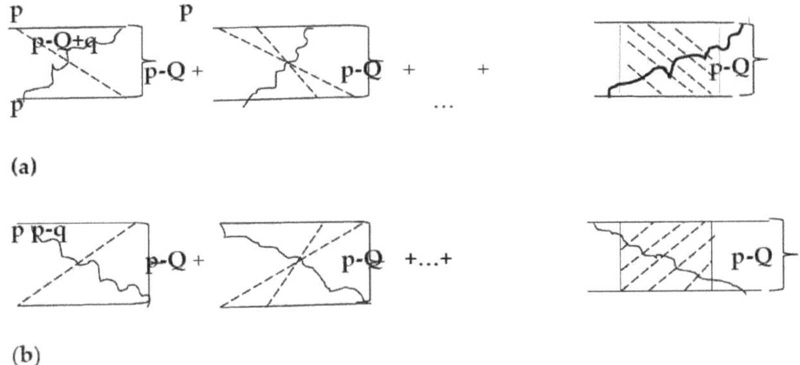

Figure 1. Crossed diagrams correspond to two, three, and so on, scattering events. The wavy line corresponds to the inelastic scattering process with plasmon excitation. Dotted lines link the same atom. (**a**) The processes, beginning from the elastic scattering events and ending by the inelastic, (**b**) processes ranging from inelastic scattering to a series of elastic scattering by atoms.

As usual, in the diagrams that match Equation (17), we will use the dotted line to connect two vertices belonging to the same atom. The lower parts of the diagrams correspond to the analytic expression, which is the complex conjugation of the expression equivalent to the upper parts. The wavy line corresponds to the plasmon. The set of ladder diagrams, correspondent to the terms contributed to the cross-section (17), are independent of the electron-scattering angle. The relation of the crossed diagram contributions and ladder diagrams determine a so-called coherency degree [25]:

$$\frac{d\sigma_{fi}}{d\Omega_Q} = \left(\frac{m_e}{2\hbar^2}\right)^{3/2} \frac{\left(p^2 - 2m_e\hbar\omega\right)^{1/2}}{2\pi^2 p} \times$$

$$\sum_q w_{mnQ} \left| \langle m, p - Q + q | \hat{G}_a (\underline{V_{at}})^N | n, p \rangle + \langle m, p - Q | (\underline{V_{at}})^N \hat{G}_a | n, p - q \rangle \right|^2 d\omega$$

$$where,$$

$$w_{mnQ} = \left| \langle m, p - q | n, p - q \rangle \langle n, p - q | V_{pl} | n, p \rangle \right|^2 \sim$$

$$\left| \langle m, p - q | V_{pl} | m, p \rangle \langle m, p | n, p \rangle \right|^2 \sim \left| \langle m, p - Q | \left(\underline{V_{pl}}\right) | m, p - Q + q \rangle \right|^2 . \tag{18}$$

Neglecting exchange effects, in the resulting expression (17), both *bra-* and *ket-* vectors become non symmetric, and we obtain the well-known result obtained in [25–27]. In contrast to the usual weak localization, one of the crossed lines in the new type of weak localization corresponds to inelastic scattering, while the others correspond to elastic interaction with randomly distributed power centers. According to formula (18), crossed diagrams together with the corresponding ladder diagrams contribute to the scattering probability factor $\mathfrak{I}(\omega, \chi)$ [20]:

$$\int d\Omega_Q \sum_q w_{mnQ}(q, \hbar\omega) \left| \langle m, p - Q + q | \hat{G}_a (\underline{V_{at}})^N | n, p \rangle + \langle m, p - Q | (\underline{V_{at}})^N \hat{G}_a | n, p - q \rangle \right|^2 d\omega \rightarrow$$

$$\sum_q w_{mnQ}(q, \hbar\omega) d\omega \left| m, p - Q + q V_{at} | n, p \rangle \right|^2 \times$$

$$\int d\Omega_Q \left| p - Q + q \hat{G}_a p - Q + q \rangle + \left| m, p - Q \hat{G}_a m, p - Q \rangle \right|^2 d\omega, \tag{19}$$

$$\mathfrak{I}(\omega, \chi) = \frac{\hbar^2}{(2\pi\hbar)^3} \int_0^\infty q \, dq \, w_{mn}(q, \hbar\omega) \int \left| G(p - Q + q, E_p) + G(p - q, E_p - \hbar\omega) \right|^2 d\Omega_Q.$$

It was shown in [24–26] that the angular dependence characteristic of a new weak localization arises due to the term in the integrand, which describes the contribution of the interference of two electron waves propagating along the same path in opposite directions:

$$G^*(p - Q + q, E_p) G(p - q, E_p - \hbar\omega) + G(p - Q + q, E_p) G^*(p - q, E_p - \hbar\omega) =$$

$$\frac{1}{E_p - E_{p-Q+q} + i\gamma} \cdot \frac{1}{E_p - E_{p-q} - \hbar\omega - i\gamma} + \frac{1}{E_p - E_{p-Q+q} - i\gamma} \cdot \frac{1}{E_p - E_{p-q} - \hbar\omega + i\gamma}, \tag{20}$$

where γ is the electron collision frequency.

This means that weak localization occurs due to such collisions of electrons in which each subsequent scattering begins earlier than the end of the previous one. Thus, a new weak localization is realized when two conditions are met simultaneously:

$$\begin{cases} E_p - E_{p-Q+q} = 0, \\ E_p - E_{p-q} - \hbar\omega = 0 \end{cases} . \tag{21}$$

If we denote $p\prime = p - Q$, then a simple kinematic approach explains the range of the scattering angle $\chi = cos^{-1} \frac{p \cdot p\prime}{p \cdot p\prime}$ typical for the new type of weak localization:

$$cos\chi = -\frac{p^2 + p\prime^2}{2pp\prime} + \frac{2(m\hbar\omega)^2}{pp\prime q^2}, . \tag{22}$$

For more details, [24,25], where it is explained why the angles typical of the new type of weak localization differ from π, and by using the simple kinematic approach, these angles can be estimated very accurately.

Considering Equation (18) taking into account the exchange contributions, we can write the following:

$$\sigma_{fi}(\chi,\omega) = \left(\frac{m_e}{2\hbar^2}\right)^{3/2} \frac{\left(p^2 - 2m_e\hbar\omega\right)^{1/2}}{2\pi^2 p} \times$$
$$\int d\Omega_Q \sum_q w_{mnQ}(q,\hbar\omega) \left|\langle m, p - Q + q|\hat{G}_a(V_{at})^{N}|n,p\rangle + \langle m, p - Q|(V_{at})^{N}\hat{G}_a|n, p - q\rangle\right|^2 =$$
$$\left(\frac{m_e}{2\hbar^2}\right)^{3/2} \frac{\left(p^2 - 2m_e\hbar\omega\right)^{1/2}}{2\pi^2 p} \sum_q w_{mnQ}(q,\hbar\omega) \left|\langle m, p - Q + q|(V_{at})^{N}|n,p\rangle\right|^2 \times$$
$$\int d\Omega_Q |\langle m, p - Q + q|\hat{G}_a|m, p - Q + q\rangle + \langle n, p - q|\hat{G}_a|n, p - q\rangle|^2. \tag{23}$$

In the case of an antisymmetric basis, the calculation of matrix elements $\langle n, p - q|\hat{G}_a|n, p - q\rangle$ requires additional comments. Instead of the resolvent operator $\hat{G}_a$, we will use a one-electron Green function $G_a^{ex}(r, r\prime)$ taking into account that exchange contributions influence the one-particle system:

$$G_a^{ex}(r, r\prime, R) = \langle n, p - q|\hat{G}_a|n, p - q\rangle_{rr\prime} = \frac{\sum_{lp} f_{nl}(r,R) f *_{ln}(r\prime, R)}{E_p - E_{p-q} - \hbar\omega + i\gamma}, \tag{24}$$

where:

$$f_{nl}(r, R) = \int dr_1 dr_2 \ldots r_n\left(\sum_v^N (-1)^{g_v}\Psi_n{}^*(r_1 - R, r_2 - R, \ldots (r - R)_v \ldots r_k - R)\psi *_{p-q}(r_v)\right) \times$$
$$\sqrt{\frac{f_{lp}}{N}}\Psi_l(r_1 - R, r_2 - R, \ldots r_v - R \ldots r_k - R)\psi_{p-q}(r), \tag{25}$$

$$f_{ln}(r\prime, R) = \int dr_1 dr_2 \ldots r_n\left(\sum_v^N (-1)^{g_v}\Psi_l{}^*(r_1 - R, r_2 - R, \ldots (r\prime - R)_v \ldots r_k - R)\psi *_{p-q}(r_v)\right) \times$$
$$\sqrt{\frac{f_{lp}}{N}}\Psi_n(r_1 - R, r_2 - R, \ldots r_v - R \ldots r_k - R)\psi_{p-q}(r\prime). \tag{26}$$

Integration over the conductor volume V means averaging over randomly distributed scatterers. Here, we used the completeness property (6). In the same way, we rewrite the matrix element of another resolvent operator:

$$G_{a(p-Q+q)}^{ex}(r, r\prime) = \langle m, p - Q + q|\hat{G}_a|m, p - Q + q\rangle_{r,r\prime} = \frac{\overline{\sum_{lp} f_{ml}(r,R) f_{lm}(r\prime,R)}_R}{E_p - E_{p-Q+q} + i\gamma}, \tag{27}$$
$$\overline{f_{ml}(r, R) f_{lm}(r\prime, R)}_R = \frac{1}{V}\int_V f_{ml}(r, R) f_{lm}(r\prime, R)dR.$$

The spectral density $\sigma_{fi}(\chi,\omega)$ of the cross-section (23) of the process of inelastic scattering of electrons on disordered media consisting of atomic defects in a metal alloy (metal alloy doped with di- or trivalent atoms), taking into account exchange effects, has the form:

$$\sigma_{fi}(\chi,\omega) = \left(\frac{m_e}{2\hbar^2}\right)^{3/2} \frac{\left(p^2 - 2m_e\hbar\omega\right)^{1/2}}{2\pi^2 p} \sum_q w_{mnQ}(q,\hbar\omega)\left|\langle m, p - Q + q|(V_{at})^{N}|n,p\rangle\right|^2 \times$$
$$\int d\Omega_Q \left|\frac{\overline{\sum_{lp} f_{ml}(r,R) f_{lm}(r\prime,R)}_R}{E_p - E_{p-Q+q} + i\gamma} + \frac{\overline{\sum_{lp} f_{nl}(r,R) f_{ln}(r\prime,R)}_R}{E_p - E_{p-q} - \hbar\omega + i\gamma}\right|^2. \tag{28}$$

Using the obtained general expression for the cross-section (18) for the case considered in [24], the alloy doped by hydrogen atoms as disordered centers, we obtain the same expression for the Green functions and for the scattering probability. It proves a limit transition of the obtained general expression (18) for the cross-section to the known case described in [24].

3. The Electron Identity Principle in Quantum Interference for Multielectron Scatterers

In our consideration, we take into account the identity of the beam electrons and the electrons belonging to the atoms of the scatterer. Firstly, we use a scatterer as a helium-like atom in the ground state ($1s^2$, ^1S) and in the excited state ($1s^1 2s^1$, ^3S).

3.1. Helium-Like Scatterer ^1S

The wave functions of the ground state of a helium-like atom and a free (incident) electron are:

$$\Phi_{He}(R - r_{11}, R - r_{22}; \xi_{11}, \xi_{22}) = (\varphi_{1s}(R - r_1)\varphi_{1s}(R - r_2)) \cdot \tfrac{1}{\sqrt{2}}(\alpha 1 \beta 2 - \beta 1 \alpha 2),$$

$$\psi_p(r) = e^{i\frac{p \cdot r}{\hbar}}, \tag{29}$$

where $\varphi_{1s}(R - r_i) = (\alpha^3/\pi)^{1/2} exp(-\alpha|R - r_i|)$, $\alpha = 27/16$.

The vector of the initial state, corresponding to the permutation $v = 0$, but antisymmetrized only accounting for the internal permutations in the helium atom, see Equation (29), has the form:

$$\left|\Phi_i^{v=0}\right) = \Phi_{He}(r_1, r_2)X_{He}(\xi_1\xi_2)\psi_e(r)\chi_e(\xi) = |n = 0, p). \tag{30}$$

We obtain the antisymmetric wave function by applying the normalized Young operator [29,32] to the wave function of the free electron–helium system as follows: Antisymmetrization of the atomic wavefunction given by Equation (30) over interatomic electrons permutations performed using the four independent Young's operators: $\omega_{11}^{[21]}$; $\omega_{12}^{[21]}$; $\omega_{21}^{[21]}$; $\omega_{22}^{[21]}$ (see Appendix A), where:

$$\Psi = \frac{1}{f_0}(\Psi_{11}(r_1, r_2, r)X_{22}(\xi_1, \xi_2, \xi) + \Psi_{12}(r_1, r_2, r)X_{21}(\xi_1, \xi_2, \xi)),$$

$$\Psi_{11}(r_1, r_2, r) = \omega_{11}^{[21]}\Phi_{He}(r_1, r_2)\psi_p(r) = \tfrac{1}{\sqrt{3}}\left(2\Phi_{He}(r_1, r_2)\psi_p(r) - \Phi_{He}(r_1, r)\psi_p(r_2) - \Phi_{He}(r, r_2)\psi_p(r_1)\right)$$

$$\Psi_{12}(r_1, r_2, r) = \omega_{12}^{[21]}\Phi_{He}(r_1, r_2)\psi_p(r) = \left(\Phi_{He}(r_1, r)\psi_p(r_2) - \Phi_{He}(r, r_2)\psi_p(r_1)\right)$$

$$X_{21}(\xi_1, \xi_2, \xi) = \omega_{21}^{[21]}X_{He}(\xi_1\xi_2)\chi_e(\xi) = \omega_{21}^{[21]}\tfrac{1}{\sqrt{2}}(\alpha_1\beta_2 - \beta_1\alpha_2)\chi_e(\xi) = \tag{31}$$

$$\tfrac{1}{\sqrt{2}}[(\alpha_1\beta - \beta_1\alpha)\chi_2(\xi) - (\alpha\beta_2 - \beta\alpha_2)\chi_1(\xi)]$$

$$X_{22}(\xi_1, \xi_2, \xi) = \omega_{22}^{[21]}X_{He}(\xi_1\xi_2)\chi_e(\xi) =$$

$$\tfrac{1}{\sqrt{3}}\tfrac{1}{\sqrt{2}}[2(\alpha_1\beta_2 - \beta_1\alpha_2)\chi_e(\xi) + (\alpha_1\beta - \beta_1\alpha)\chi_2(\xi) + (\alpha\beta_2 - \beta\alpha_2)\chi_1(\xi)],$$

where f_0 is the normalized factor, determined by:

$$\langle(\Psi_{11}(r_1, r_2, r)X_{22}(\xi_1, \xi_2, \xi) + \Psi_{12}(r_1, r_2, r)X_{21}(\xi_1, \xi_2, \xi))\big|\Phi_{He}(R - r_1, R - r_2, \xi_1, \xi_2)\psi_p(r)\chi(\xi)\rangle_R =$$

$$f_0 = \tfrac{4\pi n}{3}\left(\tfrac{2\pi\hbar}{p}\right)^3\left(1 - \tfrac{(2p/\hbar\alpha)}{1+(p/\hbar\alpha)}\right). \tag{32}$$

For this case, Equations (25) and (26) will have the forms:

$$f_{nl}(r, R) \rightarrow f_{00}(r, R) =$$

$$\int dr_1 dr_2 \tfrac{1}{\sqrt{3}}\Big\{2\varphi_{1s}(r_1 - R)\varphi_{1s}(r_2 - R)\psi *_{p-q}(r) -$$

$$\varphi_{1s}(r_1 - R)\varphi_{1s}(r - R)\psi *_{p-q}(r_2) - \varphi_{1s}(r_2 - R)\varphi_{1s}(r - R)\psi *_{p-q}(r_1)\Big\} \times$$

$$\sqrt{\tfrac{f_0}{6}}(\varphi_{1s}(R - r_1)\varphi_{1s}(R - r_2))\psi_{p-q}(r'')\tfrac{2}{\sqrt{3}} = \tag{33}$$

$$\tfrac{4}{3}\sqrt{\tfrac{f_0}{6}}\left\{e^{i\frac{(p-q)(r''-r)}{\hbar}} - \frac{8}{\left(1+\frac{|p-q|^2}{\alpha^2\hbar}\right)^2}exp\left(i\frac{(p-q)r''}{\hbar} - \alpha|r - R|\right)\right\}.$$

That is, the Green's function (28), averaged over randomly distributed centers, can be written as:

$$G^{ex}_{(p-Q+q)}(r,r\prime) = \tfrac{1}{V}\int_V dR\, G^{ex}_{a(p-Q+q)}(r,r\prime,R) \to \tfrac{1}{V}\int_V dR\,\frac{\Sigma_p\, f_{00}(r)f*_{00}(r\prime)}{E_p-E_{p-Q+q}+i\gamma} =$$

$$\left(\tfrac{4}{3}\right)^2\tfrac{f_0}{6}\tfrac{1}{V}\int_V dR \sum_p \left\{ \frac{exp\!\left(-i\frac{p(r-r\prime)}{\hbar}\right)}{E_p-E_{p-Q+q}+i\gamma}\left[1+\frac{8^2}{\left(1+\frac{|p|^2}{\alpha^2\hbar}{}^2\right)^4}\right] - \right.$$

$$\frac{8}{\left(1+\frac{|p|^2}{\alpha^2\hbar}{}^2\right)^2}\frac{exp\!\left(i\frac{p(R+r\prime-r)}{\hbar}-\alpha|r-R|\right)+exp\!\left(-i\frac{p(R-r\prime+r)}{\hbar}-\alpha|r\prime-R|\right)}{E_p-E_{p-Q+q}+i\gamma}+$$

$$\left. \frac{8^2}{\left(1+\frac{|p|^2}{\alpha^2\hbar}{}^2\right)^4}\frac{exp(-\alpha|r-R|-\alpha|r\prime-R|)}{E_p-E_{p-Q+q}+i\gamma}\right\} =$$

$$G^{ex}_{0(p-Q+q)}(r,r\prime) - G^{ex}_{1(p-Q+q)}(r,r\prime) + G^{ex}_{a(p-Q+q)}(r,r\prime),$$

(34)

which has the same construction as in [26] but with a well-defined sign in front of the second term.

Here, $G^{ex}_0(r,r\prime)$ is the ordinary Green's function, previously considered in [22,24,26], making it the main contribution to the new weak localization, corrected only by adding a normalization factor due to exchange effects. The specific exchange contribution has the form:

$$G^{ex}_{1p-Q+q}(r,r\prime) = \left(\frac{4}{3}\right)^2\frac{f_0}{6V\alpha^3}2\pi\sum_p \frac{8^2}{\left(1+\frac{p^2}{\alpha^2\hbar}{}^2\right)^4}\frac{\left(1+\frac{ip(r-r\prime)}{2\hbar}\right)e^{-i\frac{p(r-r\prime)}{\hbar}}}{E_p-E_{p-Q+q}+i\gamma}, \tag{35}$$

Since we are interested in the correlation length $|r-r\prime|\le \lambda_p$, we can neglect the higher-order terms of the parameter $\frac{p(r-r\prime)}{\hbar}$. Then, this term in the p-representation after the Fourier transform and an ordinary Green function in the p-representation have the following form:

$$G^{ex}_1(p-Q+q) = G^{ex}_1(p\prime+q) = \frac{2^{10}}{3^4}\left(1-\frac{\left(\frac{2p}{\hbar\alpha}\right)^3}{\left(1+\left(\frac{p}{\hbar\alpha}\right)^2\right)^4}\right)\frac{\left(\frac{\hbar}{\alpha p}\right)^3}{\left(1+\frac{p^2}{\alpha^2\hbar}{}^2\right)^4}\frac{(2\pi)^2 n^2}{E_p-E_{p-Q+q}+i\gamma},$$

$$G_0(p\prime+q) = \frac{1}{E_p-E_{p-Q+q}+i\gamma}. \tag{36}$$

The third term $G^{ex}_a(r,r\prime)$ describes a process that begins and ends at the same atom; therefore, it does not contribute to the quantum transfer of electrons.

Now, consider the main pre-factor Ξ in the cross-section (28):

$$\Xi = \int d\Omega_Q \left|\frac{\Sigma_{lp}\,\overline{f_{ml}(r,R)f_{lm}(r\prime,R)}_R}{E_p-E_{p-Q+q}+i\gamma} + \frac{\Sigma_{lp}\,\overline{f_{nl}(r)f_{ln}(r\prime)}_R}{E_p-E_{p-q}-\hbar\omega+i\gamma}\right|^2 =$$

$$\int d\Omega_Q \left|G^{ex}(p-Q+q) + G^{ex}(p-q)\right|^2, \tag{37}$$

where we denoted the Green's function (34) in the p-representation as $G^{ex}(p-Q+q) = G^{ex}_0(p\prime+q) - G^{ex}_1(p\prime+q)$, taking into account $p-Q=p\prime$, and in exactly the same way the Green's function of the second term in the same representation as $G^{ex}(p-q) = G^{ex}_0(p-q) - G^{ex}_1(p-q)$. Then, the factor Ξ consists of three contributions, background, "the simple new weak localization", and the exchange term:

$$\Xi = \Xi_{bg} + \Xi_0 - \Xi_{exc}. \tag{38}$$

The background term has the following form:

$$\Xi_{bg} = \int d\Omega_Q \left\{ \left| G_0^{ex}(p\prime + q) - G_1^{ex}(p\prime + q) \right|^2 + \left| G_0^{ex}(p - q) - G_1^{ex}(p - q) \right|^2 \right\} =$$
$$\int d\Omega_Q \left\{ \left| G_0^{ex}(p\prime + q) \right|^2 + \left| G_1^{ex}(p\prime + q) \right|^2 + \left| G_0^{ex}(p - q) \right|^2 + \left| G_1^{ex}(p - q) \right|^2 - \right.$$
$$\left. \left[G_0^{ex*}(p\prime + q)G_1^{ex}(p\prime + q) + G_1^{ex*}(p\prime + q)G_0^{ex}(p\prime + q) + G_0^{ex*}(p - q)G_1^{ex}(p - q) + G_1^{ex*}(p - q)G_0^{ex}(p - q) \right] \right\}. \tag{39}$$

Substituting Expression (36), we obtain:

$$\Xi_{bg} = \left(1 - \frac{\pi}{\left(1+\frac{p^2}{(\hbar\alpha)^2}\right)^6} \frac{2^{10}(\pi n)^2}{3\alpha^6} \right) \times$$
$$\left[5\left(\frac{1}{(v\prime q)^2} \ln\left(\frac{(v\prime q - \hbar\omega)^2 + (\hbar\gamma)^2}{(v\prime q + \hbar\omega)^2 + (\hbar\gamma)^2} \right) - \frac{1}{(vq)^2}\ln\left(\frac{(vq - \hbar\omega)^2 + (\hbar\gamma)^2}{(vq + \hbar\omega)^2 + (\hbar\gamma)^2} \right) \right) - \right.$$
$$\left. \left(4 - \frac{25}{\left(1+\frac{p^2}{(\hbar\alpha)^2}\right)^2} \right) \frac{\Gamma^2(5)(vq)^2}{\left(\frac{\hbar^2\alpha^2}{2m}\right)^4 \left(1+\frac{p^2}{(\hbar\alpha)^2}\right)^4} \right]. \tag{40}$$

"The ordinary new weak localization" contribution has the form:

$$\Xi_0 = \int d\Omega_Q \left\{ G_0^{ex*}(p\prime + q)G_0^{ex}(p - q) + G_0^{ex*}(p - q)G_0^{ex}(p\prime + q)^2 \right\} =$$
$$\frac{4\pi}{(qv)\hbar\gamma} \cdot \left[\arctg\left(\frac{qv + \hbar\omega_p}{\hbar\gamma} \right) + \arctg\left(\frac{qv - \hbar\omega_p}{\hbar\gamma} \right) - \frac{\gamma}{\sqrt{2\omega_p^2(1-\cos\chi) - \frac{q^2v^2}{\hbar^2}\sin^2\chi}} \right] \times$$
$$\ln\left(\frac{\hbar^2\omega_p^2 - q^2v^2\cos\chi + qv\sqrt{2\hbar^2\omega_p^2(1-\cos\chi) - q^2v^2\sin^2\chi}}{\hbar^2\omega_p^2 - q^2v^2\cos\chi - qv\sqrt{2\hbar^2\omega_p^2(1-\cos\chi) - q^2v^2\sin^2\chi}} \right). \tag{41}$$

This expression was first obtained in [20] and used in [26]. The Langmuir frequency is denoted as ω_p, and γ is the collision frequency. The specific exchange contribution in the cross-section is:

$$\Xi_{exc} = \int d\Omega_Q \left\{ G_0^{ex*}(p+q)G_1^{ex}(p-q) + G_1^{ex*}(p\prime+q)G_0^{ex}(p-q) + G_0^{ex*}(p-q)G_1^{ex}(p\prime+q) + G_1^{ex*}(p-q)G_0^{ex}(p\prime+q) - \right.$$
$$\left. \left(G_1^{ex*}(p\prime+q)G_1^{ex}(p-q) + G_1^{ex*}(p-q)G_1^{ex}(p\prime+q) \right) \right\} =$$
$$\frac{2^8 \frac{n\pi}{\alpha^3}}{\left(1+\frac{p^2}{(\hbar\alpha)^2}\right)^3} \left\{ \Xi_0\left(1 - \frac{2^5 \frac{n\pi}{\alpha^3}}{\left(1+\frac{p^2}{(\hbar\alpha)^2}\right)^5} \right) - \cos\chi\left(4 + \frac{\hbar\omega_L}{vq}\ln\left(\frac{(vq-\hbar\omega_p)^2 + (\hbar\gamma)^2}{(vq+\hbar\omega_p)^2 + (\hbar\gamma)^2} \right) \right) \frac{(2m)^2 \cdot 4\pi}{(\hbar\alpha)^4\left(1+\frac{p^2}{(\hbar\alpha)^2}\right)^2}\left(1 - \frac{2^5 \frac{n\pi}{\alpha^3}\Gamma^2(5)}{\left(1+\frac{p^2}{(\hbar\alpha)^2}\right)^5} \right) \right\}, \tag{42}$$

where $\Gamma(5) = 4!$ is a Γ-function. Then, taking into account Equations (39)–(42), the main factor determining the angular dependence of the scattering cross-section is:

$$\Xi(\chi) = \Xi_{bg} + \Xi_0\left(1 - \frac{2^8 \frac{n\pi}{\alpha^3}}{\left(1+\frac{p^2}{(\hbar\alpha)^2}\right)^3} + \frac{2^{13}\left(\frac{n\pi}{\alpha^3}\right)^2}{\left(1+\frac{p^2}{(\hbar\alpha)^2}\right)^8} \right) +$$
$$\frac{2^8 \frac{n\pi}{\alpha^3}}{\left(1+\frac{p^2}{(\hbar\alpha)^2}\right)^3}\cos\chi\left(4 + \frac{\hbar\omega_L}{vq}\ln\left(\frac{(vq-\hbar\omega_p)^2 + (\hbar\gamma)^2}{(vq+\hbar\omega_p)^2 + (\hbar\gamma)^2} \right) \right) \frac{(2m)^2 \cdot 4\pi}{(\hbar\alpha)^4\left(1+\frac{p^2}{(\hbar\alpha)^2}\right)^2}\left(1 - \frac{2^5 \frac{n\pi}{\alpha^3}\Gamma^2(5)}{\left(1+\frac{p^2}{(\hbar\alpha)^2}\right)^5} \right) = \tag{43}$$
$$\Xi_{bg} + \Xi_{0exc} + \Xi_{1exc}.$$

Thus, the cross-section $\sigma_{fi}(\chi, \omega) \propto \Xi(\chi)$. is proportional to the factor $\Xi(\chi)$ determined by Equation (43), which contains the total dependence on the scattering angle.

3.2. Helium-Like Scatterer ^3S

The wave functions of the excited state helium-like atom and a free (incident) electron are:

$$\Phi_{He}(R - r_{11}, R - r_{22}; \xi_1, \xi_2) = \frac{1}{\sqrt{2}}\left(\varphi_{1s}(R - r_1)\varphi_{2s}(R - r_2) - \varphi_{1s}(R - r_2)\varphi_{2s}(R - r_1) \right) \cdot X(\alpha 1 \alpha 2), \tag{44}$$

where:

$$\varphi_{2s}(R - r_i) = (\alpha^3/2^3\pi)^{1/2}\left(1 - \frac{\alpha|R-r_i|}{2}\right)exp\left(-\frac{\alpha|R-r_i|}{2}\right),$$
$$\alpha = 27/16. \tag{45}$$

The vector of the initial state, corresponding to the permutation $v = 0$, but antisymmetrized only accounting for the internal permutations in the helium atom, see Equation (44), has the form:

$$\left|\Phi_i^{v=0}\right) = \Phi_{He}(r_1,r_2)X_{He}(\xi_1\xi_2)\psi_e(r)\chi_e(\xi) = \left|n = 1*,p\right).$$

The antisymmetrized vector of the final state is:

$$\Psi(r_1,r_2,r_3\alpha_1\alpha_2\alpha_3) = \frac{(\alpha 1\alpha 2\alpha 3)}{f_1\sqrt{3!}}\begin{vmatrix} \phi_{1s}(R-r_1) & \phi_{2s}(R-r_1) & \psi_p(r_1) \\ \phi_{1s}(R-r_2) & \phi_{2s}(R-r_2) & \psi_p(r_2) \\ \phi_{1sa}(R-r_3) & \phi_{2s}(R-r_3) & \psi_p(r_3) \end{vmatrix} =$$
$$\frac{(\alpha 1\alpha 2\alpha 3)}{f_1\sqrt{3}}\left(\frac{1}{\sqrt{2}}[\phi_{1s}(R-r_1)\phi_{2s}(R-r_2) - \phi_{1s}(R-r_2)\phi_{2s}(R-r_1)]\cdot\psi_p(r_3) - \right.$$
$$[\phi_{1s}(R-r_1)\phi_{2s}(R-r_3) - \phi_{2s}(R-r_1)\phi_{1s}(R-r_3)]\cdot\frac{1}{\sqrt{2}}\psi_p(r_2) +$$
$$\left.[\phi_{1s}(R-r_2)\phi_{2s}(R-r_3) - \phi_{1s}(R-r_3)\phi_{2s}(R-r_2)]\cdot\frac{1}{\sqrt{2}}\psi_p(r_1)\right). \tag{46}$$

Then, an exchange normalization factor:

$$f_1 = \iiint dr_1dr_2dr_3\frac{1}{\sqrt{2}}\overline{(\varphi*_{1s}(R-r_1)\varphi*_{2s}(R-r_2) - \varphi*_{1s}(R-r_2)\varphi*_{2s}(R-r_1))\psi_p(r_3)} \times$$
$$\frac{1}{\sqrt{3}}\left(\frac{1}{\sqrt{2}}[\phi_{1s}(R-r_1)\phi_{2s}(R-r_2) - \phi_{1s}(R-r_2)\phi_{2s}(R-r_1)]\cdot\psi_p(r_3) - [\phi_{1s}(R-r_1)\phi_{2s}(R-r_3) - \phi_{2s}(R-r_1)\phi_{1s}(R-r_3)]\cdot\frac{1}{\sqrt{2}}\psi_p(r_2) + \right.$$
$$\left.[\phi_{1s}(R-r_2)\phi_{2s}(R-r_3) - \phi_{1s}(R-r_3)\phi_{2s}(R-r_2)]\cdot\frac{1}{\sqrt{2}}\psi_p(r_1)\right)_R =$$
$$\frac{1}{\sqrt{3}}\left\{1 - \left[\left|\langle\phi_{2s}|\psi_p\rangle\right|^2 + \left|\langle\phi_{1s}|\psi_p\rangle\right|^2\right]\right\}_R =$$
$$\frac{4\pi n}{3}\left(\frac{2\pi\hbar}{p}\right)^{33}\frac{1}{\sqrt{3}}\left\{1 - 2^6\left[\left(\frac{\left(\frac{2p}{\hbar}\alpha\right)^4 + 7\cdot2\left(\frac{2p}{\hbar}\alpha\right)^2 - 11}{\left(1+\left(\frac{2p}{\hbar}\alpha\right)^2\right)^4}\right)^2 + \frac{4}{\left(4+\left(\frac{2p}{\alpha\hbar}\right)^2\right)^4}\right]\right\}.$$

A one-electron amplitude with respect to the electron permutations has the form:

$$f_{11}(r,R) = \frac{(\alpha 1\alpha 2\alpha 3)}{f_1\sqrt{3}}\int dr_1dr_2\left(\frac{1}{\sqrt{2}}[\varphi_{1s}(R-r_1)\varphi_{2s}(R-r_2) - \varphi_{1s}(R-r_2)\varphi_{2s}(R-r_1)]\cdot\psi_p(r) - \right.$$
$$[\varphi_{1s}(R-r_1)\varphi_{2s}(R-r) - \varphi_{2s}(R-r_1)\varphi_{1s}(R-r)]\cdot\frac{1}{\sqrt{2}}\psi_p(r_2) +$$
$$\left.[\varphi_{1s}(R-r_2)\varphi_{2s}(R-r) - \varphi_{1s}(R-r)\varphi_{2s}(R-r_2)]\cdot\frac{1}{\sqrt{2}}\psi_p(r_1)\right) \times$$
$$\sqrt{\frac{f_1}{6}}\frac{1}{\sqrt{2}}\overline{(\varphi*_{1s}(R-r_1)\varphi*_{2s}(R-r_2) - \varphi*_{1s}(R-r_2)\varphi*_{2s}(R-r_1))\psi*_p(r'')} =$$
$$\frac{1}{\sqrt{6f_1}\sqrt{3}}\left(e^{ip(r-r'')/\hbar} - (\alpha^3/\pi)^{1/2}exp\left(-\frac{\alpha|R-r|}{2} - \frac{ipr''}{\hbar}\right) \times \right.$$
$$\left[-2^{-3/2}\left(1 - \frac{\alpha|R-r|}{2}\right)\frac{4\pi\Gamma(5)(\alpha/2)^3\left((\alpha/2)^2 - (p/\hbar)^2\right)}{\left((\alpha/2)^2 + (p/\hbar)^2\right)^4} + \right.$$
$$\left.\left.\left(2^{-3/2}\left(1 - \frac{\alpha|R-r|}{2}\right) + exp\left(-\frac{\alpha|R-r|}{2}\right)\frac{8\pi(\alpha/2)}{\left((\alpha/2)^2 + (p/\hbar)^2\right)^2}\right]\right). \tag{47}$$

These amplitudes determine a one-electron Green function while accounting for the exchange contributions:

$$
\begin{aligned}
G(r,r\prime) &= \frac{1}{18f_1} \sum_p \frac{1}{E_p - E_{p-Q+q} + i\gamma}\left(e^{ip(r-r\prime)/\hbar} - (\alpha^3/\pi)^{1/2} e^{ip(r-r\prime)/\hbar}\left(e^{-\frac{ipr}{\hbar}} + e^{\frac{ipr\prime}{\hbar}}\right)\frac{8\pi}{(\alpha/2)^3}\right) \times \\
&\left[2^{-3/2}\left(\frac{\Gamma(5)(\alpha/2)^3\left((\alpha/2)^2 - (p/\hbar)^2\right)}{\left((\alpha/2)^2 + (p/\hbar)^2\right)^4} - 2\right) + \frac{(\alpha/2)}{8\left((\alpha/2)^2 + (p/\hbar)^2\right)^2}\right]\right) + G_{aa}(r,r\prime) = \\
&\frac{1}{18f_1}\sum_p \frac{e^{ip(r-r\prime)/\hbar}}{E_p - E_{p-Q+q}+i\gamma} - \frac{1}{18f_1}\left(\sum_p \frac{e^{ip(r-r\prime)/\hbar}}{E_p - E_{p-Q+q}+i\gamma}\left(2 - \frac{ip(r-r\prime)}{\hbar}\right)8^2 n(\alpha/2)^{-3}\right) \times \\
&\left[\left(\frac{\Gamma(5)\left(1-(2p/\hbar\alpha)^2\right)}{\left(1+(2p/\hbar\alpha)^2\right)^4} - 2\right) + \frac{1}{8\left(1+(2p/\hbar\alpha)^2\right)^2}\right]\right) + G_{aa}(r,r\prime) = \\
&\frac{1}{18f_1}\left(G_0(r,r\prime) - G_1^{exc}(r,r\prime) + G_{aa}\right).
\end{aligned}
$$

$$
G_1^{exc}(r,r\prime) = \sum_p \frac{e^{ip(r-r\prime)/\hbar}}{E_p - E_{p-Q+q}+i\gamma}\left(2 - \frac{ip(r-r\prime)}{\hbar}\right)8^2 n(\alpha/2)^{-3}\left[\left(\frac{\Gamma(5)\left(1-(2p/\hbar\alpha)^2\right)}{\left(1+(2p/\hbar\alpha)^2\right)^4} - 2\right) + \frac{1}{8\left(1+(2p/\hbar\alpha)^2\right)^2}\right].
$$

(48)

The mentioned Green function in p-representation is:

$$
\begin{aligned}
G(p-Q+q) &= G(p\prime+q) = \frac{n\alpha^{-3}}{18f_1}G_0(p\prime+q)\left(1+2^{11}\pi\right) - \frac{2^{10}\pi n\alpha^{-3}}{18f_1}G_1^{exc}(p\prime+q) = \\
&\frac{n\alpha^{-3}}{18f_1}G_0((p\prime+q)\left(1+2^{11}\pi\right) - \\
&\frac{2^{10}\pi n\alpha^{-3}}{18f_1}\frac{1}{E_p - E_{p-Q+q}+i\gamma}\cdot\left[\frac{1}{8\left(1+(2|p\prime+q|/\hbar\alpha)^2\right)^2} - \frac{24}{\left(1+(2|p\prime+q|/\hbar\alpha)^2\right)^3} + \frac{24}{\left(1+(2|p\prime+q|/\hbar\alpha)^2\right)^4}\right]\right).
\end{aligned}
$$

(49)

where the general $G_0(p\prime+q)$ and the exchange $G_1^{exc}(p\prime+q)$ contributions are:

$$
G_0(p\prime+q) = \frac{1}{E_p - E_{p-Q+q}+i\gamma}
$$
$$
G_1^{exc}(p\prime+q) =
$$
$$
\frac{1}{E_p - E_{p-Q+q}+i\gamma}\cdot\left[\frac{1}{8\left(1+(2|p\prime+q|/\hbar\alpha)^2\right)^2} - \frac{24}{\left(1+(2|p\prime+q|/\hbar\alpha)^2\right)^3} + \frac{24}{\left(1+(2|p\prime+q|/\hbar\alpha)^2\right)^4}\right].
$$

(50)

The contribution Ξ in the cross-section (28) in the form (38) for our case of the excited helium atom has the same structure as Equation (40), where the background contribution is:

$$
\begin{aligned}
\Xi_{bg} &= \left(\frac{2^{10}\pi n\alpha^{-3}}{18f_1}\right)^2 \int d\Omega_Q\{4\left(|G_0^{ex}(p\prime+q)|^2 + |G_0^{ex}(p-q)|^2\right) + \left(|G_1^{ex}(p\prime+q)|^2 + |G_1^{ex}(p-q)|^2\right) - \\
&2\left[G_0^{ex*}(p\prime+q)G_1^{ex}(p\prime+q) + G_1^{ex*}(p\prime+q)G_0^{ex}(p\prime+q) + G_0^{ex*}(p-q)G_1^{ex}(p-q) + G_1^{ex*}(p-q)G_0^{ex}(p-q)\}\right].
\end{aligned}
$$

(51)

The exchange contribution (42) in this case has the form:

$$
\begin{aligned}
\Xi_{exc} &= \int d\Omega_Q\{G_0^{ex*}(p\prime+q)G_1^{ex}(p-q) + G_1^{ex*}(p\prime+q)G_0^{ex}(p-q) + G_0^{ex*}(p-q)G_1^{ex}(p\prime+q) + \\
&G_1^{ex*}(p-q)G_0^{ex}(p\prime+q) - \left(G_1^{ex*}(p\prime+q)G_1^{ex}(p-q) + G_1^{ex*}(p-q)G_1^{ex}(p\prime+q)\right)\}.
\end{aligned}
$$

(52)

The explicit form is presented in Appendix A (see Equations (A3) and (A4)). In this expression, we neglected by the second order small terms proportional to q^2, while the inelastic momentum loss is $q \ll p$. Thus, the exchange contribution to the cross-section has the form:

$$
\begin{aligned}
\Xi_{exc} &= -\left(\frac{(\hbar\alpha)^2}{8m}\right)^2 \left\{\frac{1}{8}\frac{\partial}{\partial\mathbb{R}} + 12\left(\frac{(\hbar\alpha)^2}{8m}\right)\frac{\partial^2}{\partial\mathbb{R}^2} + 4\left(\frac{(\hbar\alpha)^2}{8m}\right)^2\frac{\partial^3}{\partial\mathbb{R}^3}\right\}\left[\frac{1}{(\mathbb{R}-\hbar\omega-i\gamma)} + c.c.\right](J_0 + J_2) - \\
&\left(\frac{(\hbar\alpha)^2}{8m}\right)^2 \times \left\{\frac{1}{8}\frac{\partial}{\partial\mathbb{R}'} + 12\frac{(\hbar\alpha)^2}{8m}\frac{\partial^2}{\partial\mathbb{R}'^2} + 4\left(\frac{(\hbar\alpha)^2}{8m}\right)^2\frac{\partial^3}{\partial\mathbb{R}'^3}\right\}\left[\frac{1}{\mathbb{R}'+\hbar\omega+i\gamma} + c.c.\right](J_0 + J_1) - \\
&\left(\frac{(\hbar\alpha)^2}{8m}\right)^4\left[\frac{1}{8}\frac{\partial}{\partial\mathbb{R}'} + 12\left(\frac{(\hbar\alpha)^2}{8m}\right)\frac{\partial^2}{\partial\mathbb{R}'^2} + 4\left(\frac{(\hbar\alpha)^2}{8m}\right)^2\frac{\partial^3}{\partial\mathbb{R}'^3}\right]\times \\
&\left[\frac{1}{8}\frac{\partial}{\partial\mathbb{R}} + 12\left(\frac{(\hbar\alpha)^2}{8m}\right)\frac{\partial^2}{\partial\mathbb{R}^2} + 4\left(\frac{(\hbar\alpha)^2}{8m}\right)^2\frac{\partial^3}{\partial\mathbb{R}^3}\right]\times \\
&\frac{1}{\mathbb{R}-i\gamma}\frac{1}{\mathbb{R}'-i\gamma}\{J_0 + J_1 + J_2 + J_3\}.,
\end{aligned}
\tag{53}
$$

where the contribution $J_0 = \Xi_0$ is determined by Equation (41), c.c. is denoted by a complex conjugate expression. Here:

$$
\begin{aligned}
\mathbb{R} &= \left(\frac{(\hbar\alpha)^2}{8m} + E_p\right) \\
\mathbb{R}' &= \mathbb{R} - \hbar\omega = \left(\frac{(\hbar\alpha)^2}{8m} + E_p - \hbar\omega\right).
\end{aligned}
\tag{54}
$$

The total exchange contribution consists of four terms:

$$
\Xi_{exc} = \Xi_{exc0} + \Xi_{exc1} + \Xi_{exc2} + \Xi_{exc3},
\tag{55}
$$

where each term is determined in detail in Appendix A, and the results of the calculations for each term are the following:

(1) The usual term "new weak localization" Ξ_{exc0}, which retains the angular dependence of the kinematic model with a factor depending on the exchange contributions:

$$
\begin{aligned}
\Xi_{exc\_0} = &\left\{\frac{1}{2^6}\left(\frac{(\hbar\alpha)^2}{2m}\right)^2\frac{1}{(\mathbb{R})^2}\left[\frac{(\mathbb{R})^2}{(\mathbb{R}-\hbar\omega)^2} + 1\right] - 12\left(\frac{(\hbar\alpha)^2}{2m}\right)^3\frac{1}{(\mathbb{R})^3}\left[\frac{(\mathbb{R})^3}{(\mathbb{R}-\hbar\omega)^3} + 1\right] + \right. \\
&3\left(\frac{(\hbar\alpha)^2}{2m}\right)^4\frac{1}{(\mathbb{R})^4}\left[\frac{(\mathbb{R})^4}{(\mathbb{R}-\hbar\omega)^4} + 1\right] - \frac{1}{2^{14}}\left(\frac{(\hbar\alpha)^2}{2m}\right)^4\frac{1}{(\mathbb{R}-\hbar\omega)^2(\mathbb{R})^2} + \\
&\frac{3}{2^{10}}\left(\frac{(\hbar\alpha)^2}{2m}\right)^5\frac{1}{(\mathbb{R})^5}\left(\frac{(\mathbb{R})^2}{(\mathbb{R}-\hbar\omega)^2} + \frac{(\mathbb{R})^3}{(\mathbb{R}-\hbar\omega)^3}\right) - \frac{9.}{2^6}\left(\frac{(\hbar\alpha)^2}{2m}\right)^6\frac{1}{(\mathbb{R})^6}\frac{(\mathbb{R})^3}{(\mathbb{R}-\hbar\omega)^3} - \\
&\frac{3}{2^{13}}\left(\frac{(\hbar\alpha)^2}{2m}\right)^6\frac{1}{(\mathbb{R})^6}\left(\frac{(\mathbb{R})^2}{(\mathbb{R}-\hbar\omega)^2} + \frac{(\mathbb{R})^4}{(\mathbb{R}-\hbar\omega)^4}\right) + \\
&\left.\frac{9}{2^9}\left(\frac{(\hbar\alpha)^2}{2m}\right)^7\frac{1}{(\mathbb{R})^7}\left(\frac{(\mathbb{R})^4}{(\mathbb{R}-\hbar\omega)^4} + \frac{(\mathbb{R})^3}{(\mathbb{R}-\hbar\omega)^3}\right) - \frac{9}{2^{12}}\left(\frac{(\hbar\alpha)^2}{2m}\right)^8\frac{1}{(\mathbb{R}-\hbar\omega)^4(\mathbb{R})^4}\right\}J_0.
\end{aligned}
\tag{56}
$$

(2) There are three typical exchange contributions, which differ from each other, each of which contains, in addition to the background exchange terms, a term proportional to $\cos\chi$

$$
\begin{aligned}
\Xi_{exc1} = &\frac{1}{8}\frac{\pi}{\mathbb{R}^4}\left(\frac{(\hbar\alpha)^2}{2m}\right)^2\left\{1 + \frac{1}{2^5}\left[\frac{1}{8}\frac{\left(\frac{(\hbar\alpha)^2}{2m}\right)^2}{(\mathbb{R}-\hbar\omega)^2} - 6\frac{\left(\frac{(\hbar\alpha)^2}{2m}\right)^3}{(\mathbb{R}-\hbar\omega)^3} + \frac{3}{4}\frac{\left(\frac{(\hbar\alpha)^2}{2m}\right)^4}{(\mathbb{R}-\hbar\omega)^4}\right]\right\}\times \\
&\left\{\frac{\mathbb{R}}{vq}\ln\left(\frac{(vq-\hbar\omega_p)^2+(\hbar\gamma)^2}{(vq+\hbar\omega_p)^2+(\hbar\gamma)^2}\right)\left[\frac{1}{8}\left(2 + \frac{3\hbar\omega}{\mathbb{R}}\right) - 3\frac{(\hbar\alpha)^2}{2m}\left(\frac{6}{\mathbb{R}} + \frac{12\hbar\omega}{\mathbb{R}^2}\right) + \left(\frac{(\hbar\alpha)^2}{2m}\right)^2\left(\frac{6}{\mathbb{R}^2} + \frac{15\hbar\omega}{\mathbb{R}^3}\right)\right] + \right. \\
&\left.3\frac{v'}{v}\cos\chi\left(2 + \frac{\hbar\omega}{2v\cdot q}\ln\left(\frac{(vq-\hbar\omega_p)^2+(\hbar\gamma)^2}{(vq+\hbar\omega_p)^2+(\hbar\gamma)^2}\right)\right)\left[\frac{1}{4} - 24\frac{\left(\frac{(\hbar\alpha)^2}{2m}\right)}{\mathbb{R}} + 10\frac{\left(\frac{(\hbar\alpha)^2}{2m}\right)^2}{\mathbb{R}^2}\right]\right\}.
\end{aligned}
\tag{57}
$$

$$\Xi_{exc2} = \frac{1}{8}\left(\frac{\pi}{\mathbb{R}^4}\right)\left(\frac{(\hbar\alpha)^2}{2m}\right)^2 \left\{ 1 + \frac{1}{2^5} \frac{\left(\frac{(\hbar\alpha)^2}{2m}\right)^2}{(\mathbb{R}-\hbar\omega)^2}\left[\frac{1}{8} - 6\frac{\left(\frac{(\hbar\alpha)^2}{2m}\right)}{(\mathbb{R}-\hbar\omega)} + \frac{3}{4}\frac{\left(\frac{(\hbar\alpha)^2}{2m}\right)^2}{(\mathbb{R}-\hbar\omega)^2}\right]\right\} \times$$

$$\left\{\frac{2\mathbb{R}}{v\prime q}\ln\left(\frac{(v\prime q-\hbar\omega_p)^2+(\hbar\gamma)^2}{(v\prime q+\hbar\omega_p)^2+(\hbar\gamma)^2}\right)\left[\frac{1}{8} - 9\frac{\frac{(\hbar\alpha)^2}{2m}}{\mathbb{R}} + 3\frac{\left(\frac{(\hbar\alpha)^2}{2m}\right)^2}{\mathbb{R}^2}\right] + \right.$$

$$\left. 3\frac{v}{v\prime}\cos\chi\left(2 + \frac{\hbar\omega+i\gamma}{2v\cdot q}\ln\left(\frac{(v\prime q-\hbar\omega_p)^2+(\hbar\gamma)^2}{(v\prime q+\hbar\omega_p)^2+(\hbar\gamma)^2}\right)\right)\left[\frac{1}{4} - \left(\frac{(\hbar\alpha)^2}{2m}\right)\frac{24}{\mathbb{R}} + \left(\frac{(\hbar\alpha)^2}{2m}\right)^2\frac{10}{\mathbb{R}^2}\right]\right\}. \tag{58}$$

$$\Xi_{exc3} = -\pi\cdot\cos\chi\frac{1}{(\mathbb{R}-\hbar\omega)^2}\frac{\left(\frac{(\hbar\alpha)^2}{2m}\right)^4}{(4\mathbb{R})^4}\left[-\frac{1}{2} + 24\frac{\left(\frac{(\hbar\alpha)^2}{2m}\right)}{(\mathbb{R}-\hbar\omega)} + 3\frac{\left(\frac{(\hbar\alpha)^2}{2m}\right)^2}{(\mathbb{R}-\hbar\omega)^2}\right] \times$$

$$\left\{\left[-\frac{1}{8}\left(3 + \frac{4\hbar\omega}{\mathbb{R}}\right) + \frac{5}{3}\frac{(vq)(v\prime q)}{\mathbb{R}^2}\right] + 6\frac{\left(\frac{(\hbar\alpha)^2}{2m}\right)}{\mathbb{R}}\left[\left(6 + \frac{10\hbar\omega}{\mathbb{R}}\right) - 5\frac{(vq)(v\prime q)}{\mathbb{R}^2}\right] + \right.$$

$$\left. \frac{5}{2}\frac{\left(\frac{(\hbar\alpha)^2}{2m}\right)^2}{\mathbb{R}^2}\left[-6\left(1 + \frac{2\hbar\omega}{\mathbb{R}}\right) + 7\frac{(vq)(v\prime q)}{\mathbb{R}^2}\right]\right\}. \tag{59}$$

These three exchange terms (57)–(59) depend on the initial atomic states of the scatterers, the number of electrons, and their multiplicity, but this dependence is expressed in the value of the background contribution and the value of the factor in front of $\cos\chi$. Variance in the number of electrons for different scatterers and their states does not lead to a new angular dependence in the scattering cross-section under conditions of a new weak localization.

In our work, we consider the identity of the incident electrons and electrons on helium-like scatterers. Helium-doped Zi-Ni alloys were used as a disordered medium. Doping the samples with helium increases disorder and enhances quantum interference. For our estimates, we take $\omega_p \sim 10^{15}$, then the energy losses due to inelastic scattering will be ~1 eV, then the contribution from the usual weak localization of a new type, determined by Equation (42), to the factor in the cross-section is shown in Figure 2a. It is easy to see the fluctuations of the factor with increasing energy of the incident electron (the value determined by Equation (52)) and the singularity at an angle of 2.9 rad, which corresponds to 166 angular degrees. This angle is in a good agreement with the results of the kinematic model Equation (22):

$$\cos\chi = -\frac{E_p + \left(E_p - \hbar\omega\right)}{2\sqrt{E_p\left(E_p - \hbar\omega\right)}} + 2\frac{(\hbar\omega)^2}{vv\prime q^2} \approx 0.98, \Rightarrow \chi \approx 2{,}94, \tag{60}$$

where it was taken $\frac{\hbar\omega}{vq} \approx 0.1$, and $\frac{\hbar\omega}{E_p} \approx 0.05$. Taking into account the identity of incident electrons with atomic electrons increases the contribution Ξ_{exc0} in Equation (57) by 3–15 times dependently from the incident electron energy and does not change the singularity angle (see Figure 2b). True, it should be noted that a special contribution of exchange effects occurs at low energies, of the order of $E_p \sim 2$ (a.u.) atomic energy units, which is natural, since the de Broglie wavelength of an incident electron becomes comparable to the wavelength of an atomic electron. Under such conditions, exchange effects become dominant. Both exchange eigenvalues Ξ_{exc1}, Ξ_{exc2}, proportional to $\cos\chi$, give the same contribution as shown in Figure 2c. Naturally, it makes the main contribution, only at low energies of incident electrons, less than 3 eV. The third exchange term also proportional to $\cos\chi$ makes the main exchange contribution in this effect. It is shown in Figure 2d. Then, the total angle factor in the scattering cross-section has the form shown in Figure 2e. Figure 2f,h show the shape of the peak for the incident electron energy ~2 a.u. for the intrinsic contribution J_0 (41), and its increasing exchange effects and for the total factor in the cross-section taking into account the exchange terms. It can be argued that the exchange effects do not change the shape and angular position of this peak. The peak size depends on the energy of the incident electrons due to the exchange effects associated with the ratio of the de Broglie wavelength to the size of the atom, as mentioned above. Figure 3 shows all contributions to the angle factor of the cross-section for the singlet state of the helium atom. Here, the intrinsic contribution J_0 (41) is the same, but the exchange coefficient differs from that shown in Figure 2b and is determined by the exchange effects from the overlap of the wave function of freely falling electrons with the singlet state of the helium atom. The exchange factor of the singlet state is shown in Figure 3a, and it is 1.5 times less then the triplet's exchange factor. The eigent exchange contribution Ξ_{exc1}, proportional to $\cos\chi$, presented by Equation (44), is shown in Figure 3b, and it has the same shape as the Ξ_{exc3} of the triplet state but is five times less compared with the last. The total angle factor to the cross-section, including the exchange effects from the singlet state, is shown in Figure 3c.

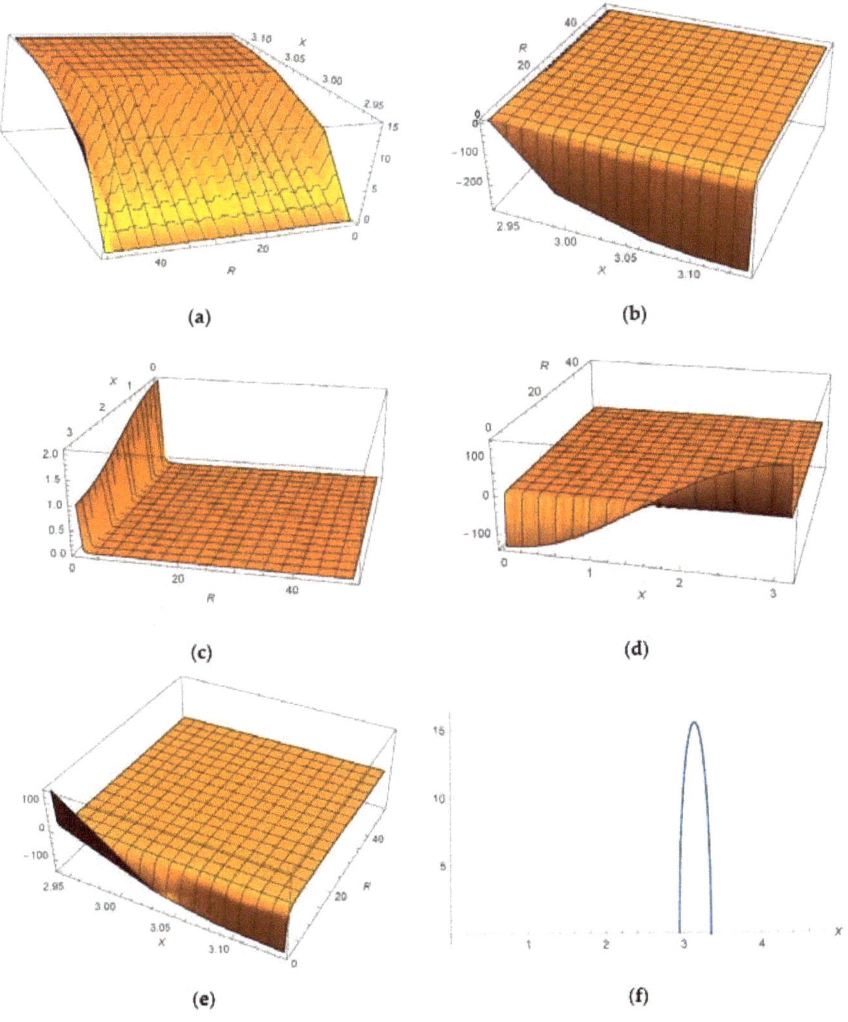

Figure 2. *Cont.*

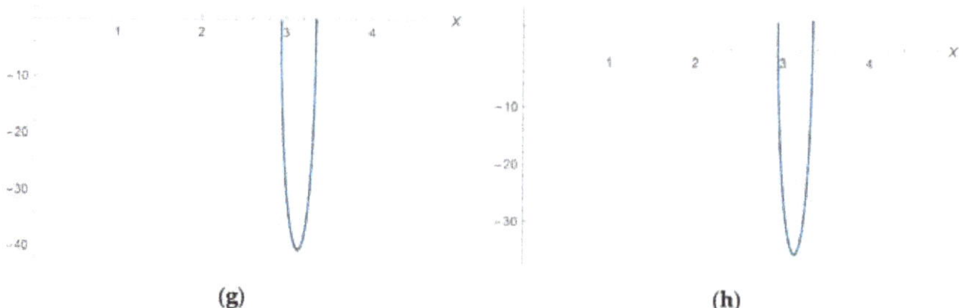

(g)   (h)

Figure 2. This is a figure of the contributions to the angular factor in the cross-section for electron scattering by a helium atom in the triplet state. Here, one axis corresponds to the energy of the incident electron, R, measured from the energy of the bound state in the atomic units (a.u.). The second axis corresponds to scattering angle X, measured in radian (rad.). The third axis corresponds to the measureless factor. (**a**) A contribution J_0 from the usual weak localization of a new type, determined by Equation (42), to the factor in the cross-section; (**b**) corresponds to the term Ξ_{exc0}, presented by Equation (58); (**c**) Both exchange eigenvalues Ξ_{exc1}, Ξ_{exc2}, proportional to $\cos\chi$, give the same contribution; (**d**) The third "exchange-exchange" term Ξ_{exc3} is also proportional to $\cos\chi$. (**e**) The total angle factor in the scattering cross-section; (**f**) The peak for the incident electron energy ~2 a.u. for the intrinsic contribution J_0 (42); (**g**) The increase of the peak by exchange effects; (**h**) The total factor in the cross-section taking into account the exchange terms.

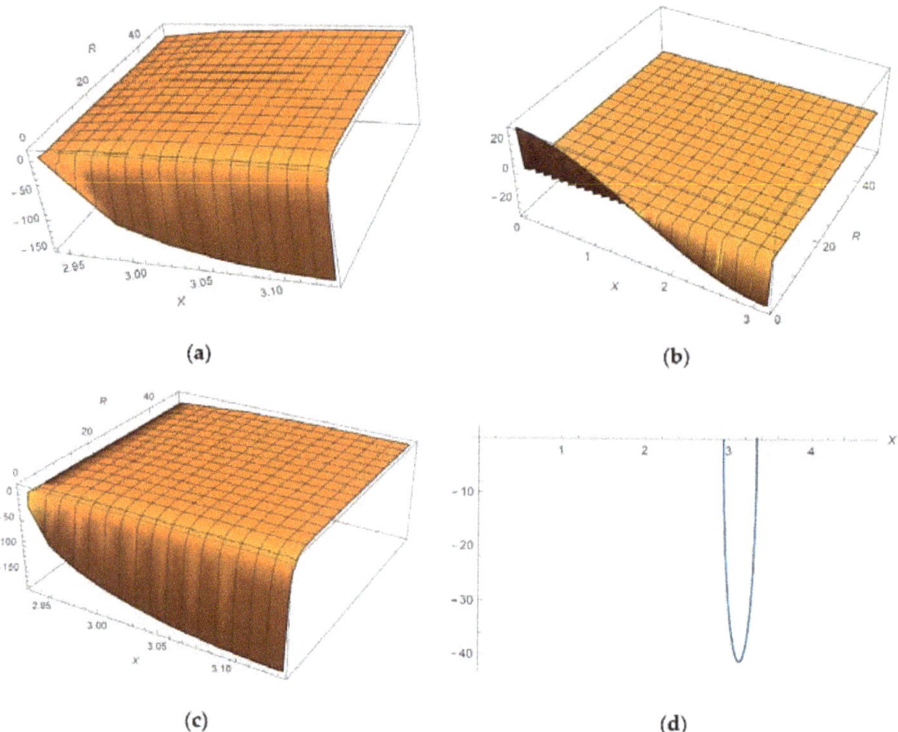

(a)   (b)

(c)   (d)

Figure 3. *Cont.*

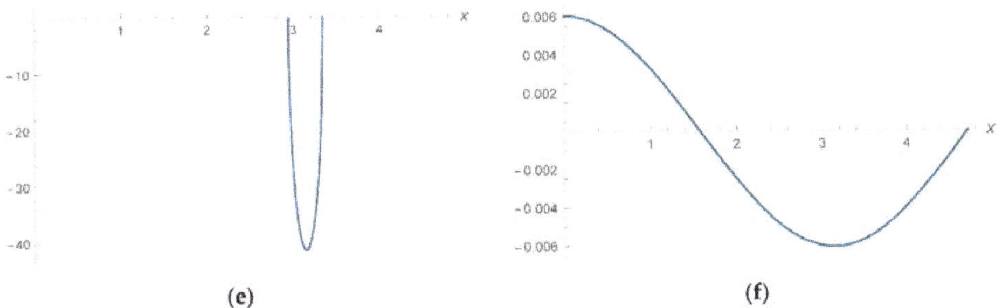

(e)   (f)

Figure 3. This is a figure of the contributions to the angular factor in the cross-section for electron scattering by a helium atom in the singlet state. Here, one axis corresponds to the energy of the incident electron, R, measured from the energy of the bound state (in the atomic units (a.u.)) and the second axis corresponds to scattering angle X, measured in radians (rad.). The third axis corresponds to the measureless factor. (**a**) A contribution Ξ_{0exc} from the usual weak localization of a new type, determined by Equation (44), to the factor in the cross-section; (**b**) corresponds to the eigent exchange contribution Ξ_{exc1}, proportional to $\cos\chi$ and presented by the Equation (44); (**c**) The total angle factor in the scattering cross-section; (**d**). The increase of the peak by exchange effects peak for the incident electron energy ~2 a.u. (42); (**e**) The total factor in the cross-section taking into account the exchange terms; (**f**) The exchange contribution Ξ_{exc1}, proportional to $\cos\chi$.

4. Discussion

Investigation of the influence of exchange effects on the process of coherent backscattering of electrons by disordered media is a very important task both for weak localization of the Anderson type and for a new type of weak localization with an inelastic process. For a new type of weak localization, this problem stood and was solved theoretically only for the case of hydrogen-like scatterers in metal alloys [26], where an additional angular dependence was mentioned, proportional to $\cos\chi$ in the cross-section. It is now clear that this new dependence is not clearly visible, in contrast to the angular dependence of the new weak localization, explained by the kinematic consideration. The latter has a sharper peak and remains invariable even in the range of beam energies, where exchange effects are strong. In this work, we investigated the exchange effects from the first principle consideration for general case. We used the formulas for the S-matrix and scattering cross-section developed generally [27] while taking into account the exchange contributions. We developed the Green function (13), taking into account the wave function overlapping due to permutation symmetry. After analyzing all exchange contributions in the example of helium-like scatterers in metallic alloys, we come to the conclusion that exchange effects do not change the angle dependence in the new weak localization but could change the size of the peak very dramatically. A detailed comparison of the influence of exchange effects on the new weak localization for the singlet and triplet states of the helium atom shows exactly that the role of exchange effects in the case of a singlet is negligible. While, for the triplet state, it is decisive, especially for those values of the energy of incident electrons when de Broglie's waves are commensurate with the atomic.

5. Conclusions

In our work, we focused on two aspects of the theoretical description of non-Anderson weak localization of electrons in disordered media. First, we made an attempt to create a general method for taking into account the exchange of electrons between the incident beam and the belonging of electrons to atomic scatterers, using the formalism of the invariant exchange perturbation theory. We applied the general expressions for the scattering amplitude and scattering probability with the allowance for permutation symmetry obtained in [27,28] to the problem of the process of electron scattering by disordered media under new weak localization. The result of this development of the theory is expressed by Equations (27) and (28). Second, we applied the developed general equations to a specific system: a scattering electron beam on disordered helium-like scatterers in Zi-Ni alloys doped with helium. We performed a detailed analytical calculation of the angle factor in the scattering cross-section for this system. We showed that the exchange contribution in the angle dependence of the electrons' cross-section under the "new weak localization" condition appears in two apostasies, in the additional angle dependence, proportional to $\cos\chi$, and in the exchange factor to the "kinematic" angle dependence. In our opinion, such consideration would also be useful for the so-called anti-Anderson localization considered experimentally and theoretically in the works [33–36] for the different disordered systems.

Author Contributions: Conceptualization, E.V.O.; methodology, E.V.O.; software, E.V.O. and F.E.O.; validation, E.V.O. and F.E.O.; formal analysis, E.V.O. and F.E.O.; investigation, E.V.O. and F.E.O.; resources, F.E.O.; data curation, F.E.O.; writing—original draft preparation, E.V.O.; writing—review and editing, E.V.O. and F.E.O.; visualization, F.E.O. All authors have read and agreed to the published version of the manuscript.

Funding: This research received no external funding.

Acknowledgments: The authors are grateful to Alexey I. Borovkov, Vice-Rector for Innovative Projects at Peter the Great St. Petersburg Polytechnic University, for financial support of the presented publication.

Conflicts of Interest: The authors declare no conflict of interest.

Appendix A. Mathematical Additions

Appendix A.1. The Proof of Completeness Property of Antisymmetric Basis

It is easy to show this property by the acting operator (6) on to any antisymmetric vector $|m, p\prime\rangle$ and by using Equation (5):

$$\left(\sum_{n,p}|n,p\rangle\frac{f_{np}}{N}\langle n,p|\right)|m,p\prime\rangle = \sum_{n,p}|n,p\rangle\frac{f_{np}}{N}\langle n,p|m,p\prime\rangle = \sum_{v}\sum_{n,p}(-1)^{g_v}\frac{1}{\boxed{f_{np}}}\frac{\boxed{f_{np}}}{N}|n,p\rangle_v\langle n,p|m,p\prime\rangle =$$

$$\sum_{v}\sum_{n,p}\boxed{(-1)^{g_v}}\frac{1}{\boxed{f_{np}}}\frac{\boxed{f_{np}}}{N}|n,p\rangle_v\langle n,p|_v m,p\prime\rangle\boxed{(-1)^{g_v}} = \sum_{v}\frac{1}{N}\sum_{n,p}|n,p\rangle_v\langle n,p|_v m,p\prime\rangle = \frac{1}{N}\sum_{v}|m,p\prime\rangle = |m,p\prime\rangle$$

(A1)

Appendix A.2. Young Operators

$$\begin{aligned}
\omega_{11}^{[21]} &= \tfrac{1}{\sqrt{12}}(2 + 2P_{12} - P_{23} - P_{13} - P_{123} - P_{132}),\\
\omega_{12}^{[21]} &= \tfrac{1}{2}(P_{23} - P_{13} + P_{123} - P_{132}),\\
\omega_{21}^{[21]} &= \tfrac{1}{2}(P_{23} - P_{13} - P_{123} + P_{132})\\
\omega_{22}^{[21]} &= \tfrac{1}{\sqrt{12}}(2 - 2P_{12} + P_{23} + P_{13} - P_{123} - P_{132}).
\end{aligned}$$

(A2)

Appendix A.3. The Explicit Form of the Cross-Section Pre-Factor $\Xi(\chi)$ for Expression (52)

Here, we took into account the obvious relations:

$$\begin{aligned}
E_p - E_{p-Q+q} &= \tfrac{1}{2m}\left(p^2 - p\prime^2 - 2p\prime q - q^2\right) = \hbar\omega - v\prime q,\\
E_p - E_{p-q} &= \tfrac{1}{2m}\left(p^2 - p^2 + 2p\cdot q - q^2 - \hbar\omega\right) = v\cdot q - \hbar\omega,\\
\left(2|p-q|/\hbar\alpha\right)^2 &= \left(\tfrac{2}{\hbar\alpha}\right)^2\tfrac{2m}{2m}\left(p^2 - 2p\cdot q + q^2\right) = 4\frac{2m}{(\hbar\alpha)^2}\left(E_p - v\cdot q + \boxed{q^2}\right),\\
\left(2|p\prime+q|/\hbar\alpha\right)^2 &= \left(\tfrac{2}{\hbar\alpha}\right)^2\tfrac{2m}{2m}\left(p\prime^2 + 2p\prime\cdot q + q^2\right) = 4\frac{2m}{(\hbar\alpha)^2}\left(E_p - \hbar\omega + v\prime\cdot q + \boxed{q^2}\right),
\end{aligned}$$

(A3)

$$
\begin{aligned}
\Xi_{exc} = \int d\Omega_Q \Bigg\{ &\frac{1}{\hbar\omega - v'q + i\gamma} \cdot \frac{1}{v\cdot q - \hbar\omega - i\gamma}\left[\frac{1}{8\left(1+(2|p-q|/\hbar\alpha)^2\right)^2} - \frac{24}{\left(1+(2|p-q|/\hbar\alpha)^2\right)^3} + \frac{24}{\left(1+(2|p-q|/\hbar\alpha)^2\right)^4}\right] + \\
&\frac{1}{v\cdot q - \hbar\omega - i\gamma} \cdot \frac{1}{\hbar\omega - v'q + i\gamma}\left[\frac{1}{8\left(1+(2|p'+q|/\hbar\alpha)^2\right)^2} - \frac{24}{\left(1+(2|p'+q|/\hbar\alpha)^2\right)^3} + \frac{24}{\left(1+(2|p'+q|/\hbar\alpha)^2\right)^4}\right] + \\
&\frac{1}{v\cdot q - \hbar\omega + i\gamma} \cdot \frac{1}{\hbar\omega - v'q - i\gamma}\left[\frac{1}{8\left(1+4\frac{2m}{(\hbar\alpha)^2}(E_p-\hbar\omega+v'\cdot q)\right)^2} - \frac{24}{\left(1+4\frac{2m}{(\hbar\alpha)^2}(E_p-\hbar\omega+v'\cdot q)\right)^3} + \frac{24}{\left(1+4\frac{2m}{(\hbar\alpha)^2}(E_p-\hbar\omega+v'\cdot q)\right)^4}\right] + \\
&\frac{1}{\hbar\omega - v'q - i\gamma}\frac{1}{v\cdot q - \hbar\omega + i\gamma}\left[\frac{1}{8\left(1+4\frac{2m}{(\hbar\alpha)^2}\left(E_p-v\cdot q+\boxed{q^2}\right)\right)^2} - \frac{24}{\left(1+4\frac{2m}{(\hbar\alpha)^2}(E_p-v\cdot q)\right)^3} + \frac{24}{\left(1+4\frac{2m}{(\hbar\alpha)^2}(E_p-v\cdot q)\right)^4}\right] - \\
&\frac{1}{\hbar\omega - v'q + i\gamma}\left[\frac{1}{8\left(1+(2|p'+q|/\hbar\alpha)^2\right)^2} - \frac{24}{\left(1+(2|p'+q|/\hbar\alpha)^2\right)^3} + \frac{24}{\left(1+(2|p'+q|/\hbar\alpha)^2\right)^4}\right] \times \\
&\frac{1}{v\cdot q - \hbar\omega - i\gamma}\left[\frac{1}{8\left(1+(2|p-q|/\hbar\alpha)^2\right)^2} - \frac{24}{\left(1+(2|p-q|/\hbar\alpha)^2\right)^3} + \frac{24}{\left(1+(2|p-q|/\hbar\alpha)^2\right)^4}\right] - \\
&\frac{1}{v\cdot q - \hbar\omega + i\gamma}\left[\frac{1}{8\left(1+4\frac{2m}{(\hbar\alpha)^2}\left(E_p-v\cdot q+\boxed{q^2}\right)\right)^2} - \frac{24}{\left(1+4\frac{2m}{(\hbar\alpha)^2}(E_p-v\cdot q)\right)^3} + \frac{24}{\left(1+4\frac{2m}{(\hbar\alpha)^2}(E_p-v\cdot q)\right)^4}\right] \times \\
&\frac{1}{\hbar\omega - v'q - i\gamma}\left[\frac{1}{8\left(1+4\frac{2m}{(\hbar\alpha)^2}(E_p-\hbar\omega+v'\cdot q)\right)^2} - \frac{24}{\left(1+4\frac{2m}{(\hbar\alpha)^2}(E_p-\hbar\omega+v'\cdot q)\right)^3} + \frac{24}{\left(1+4\frac{2m}{(\hbar\alpha)^2}(E_p-\hbar\omega+v'\cdot q)\right)^4}\right] \Bigg\},
\end{aligned}
\tag{A4}
$$

In this expression, we neglected by the second order small terms proportional to q^2, while inelastic momentum loss is $q << p$.

Appendix A.4. The Integrals Used to Derive the Expression for the Factor $\Xi(\chi)$

Here are the integrals used to derive the expression for the factor $\Xi(\chi)$ ((39) and (57)), which contains the angular dependence introduced into Expression (28) for the cross-section:

$$
\begin{aligned}
J_0 = \Xi_0 = \int d\Omega_Q &\left\{ G_0^{ex*}(p'+q)G_0^{ex}(p-q) + G_0^{ex*}(p-q)G_0^{ex}(p'+q)^2 \right\} = \\
&\frac{4\pi\hbar\omega^2 p}{qv\gamma} \cdot \left[arctg\left(\frac{qv+\hbar\omega_p}{\hbar\gamma}\right) + arctg\left(\frac{qv-\hbar\omega_p}{\hbar\gamma}\right) - \frac{\gamma}{\sqrt{2\omega_p^2(1-\cos\chi)-\frac{q^2v^2}{\hbar^2}\sin^2\chi}}\right] \times \\
&ln\left(\frac{\hbar^2\omega_p^2-q^2v^2\cos\chi+qv\sqrt{2\hbar^2\omega_p^2(1-\cos\chi)-q^2v^2\sin^2\chi}}{\hbar^2\omega_p^2-q^2v^2\cos\chi-qv\sqrt{2\hbar^2\omega_p^2(1-\cos\chi)-q^2v^2\sin^2\chi}}\right).
\end{aligned}
\tag{A5}
$$

$$
\begin{aligned}
J_1 = \int d\Omega_q &\frac{1}{v\cdot q - \hbar\omega + i\gamma}\frac{1}{\left(\left(\frac{(\hbar\alpha)^2}{8m}+E_p-\hbar\omega\right)+v'\cdot q\right)} = \\
&\frac{1}{\left(\frac{(\hbar\alpha)^2}{8m}+E_p\right)}\left\{\left(1+\frac{\hbar\omega}{\left(\frac{(\hbar\alpha)^2}{8m}+E_p\right)}\right)\frac{2\pi}{vq}ln\left|\frac{v\cdot q-\hbar\omega-i\gamma}{v\cdot q+\hbar\omega+i\gamma}\right| - \frac{2\pi}{vq}\frac{v'q}{\left(\frac{(\hbar\alpha)^2}{8m}+E_p\right)}\cos\chi\left(2+zln\left|\frac{v\cdot q-\hbar\omega-i\gamma}{v\cdot q+\hbar\omega+i\gamma}\right|\right)\right\}
\end{aligned}
\tag{A6}
$$

$$
\begin{aligned}
J_2 = \int d\Omega_{Q'} &\frac{1}{\hbar\omega - v'q + i\gamma}\frac{1}{\left(\left(\frac{(\hbar\alpha)^2}{8m}+E_p\right)-v\cdot q\right)} = \frac{-1}{\left(\frac{(\hbar\alpha)^2}{8m}+E_p\right)}\left\{\frac{2\pi}{v'q}ln\left|\frac{v'\cdot q-\hbar\omega-i\gamma}{v'\cdot q+\hbar\omega+i\gamma}\right| + \right. \\
&\left. \frac{vq}{v'q}\cos\chi\frac{2\pi}{\left(\frac{(\hbar\alpha)^2}{8m}+E_p\right)}\left(2+z'ln\left|\frac{v'\cdot q-\hbar\omega-i\gamma}{v'\cdot q+\hbar\omega+i\gamma}\right|\right)\right\}, \\
&z' = \frac{\hbar\omega+i\gamma}{v'\cdot q}.
\end{aligned}
\tag{A7}
$$

$$J_3 = \int d\Omega_{Q'} \frac{1}{\left(\left(\frac{(\hbar\alpha)^2}{8m}+E_p\right)-v\cdot q\right)} \frac{1}{\left(\left(\frac{(\hbar\alpha)^2}{8m}+E_p-\hbar\omega\right)+v'\cdot q\right)} =$$

$$\frac{4\pi\cdot\cos\chi}{\left(\frac{(\hbar\alpha)^2}{8m}+E_p\right)^2}\left[\left(1+\frac{\hbar\omega}{\left(\frac{(\hbar\alpha)^2}{8m}+E_p\right)}\right)-\frac{1}{3}\frac{(vq)(v'q)}{\left(\frac{(\hbar\alpha)^2}{8m}+E_p\right)^2}\right] \tag{A8}$$

Here, the c.c. complex conjugate expression is denoted:

$$\begin{aligned}
\Xi_{exc_0} =& \left\{\frac{1}{2^6}\left(\frac{(\hbar\alpha)^2}{2m}\right)^2\left[\frac{1}{(\mathbb{R}-\hbar\omega)^2}+\frac{1}{(\mathbb{R})^2}\right]-12\left(\frac{(\hbar\alpha)^2}{2m}\right)^3\left[\frac{1}{(\mathbb{R}-\hbar\omega)^3}+\frac{1}{(\mathbb{R})^3}\right]+\right. \\
& 3\left(\frac{(\hbar\alpha)^2}{2m}\right)^4\left[\frac{1}{(\mathbb{R}-\hbar\omega)^4}+\frac{1}{(\mathbb{R})^4}\right]-\frac{1}{2^{14}}\left(\frac{(\hbar\alpha)^2}{2m}\right)^4\frac{1}{(\mathbb{R}-\hbar\omega)^2(\mathbb{R})^2}+ \\
& \frac{3}{2^{10}}\left(\frac{(\hbar\alpha)^2}{2m}\right)^5\left(\frac{1}{(\mathbb{R}-\hbar\omega)^2(\mathbb{R})^3}+\frac{1}{(\mathbb{R}-\hbar\omega)^3(\mathbb{R})^2}\right)-\frac{9\cdot}{2^6}\left(\frac{(\hbar\alpha)^2}{2m}\right)^6\frac{1}{(\mathbb{R}-\hbar\omega)^3(\mathbb{R})^3}- \\
& \frac{3}{2^{13}}\left(\frac{(\hbar\alpha)^2}{2m}\right)^6\left(\frac{1}{(\mathbb{R}-\hbar\omega)^2(\mathbb{R})^4}+\frac{1}{(\mathbb{R}-\hbar\omega)^4(\mathbb{R})^2}\right)+ \\
& \left.\frac{9}{2^9}\left(\frac{(\hbar\alpha)^2}{2m}\right)^7\left(\frac{1}{(\mathbb{R}-\hbar\omega)^4(\mathbb{R})^3}+\frac{1}{(\mathbb{R}-\hbar\omega)^3(\mathbb{R})^4}\right)-\frac{9}{2^{12}}\left(\frac{(\hbar\alpha)^2}{2m}\right)^8\frac{1}{(\mathbb{R}-\hbar\omega)^4(\mathbb{R})^4}\right\}J_0.
\end{aligned} \tag{A9}$$

$$\begin{aligned}
\Xi_{exc1} =& -\left(\frac{(\hbar\alpha)^2}{8m}\right)^2\times\left\{\frac{1}{8}\frac{\partial}{\partial\mathbb{R}'}+12\frac{(\hbar\alpha)^2}{8m}\frac{\partial^2}{\partial\mathbb{R}'^2}+4\left(\frac{(\hbar\alpha)^2}{8m}\right)^2\frac{\partial^3}{\partial\mathbb{R}'^3}\right\}\left[\frac{1}{\mathbb{R}'+\hbar\omega+i\gamma}+c.c.\right]J_1- \\
& \left(\frac{(\hbar\alpha)^2}{8m}\right)^4\left[\frac{1}{8}\frac{\partial}{\partial\mathbb{R}'}+12\left(\frac{(\hbar\alpha)^2}{8m}\right)\frac{\partial^2}{\partial\mathbb{R}'^2}+4\left(\frac{(\hbar\alpha)^2}{8m}\right)^2\frac{\partial^3}{\partial\mathbb{R}'^3}\right]\times \\
& \left[\frac{1}{8}\frac{\partial}{\partial\mathbb{R}}+12\left(\frac{(\hbar\alpha)^2}{8m}\right)\frac{\partial^2}{\partial\mathbb{R}^2}+4\left(\frac{(\hbar\alpha)^2}{8m}\right)^2\frac{\partial^3}{\partial\mathbb{R}^3}\right]\frac{1}{\mathbb{R}-i\gamma}\frac{1}{\mathbb{R}'-i\gamma}J_1= \\
& \frac{1}{8}\left(\frac{(\hbar\alpha)^2}{2m}\right)^2\left\{1+\frac{1}{2^5}\left[\frac{1}{8}\frac{\left(\frac{(\hbar\alpha)^2}{2m}\right)^2}{(\mathbb{R}-\hbar\omega)^2}-6\frac{\left(\frac{(\hbar\alpha)^2}{2m}\right)^3}{(\mathbb{R}-\hbar\omega)^3}+\frac{3}{4}\frac{\left(\frac{(\hbar\alpha)^2}{2m}\right)^4}{(\mathbb{R}-\hbar\omega)^4}\right]\right\}\times \\
& \left\{\frac{\pi}{vq}\frac{1}{\mathbb{R}^3}ln\left(\frac{(vq-\hbar\omega_p)^2+(\hbar\gamma)^2}{(vq+\hbar\omega_p)^2+(\hbar\gamma)^2}\right)\left[\frac{1}{8}\left(2+\frac{3\hbar\omega}{\mathbb{R}}\right)-3\frac{(\hbar\alpha)^2}{2m}\left(\frac{6}{\mathbb{R}}+\frac{12\hbar\omega}{\mathbb{R}^2}\right)+\left(\frac{(\hbar\alpha)^2}{2m}\right)^2\left(\frac{6}{\mathbb{R}^2}+\frac{15\hbar\omega}{\mathbb{R}^3}\right)\right]+\right. \\
& \left.\frac{3\pi}{\mathbb{R}^4}\frac{v'}{v}\cos\chi\left(2+\frac{\hbar\omega}{2v\cdot q}ln\left(\frac{(vq-\hbar\omega_p)^2+(\hbar\gamma)^2}{(vq+\hbar\omega_p)^2+(\hbar\gamma)^2}\right)\right)\left[\frac{1}{4}-24\frac{\left(\frac{(\hbar\alpha)^2}{2m}\right)}{\mathbb{R}}+10\frac{\left(\frac{(\hbar\alpha)^2}{2m}\right)^2}{\mathbb{R}^2}\right]\right\}.
\end{aligned} \tag{A10}$$

$$\begin{aligned}
\Xi_{exc2} =& \frac{1}{8}\left(\frac{(\hbar\alpha)^2}{2m}\right)^2\left\{1+\frac{1}{2^5}\left(\frac{(\hbar\alpha)^2}{2m}\right)^2\left[\frac{1}{8}\frac{1}{(\mathbb{R}-\hbar\omega)^2}-6\left(\frac{(\hbar\alpha)^2}{2m}\right)\frac{1}{(\mathbb{R}-\hbar\omega)^3}+\frac{3}{4}\left(\frac{(\hbar\alpha)^2}{2m}\right)^2\frac{1}{(\mathbb{R}-\hbar\omega)^4}\right]\right\}\times \\
& \left\{\frac{\pi}{v'q}ln\left(\frac{(v'q-\hbar\omega_p)^2+(\hbar\gamma)^2}{(v'q+\hbar\omega_p)^2+(\hbar\gamma)^2}\right)\left[\frac{1}{8}\left(\frac{2}{\mathbb{R}^3}\right)-3\frac{(\hbar\alpha)^2}{2m}\left(\frac{6}{\mathbb{R}^4}\right)+\left(\frac{(\hbar\alpha)^2}{2m}\right)^2\left(\frac{6}{\mathbb{R}^5}\right)\right]+\right. \\
& \frac{\pi}{v'q}ln\left(\frac{(v'q-\hbar\omega_p)^2+(\hbar\gamma)^2}{(v'q+\hbar\omega_p)^2+(\hbar\gamma)^2}\right)\left[\frac{1}{8}\left(\frac{2}{\mathbb{R}^3}\right)-3\frac{(\hbar\alpha)^2}{2m}\left(\frac{6}{\mathbb{R}^4}\right)+\left(\frac{(\hbar\alpha)^2}{2m}\right)^2\left(\frac{6}{\mathbb{R}^5}\right)\right]+ \\
& \left.\frac{v}{v'}\cos\chi\left(2+\frac{\hbar\omega+i\gamma}{2v\cdot q}ln\left(\frac{(v'q-\hbar\omega_p)^2+(\hbar\gamma)^2}{(v'q+\hbar\omega_p)^2+(\hbar\gamma)^2}\right)\right)\left[+\frac{3\pi}{4\mathbb{R}^4}-\left(\frac{(\hbar\alpha)^2}{2m}\right)\frac{72\pi}{\mathbb{R}^5}+\left(\frac{(\hbar\alpha)^2}{2m}\right)^2\frac{30\pi}{\mathbb{R}^6}\right]\right\}
\end{aligned} \tag{A11}$$

$$
\Xi_{exc3} = -\left(\frac{(\hbar\alpha)^2}{8m}\right)^4 \left[\frac{1}{8}\frac{\partial}{\partial\mathbb{R}'} + 12\left(\frac{(\hbar\alpha)^2}{8m}\right)\frac{\partial^2}{\partial\mathbb{R}'^2} + 4\left(\frac{(\hbar\alpha)^2}{8m}\right)^2\frac{\partial^3}{\partial\mathbb{R}'^3}\right] \times
$$

$$
\left[\frac{1}{8}\frac{\partial}{\partial\mathbb{R}} + 12\left(\frac{(\hbar\alpha)^2}{8m}\right)\frac{\partial^2}{\partial\mathbb{R}^2} + 4\left(\frac{(\hbar\alpha)^2}{8m}\right)^2\frac{\partial^3}{\partial\mathbb{R}^3}\right]\frac{1}{\mathbb{R}-i\gamma}\frac{1}{\mathbb{R}'-i\gamma}J_3 =
$$

$$
-\left(\frac{(\hbar\alpha)^2}{8m}\right)^4 \left[\frac{1}{8}\frac{\partial}{\partial\mathbb{R}'} + 12\left(\frac{(\hbar\alpha)^2}{8m}\right)\frac{\partial^2}{\partial\mathbb{R}'^2} + 4\left(\frac{(\hbar\alpha)^2}{8m}\right)^2\frac{\partial^3}{\partial\mathbb{R}'^3}\right] \times
$$

$$
\left[\frac{1}{8}\frac{\partial}{\partial\mathbb{R}} + 12\left(\frac{(\hbar\alpha)^2}{8m}\right)\frac{\partial^2}{\partial\mathbb{R}^2} + 4\left(\frac{(\hbar\alpha)^2}{8m}\right)^2\frac{\partial^3}{\partial\mathbb{R}^3}\right]\frac{1}{\mathbb{R}-i\gamma}\frac{1}{\mathbb{R}'-i\gamma}\frac{4\pi\cdot cos\chi}{\mathbb{R}^2}\left[\left(1+\frac{\hbar\omega}{\mathbb{R}}\right) - \frac{1}{3}\frac{(vq)(v'q)}{\mathbb{R}^2}\right] =
$$

$$
-\pi\cdot cos\chi\frac{1}{(\mathbb{R}-\hbar\omega)^2}\frac{\left(\frac{(\hbar\alpha)^2}{2m}\right)^4}{(4\mathbb{R})^4}\left[-\frac{1}{2} + 24\frac{\left(\frac{(\hbar\alpha)^2}{2m}\right)}{(\mathbb{R}-\hbar\omega)} + 3\frac{\left(\frac{(\hbar\alpha)^2}{2m}\right)^2}{(\mathbb{R}-\hbar\omega)^2}\right] \times
$$

$$
\left\{\left[-\frac{1}{8}\left(3+\frac{4\hbar\omega}{\mathbb{R}}\right) + \frac{5}{3}\frac{(vq)(v'q)}{\mathbb{R}^2}\right] + 6\frac{\left(\frac{(\hbar\alpha)^2}{2m}\right)}{\mathbb{R}}\left[\left(6+\frac{10\hbar\omega}{\mathbb{R}}\right) - 5\frac{(vq)(v'q)}{\mathbb{R}^2}\right] + \right.
$$

$$
\left. \frac{5}{2}\frac{\left(\frac{(\hbar\alpha)^2}{2m}\right)^2}{\mathbb{R}^2}\left[-6\left(1+\frac{2\hbar\omega}{\mathbb{R}}\right) + 7\frac{(vq)(v'q)}{\mathbb{R}^2}\right]\right\}
$$

(A12)

References

1. Pollak, M.; Ortuño, M.; Frydman, A. Disordered electronic systems. In *The Electron Glass*; Cambridge University Press: Cambridge, UK, 2013; pp. 7–39. [CrossRef]
2. Bergmann, G. Weak localization in thin films. A time-of-light experiment with conduction electrons. *Phys.Rep.* **1984**, *107*, 1–58.
3. Rammer, J. Quantum transport theory of electrons in solids: A single-particle approach. *Rev. Mod. Phys.* **1991**, *63*, 781–817. [CrossRef]
4. Kramer, B.; MacKinnon, A. Localization: Theory and experiment. *Rep. Prog. Phys.* **1993**, *56*, 1469.
5. Aronov, A. Fourteen years of quantum interference in disordered metals. *Phys. Scr.* **1993**, *1993*, 28–33. [CrossRef]
6. Van Der Mark, M.B.; Van Albada, M.P.; Lagendijk, A. Light scattering in strongly scattering media: Multiple scattering and weak localization. *Phys. Rev. B* **1988**, *37*, 3575–3592. [CrossRef]
7. Kravzov, Y.A. Propagation of electromagnetic waves through a turbulent atmosphere. *Rep. Prog. Phys.* **1992**, *55*, 39–51.
8. Kaveh, M. Electron and optical wave phenomena: Some new statistical properties. *Physica B* **1991**, *17*, 1–5.
9. Sheng, P. *Scattering and Localization of Classical Waves in Random Media*; World Scientific Pub. Co. Pte. Lt.: Singapore, 1990; p. 648.
10. Bergmann, G. Physical interpretation of weak localization: A time-of-flight experiment with conduction electrons. *Phys. Rev. B* **1983**, *28*, 2914–2920. [CrossRef]
11. Berkovits, R.; Kaveh, M. Back scattering of high-energy electrons from disordered media: Antienhancement due to spin-orbit interaction. *Phys. Rev. B* **1988**, *37*, 584–587.
12. Berkovits, R.; Eliyahu, D.; Kaveh, M. Enhanced backscattering of electrons in a magnetic field. *Phys. Rev. B* **1990**, *41*, 407–412. [CrossRef]
13. Gorodnichev, E.E.; Dudarev, S.L.; Rogozkin, D.B. Polarization effects in coherent backscattering of particles from random media. *JETP* **1990**, *70*, 853–863.
14. Igarashi, J.-I. Coherent backscattering of neutrons. *Phys. Rev. B* **1987**, *35*, 8894–8897. [CrossRef]
15. Washburn, S.; Webb, R.A. Aharonov-Bohm effect in normal metal. Quatnum coherence and transport. *Adv. Phys.* **1986**, *35*, 375–422.
16. Kouwenhoven, L.P.; Harmans, C.J.P.M.; Timmering, C.E.; Hekking, F.W.J.; Van Wees, B.J.; Foxon, C.T. Transport through a finite one-dimensional crystal. *Phys. Rev. Lett.* **1990**, *65*, 361–364. [CrossRef] [PubMed]
17. Maschke, K.; Schreiber, M. Unified description of coherent and dissipative electron transport. *Phys. Rev. B* **1991**, *44*, 3835–3841. [CrossRef]
18. Kanzieper, E.A.; Freilikher, V. Coherent phenomena in inelastic backscattering of electrons from disordered media. *Phys. Rev. B* **1995**, *51*, 2759–2769.

19. Tsang, L.; Ishimaru, A. Backscattering enhancement of random discrete scatterers. *J. Opt. Soc. Am. A* **1984**, *1*, 836–839.
20. Libenson, B.N.; Platonov, K.Y.; Rumyantsev, V.V. New type of weak localization of electrons in disordered media. *Sov. Phys. JETP* **1992**, *74*, 326–333.
21. Lebed', A.A.; Padusenko, E.A.; Roshchupkin, S.P.; Dubov, V.V. Resonant parametric interference effect in spontaneous bremsstrahlung of an electron in the field of a nucleus and two pulsed laser waves. *Phys. Rev. A* **2018**, *97*, 043404. [CrossRef]
22. Dubov, A.; Dubov, V.V.; Roshchupkin, S.P. Resonant high-energy bremsstrahlung of ultrarelativistic electrons in the field of a nucleus and a weak electromagnetic wave. *Laser Phys. Lett.* **2020**, *17*, 045301. [CrossRef]
23. Rumyantsev, V.V.; Doubov, V.V. Quantum transport of electrons scattered inelastically from disordered media. *Phys. Rev. B* **1994**, *49*, 8643–8654. [CrossRef] [PubMed]
24. Rumyantsev, V.; Orlenko, E.; Libenson, B. Nature of coherent enhancement of inelastic electron scattering from disordered media. *Eur. Phys. J. B* **1997**, *103*, 53–60. [CrossRef]
25. Rumyantsev, V.V.; Orlenko, E.V.; Libenson, B.N. On the theory of quantum interference between inelastic and elastic electron scattering events. *J. Exp. Theor. Phys.* **1997**, *84*, 552–559. [CrossRef]
26. Orlenko, E.V.; Rumyantsev, V.V. The effect of particle identity on a new type of weak localization. *J. Phys. Condens. Matter* **1995**, *7*, 3557–3564. [CrossRef]
27. Orlenko, E.V. Exchange Perturbation theory. In *Perturbation Theory: Advances in Research and Applications*, *Nova*; Science Publishers, Inc.: New York, NY, USA, 2018; pp. 1–59.
28. Orlenko, E.; Latychevskaia, T.; Evstafev, A.V.; Orlenko, F.E. Invariant time-dependent exchange perturbation theory and its application to the particles collision problem. *Theor. Chem. Acc.* **2015**, *134*. [CrossRef]
29. Davydov, A.S. *Quantum Mechanics*, 2nd ed.; Pergamon Press: Oxford, UK; New York, NY, USA, 1976; p. 680.
30. Libenson, B.N. Trinal nature of the excitation of bulk plasmons by a fast charged particle at grazing incidence on the interface between vacuum and a metal with strong spatial dispersion. *J. Exp. Theor. Phys.* **2018**, *127*, 37–47. [CrossRef]
31. Libenson, B.N. Effect of metallic surface microroughness on the probability and spectrum of plasmon excitation by a fast charged particle moving parallel to the surface. *J. Exp. Theor. Phys.* **2012**, *114*, 194–204. [CrossRef]
32. Dixon, R.N. Symmetry of many electron systems. *Phys. Bull.* **1975**, *26*, 546. [CrossRef]
33. Wang, H.; Liu, H.; Chang, C.-Z.; Zuo, H.; Zhao, Y.; Sun, Y.; Xia, Z.; He, K.; Ma, X.; Xie, X.C.; et al. Crossover between weak antilocalization and weak localization of bulk states in ultrathin Bi2Se3 films. *Sci. Rep.* **2014**, *4*, 5817. [CrossRef]
34. Lu, H.Z.; Shen, S.Q. Finite-Temperature Conductivity and Magnetoconductivity of Topological Insulators. *Phys. Rev. Lett.* **2014**, *112*, 146601. [CrossRef]
35. Hasan, M.Z.; Kane, C.L. Colloquium: Topological insulators. *Rev. Mod. Phys.* **2010**, *82*, 3045–3067. [CrossRef]
36. Gornyi, I.V.; Kachorovskii, V.Y.; Ostrovsky, P.M. Interference-induced magnetoresistance in HgTe quantum wells. *Phys. Rev. B* **2014**, *90*, 085401. [CrossRef]

Publisher's Note: MDPI stays neutral with regard to jurisdictional claims in published maps and institutional affiliations.

Article

An Invariant Characterization of the Levi-Civita Spacetimes

Cooper K. Watson [1,2,*], William Julius[1,2], Matthew Gorban [1,2], David D. McNutt [3], Eric W. Davis [1] and Gerald B. Cleaver [1,2]

1   Early Universe Cosmology and Strings (EUCOS) Group, Center for Astrophysics, Space Physics and Engineering Research (CASPER), Baylor University, Waco, TX 76798, USA; William_Julius1@baylor.edu (W.J.); Matthew_Gorban1@baylor.edu (M.G.); Eric_W_Davis@baylor.edu (E.W.D.); Gerald_Cleaver@baylor.edu (G.B.C.)
2   Department of Physics, Baylor University, Waco, TX 76798, USA
3   Faculty of Science and Technology, University of Stavanger, 4036 Stavanger, Norway; david.d.mcnutt@uis.no
*   Correspondence: Cooper_Watson@baylor.edu

Abstract: In the years 1917–1919 Tullio Levi-Civita published a number of papers presenting new solutions to Einstein's equations. This work, while partially translated, remains largely inaccessible to English speaking researchers. In this paper we review these solutions, and present them in a modern readable manner. We will also compute both Cartan–Karlhede and Carminati–Mclenaghan invariants such that these solutions are invariantly characterized by two distinct methods. These methods will allow for these solutions to be totally and invariantly characterized. Because of the variety of solutions considered here, this paper will also be a useful reference for those seeking to learn to apply the Cartan–Karlhede algorithm in practice.

Keywords: Levi-Civita metric; general relativity; curvature invariant

Citation: Watson, C.K.; Julius, W.; Gorban, M.; McNutt, D.D.; Davis, E.W.; Cleaver, G.B. An Invariant Characterization of the Levi-Civita Spacetimes. *Symmetry* **2021**, *13*, 1469. https://doi.org/10.3390/sym13081469

Academic Editor: Roberto Passante

Received: 23 July 2021
Accepted: 4 August 2021
Published: 11 August 2021

Publisher's Note: MDPI stays neutral with regard to jurisdictional claims in published maps and institutional affiliations.

1. Introduction

In the years 1917–1919 Tullio Levi-Civita (LC) published nearly a dozen papers introducing and analyzing a variety of new solutions to Einstein's field equations (collected works in Italian available in Volume IV at [1]). Recently, several key papers have been republished in English, including two of Levi-Civita's original papers [2,3], and [4] containing an overview of several solutions not included in any of the other translations. In [2], a homogeneous Einstein–Maxwell spacetime is derived; in [3], a spacetime with a potential analogous to the logarithmic Newtonian gravitational potential is derived; and [4] discusses derivation of several degenerate vacuum spacetimes. There are additional spacetime solutions in literature not translated into English [5,6] which are similar, but distinct from the other degenerate vacuum solutions.

The age and structure of these papers has resulted in more contemporary works citing these papers in confusing or incorrect ways. Here, we clarify the structure of works on exact solutions that are of interest. First, a homogeneous solution was published in 1917 as a standalone paper [7]. Then, in the years from 1917 to 1919, a series of nine notes were published starting with [8] and ending with [9]. It is not uncommon to find the different papers in this series cited by the general heading of the entire series or by referencing to only the first article in the series.

Additionally, we will provide an invariant (local) characterization of these solutions via two different methods. First, we will utilize the Cartan–Karlhede (CK) algorithm [10] to generate an invariant coframe and the corresponding scalar quantities which uniquely characterize these spacetimes. The variety of solutions considered in this work will result in the CK algorithm running in several markedly different ways. We will present an overview of the algorithm itself, as well as a comprehensive guide that fully outlines and computes each step of the CK algorithm for the different spacetimes. Thus, this paper should serve as a useful resource for those attempting to learn to apply the CK algorithm in practice.

233

Second, we calculate the Carminati–McLenaghan (CM) [11] scalar invariants to construct a coframe independent classification of each solution. The set of CM invariants is advantageous as they are of the lowest possible degree and are generally the minimal independent set for any valid Petrov and Segre type spacetime. All spacetimes considered herein are of a Segre type, such that only CM invariants are needed, i.e., we do not need the extended set of invariants given in [12]. In fact, in several of the cases considered the spacetime is sufficiently special, such that only a subset of the CM invariants are needed [13]. We do note explicitly that the CM invariants will only uniquely characterize these solutions to zeroth order (in derivatives), but such invariants are useful for distinguishing LC solutions. In several cases, we will also present "$\mathcal{I}$" invariants [14], as these invariants are distinct from the CM invariants and may contain information regarding algebraically special surfaces [14,15]. For completeness, we note that all spacetimes considered here are $\mathcal{I}$ non-degenerate, as the only case considered with constant scalar invariants is homogeneous [16]. We also note that the CK algorithm will always generate a complete classification of the spacetime, thus there is no possibility this may fail for the specific cases considered here.

We will also consider several generalizations of these solutions in cases where our methods of characterization extend directly, and in an instructive manner. In particular, we are interested in the work given by [17], which generalizes the solution in [2,7] to one which is not conformally flat, and in [18] which generalizes the work in [3,9] to a solution which is not generally static.

Throughout, the $(-1,1,1,1)$ signature convention will be used. Greek indices will be taken to run from 0 to 3, and follow the Einstein summation convention. Parentheses (brackets) around indices will denote the usual (anti)-symmetrized indices. Partial derivatives in the form $\frac{\partial}{\partial x}$ will always be taken to be covariant, and null tetrads will always be listed in covariant form. The null vectors $\{l, k, m, \bar{m}\}$ will be taken to have normalization, such that $l_\mu k^\mu = -1$ and $\bar{m}_\nu m^\nu = 1$.

2. Overview of the CK Algorithm

Here, we provide an overview of the practical CK algorithm used throughout this paper. For a review of the theoretic underpinning of the general Cartan process, see [19]. For a review of this process's application to general relativity, refer to [10,20,21]. In this algorithm we will use q to denote the order of differentiation, which tracks the current iteration of the algorithm. In the steps given here, we will also depart from the "standard" description of the algorithm by treating the $q = 0$ order step as a distinct "initialization" step and all steps with $q \geq 1$ as the repeated part of the algorithm. We do this as the zeroth order step is the only step in which we will require full knowledge of the algebraic type of the tensors considered as it will usually be the step at which the parameters of the isotropy group are fixed the most.

The CK algorithm will run as follows:

1. Take the order of differentiation to be $q = 0$;
2. Determine the Petrov and Segre types of the spacetime. Practical algorithms for this can be found in [22,23], respectively. These types will be used to determine the possible invariant forms to use at zeroth order;
3. Construct a null tetrad for the spacetime;
4. Calculate the components of the Riemann tensor along the current null tetrad. It will be useful to split the Riemann tensor into its irreducible parts;
5. Using the known Petrov and Segre types, along with the forms of the curvature computed above, determine an invariantly defined frame for the spacetime which fixes the frame as much as possible. Here we will make use of the invariant forms given in [24]. We also note that while it is possible to start with any frame and determine the transform which brings it to its standard form, we will usually try to determine a frame which is in (or as close as possible to) an invariant form at zeroth order. At this step, one will often have to select, by hand, if one is setting the Ricci

or Weyl tensor into an invariant form, as it is not generally possible to find a frame which fixes both tensors into their canonical form;

6. Using this canonical form, determine the number of functionally independent terms which are now invariantly defined by the given frame. One method of doing this is constructing the Jacobian for the functions and determining its rank;

7. Set $q = 1$;

8. Calculate the qth derivatives of the tensor which has been set into an invariant form;

9. Determine the isotropy group which leaves these derivatives invariant. This group will be a subgroup of the isotropy group at order $q - 1$, and thus one only needs to check how the qth derivatives transform under the $(q - 1)$st isotropy group and find the new maximal invariant subgroup;

10. If the new invariant subgroup is smaller than the previous group, fix the transformation parameters such that the derivatives are in an invariant form;

11. Determine the number of new functionally independent terms appearing at order q;

12. If the isotropy group and number of functionally independent terms has not changed from the $q - 1$ step, the algorithm terminates. The full set of CK invariants are all of the derivative components computed thus far. If the isotropy group or functionally independent terms has changed, then set $q = q + 1$ and return to step 8.

3. The Homogeneous Levi-Civita Solution (1917)

In 1917, Levi-Civita presented a solution to Einstein's field equations which described a space permeated by a homogeneous, non-null Maxwell field [2,7]. This solution was later rediscovered independently (and nearly simultaneously) in [17,25], and, as such, is often called the Bertotti–Robinson metric in literature. It was [17] that presented a slight generalization (discussed below) which has a non-vanishing cosmological constant. It was shown in [26] that this metric was generally singularity free. This spacetime can also be shown to be a limiting case of the more general Petrov solution [27].

Later considerations of more general Maxwell spacetimes have also revealed several interesting properties regarding this solution. First, this is the only homogeneous non-null Maxwell solution [28]. Generalizations where the Maxwell field does not share the homogeneous symmetry also give interesting solutions (not discussed here) which are algebraically more general [29,30].

3.1. Forms of the Metric and Nature of the Coordinates

In present literature, there are at least six equivalent forms of the homogeneous 1917 Levi-Civita line element. Here, we will introduce these different forms and discuss the relations between them when possible and mention any relevant coordinate artifacts that might be present. We will also present several new forms of this metric and discuss several cases in which no coordinate transforms exist in the literature, nor can it be derived without the use of complex transforms.

The original form of the line element [2] was given in cylindrical coordinates as:

$$ds^2 = -\left(c_1 e^{z/a} + c_2 e^{-z/a}\right)^2 dt^2 + d\rho^2 + a^2 \sin(\rho/a)^2 d\phi^2 + dz^2, \tag{1}$$

where c_1 and c_2 are integration constants originally derived by Levi-Civita and a is a constant usually associated with an electric or magnetic field strength [2,7]. If c_1 and c_2 are both non-zero and of opposite sign, this metric will be singular for $z = \frac{a}{2}\ln\left|\frac{c_2}{c_1}\right|$. Additionally, this metric is singular for $\rho = n\pi a$, where $n \in \mathbb{Z}$.

One may define an angular parameter $\theta = \rho/a$, such that (1) can be rewritten in the form given in [31] as:

$$ds^2 = -\left(c_1 e^{z/a} + c_2 e^{-z/a}\right)^2 dt^2 + dz^2 + a^2 d\Omega^2, \tag{2}$$

where $d\Omega^2$ is the line element for the 2-sphere. We explicitly note that $\theta \in \mathbb{R}$. This metric can also be be put into a secondary set of cylindrical coordinates [31] by taking $\rho'/a = \sin\theta$ giving:

$$ds^2 = -\left(c_1 e^{z/a} + c_2 e^{-z/a}\right)^2 dt^2 + \frac{d\rho'^2}{\left(1 - (\rho'/a)^2\right)} + \rho'^2 d\phi^2 + dz^2, \tag{3}$$

We note here that $\rho' \in [0, a]$. In these coordinates the countably infinite singularities in the original radial coordinate have been reduced to only $\rho = 0$ and $\rho = a$.

We may once again rewrite (1), this time in the associated Cartesian coordinates previously exploited in [32], as:

$$ds^2 = -\left(c_1 e^{z/a} + c_2 e^{-z/a}\right)^2 dt^2 + dx'^2 + dy'^2 + dz^2 + \frac{(x'\,dx' + y'\,dy')^2}{a^2 - (x'^2 + y'^2)}, \tag{4}$$

where we explicitly note that the primed coordinates x' and y' are restricted to take values subject to $x'^2 + y'^2 \leq a^2$.

It is also possible to eliminate c_1 and c_2 via a transform on both t and z. By scaling t and translating z, the original metric can be rewritten into one of two equivalent forms:

$$ds^2 = -\sinh(z)^2 dt^2 + a^2 dz^2 + a^2 d\Omega^2 \tag{5}$$

or

$$ds^2 = -\cosh(z)^2 dt^2 + a^2 dz^2 + a^2 d\Omega^2, \tag{6}$$

where, for real coordinate transforms, one can get either a *sinh* or *cosh* solution depending on the relative signs of c_1 and c_2. Interestingly, it appears that the only method of connecting these two *equivalent* solutions is via complex transformations on both t and z, although it is not presently understood why this method works. We do note that this property of certain complex transformations, resulting in the same solution in different forms, is remarkably similar to the application of the Newman–Janis trick when applied to Minkowski space [33].

Here we also present a new form of this metric which is not related to previous solutions by any known transform, real or complex:

$$ds^2 = -e^{2z} dt^2 + a^2 dz^2 + a^2 d\Omega^2. \tag{7}$$

Using this form and making the coordinate transform $r = e^{-z}$ (along with a rescaling in t) we get the form seen in [20], given as:

$$ds^2 = \frac{a^2}{r^2}\left(dr^2 - dt^2\right) + a^2 d\Omega^2. \tag{8}$$

This reference also presents the form

$$ds^2 = -\left(1 + \frac{z^2}{a^2}\right)dt^2 + \left(1 - \frac{y^2}{a^2}\right)dx^2 + \left(1 - \frac{y^2}{a^2}\right)^{-1}dy^2 + \left(1 + \frac{z^2}{a^2}\right)^{-1}dz^2, \tag{9}$$

which we see is a special case of the more general metric considered in (17).

Throughout the rest of this paper we will work with the metric and coordinates given by (8). A convenient choice of covariant null tetrad is given by:

$$l = \frac{a}{\sqrt{2}r}\left(\frac{\partial}{\partial r} - \frac{\partial}{\partial t}\right), \quad k = -\frac{a}{\sqrt{2}r}\left(\frac{\partial}{\partial r} + \frac{\partial}{\partial t}\right),$$
$$m = \frac{a}{\sqrt{2}}\left(\frac{\partial}{\partial\theta} + i\sin(\theta)\frac{\partial}{\partial\phi}\right), \quad \bar{m} = \frac{a}{\sqrt{2}}\left(\frac{\partial}{\partial\theta} - i\sin(\theta)\frac{\partial}{\partial\phi}\right). \tag{10}$$

3.2. Curvature Invariants and CK Classification

Using the null tetrad defined in (10), one may calculate the only non-vanishing curvature component

$$\Phi_{00} = \frac{1}{2a^2}. \tag{11}$$

This constant solution is therefore both conformally flat and Ricci flat. Since this curvature component is constant and is an invariantly defined frame, (11) is the only non-vanishing CK invariant [24].

Additionally, this result fixes all CM invariants to be either zero or of the form ca^{2n} for $a, c \in \mathbb{R}$ and $n \in \mathbb{Z}$. For example, the only two non-vanishing CM invariants are

$$r_1 = 2\sqrt{r_3} = \frac{1}{a^4}, \tag{12}$$

where only r_1 is independent since the spacetime is of warped product type B_2 [13].

3.3. Regarding Electromagnetic "Wormholes"

It has been suggested that this solution may in some sense constitute a wormhole supported by electromagnetic stress [34,35]. It was shown in [31,36,37] that, despite there being a coordinate singularity at $\rho' = a$ in (3), this does not correspond to a wormhole throat, as an appropriate choice of coordinate transform can be made, such that the spatial part of the metric becomes that of a hypercylinder. Here, we will highlight two different methods of characterizing this surface, which provide a secondary method of determining that this solution is not a wormhole.

Working with the coordinates and metric given by (3), we choose the following null tetrad to analyze the surface $\rho = a$, where we have dropped the prime out of convenience,

$$l = \frac{1}{\sqrt{2}} \left(-\left| c_1 e^{z/a} + c_2 e^{-z/a} \right| \frac{\partial}{\partial t} + \frac{a}{\sqrt{a^2 - \rho^2}} \frac{\partial}{\partial \rho} \right),$$

$$k = \frac{1}{\sqrt{2}} \left(-\left| c_1 e^{z/a} + c_2 e^{-z/a} \right| \frac{\partial}{\partial t} - \frac{a}{\sqrt{a^2 - \rho^2}} \frac{\partial}{\partial \rho} \right), \tag{13}$$

$$m = \frac{1}{\sqrt{2}} \left(\rho \frac{\partial}{\partial \phi} + i \frac{\partial}{\partial z} \right), \quad \bar{m} = \frac{1}{\sqrt{2}} \left(\rho \frac{\partial}{\partial \phi} - i \frac{\partial}{\partial z} \right).$$

The null expansions along l and k are, respectively,

$$\theta_{(l)} = \bar{q}^{\mu\nu} \nabla_\mu l_\nu \quad \text{and} \quad \theta_{(k)} = \bar{q}^{\mu\nu} \nabla_\mu k_\nu , \tag{14}$$

where $\bar{q}_{\mu\nu} = g_{\mu\nu} + 2l_{(\mu} n_{\nu)}$ is a local, induced two-metric and ∇_μ is the standard covariant derivative. In terms of the given null frame, these are explicitly:

$$\theta_{(l)} = -\theta_{(n)} = -\frac{\sqrt{a^2 - \rho^2}}{\sqrt{2}a\rho}. \tag{15}$$

It can be seen that the surface $\rho = a$ is indeed a surface on which the expansion of these null directions vanish. For this surface to correspond to a wormhole throat, the derivative of the expansion projected along the null direction must be positive on the surface [38]. Computing these terms at the surface we can show that

$$l^\mu \nabla_\mu \theta_{(l)} = k^\mu \nabla_\mu \theta_{(k)} = -\frac{1}{2a^2}. \tag{16}$$

Thus, this surface cannot correspond to a wormhole throat. It is in fact only a maximal surface relating to the given range of coordinates [39,40]. This surface is also not detectable by any scalar invariants (since all invariants are constant) and, thus, is not a geometric surface, as would be expected for a wormhole throat [15].

3.4. Bertotti Generalization

In [17], a generalization of the 1917 solution is used, with the line element taking the form of

$$ds^2 = -\left(1 + \frac{y^2}{r_+^2}\right) dt^2 + \left(1 - \frac{z^2}{r_-^2}\right) dx^2 + \left(1 + \frac{y^2}{r_+^2}\right)^{-1} dy^2 + \left(1 - \frac{z^2}{r_-^2}\right)^{-1} dz^2 , \quad (17)$$

where x, y, z are Cartesian coordinates. If $r_+ \neq r_-$, this solution is no longer conformally flat or Ricci flat. In the case where $r_+ = r_- = a$, the spacetime becomes conformally flat and reduces to (9).

Using the null tetrad:

$$l = \frac{1}{\sqrt{2}}\left(-\sqrt{1 + \frac{x^2}{r_+^2}}\frac{\partial}{\partial t} + \frac{1}{\sqrt{1 + \frac{x^2}{r_+^2}}}\frac{\partial}{\partial x}\right), \quad k = \frac{1}{\sqrt{2}}\left(-\sqrt{1 + \frac{x^2}{r_+^2}}\frac{\partial}{\partial t} - \frac{1}{\sqrt{1 + \frac{x^2}{r_+^2}}}\frac{\partial}{\partial x}\right),$$

$$m = \frac{1}{\sqrt{2}}\left(\sqrt{1 - \frac{z^2}{r_-^2}}\frac{\partial}{\partial y} + \frac{i}{\sqrt{1 - \frac{z^2}{r_-^2}}}\frac{\partial}{\partial z}\right), \quad \bar{m} = \frac{1}{\sqrt{2}}\left(\sqrt{1 - \frac{z^2}{r_-^2}}\frac{\partial}{\partial y} - \frac{i}{\sqrt{1 - \frac{z^2}{r_-^2}}}\frac{\partial}{\partial z}\right),$$

$$(18)$$

the only non-vanishing curvature components are found to be

$$R = 2\left(\frac{1}{r_-^2} - \frac{1}{r_+^2}\right),$$

$$\Phi_{11} = \frac{1}{4}\left(\frac{1}{r_+^2} + \frac{1}{r_-^2}\right), \quad \Psi_2 = \frac{1}{6}\left(\frac{1}{r_+^2} - \frac{1}{r_-^2}\right),$$

$$(19)$$

which are again all constants. Since both terms in (19) are constant, and this is already an invariantly defined frame to zeroth order, these are the only three non-vanishing Cartan invariants.

Since this spacetime is a warped product spacetime of type B_2, the four invariants given below are the complete set [13]. These are:

$$R = 2\left(\frac{1}{r_-^2} - \frac{1}{r_+^2}\right), \quad r_1 = \frac{1}{4}\left(\frac{1}{r_-^2} + \frac{1}{r_+^2}\right)^2 , \quad r_2 = 0, \quad w_2 = \frac{1}{36}\left(\frac{1}{r_-^2} - \frac{1}{r_+^2}\right)^3 . \quad (20)$$

4. The Cylindrical Levi-Civita Solution (1919)

In [3,9], a solution to Einstein's field equations is presented that serves as the analog to the Newtonian logarithmic potential. In [41], the CK invariants have already been computed, and in [42] these invariants were used to show that this solution can be found as a limiting subcase of the γ (or "Zipoy-Voorhees") [43,44] solution. Here, for completeness, we will independently compute the CK invariants.

This spacetime solution has also been of interest as an exterior vacuum solution to various physical sources, see [41,45]. There has also been interest in global (topological) properties of this solution as certain parameters (which do no affect local properties) are related to cosmic strings, see [45–47] for discussion of these properties as they pertain to this solution, and see [48] for a general review of cosmic strings.

4.1. Forms of the Metric and Nature of the Coordinates

The line element for this spacetime is given by

$$ds^2 = -e^{2\nu}dt^2 + e^{-2\nu}\left(e^{2\lambda}\left(d\rho^2 + dz^2\right) + r^2 d\phi^2\right), \tag{21}$$

in standard cylindrical coordinates, where $\nu(\rho,z)$ is a solution to

$$\frac{1}{\rho}\frac{\partial}{\partial\rho}\left(\rho\frac{\partial\nu}{\partial\rho}\right) + \frac{\partial^2\nu}{\partial z^2} = 0, \tag{22}$$

and λ is defined, up to a constant of integration, by the differential relation:

$$d\lambda = \rho\left(\frac{\partial\nu}{\partial\rho}^2 - \frac{\partial\nu}{\partial z}^2\right)dr + 2\rho\nu_1\nu_2 dz. \tag{23}$$

Once solved, particular solutions for ν and λ are given by

$$e^{-\nu} = \left(\frac{\rho_0}{\rho}\right)^h, \quad e^{\lambda-\nu} = \left(\frac{\rho}{\rho_0}\right)^{h^2-h}, \tag{24}$$

where h is an arbitrary real constant. Note that, with the form of line element given by (21), it is possible to choose coordinates, such that ρ_0 may be eliminated from the metric, although this will result in ϕ not being parameterized from $(0, 2\pi)$. This will correspond to a global angular defect discussed in [47,49]. Since the metric is independent of the angular coordinate, such a reparameterization will not necessarily affect the local classification given by either CM or CK invariants. Explicitly, we will take this metric to be

$$ds^2 = -\rho^{2h}dt^2 + \rho^{2(h^2-h)}\left(d\rho^2 + dz^2\right) + \rho^{2(1-h)}dx_3^2, \tag{25}$$

and use it for all calculations going forward. We note, once again for clarity, that $x_3 \in [0, a]$ and a is determined by the specific value of r_0 taken above.

Going forward we will take a null frame given by:

$$l = \frac{1}{\sqrt{2}}\left(-\rho^h\frac{\partial}{\partial t} + \rho^{h^2-h}\frac{\partial}{\partial\rho}\right), \quad k = \frac{1}{\sqrt{2}}\left(-\rho^h\frac{\partial}{\partial t} - \rho^{h^2-h}\frac{\partial}{\partial\rho}\right),$$

$$m = \frac{1}{\sqrt{2}}\left(\rho^{h^2-h}\frac{\partial}{\partial z} + i\rho^{h-1}\frac{\partial}{\partial x_3}\right), \quad \bar{m} = \frac{1}{\sqrt{2}}\left(\rho^{h^2-h}\frac{\partial}{\partial z} - i\rho^{h-1}\frac{\partial}{\partial x_3}\right). \tag{26}$$

4.2. Curvature Invariants and CK Classification

In the above frame, the only non-vanishing components of curvature are

$$\Psi_0 = \Psi_4 = \frac{1}{2}h\left(h^2 - 1\right)\rho^{-2(h^2-h+1)} \quad \text{and} \quad \Psi_2 = -\frac{1}{2}h(h-1)^2\rho^{-2(h^2-h+1)}. \tag{27}$$

This frame is invariantly defined and thus these are the zeroth order CK invariants. Additionally, this frame fixes out all isotropy. Taking the covariant derivatives, the only non-vanishing terms are

$$C_{kmkm;k} = -C_{\bar{m}l\bar{m}l;l} = \frac{(h^5 - h)\rho^{-3(1+h^2-h)}}{\sqrt{2}},$$

$$C_{kmkm;l} = C_{kmkl;m} = -C_{kl\bar{m}l;\bar{m}} = -C_{\bar{m}l\bar{m}l;\bar{m}} = -\frac{(h^2+h)(h-1)^3\rho^{-3(1+h^2-h)}}{\sqrt{2}}, \tag{28}$$

$$C_{kmkl;\bar{m}} = C_{kmkl;k} = -C_{km\bar{m}l;l} = -C_{kl\bar{m}l;m} = -\frac{(h-1)^2\left(h^3 - h^2 + h\right)\rho^{-3(1+h^2-h)}}{\sqrt{2}},$$

none of which are functionally independent from the zeroth order invariants. Thus, the algorithm terminates here.

We do note that for certain special cases this spacetime is of type D $\left(h = -1, \frac{1}{2}, 2\right)$ [20,42]. In these cases, the algorithm must be approached separately as the frame given is only invariantly defined when $h = -1$. Additionally, since the Weyl tensor is now more algebraically special, there will be boost and spin isotropy remaining at zeroth order. This means that, in general, one will need to compute higher order derivatives to fully classify these special cases. Here, we will work each of these three cases explicitly.

When $h = -1$, the frame given by (25) becomes

$$l = \frac{1}{\sqrt{2}}\left(-\rho^{-1}\frac{\partial}{\partial t} + \rho^2\frac{\partial}{\partial \rho}\right), \quad k = \frac{1}{\sqrt{2}}\left(-\rho^{-1}\frac{\partial}{\partial t} - \rho^2\frac{\partial}{\partial \rho}\right),$$
$$m = \frac{1}{\sqrt{2}}\left(\rho^2\frac{\partial}{\partial z} + i\rho^2\frac{\partial}{\partial x_3}\right), \quad \bar{m} = \frac{1}{\sqrt{2}}\left(\rho^2\frac{\partial}{\partial z} - i\rho^2\frac{\partial}{\partial x_3}\right). \tag{29}$$

In this frame, the only non-vanishing curvature component is

$$\Psi_2 = 2\rho^{-6}, \tag{30}$$

where we have one functionally independent term, and remaining isotropy

$$\begin{pmatrix} \alpha & 0 \\ 0 & \alpha^{-1} \end{pmatrix}, \text{ with } \alpha \in \mathbb{C} \text{ and } \begin{pmatrix} 0 & 1 \\ -1 & 0 \end{pmatrix}. \tag{31}$$

At first order, the non-vanishing derivatives are

$$C_{kmkl;\bar{m}} = C_{km\bar{m}l;k} = -C_{km\bar{m}l;l} = -C_{kl\bar{m}l;m} = 6\sqrt{2}\rho^{-9}, \tag{32}$$

which are not functionally independent of the zeroth order components. At first order, the isotropy group is reduced to just

$$\begin{pmatrix} e^{i\theta} & 0 \\ 0 & e^{-i\theta} \end{pmatrix}, \text{ with } \theta \in \mathbb{R}. \tag{33}$$

The non-vanishing second derivative components are

$$C_{kmkm;\bar{m}\bar{m}} = C_{kmkl;k\bar{m}} = C_{kmkl;\bar{m}k} = C_{km\bar{m}l;kk} = C_{km\bar{m}l;ll} = C_{kl\bar{m}l;lm} = C_{kl\bar{m}l;ml}$$
$$= C_{\bar{m}l\bar{m}l;mm} = -C_{kl\bar{m}l;\bar{m}m} = -C_{kmkl;l\bar{m}} = -C_{km\bar{m}l;\bar{m}m} = -C_{kl\bar{m}l;km}$$
$$= -\frac{4}{5}C_{kmkl;\bar{m}l} = -\frac{4}{5}C_{km\bar{m}l;kl} = -\frac{4}{5}C_{km\bar{m}l;lk} = -\frac{4}{5}C_{kl\bar{m}l;mk} = \frac{48}{\rho^{12}}, \tag{34}$$

which do not reduce the above isotropy. Thus, the algorithm stops at second order.

In the following two cases, the functional independence follows exactly as above. At zeroth order the isotropy is identical, but in both cases, at first order, it is reduced to

$$\begin{pmatrix} \beta & 0 \\ 0 & \beta^{-1} \end{pmatrix}, \text{ with } \beta \in \mathbb{R}. \tag{35}$$

For $h = \frac{1}{2}$, using the frame:

$$l = \frac{1}{\sqrt{2}}\left(-\sqrt{\rho}\frac{\partial}{\partial t} + \sqrt{\rho}\frac{\partial}{\partial x_3}\right), \quad k = \frac{1}{\sqrt{2}}\left(-\sqrt{\rho}\frac{\partial}{\partial t} - \sqrt{\rho}\frac{\partial}{\partial x_3}\right),$$
$$m = \frac{1}{\sqrt{2}}\left(\rho^{-1/4}\frac{\partial}{\partial \rho} + i\rho^{-1/4}\frac{\partial}{\partial z}\right), \quad \bar{m} = \frac{1}{\sqrt{2}}\left(\rho^{-1/4}\frac{\partial}{\partial \rho} - i\rho^{-1/4}\frac{\partial}{\partial z}\right), \tag{36}$$

we have the CK invariants:

$$\Psi_2 = \frac{1}{8}\rho^{-3/2},$$

$$C_{kmkl;l} = C_{km\bar{m}l;m} = C_{km\bar{m}l;\bar{m}} = C_{kl\bar{m}l;k} = -\frac{3}{16\sqrt{2}}\rho^{-9/4},$$

$$C_{kmkm;ll} = C_{kmkl;lm} = C_{kmkl;ml} = C_{kmkl;\bar{m}l} = C_{km\bar{m}l;kl} = C_{km\bar{m}l;lk}$$

$$= C_{km\bar{m}l;mm} = C_{km\bar{m}l;\bar{m}\bar{m}} = C_{kl\bar{m}l;k\bar{m}} = C_{kl\bar{m}l;mk} = C_{kl\bar{m}l;\bar{m}k} = C_{\bar{m}l\bar{m}l;kk}$$

$$= \frac{4}{5}C_{kmkl;l\bar{m}} = \frac{4}{5}C_{km\bar{m}l;m\bar{m}} = \frac{4}{5}C_{km\bar{m}l\bar{m}m} = \frac{4}{5}C_{kl\bar{m}l;km} = \frac{3}{16}\rho^{-3}. \tag{37}$$

For $h = 2$, using the frame:

$$l = \frac{1}{\sqrt{2}}\left(-\rho^2\frac{\partial}{\partial t} + \rho^2\frac{\partial}{\partial z}\right), \quad k = \frac{1}{\sqrt{2}}\left(-\rho^2\frac{\partial}{\partial t} - \rho^2\frac{\partial}{\partial z}\right),$$

$$m = \frac{1}{\sqrt{2}}\left(\rho^2\frac{\partial}{\partial\rho} + \frac{i}{\rho}\frac{\partial}{\partial x_3}\right), \quad \bar{m} = \frac{1}{\sqrt{2}}\left(\rho^2\frac{\partial}{\partial\rho} - \frac{i}{\rho}\frac{\partial}{\partial x_3}\right), \tag{38}$$

we have the CK invariants:

$$\Psi_2 = 2\rho^{-6},$$

$$C_{kmkl;l} = C_{km\bar{m}l;m} = C_{km\bar{m}l;\bar{m}} = C_{kl\bar{m}l;k} = -6\sqrt{2}\rho^{-9},$$

$$C_{kmkm;ll} = C_{kmkl;lm} = C_{kmkl;ml} = C_{kmkl;\bar{m}l} = C_{km\bar{m}l;kl} = C_{km\bar{m}l;lk} \tag{39}$$

$$= C_{km\bar{m}l;mm} = C_{km\bar{m}l;\bar{m}\bar{m}} = C_{kl\bar{m}l;k\bar{m}} = C_{kl\bar{m}l;mk} = C_{kl\bar{m}l;\bar{m}k} = C_{\bar{m}l\bar{m}l;kk}$$

$$= \frac{4}{5}C_{kmkl;l\bar{m}} = \frac{4}{5}C_{km\bar{m}l;mm} = \frac{4}{5}C_{km\bar{m}l;\bar{m}m} = \frac{4}{5}C_{kl\bar{m}l;kl} = 48\rho^{-12}.$$

Here, we can see that the $h = 1/2$ and $h = 2$ solutions are in fact identical as their CK invariants can be found to be compatible.

In the case where $h = 0, 1$, the spacetime is (locally) flat, and, thus, all invariants will vanish, although globally there are topological properties mentioned above not captured by this approach.

The only two non-vanishing CM invariants are

$$w_1 = 2\left(h^2 - h\right)^2\left(h^2 - h + 1\right)\rho^{-4\left(h^2-h+1\right)}, \quad w_2 = -3\left(h^2 - h\right)^4\rho^{-6\left(h^2-h+1\right)}, \tag{40}$$

which reduces as expected for the special values of h.

4.3. Kasner Generalization

In [18], a generalization of the 1919 solution was given as

$$ds^2 = -r^{2D}dt^2 + t^{2A}dr^2 + r^{2E}t^{2B}dz^2 + \alpha^2 r^{2F}t^{2C}d\phi^2, \tag{41}$$

which is in general stationary (rather than static). The constants in this solution are related via the Kasner constraints

$$A + B + C = A^2 + B^2 + C^2 = 1, \quad D + E + F = D^2 + E^2 + F^2 = 1, \tag{42}$$

and take specific values given by:

$$A = \frac{2s + H}{S + H}, \quad B = \frac{(2s - 1)(2s + \epsilon(2s - 1))}{S + H}, \quad C = \frac{(1 - 2s)(1 + \epsilon(2s - 1))}{S + H}, \tag{43}$$

$$D = \frac{2s}{S}, \quad E = \frac{2s(2s - 1)}{S}, \quad F = \frac{1 - 2s}{S},$$

where s is a real constant, $\epsilon = \pm 1$, and

$$H = \epsilon\left(4s^2 - 1\right) + (1 - 2s)^2, \quad S = 4s^2 - 2s + 1. \tag{44}$$

Computing the curvature with respect to the null tetrad

$$l = \frac{1}{\sqrt{2}}\left(-r^D \frac{\partial}{\partial t} + t^A \frac{\partial}{\partial r}\right), \quad k = \frac{1}{\sqrt{2}}\left(-r^D \frac{\partial}{\partial t} - t^A \frac{\partial}{\partial r}\right),$$

$$m = \frac{1}{\sqrt{2}}\left(r^E t^B \frac{\partial}{\partial z} + i\alpha^2 r^F t^C \frac{\partial}{\partial \phi}\right), \quad \bar{m} = \frac{1}{\sqrt{2}}\left(r^E t^B \frac{\partial}{\partial z} - i\alpha^2 r^F t^C \frac{\partial}{\partial \phi}\right), \tag{45}$$

we have the non-vanishing Ricci components:

$$\Phi_{00} = -\frac{1}{2}(B(D - E) + C(D - F) + A(E + F))t^{-2+E+F}r^{-2+B+C},$$

$$\Phi_{22} = \frac{1}{2}(B + C + E - 3BE - 2CE + F - 2BF - 3CF)r^{-2+E+F}t^{-2+B+C}, \tag{46}$$

and

$$\Psi_0 = \frac{1}{2r^2 t^2}\Big(C(-2 + B + 2C)r^{2(E+F)}$$

$$+(C - E + B(3E - 1) + F - 3CF)r^{E+F}t^{B+C} + F(-2 + E + 2F)t^{2(B+C)}\Big),$$

$$\Psi_2 = \frac{1}{2}\left(-\frac{BCr^{-2D}}{t^2} + \frac{EFt^{-2A}}{r^2}\right), \tag{47}$$

$$\Psi_4 = \frac{1}{2r^2 t^2}\Big(C(-2 + B + 2C)r^{2(E+F)}$$

$$+(B + E - 3BE - F + C(3F - 1))r^{E+F}t^{B+C} + F(-2 + E + 2F)t^{2(B+C)}\Big).$$

This frame is not canonical, as this spacetime is Petrov type I, and $\Psi_0 \neq \Psi_4$, but the boost

$$\begin{pmatrix} (\Psi_0/\Psi_4)^{1/8} & 0 \\ 0 & (\Psi_4/\Psi_0)^{1/8} \end{pmatrix}, \tag{48}$$

will bring it to the canonical form

$$\Psi_0' = \Psi_4' = \sqrt{\Psi_0 \Psi_4},$$
$$\Psi_2' = \Psi_2. \tag{49}$$

For compactness we will write curvature components of the new canonical frame in terms of the old frame, where the new frame will be primed and the old frame unprimed. At zeroth order we have no remaining isotropy and have only two functionally independent terms. The first order derivatives, shown below, cannot reduce the isotropy further and contain no new functionally independent terms, and so the algorithm terminates.

$$C_{kmkm;k} = \frac{1}{2\sqrt{2}r^3 t^3}\Big(2C\left(4 - 3B - 5C + 2BC + C^2\right)r^{3(E+F)} + (4(E - F)$$

$$+ B(1 - 6E + C(7 - 4E - 13F) + 3F) + C(-15 + 14E + C(14 - 17E - 17F)$$

$$+ 23F))r^{2(E+F)}t^{B+C} + (-E + 15F - 7F(E + 2F) + C(4 + E(-3 + 13F)$$

$$+ F(-23 + 17F)) + B(-4 + E(6 + 4F) + F(-14 + 17F)))r^{E+F}t^{2(B+C)}$$

$$- 2F\left(4 - 3E - 5F + 2EF + F^2\right)t^{3(B+C)}\Big)\left(\frac{\Psi_4}{\Psi_0}\right)^{3/4},$$

$$C_{kmkm;l} = -\frac{1}{2\sqrt{2}r^3t^3}\Big(2C((-1+C)C+B(-1+2C))r^{3(E+F)} + (C(-1+2E$$
$$+C(2-3E-3F)+F)+B(-1+C+2E+F-3CF))r^{2(E+F)}t^{B+C}$$
$$+(E(-1+2B+C+F-3CF)+F(-1+2B+C+2F-3(B+C)F))r^{E+F}t^{2(B+C)}$$
$$+2F((-1+F)F+E(-1+2F))t^{3(B+C)}\Big)\left(\frac{\Psi_4}{\Psi_0}\right)^{1/4},$$

$$C_{kmkl;m} = \frac{1}{4\sqrt{2}r^3t^3}\Big(-4C((-1+C)C+B(-1+2C))r^{3(E+F)}$$
$$+(-B(-1+C+2E+4CE+F-5CF)+C(1+F+C(-2+E+F)))$$
$$\times r^{2(E+F)}t^{B+C} + (F(1+C+(-2+B+C)F)-E(-1+C+F-5CF$$
$$+B(2+4F)))r^{E+F}t^{2(B+C)} - 4F((-1+F)F+E(-1+2F))t^{3(B+C)}\Big)\left(\frac{\Psi_4}{\Psi_0}\right)^{1/4},$$

$$C_{kmkl;\bar{m}} = \frac{1}{4\sqrt{2}}r^{-3+E+F}t^{-3+B+C}\big((B+C-2BE+(-2+C)CE$$
$$-BF-C(1+C)F+BC(-3+2E+F))r^{E+F} + 4BCr^{2(E+F)}t^{-B-C}$$
$$+(E-CE+2BE(-1+F)+F+(-3+C)EF+B(-2+F)F-CF(1+F))$$
$$\times t^{B+C} + 4EFr^{-E-F}t^{2(B+C)}\big)\left(\frac{\Psi_4}{\Psi_0}\right)^{1/4},$$

$$C_{km\bar{m}l;k} = \frac{1}{\sqrt{2}}r^{-3(1+D)}t^{-3(1+A)}\Big(EFr^{3D}t^3 + BCr^3t^{3A}$$
$$+(-1+B+C)EFr^{1+2D}t^{2+A} + BC(-1+E+F)r^{2+D}t^{1+2A}\Big)\left(\frac{\Psi_4}{\Psi_0}\right)^{1/4},$$

$$C_{km\bar{m}l;l} = \frac{r^{-3(1+D)}t^{-3(1+A)}}{\sqrt{2}\left(\frac{\Psi_4}{\Psi_0}\right)^{1/4}}\Big(-EFr^{3D}t^3 + BCr^3t^{3A}$$
$$+(-1+B+C)EFr^{1+2D}t^{2+A} - BC(-1+E+F)r^{2+D}t^{1+2A}\Big),$$

$$C_{kl\bar{m}l;m} = \frac{r^{-3+E+F}t^{-3+B+C}}{4\sqrt{2}\left(\frac{\Psi_4}{\Psi_0}\right)^{1/4}}\big(-(B+C-2BE+(-2+C)CE-BF-C(1+C)F$$
$$+BC(-3+2E+F))r^{E+F} + 4BCr^{2(E+F)}t^{-B-C} + (E-CE+2BE(-1+F)$$
$$+F+(-3+C)EF+B(-2+F)F-CF(1+F))t^{B+C} - 4EFr^{-E-F}t^{2(B+C)}\big),$$

$$C_{kl\bar{m}l;\bar{m}} = \frac{1}{4\sqrt{2}r^3t^3\left(\frac{\Psi_4}{\Psi_0}\right)^{1/4}}\Big(-4C((-1+C)C+B(-1+2C))r^{3(E+F)}$$
$$+(B(-1+C+2E+4CE+F-5CF)-C(1+F+C(-2+E+F)))r^{2(E+F)}t^{B+C}$$
$$+(F(1+C+(-2+B+C)F)-E(-1+C+F-5CF+B(2+4F)))r^{E+F}t^{2(B+C)}$$
$$+4F((-1+F)F+E(-1+2F))t^{3(B+C)}\Big),$$

$$C_{\bar{m}l\bar{m}l;k} = \frac{1}{2\sqrt{2}r^3t^3\left(\frac{\Psi_4}{\Psi_0}\right)^{1/4}}\Big(-2C((-1+C)C+B(-1+2C))r^{3(E+F)}$$

$$+(C(-1+2E+C(2-3E-3F)+F)+B(-1+C+2E+F-3CF))r^{2(E+F)}t^{B+C}$$

$$+(-E(-1+2B+C+F-3CF)+F(1-2B-C-2F+3(B+C)F))r^{E+F}t^{2(B+C)}$$

$$+2F((-1+F)F+E(-1+2F))t^{3(B+C)}\Big),$$

$$C_{\bar{m}l\bar{m}l;l} = \frac{1}{2\sqrt{2}r^3t^3\left(\frac{\Psi_4}{\Psi_0}\right)^{3/4}}\Big(2C\Big(4-3B-5C+2BC+C^2\Big)r^{3(E+F)}$$

$$+(-4E+4F+B(-1+6E-3F+C(-7+4E+13F))+C(15-14E-23F$$

$$+C(-14+17E+17F)))r^{2(E+F)}t^{B+C}+(-E+15F-7F(E+2F)$$

$$+C(4+E(-3+13F)+F(-23+17F))+B(-4+E(6+4F)$$

$$+F(-14+17F)))r^{E+F}t^{2(B+C)}+2F\Big(4-3E-5F+2EF+F^2\Big)t^{3(B+C)}\Big).$$

$$\tag{50}$$

Here, the non-vanishing CM invariants are:

$$r_1 = \sqrt{2}r_3 = -\frac{1}{2}(B+C+E-3BE-2CE+F-2BF-3CF)$$

$$\times(B(D-E)+C(D-F)+A(E+F))r^{2(E+F-2)}t^{2(B+C-2)},$$

$$w_1 = \frac{1}{2r^4t^4}\Big(3\Big(BCr^{2(E+F)}-EFt^{2(B+C)}\Big)^2+\Big(C(B+2C-2)r^{2(E+F)}$$

$$+(C-E+B(3E-1)+F-3CF)r^{E+F}t^{B+C}+F(E+2F-2)t^{2(B+C)}\Big)$$

$$\times\Big(C(B+2C-2)r^{2(E+F)}+(B+E-3BE-F+C(3F-1))r^{E+F}t^{B+C}+$$

$$+F(E+2F-2)t^{2(B+C)}\Big)\Big),$$

$$w_2 = \frac{3}{4r^6t^6}\Big(\Big(BCr^{2(E+F)}-EFt^{2(B+C)}\Big)^3-\Big(BCr^{2(E+F)}-EFt^{2(B+C)}\Big)$$

$$\times\Big(C(-2+B+2C)r^{2(E+F)}+(C-E+B(-1+3E)$$

$$+F-3CF)r^{E+F}t^{B+C}+F(-2+E+2F)t^{2(B+C)}\Big)$$

$$\times\Big(C(-2+B+2C)r^{2(E+F)}+(B+E-3BE-F+C(-1+3F))r^{E+F}t^{B+C}$$

$$+F(-2+E+2F)t^{2(B+C)}\Big)\Big),$$

$$m_1 = \frac{1}{4}(B(E-D)+C(F-D)-A(E+F))$$

$$\times(-E-F+B(-1+3E+2F)+C(-1+2E+3F))r^{4(-2+E+F)}t^{4(-2+B+C)}$$

$$\times\Big(-EFr^{2D}t^2+BCr^2t^{2A}\Big),$$

$$m_2 = \frac{1}{8}(B + C + E - 3BE - 2CE + F - 2BF - 3CF)$$
$$\times (B(D - E) + C(D - F) + A(E + F))r^{2(-4+E+F)}t^{2(-4+B+C)}$$
$$\times \left(-\left(BCr^{2(E+F)} - EFt^{2(B+C)}\right)^2 - \left(C(-2 + B + 2C)r^{2(E+F)}\right.\right.$$
$$+(C - E + B(-1 + 3E) + F - 3CF)r^{E+F}t^{B+C} + F(-2 + E + 2F)t^{2(B+C)}\Big)$$
$$\times \left(C(-2 + B + 2C)r^{2(E+F)} + (B + E - 3BE - F + C(-1 + 3F))r^{E+F}t^{B+C}\right.$$
$$\left.\left.+F(-2 + E + 2F)t^{2(B+C)}\right)\right),$$

$$m_3 = \frac{1}{16}r^{2(E+F-4)}t^{2(B+C-4)}\left(-2(B(E - D) + C(F - D) - A(E + F))\right.$$
$$\times(-E - F + B(-1 + 3E + 2F) + C(-1 + 2E + 3F))r^{-4D}t^{-4A}\left(EFr^{2D}t^2 - BCr^2t^{2A}\right)^2$$
$$+(B + C + E - 3BE - 2CE + F - 2BF - 3CF)^2$$
$$\times \left(2C^2r^{2(E+F)} + Cr^{E+F}\left((B - 2)r^{E+F} + (1 - 3F)t^{B+C}\right)\right.$$
$$+t^{B+C}\left(B(3E - 1)r^{E+F} + F\left(r^{E+F} + 2(F - 1)t^{B+C}\right) + E\left(-r^{E+F} + Ft^{B+C}\right)\right)\Big)^2$$
$$+(B(D - E) + C(D - F) + A(E + F))^2$$
$$\times \left(2C^2r^{2(E+F)} + Cr^{E+F}\left((B - 2)r^{E+F} + (3F - 1)t^{B+C}\right)\right.$$
$$\left.+t^{B+C}\left(-B(3E - 1)r^{E+F} - Fr^{E+F} + 2(F - 1)Ft^{B+C} + E\left(r^{E+F} + Ft^{B+C}\right)\right)\right)^2\Big),$$

$$m_5 = \frac{1}{8}r^{2(E+F-4)}t^{2(B+C-4)}\left(\frac{1}{2}(B(E - D) + C(F - D) - A(E + F))\right.$$
$$\times(-E - F + B(3E + 2F - 1) + C(2E + 3F - 1))r^{-2-6D}t^{-2-6A}\left(BCr^2t^{2A} - EFr^{2D}t^2\right)^3$$
$$+\frac{1}{2}(B + C + E - 3BE - 2CE + F - 2BF - 3CF)^2r^{2(E+F-2)}t^{2(B+C-2)}$$
$$\times \left(EFr^{2D}t^2 - BCr^2t^{2A}\right)\left(2C^2r^{2(E+F)} + Cr^{E+F}\left((B - 2)r^{E+F} + (1 - 3F)t^{B+C}\right)\right.$$
$$\left.+t^{B+C}\left(B(3E - 1)r^{E+F} + F\left(r^{E+F} + 2(F - 1)t^{B+C}\right) + E\left(-r^{E+F} + Ft^{B+C}\right)\right)\right)^2$$
$$+\frac{1}{2}(B(E - D) + C(F - D) - A(E + F))(-E - F + B(3E + 2F - 1)$$
$$+C(2E + 3F - 1))r^{-2(1+D)}t^{-2(1+A)}\left(-EFr^{2D}t^2 + BCr^2t^{2A}\right)\left(2C^2r^{2(E+F)}\right.$$
$$+Cr^{E+F}\left((B - 2)r^{E+F} + (1 - 3F)t^{B+C}\right) + t^{B+C}\left(B(3E - 1)r^{E+F}\right.$$
$$\left.+F\left(r^{E+F} + 2(-1 + F)t^{B+C}\right) + E\left(-r^{E+F} + Ft^{B+C}\right)\right))$$
$$\times \left(2C^2r^{2(E+F)} + Cr^{E+F}\left((-2 + B)r^{E+F} + (3F - 1)t^{B+C}\right) + t^{B+C}\left(-B(3E - 1)r^{E+F}\right.\right.$$
$$\left.\left.-Fr^{E+F} + 2(F - 1)Ft^{B+C} + E\left(r^{E+F} + Ft^{B+C}\right)\right)\right)$$
$$-\frac{1}{2}(B(D - E) + C(D - F) + A(E + F))^2r^{2(-2+E+F)}t^{2(-2+B+C)}$$
$$\times \left(-EFr^{2D}t^2 + BCr^2t^{2A}\right)\left(2C^2r^{2(E+F)} + Cr^{E+F}\left((-2 + B)r^{E+F} + (-1 + 3F)t^{B+C}\right)\right.$$
$$\left.\left.+t^{B+C}\left(-B(3E - 1)r^{E+F} - Fr^{E+F} + 2(F - 1)Ft^{B+C} + E\left(r^{E+F} + Ft^{B+C}\right)\right)\right)^2\right),$$

(51)

where we point out that the vanishing of the Ricci invariant r_2 and the mixed invariant m_4 are useful for identifying this spacetime.

5. The Longitudinal, Quadrantal, and Oblique Levi-Civita Solutions (1918)

There are three classes of solutions that are less commonly discussed in the literature than the proceeding two we presented. These solutions are derived in [5,6,50] and all three are reviewed and summarized in [6]. These three papers are not published in English and the latter two remain nearly uncited directly, thus we will carefully review these solutions in more depth. Throughout this section we will also present the non-vanishing $\mathcal{I}$ invariants (see [14]) as these scalars contain information not present in the CM invariants, due to being constructed from derivatives of the Weyl tensor.

5.1. The Longitudinal Solutions

Here we characterize the solutions described in [50], which Levi-Civita refers to as longitudinal solutions, described by the metric

$$ds^2 = -(\mu - \epsilon\eta)dt^2 + \frac{d\sigma^2}{\eta^2} + \frac{d\eta^2}{K_0\eta^4(\mu - \epsilon\eta)}, \tag{52}$$

where $d\sigma^2$ is the line element for a two-dimensional space with constant Gaussian curvature and μ, η are coordinates which run over intervals such that the spacetime is consistent with the chosen signature convention, μ is a real constant, K_0 is a positive constant, and $\epsilon = \pm 1$.

This solution can be split into three distinct subcases depending on the sign of μ. Here, we will move to work with the coordinates given by [4], as they are closer to those typically employed in modern references and explicitly reduce the number of free constants to one. The three forms given are

$$(\mu > 0): \quad ds^2 = -\left(1 - \frac{2m}{r}\right)dt^2 + \left(1 - \frac{2m}{r}\right)^{-1}dr^2 + r^2\left(d\theta^2 + \sin(\theta)^2 d\phi^2\right), \tag{53}$$

where $m > 0$ and $2m < r < \infty$, or $m < 0$ and $0 < r < \infty$,

$$(\mu < 0): \quad ds^2 = -\left(\frac{2m}{z} - 1\right)dt^2 + \left(\frac{2m}{z} - 1\right)^{-1}dz^2 + z^2\left(dr^2 + \sinh(r)^2 d\phi^2\right), \tag{54}$$

where $m > 0$ and $0 < z < 2m$,

$$(\mu = 0): \quad ds^2 = -\frac{dt^2}{z} + zdz^2 + z^2\left(dr^2 + r^2 d\phi^2\right), \tag{55}$$

where $z > 0$.

For completeness, we note that (55) is not the only solution for $\mu = 0$ (we also note that the degenerate static vacuum fields are also listed in Table 2–3.1 in [51], in which Equation (53) is classified as "A1", (54) is classified as "A2", and (55) is classified as "A3", where $b = 2m$). In particular, this solution takes Gaussian flat two-spaces to be a two-dimensional plane. An equally valid choice would be to take these two spaces to be cylinders,

$$ds^2 = -\frac{dt^2}{z} + zdz^2 + z^2\left(d\rho^2 + a^2 d\phi^2\right), \tag{56}$$

but these solutions are only different globally, and, thus, both the CK algorithm and CM invariants will not be able to detect this difference.

The solution given by (53) is the Schwarzschild solution which has been invariantly characterized via the CK algorithm in [10], via CM invariants in [52], and via $\mathcal{I}$ invariants as a subcase in [14]. The solutions given by (54) and (55) are distinct and, as such, we will explicitly state CK and scalar invariants.

For (54), the null frame:

$$l = \frac{1}{\sqrt{2}}\left(-\sqrt{\frac{2m}{z} - 1}\,\frac{\partial}{\partial t} + \frac{1}{\sqrt{\frac{2m}{z} - 1}}\,\frac{\partial}{\partial z}\right),$$

$$k = \frac{1}{\sqrt{2}}\left(-\sqrt{\frac{2m}{z} - 1}\,\frac{\partial}{\partial t} - \frac{1}{\sqrt{\frac{2m}{z} - 1}}\,\frac{\partial}{\partial z}\right),$$

$$m = \frac{1}{\sqrt{2}}\left(z\frac{\partial}{\partial r} + iz\sinh(r)\frac{\partial}{\partial \phi}\right), \quad \bar{m} = \frac{1}{\sqrt{2}}\left(z\frac{\partial}{\partial r} - iz\sinh(r)\frac{\partial}{\partial \phi}\right),$$

$$(57)$$

gives the only non-vanishing curvature component to be

$$\Psi_2 = mz^{-3}. \tag{58}$$

This frame is, therefore, invariant to zeroth order, and the remaining isotropy is given by (31). The independent, first order, non-zero derivatives (components) are

$$C_{kmkl;\bar{m}} = C_{km\bar{m}l;k} = -C_{km\bar{m}l;l} = -C_{kl\bar{m}l;m} = \frac{3m}{\sqrt{2}z^4}\sqrt{\frac{2m}{z} - 1}, \tag{59}$$

which are not functionally independent of the zeroth order components. At first order, the isotropy group is reduced to just (33). Since the isotropy group has been reduced at first order, the algorithm proceeds to second order, where the independent non-vanishing terms are

$$C_{kmkl;l\bar{m}} = C_{km\bar{m}l;m\bar{m}} = C_{km\bar{m}l;m\bar{m}} = C_{kl\bar{m}l;km} = -C_{kmkm;\bar{m}\bar{m}} = -C_{kmkl;k\bar{m}} = -C_{kmkl;\bar{m}k}$$

$$= -C_{km\bar{m}l;kk} = -C_{km\bar{m}l;ll} = -C_{kl\bar{m}l;lm} = -C_{kl\bar{m}l;ml} = -C_{\bar{m}l\bar{m}l;mm} = \frac{3m(2z - 4m)}{z^6}, \tag{60}$$

$$C_{kmkl;\bar{m}l} = C_{km\bar{m}l;kl} = C_{km\bar{m}l;lk} = C_{kl\bar{m}l;mk} = \frac{3m(2z - 5m)}{z^6},$$

which possess the same isotropy and produces no new functionally independent terms. Thus the zeroth, first, and second order terms are the CK invariants needed to fully characterize this spacetime.

In this case, the non-vanishing CM invariants are:

$$w_1 = -\frac{1}{6\sqrt{w_2}} = -\frac{z^3}{m}, \tag{61}$$

and the non-vanishing $\mathcal{I}$ invariants are

$$I_1 = 48\frac{m^2}{z^6}, \quad I_3 = 720\frac{m^2(2m - z)}{z^9}, \quad I_5 = 82944\frac{m^4(2m - z)}{z^{15}}. \tag{62}$$

We note in this case that the I_3 and I_5 invariants both vanish for the extremal value $z = 2m$, which indicates this hypersurface is invariantly defined in much the same way that the Schwarzschild solution's event horizon is.

For the spacetime given by (55), the algorithm proceeds identically to the preceding case with the null frame given by

$$l = \frac{1}{\sqrt{2}}\left(-\frac{1}{\sqrt{z}}\frac{\partial}{\partial t} + \sqrt{z}\frac{\partial}{\partial z}\right), \quad k = \frac{1}{\sqrt{2}}\left(-\frac{1}{\sqrt{z}}\frac{\partial}{\partial t} - \sqrt{z}\frac{\partial}{\partial z}\right),$$

$$m = \frac{1}{\sqrt{2}}\left(z\frac{\partial}{\partial r} + izr\frac{\partial}{\partial \phi}\right), \quad \bar{m} = \frac{1}{\sqrt{2}}\left(z\frac{\partial}{\partial r} - izr\frac{\partial}{\partial \phi}\right),$$

$$(63)$$

which gives CK invariants at zeroth, first, and second order as:

$$\Psi_2 = \frac{1}{2}z^{-3},$$

$$C_{kmkl;\bar{m}} = C_{km\bar{m}l;k} = -C_{km\bar{m}l;l} = -C_{kl\bar{m}l;m} = \frac{3}{2\sqrt{2}}z^{-9/2},$$

$$
\begin{aligned}
C_{kmkm;\bar{m}\bar{m}} &= C_{kmkl;k\bar{m}} = C_{kmkl;\bar{m}k} = C_{km\bar{m}l;kk} = C_{km\bar{m}l;ll} = C_{kl\bar{m}l;lm} = C_{kl\bar{m}l;ml} \\
&= C_{\bar{m}l\bar{m}l;mm} = -C_{kmkl;l\bar{m}} = -C_{km\bar{m}l;m\bar{m}} = -C_{km\bar{m}l;m\bar{m}} = -C_{kl\bar{m}l;km} \\
&= -\frac{4}{5}C_{kmkl;\bar{m}l} = -\frac{4}{5}C_{km\bar{m}l;kl} = -\frac{4}{5}C_{km\bar{m}l;lk} = -\frac{4}{5}C_{kl\bar{m}l;mk} = \frac{3}{z^6}.
\end{aligned}
$$

(64)

A particularly interesting remark is that this special case coincides (at least locally) with (21) where $h = -1$.

We also have the following non-vanishing CM invariants:

$$w_1 = \frac{3}{2z^6}, \quad w_2 = -\frac{3}{4z^9},$$

(65)

and $\mathcal{I}$ invariants

$$I_1 = \frac{12}{z^6}, \quad I_3 = \frac{180}{z^9}, \quad I_5 = \frac{5184}{z^{15}}.$$

(66)

5.2. Quadrantal Solutions

In [5], Levi-Civita gave a second set of vacuum solutions. These are similar to, but distinct from, the longitudinal solutions (which he called quadrantal solutions). He gave the following line element

$$ds^2 = -\frac{e^{2\zeta(\psi)}}{\xi^2}dl^2 + \frac{1}{K_0\xi^2}\left(\frac{d\xi^2}{\Xi(\xi)} + \Xi(\xi)d\phi^2 + d\psi^2\right),$$

(67)

where $\Xi(\xi) = \mu\xi^3 + \epsilon\xi^2$, K_0 is a positive constant, μ is a real constant, and $\epsilon = \pm 1$. Note that the coordinate range of ξ is restricted to the subset of $\mathbb{R}^+$ such that $\Xi(\xi) > 0$. The undetermined function $\zeta(\psi)$ is defined by the differential equation

$$\frac{\partial^2}{\partial\psi^2}e^{\zeta(\psi)} + \mu e^{\zeta(\psi)} = 0,$$

(68)

which will have solutions of the form

$$
\begin{aligned}
\mu > 0: \quad & e^{\zeta(\psi)} = \cos(\sqrt{\mu}\psi), \\
\mu < 0: \quad & e^{\zeta(\psi)} = \cosh(\sqrt{|\mu|}\psi), \\
\mu = 0: \quad & e^{\zeta(\psi)} = \psi.
\end{aligned}
$$

(69)

Working in the null frame

$$
\begin{aligned}
l &= \frac{1}{\sqrt{2}}\left(-\frac{e^{\zeta(\psi)}}{\xi}\frac{\partial}{\partial t} + \frac{1}{\sqrt{K_0}\xi}\frac{\partial}{\partial\psi}\right), \quad
k = \frac{1}{\sqrt{2}}\left(-\frac{e^{\zeta(\psi)}}{\xi}\frac{\partial}{\partial t} - \frac{1}{\sqrt{K_0}\xi}\frac{\partial}{\partial\psi}\right), \\
m &= \frac{1}{\sqrt{2}}\left(\frac{1}{\xi\sqrt{K_0\Xi(\xi)}}\frac{\partial}{\partial\xi} + \frac{i}{\xi}\sqrt{\frac{\Xi(\xi)}{K_0}}\frac{\partial}{\partial\psi}\right), \quad
\bar{m} = \frac{1}{\sqrt{2}}\left(\frac{1}{\xi\sqrt{K_0\Xi(\xi)}}\frac{\partial}{\partial\xi} - \frac{i}{\xi}\sqrt{\frac{\Xi(\xi)}{K_0}}\frac{\partial}{\partial\psi}\right),
\end{aligned}
$$

(70)

the only non-vanishing curvature component is

$$\Psi_2 = \frac{1}{2}\epsilon K_0\xi^3,$$

(71)

which is in an invariant form to first order. The remaining isotropy is given by (31). The independent first derivative components are

$$C_{kmkl;l} = C_{km\bar{m}l;m} = C_{km\bar{m}l;\bar{m}} = C_{kl\bar{m}l;k} = \frac{3\epsilon K_0^{\frac{3}{2}} \zeta^4}{2\sqrt{2}} \sqrt{\mu + \epsilon\zeta}, \tag{72}$$

which are not functionally independent of Ψ_2 and the isotropy is reduced to (35). The second derivative components are

$$C_{kmkm;ll} = C_{kmkl;lm} = C_{kmkl;ml} = C_{kmkl;\bar{m}l} = C_{km\bar{m}l;kl} = C_{km\bar{m}l;lk} = C_{km\bar{m}l;mm}$$
$$= C_{km\bar{m}l;\bar{m}\bar{m}} = C_{kl\bar{m}l;k\bar{m}} = C_{kl\bar{m}l;mk} = C_{kl\bar{m}l;\bar{m}k} = C_{\bar{m}l\bar{m}l;kk} = 3\epsilon K_0^2 \zeta^5 (\mu + \epsilon\zeta), \tag{73}$$
$$C_{kmkl;l\bar{m}} = C_{km\bar{m}l;mm} = C_{km\bar{m}l;m\bar{m}} = C_{kl\bar{m}l;km} = \frac{3}{4}\epsilon K_0^2 \zeta^5 (4\mu + 5\epsilon\zeta),$$

which have the same isotropy and introduce no new functionally independent terms. Thus the algorithm terminates here. In the special case where $\mu = 0$ and $\epsilon = 1$, this solution will coincide with (21) where $h = 1/2$ or $h = 2$.

In this case, the non-vanishing CM invariants are given by

$$w_1 = \frac{3}{2}K_0^2 \zeta^6, \quad w_2 = -\frac{3}{4}\epsilon K_0^3 \zeta^9, \tag{74}$$

and the non-vanishing $\mathcal{I}$ invariants are given by

$$I_1 = 12K_0^2 \zeta^6, \quad I_3 = 180K_0^3 \zeta^8 (\mu + \epsilon\zeta), \quad I_5 = 5184K_0^5 \zeta^{14} (\mu + \epsilon\zeta). \tag{75}$$

Once again the derivative $\mathcal{I}$ invariants will vanish at the extremal value of the coordinate ζ (if that coordinate is restricted due to the signs of μ and ϵ).

We note that, despite this solution appearing very similar to the longitudinal solutions, the fact that the first order isotropy groups are distinct guarantees that these are inequivalent solutions.

5.3. Oblique Solutions

In [6], a third vacuum solution was given which Levi-Civita described as oblique. This solution is characterized by the line element:

$$ds^2 = -\frac{H(\eta)}{(\zeta + \eta)^2}dt^2 + \frac{1}{K_0\zeta^2}\left(\frac{d\zeta^2}{\Xi(\zeta)} + \frac{d\eta^2}{H(\eta)} + \Xi(\zeta)d\phi^2\right), \tag{76}$$

where K_0 is a positive constant and H and Ξ are given by

$$\Xi(\zeta) = 4\zeta^3 - g_2\zeta - g_3, \quad H(\eta) = 4\eta^3 - g_2\eta + g_3, \tag{77}$$

with g_2 and g_3 being real numbers. Working in the tetrad

$$l = \frac{1}{\sqrt{2}}\left(-\frac{\sqrt{g_3 - g_2\eta + 4\eta^3}}{\eta + \zeta}\frac{\partial}{\partial t} + \frac{1}{(\eta + \zeta)\sqrt{K_0(g_3 - g_2\eta + 4\eta^3)}}\frac{\partial}{\partial \eta}\right),$$

$$k = \frac{1}{\sqrt{2}}\left(-\frac{\sqrt{g_3 - g_2\eta + 4\eta^3}}{\eta + \zeta}\frac{\partial}{\partial t} - \frac{1}{(\eta + \zeta)\sqrt{K_0(g_3 - g_2\eta + 4\eta^3)}}\frac{\partial}{\partial \eta}\right),$$

$$m = \frac{1}{\sqrt{2K_0}}\left(\frac{1}{(\eta + \zeta)\sqrt{(-g_3 - g_2\zeta + 4\zeta^3)}}\frac{\partial}{\partial \zeta} + i\frac{\sqrt{-g_3 - g_2\zeta + 4\zeta^3}}{(\eta + \zeta)}\frac{\partial}{\partial \phi}\right),$$

$$\bar{m} = \frac{1}{\sqrt{2K_0}}\left(\frac{1}{(\eta + \zeta)\sqrt{(-g_3 - g_2\zeta + 4\zeta^3)}}\frac{\partial}{\partial \zeta} - i\frac{\sqrt{-g_3 - g_2\zeta + 4\zeta^3}}{(\eta + \zeta)}\frac{\partial}{\partial \phi}\right), \tag{78}$$

the only non-vanishing curvature component is

$$\Psi_2 = 2K_0(\eta + \xi)^3. \tag{79}$$

This frame is canonical and has remaining isotropy given by (31) as before. The first order derivative components are

$$
\begin{aligned}
C_{kmkl;l} &= C_{km\bar{m}l;m} = C_{km\bar{m}l;\bar{m}} = C_{kl\bar{m}l;k} = 3\sqrt{2}\sqrt{K_0^3(\eta+\xi)^6(-g_3 - g_2\xi + 4\xi^3)}, \\
C_{km\bar{m}l;l} &= C_{kl\bar{m}l;m} = -C_{kmkl;\bar{m}} = -C_{km\bar{m}l;k} = 3\sqrt{2}\sqrt{K_0^3(\eta+\xi)^6(g_3 - g_2\eta + 4\eta^3)}.
\end{aligned}
\tag{80}
$$

These terms introduce one new functionally independent term. Additionally, since the derivatives are mixed between those seen in (59) and (72), we see that all isotropy will be fixed out at first order. The second order derivatives are

$$
\begin{aligned}
C_{kmkm;ll} &= C_{kmkl;lm} = C_{kmkl;ml} = C_{km\bar{m}l;mm} = C_{km\bar{m}l;\bar{m}\bar{m}} = C_{kl\bar{m}l;k\bar{m}} = C_{kl\bar{m}l;\bar{m}k} = C_{\bar{m}l\bar{m}l;kk} \\
&= 12K_0^2(\eta+\xi)^3\left(-g_3 - g_2\xi + 4\xi^3\right), \\[4pt]
C_{kmkm;\bar{m}\bar{m}} &= C_{kmkl;k\bar{m}} = C_{kmkl;\bar{m}k} = C_{km\bar{m}l;kk} = C_{km\bar{m}l;ll} = C_{kl\bar{m}l;lm} = C_{kl\bar{m}l;ml} = C_{\bar{m}l\bar{m}l;mm} \\
&= 12K_0^2(\eta+\xi)^3\left(g_3 - g_2\eta + 4\eta^3\right), \\[4pt]
C_{kl\bar{m}l;mm} &= C_{kl\bar{m}l;\bar{m}m} = C_{\bar{m}l\bar{m}l;km} = C_{\bar{m}l\bar{m}l;mk} = C_{kl\bar{m}l;lk} = C_{kl\bar{m}l;kl} = C_{km\bar{m}l;\bar{m}l} = C_{km\bar{m}l;l\bar{m}} \\
&= -C_{kmkm;l\bar{m}} = -C_{kmkm;\bar{m}\bar{m}} = -C_{kmkl;kl} = -C_{kmkl;lk} \\
&= -C_{km\bar{m}l;km} = -C_{kmkl;\bar{m}m} = -C_{km\bar{m}l;mk} = -C_{kmkl;m\bar{m}} \\
&= \frac{3}{2}C_{kmkl;ll} = \frac{3}{2}C_{km\bar{m}l;lm} = \frac{3}{2}C_{km\bar{m}l;ml} = \frac{3}{2}C_{kl\bar{m}l;mm} \\
&= -\frac{3}{2}C_{kmkl;\bar{m}\bar{m}} = -\frac{3}{2}C_{km\bar{m}l;k\bar{m}} = -\frac{3}{2}C_{km\bar{m}l;\bar{m}k} = -\frac{3}{2}C_{kl\bar{m}l;kk} \\
&= 12K_0^2(\eta+\xi)^3\sqrt{(g_3 - g_2\eta + 4\eta^3)(-g_3 - g_2\xi + 4\xi^3)}, \\[4pt]
C_{kmkl;l\bar{m}} &= C_{km\bar{m}l;mm} = C_{km\bar{m}l;\bar{m}m} = C_{kl\bar{m}l;km} \\
&= 3K_0^2(\eta+\xi)^3\left(-6g_3 + 3g_2\eta - 3g_2\xi - 16\eta^3 + 12\eta\xi^2 + 20\xi^3\right), \\[4pt]
C_{kmkl;\bar{m}l} &= C_{km\bar{m}l;kl} = C_{km\bar{m}l;lk} = C_{kl\bar{m}l;mk} \\
&= 3K_0^2(\eta+\xi)^3\left(-6g_3 + 3g_2\eta - 3g_2\xi + 16\xi^3 - 12\xi\eta^2 - 20\eta^3\right),
\end{aligned}
\tag{81}
$$

which introduce no new functionally independent terms, and, thus, the algorithm stops here.

The non-vanishing CM invariants are

$$w_1 = -\frac{1}{\sqrt{6}}w_3^{2/3} = 24K_0^2(\eta + \xi)^6, \tag{82}$$

and the non-vanishing $\mathcal{I}$ invariants are

$$
\begin{aligned}
I_1 &= 192K_0(\eta + \xi), \quad I_3 = -2880K_0^3(\eta+\xi)^7\left(g_2 - 4\left(\eta^2 - \eta\xi + \xi^2\right)\right), \\
I_5 &= -1327104K_0^5(\eta+\xi)^{13}\left(g_2 - 4\left(\eta^2 - \eta\xi + \xi^2\right)\right).
\end{aligned}
\tag{83}
$$

Note that there are cases where the $\mathcal{I}$ invariants will vanish and detect some special surface, although the physical meaning of these surfaces has yet to be understood.

6. Conclusions

In this paper, we discussed a number of historical solutions originally presented by Tullio Levi-Civita, outlining these solutions in a modern fashion. We also presented two different invariant characterizations of these solutions, using both the CK algorithm and CM invariants. As these methods are fully coordinate invariant they present a unique method of identifying these solutions for future work.

The CK algorithm is used here to generate invariantly defined frames, along with curvature components projected along those frames. These invariant quantities are used to classify these spacetimes and provide insight into certain special cases which can now be seen to be identical. The CK invariants are of particular interest, as this method will extend to higher dimensional generalizations of these solutions in a natural way [19,53]. This paper should also serve as a useful reference for those who want to learn to use this algorithm in their own work, as the numerous cases presented here allow for one to see how the algorithm is applied under a number of different circumstances.

The CM invariants, and in some cases $\mathcal{I}$ invariants, for these solutions are also constructed. These invariants provide a coframe independent classification of these solutions (to zeroth order), and also offer a useful characterization in cases where the CK invariants become complicated. In particular, the vanishing of certain invariants can be used to distinguish solutions that might otherwise appear deceivingly similar. The $\mathcal{I}$ invariants are also of interest, as there are cases in which these invariants vanish for certain coordinate values. This indicates that there may be interesting, and invariantly defined, surfaces present in these solutions which have not yet been analyzed.

As gravitational wave astronomy moves into the limelight of modern physics, solutions to Einstein's General Theory of Relativity have become increasingly important as a field of study even as an example for proof-by-contradiction. The high non-linearity of Einstein's field equations impose difficulties on finding solutions and an even greater difficulty in the interpretation of them. It is possible that invariantly analyzing these solutions could prove useful in understanding gravitational wave signals from novel sources. Though the 1917 Levi-Civita solution was not a wormhole, cylindrical symmetry seems to be a way to avoid topological censorship and consequently give hope to obtaining phantom-free wormholes, asymptotically flat in the radial direction [45]. Of interest is the possibility that Levi-Civita's metric could provide the pathway towards the engineering of artificial gravitational fields for human spaceflight.

Author Contributions: The individual contributions of the authors are: conceptualization, C.K.W. and W.J.; methodology, C.K.W. and W.J.; software, C.K.W. and W.J.; validation, D.D.M.; formal analysis, C.K.W. and W.J.; investigation, C.K.W. and W.J.; resources, E.W.D. and G.B.C.; data curation, C.K.W., W.J.; writing—original draft preparation, C.K.W. and W.J.; writing—review and editing, M.G., D.D.M., E.W.D., and G.B.C; supervision, G.B.C.; project administration, E.W.D. and G.B.C. All authors have read and agreed to the published version of the manuscript.

Funding: This research received no external funding.

Institutional Review Board Statement: Not applicable.

Informed Consent Statement: Not applicable.

Data Availability Statement: Not applicable.

Acknowledgments: The authors would like to thank Anzhong Wang for beneficial discussions. The authors would like to thank the reviewers at MDPI symmetry.

Conflicts of Interest: The authors declare no conflict of interest.

References

1. Opere Matematiche. Available online: http://mathematica.sns.it/opere/433/ (accessed on 1 July 202).
2. Levi-Civita, T. Republication of: The physical reality of some normal spaces of Bianchi. *Gen. Relativ. Gravit.* **2011**, *43*, 2307–2320. [CrossRef]
3. Levi-Civita, T. Republication of: Einsteinian ds 2 in Newtonian fields. IX: The analog of the logarithmic potential. *Gen. Relativ. Gravit.* **2011**, *43*, 2321–2330. [CrossRef]
4. Jordan, P.; Ehlers, J.; Kundt, W. Republication of: Exact solutions of the field equations of the general theory of relativity. *Gen. Relativ. Gravit.* **2009**, *41*, 2191–2280. [CrossRef]

5.   Levi-Civita, T. *ds²* Einsteiniani in campi Newtoniani. 6. Il sottocaso B2: Soluzioni quadrantali ($\eta = 0$). *Tip. Della R. Accad. Lincei* **1918**, *27*, 283–292.

6.   Levi-Civita, T. *ds²* Einsteiniani in campi Newtoniani. 7. Il sottocaso B2: Soluzioni oblique. *Tip. Della R. Accad. Lincei.* **1918**, *27*, 343–351.

7.   Levi-Civita, T. Realta fisica di alcuni spazi normali del Bianchi. *Tip. Della R. Accad. Lincei.* **1917**, *26*, 519–531.

8.   Levi-Civita, T. *ds²* Einsteiniani in campi Newtoniani. 1. Generalitá e prima approssimazione. *Tip. Della R. Accad. Lincei.* **1917**, *26*, 307–317.

9.   Levi-Civita, T. *ds²* Einsteiniani in campi Newtoniani. 9. L'Analogo del potenziale logaritmico. *Tip. Della R. Accad. Lincei.* **1919**, *28*, 101–109.

10.   Karlhede, A. A review of the geometrical equivalence of metrics in general relativity. *Gen. Relativ. Gravit.* **1980**, *12*, 693–707. [CrossRef]

11.   Carminati, J.; McLenaghan, R.G. Algebraic invariants of the Riemann tensor in a four-dimensional Lorentzian space. *J. Math Phys.* **1991**, *32*, 3135–3140. [CrossRef]

12.   Zakhary, E.; McIntosh, C.B.G. A Complete Set of Riemann Invariants. *Gen. Relativ. Gravit.* **1997**, *29*, 539–581. [CrossRef]

13.   Santosuosso, K.; Pollney, D.; Pelavas, N.; Musgrave, P.; Lake, K. Invariants of the Riemann tensor for class B warped product space-times. *Comput. Phys. Commun.* **1998**, *115*, 381–394. [CrossRef]

14.   Abdelqader, M.; Lake, K. Invariant characterization of the Kerr spacetime: Locating the horizon and measuring the mass and spin of rotating black holes using curvature invariants. *Phys. Rev. D* **2015**, *91*, 084017. [CrossRef]

15.   McNutt, D.D.; Julius, W.; Gorban, M.; Mattingly, B.; Brown, P.; Cleaver, G. Geometric surfaces: An invariant characterization of spherically symmetric black hole horizons and wormhole throats. *Phys. Rev. D* **2021**, *103*, 124024. [CrossRef]

16.   Coley, A.; Hervik, S.; Pelavas, N. Spacetimes characterized by their scalar curvature invariants. *Class. Quantum Gravity* **2009**, *26*. [CrossRef]

17.   Bertotti, B. Uniform Electromagnetic Field in the Theory of General Relativity. *Phys. Rev.* **1959**, *116*, 1331–1333. [CrossRef]

18.   Delice, O. Kasner generalization of Levi-Civita spacetime. *arXiv* **2004**, arXiv:gr-qc/0411011.

19.   Olver, P.J. *Equivalence, Invariants and Symmetry*; Cambridge University Press: Cambridge, UK, 1995. [CrossRef]

20.   Stephani, H.; Kramer, D.; MacCallum, M.A.H.; Hoenselaers, C.; Herlt, E. *Exact Solutions of Einstein's Field Equations*; Cambridge University Press: Cambridge, UK, 2009. [CrossRef]

21.   MacCallum, M.A.H.; McLenaghan, R.G. *Algebraic Computing in General Relativity: Lecture Notes from the First Brazilian School on Computer Algebra*; Oxford University Press: Oxford, UK, 1994; Volume 2.

22.   Zakhary, E.; Vu, K.T.; Carminati, J. A New Algorithm for the Petrov Classification of the Weyl Tensor. *Gen. Relativ. Gravit.* **2003**, *35*, 1223–1242. [CrossRef]

23.   Zakhary, E.; Carminati, J. A New Algorithm for the Segre Classification of the Trace-Free Ricci Tensor. *Gen. Relativ. Gravit.* **2004**, *36*, 1015–1038. [CrossRef]

24.   Pollney, D.; Skea, J.E.F.; A D'Inverno, R. Classifying geometries in general relativity: I. Standard forms for symmetric spinors. *Class. Quantum Gravity* **2000**, *17*, 643–663. [CrossRef]

25.   Robinson, I. A solution of the Maxwell-Einstein equations. *Bull. Acad. Pol. Sci. Ser. Sci. Math. Astron. Phys.* **1959**, *7*, 351.

26.   Dolan, P. A singularity free solution of the Maxwell-Einstein equations. *Commun. Math. Phys.* **1968**, *9*, 161–168. [CrossRef]

27.   Bonnor, W. A source for Petrov's homogeneous vacuum space-time. *Phys. Lett. A* **1979**, *75*, 25–26. [CrossRef]

28.   Kramer, D. Homogeneous Einstein—Maxwell fields. *Acta Phys. Acad. Sci. Hung.* **1978**, *44*, 353–356. [CrossRef]

29.   McLenaghan, R.G. A new solution of the Einstein–Maxwell equations. *J. Math. Phys.* **1975**, *16*, 2306–2312. [CrossRef]

30.   Tariq, N.; Tupper, B.O.J. A class of algebraically general solutions of the Einstein-Maxwell equations for non-null electromagnetic fields. *Gen. Relativ. Gravit.* **1975**, *6*, 345–360. [CrossRef]

31.   Puthoff, H.E.; Davis, E.W.; Maccone, C. Levi-Civita effect in the polarizable vacuum (PV) representation of general relativity. *Gen. Relativ. Gravit.* **2005**, *37*, 483–489. [CrossRef]

32.   Pauli, W. *Theory of Relativity*; Dover Publications: Mineola, NY, USA, 1981; pp. 171–172.

33.   Rajan, D. Complex spacetimes and the Newman-Janis trick. *arXiv* **2016**, arXiv:1601.03862.

34.   Maccone, C. Interstellar travel through magnetic wormholes. *J. Br. Interplanet. Soc.* **1995**, *48*, 453–458.

35.   Maccone, C. SETI via wormholes. *Acta Astronaut.* **2000**, *46*, 633–639. [CrossRef]

36.   Davis, E.W. Wormhole Induction Propulsion (WHIP). In Proceedings of the NASA Breakthrough Propulsion Physics Workshop, NASA Lewis Research Center, Cleveland, OH, USA, 12–14 August 1997; pp. 157–164.

37.   Davis, E.W. Interstellar Travel by Means of Wormhole Induction Propulsion (WHIP). In *AIP Conference Proceedings*; American Institute of Physics: College Park, MD, USA, 1998; Volume 420, pp. 1502–1508. [CrossRef]

38.   Hochberg, D.; Visser, M. Dynamic wormholes, antitrapped surfaces, and energy conditions. *Phys. Rev. D* **1998**, *58*, 044021. [CrossRef]

39.   Landis, G.A. Magnetic Wormholes and the Levi-Civita solution to the Einstein equation. *J. Br. Interplanet. Soc.* **1997**, *50*, 155–157.

40.   Senovilla, J.M.M. Trapped surfaces. In *Black Holes: New Horizons*; Springer: Hackensack, NJ, USA, 2013; pp. 203–234. [CrossRef]

41.   Da Silva, M.F.A.; Herrera, L.; Paiva, F.M.; Santos, N.O. The Levi–Civita space–time. *J. Math. Phys.* **1995**, *36*, 3625–3631. [CrossRef]

42.   Herrera, L.; Paiva, F.M.; Santos, N.O. The Levi-Civita space–time as a limiting case of the γ space–time. *J. Math. Phys.* **1999**, *40*, 4064–4071. [CrossRef]

43. Zipoy, D.M. Topology of some spheroidal metrics. *J. Math. Phys.* **1966**, *7*, 1137–1143. [CrossRef]
44. Voorhees, B.H. Static axially symmetric gravitational fields. *Phys. Rev. D* **1970**, *2*, 2119. [CrossRef]
45. Bronnikov, K.A.; Santos, N.O.; Wang, A. Cylindrical systems in general relativity. *Class. Quantum Gravity* **2020**, *37*, 113002. [CrossRef]
46. Wang, A.Z.; Da Silva, M.F.A.; Santos, N.O. On parameters of the Levi-Civita solution. *Class. Quantum Gravity* **1997**, *14*, 2417. [CrossRef]
47. Krisch, J.P.; Glass, E.N. Levi-Civita cylinders with fractional angular deficit. *J. Math. Phys.* **2011**, *52*, 052503. [CrossRef]
48. Hindmarsh, M.B.; Kibble, T.W.B. Cosmic strings. *Rep. Prog. Phys.* **1995**, *58*, 477. [CrossRef]
49. MacCallum, M.A.H. Editorial note to: T. Levi-Civita, The physical reality of some normal spaces of Bianchi and to: Einsteinian ds^2 in Newtonian fields. IX: The analog of the logarithmic potential. *Gen. Relativ. Gravit.* **2011**, *43*, 2297–2306. [CrossRef]
50. Levi-Civita, T. ds^2 Einsteiniani in campi Newtoniani. 5. Il sottocaso B2: Soluzioni longitudinali ($\xi = 0$). *Tip. Della R. Accad. Lincei.* **1918**, *27*, 240–248.
51. Witten, L. *Gravitation: An Introduction to Current Research*; Wiley: New York, NY, USA, 1962.
52. Mattingly, B.; Kar, A.; Julius, W.; Gorban, M.; Watson, C.; Ali, M.D.; Baas, A.; Elmore, C.; Shakerin, B.; Davis, E.; et al. Curvature Invariants for Lorentzian Traversable Wormholes. *Universe* **2020**, *6*, 11. [CrossRef]
53. McNutt, D.D.; Coley, A.A.; Forget, A. The Cartan algorithm in five dimensions. *J. Math. Phys.* **2017**, *58*, 032502. [CrossRef]

Article

Inflation with Scalar-Tensor Theory of Gravity

Dalia Saha *, Susmita Sanyal and Abhik Kumar Sanyal

Department of Physics, Jangipur College, Murshidabad 742213, West Bengal, India;
susmitasanyal@yahoo.com (S.S.); sanyal_ak@yahoo.com (A.K.S.)
* Correspondence: daliasahamandal1983@gmail.com

Received: 2 June 2020; Accepted: 9 July 2020; Published: 1 August 2020

Abstract: The latest released data from Planck in 2018 put up tighter constraints on inflationary parameters. In the present article, the in-built symmetry of the non-minimally coupled scalar-tensor theory of gravity is used to fix the coupling parameter, the functional Brans–Dicke parameter, and the potential of the theory. It is found that all the three different power-law potentials and one exponential pass these constraints comfortably, and also gracefully exit from inflation.

1. Introduction

The standard (FLRW) model of cosmology based on the basic assumption of homogeneity and isotropy, known as the 'cosmological principle', has successfully been able to explain several very important issues in connection with the evolution of the universe. First of all, it predicts the observed expansion of the universe being supported by the Hubble's law. It also postulates the existence of cosmic microwave background radiation (CMBR), formed since recombination when the electrons combined to form atoms, allowing photons to free stream, with extreme precession, being verified by Penzias and Wilson for the first time [1]. It further predicts with absolute precession the abundance of the light atomic nuclei ($\frac{^4\text{He}}{\text{H}} \sim 0.25$, $\frac{^2\text{D}}{\text{H}} \sim 10^{-3}$, $\frac{^3\text{He}}{\text{H}} \sim 10^{-4}$, $\frac{^7\text{Li}}{\text{H}} \sim 10^{-9}$, by mass and not by number) observed in the present universe [2–4]. Finally, assuming the presence of the seeds of perturbation in the early universe, it can explain the observed present structure of the universe. Despite such tremendous success, the model inevitably suffers from a plethora of pathologies. The problems at a glance are the following [5,6].

1. 'The singularity problem': Extrapolating the FLRW solutions back in time one encounters an unavoidable singularity, since all the physical parameters viz. the energy density (ρ), the thermodynamic pressure (p), the Ricci scalar (R), the Kretschmann scalar ($R_{\alpha\beta\gamma\delta}R^{\alpha\beta\gamma\delta}$) etc. diverge.

2a. 'The flatness problem': The model does not provide any explanation to the observed value of the density parameter $\Omega \approx 1$, which depicts that the universe is spatially flat.

2b. 'The horizon problem': It also can not provide any reason to the observed tremendous isotropy of the CMBR being split in 1.4×10^4 patches of the sky, that were never causally connected before emission of the CMBR.

2c. 'The structure formation problem': It does not also provide any clue to the seeds of perturbation responsible for the structure formation.

3. 'The dark energy problem': Finally, the standard FLRW model does not fit the redshift versus luminosity-distance curve plotted in view of the observed SN1a (Supernova type a) data.

In connection with the first problem, viz. the so called 'Big-Bang singularity', and also to understand the underlying physics of 'Black-Hole' being associated with Schwarzschild singularity, it has been realized long ago that 'General Theory of Relativity' (GTR) must have to be replaced by a quantum theory of gravity when and where gravity is strong enough. However, GTR is not renormalizable and a renormalized theory requires to include higher-order curvature invariant terms in the gravitational action [7]. Despite serious and intense research over several decades and formulation of new high energy physical theories like superstring [8,9] and supergravity [10,11] theories, a viable quantum theory of gravity is still far from being realized. In connection with last problem, a host of research is in progress over last two decades [12]. It has been realized that to fit the observed redshift versus luminosity-distance curve, it is either required to take into account some form of exotic matter in addition to the barotropic fluid (ordinary plus the cold dark matter) which violates the strong energy condition ($\rho + p \geq 0$, $\rho + 3p \geq 0$), and is dubbed as 'dark energy' (since it interacts none other than with the gravitational field) or to modify the theory of gravity by including additional curvature scalars in the Einstein–Hilbert action, known as 'the modified theory of gravity' [12]. It has been observed that both the possibilities lead to present accelerated expansion of the universe. The problem is thus rephrased as: why the universe undergoes an accelerated expansion at present? The pathology 2, in connection with the flatness, horizon and structure formation problems has however been solved under the hypothesis called 'Inflation' [5], which is our present concern.

Under the purview of cosmological principle, i.e., taking into account Robertson–Walker line element,

$$ds^2 = -dt^2 + a^2(t) \left[\frac{dr^2}{1 - kr^2} + r^2(d\theta^2 + sin^2\theta d\phi^2) \right], \tag{1}$$

the co-moving distance (the present-day proper distance) traversed by light between cosmic time t_1 and t_2 in an expanding universe may be expressed as, $d_0(t_i, t_f) = a_0 \int_{t_i}^{t_f} \frac{dt}{a(t)}$, where $a(t)$ is called the scale factor. The co-moving size of the particle horizon at the last-scattering surface of CMBR ($a_f = a_{lss}$) corresponds to $d_0 \sim 100$ Mpc, or approximately 1^0 (one degree) on the CMB sky today. In the decelerated radiation dominated era of the standard model of cosmology (FLRW model), for which $a \propto \sqrt{t}$ the integrand, $(\dot{a})^{-1} \sim 2\sqrt{t}$ decreases towards the past, and there exists a finite co-moving distance traversed by light since the Big Bang ($a_i \rightarrow 0$), called the particle horizon. The hypothesis of inflation [13,14] postulates a period of accelerated expansion, $\ddot{a} > 0$, in the very early universe ($t \sim 10^{-36}$ s), prior to the hot Big-Bang era, administering certain initial conditions [15–21]. During a period of inflation e.g., a de-Sitter universe ($a \propto e^{\Lambda t}$) driven by a cosmological constant (say), $(\dot{a})^{-1} \sim (\Lambda e^{\Lambda t})^{-1}$ increases towards the past, and hence the integral diverges as ($a_i \rightarrow 0$). This allows an arbitrarily large causal horizon dependent only upon the duration of the accelerated expansion. Assuming that the universe inflates with a finite Hubble rate H_i, (instead of a constant exponent Λ) ending with $H_f < H_i$, we may have, $d_0(t_i, t_f) > (\frac{a_i}{a_f})H_i^{-1}(e^N - 1)$ where $N = \ln\left(\frac{a_f}{a_i}\right)$ is measured in terms of the logarithmic expansion (or 'e-folds'), and describes the duration of inflation. It has been found that a 40–60 e-folds of inflation can encompass our entire observable universe today, and thus solves the horizon and the flatness problem discussed earlier. In some situations, e-fold may range between $25 \leq N \leq 70$, depending on the model under consideration.

A false vacuum state can drive an exponential expansion, corresponding to a de-Sitter space-time with a constant Hubble rate on spatially-flat hypersurfaces. However, a graceful exit from such exponential expansion requires a phase transition to the true vacuum state. A second-order phase transition [22,23], under the slow roll condition of the scalar field (that can also drive the inflation instead of the cosmological constant), potentially leads to a smooth classical exit from the vacuum-dominated phase. Further, the quantum fluctuations of the scalar field, which essentially are the origin of the structures seen in the universe today, provides a source of almost scale-invariant density fluctuations [24–28], as detected in the CMBR. Accelerated expansion and primordial perturbations can also be produced in some modified theories of gravity (e.g., [13,29] and also in a host of models presently available in the literature), which introduce additional non-minimally coupled degrees of freedom. Such inflationary models are conveniently studied by transforming variables to the so-called 'Einstein frame', in which Einstein's equations apply with minimally coupled scalar fields [30,31], which we shall deal with, in the present manuscript.

Non-minimal coupling with the scalar field ϕ is unavoidable in a quantum theory, since such coupling is generated by quantum corrections, even if it is primarily absent in the classical action. Particularly, it is required by the renormalization properties of the theory in curved space-time background. Horndeski' theory [32] presents the most general scalar-tensor theory of gravity ensuring no more than second-order field equations to avoid ghost instability due to Ostrogradski theorem. Recently, in view of a general conserved current, obtained under suitable manipulation of the field Equations [33–36], a non-minimally coupled scalar-tensor theory of gravity, being a special case of Horndeski' model, has been studied extensively in connection with the cosmological evolution, starting from the very early stage (Inflationary regime) to the late-stage (presently accelerated matter-dominated era) via a radiation dominated era [37]. It has been found that such a theory admits a viable inflationary regime, since the inflationary parameters viz. the scalar-tensor ratio (r), and the spectral index n_s lie well within the limits of the constraints imposed by Planck's data, released in 2014 [38] and in 2016 [39]. Furthermore, the model passes through a Friedmann-like radiation era ($a \propto \sqrt{t}, q = 1$) and also an early stage of long Friedmann-like decelerating matter dominated era ($a \propto t^{\frac{2}{3}}$, $q = \frac{1}{2}$) till $z \approx 4$, where a, q and z denote the scale factor, the deceleration parameter and the red-shift respectively. The universe was also found to enter a recent accelerated expansion at a red-shift, $z \approx 0.75$, which is very much at par with recent observations. Further, the present numerical values of the cosmological parameters obtained in the process are also quite absorbing, since it revealed the age of the universe ($13.86 < t_0 < 14.26$) Gyr, the present value of the Hubble parameter ($69.24 < H_0 < 69.96$) Km.s^{-1}Mpc^{-1}, so that $0.991 < H_0 t_0 < 1.01$ fit with the observation with appreciable precision. Numerical analysis also reveals that the state finder $\{r, s\} = \{1, 0\}$, which establishes the correspondence of the present model with the standard ΛCDM universe. Last but not the least important outcome is: considering the CMBR temperature at decoupling ($z \sim 1080$) to be 3000 K, required for recombination, its present value is found to be 2.7255 K, which again fits the observation with extremely high precision. Thus, non-minimally coupled scalar-tensor theory of gravity appears to serve as a reasonably fair candidate for describing the evolution history of our observable universe, beyond quantum domain.

In the mean time, new Planck's data have been released [40,41], which imposed even tighter constraints on the inflationary parameters. In this manuscript, we therefore pose if the theory [37] admits these new constraints. However, earlier we considered a particular form of coupling parameter along with the potential in the form $V(\phi) = V_0 \phi^4 - B\phi^2$, where, V_0 and B are constants [37]. Here instead, we choose different forms of the coupling parameters and also different potentials to study the inflationary regime. In Section 2, we describe the model, write down the field equations, find the parameters involved in the theory in view of a general conserved current. We also present the scalar-tensor equivalent form of the action in Einstein's frame to find the inflationary parameters. In Section 3, we choose different forms of the coupling parameters and associated potentials to test the viability of the model in view of the latest released data from Planck [40,41]. We conclude in Section 4.

2. The Model, Conserved Current, Scalar-Tensor Equivalence and Inflationary Parameters

As mentioned, here we concentrate upon pure non-minimally coupled scalar-tensor theory of gravity, for which the action is expressed in the form,

$$A = \int \left[f(\phi)R - \frac{\omega(\phi)}{\phi} \phi_{,\mu} \phi^{,\mu} - V(\phi) - \mathcal{L}_m \right] \sqrt{-g}\, d^4 x, \tag{2}$$

where, $\mathcal{L}_m$ is the matter Lagrangian density, $f(\phi)$ is the coupling parameter, while, $\omega(\phi)$ is the variable Brans–Dicke parameter. As mentioned, action (2) is a special case of the generalized Horndeski' model [32] having an action S_H given by,

$$S_H = \int d^4 x \sqrt{-g} \mathcal{L}_H = \int d^4 x \sqrt{-g} \left(\mathcal{L}_2 + \mathcal{L}_3 + \mathcal{L}_4 + \mathcal{L}_5 \right), \tag{3}$$

where, $\mathcal{L}_H$ is the Lagrangian density which is the sum of a quadratic, cubic, quartic and quintic terms:

$$\mathcal{L}_2 = K(\phi, X); \tag{4a}$$

$$\mathcal{L}_3 = -G_3(\phi, X)\Box\phi; \tag{4b}$$

$$\mathcal{L}_4 = G_4(\phi, X)R + G_{4X}[(\Box\phi)^2 - (\nabla_\mu\nabla_\nu\phi)^2]; \tag{4c}$$

$$\mathcal{L}_5 = G_5(\phi, X)G_{\mu\nu}\nabla^\mu\nabla^\nu\phi - \frac{G_{5X}}{6}[(\Box\phi)^3 - 3(\Box\phi)(\nabla_\mu\nabla_\nu\phi)^2 + 2(\nabla_\mu\nabla_\nu\phi)^3]. \tag{4d}$$

In the above, K and $G_{3,4,5}$ are generic functions of ϕ and X, while, the kinetic term $X := -\frac{\partial_\mu\phi\partial^\mu\phi}{2}$, R is the Ricci tensor, $G_{\mu\nu}$ is the Einstein tensor, $(\nabla_\mu\nabla_\nu\phi)^2 = \nabla_\mu\nabla_\nu\phi\nabla^\mu\nabla^\nu\phi$ and $G_{iX} = \frac{\partial G_i}{\partial X}$. It is important to mention that, gravity becomes dynamical only through mixing with a scalar field, a phenomenon dubbed as kinetic gravity braiding. General relativity is recovered by setting $K = G_3 = G_5 = 0$ and $G_4 = \frac{M_p^2}{2}$. Note that, when $G_4 = \frac{M_p^2}{2}$, $\mathcal{L}_4$ reduces to the Einstein–Hilbert term. We also obtain a non minimal coupling in the form $f(\phi)R$ from $\mathcal{L}_4$, by setting $G_4 = f(\phi)$. The Horndeski Theory and beyond, have been studied by numerous people from different perspectives [42–50]. In the context of inflation, the generalised (kinetically driven and potentially driven slow-roll) G-inflation has been studied to some details, although there had been no attempt to fit inflationary parameters with the observed data [42]. Additionally, Higgs G-inflation [43], inflation with Scalar-Tensor Horndeski Model [44] and K-Essence non-minimally coupled Gauss–Bonnet invariant for inflation [45], appear in the literature, as some Special cases of Horndeski model, since, it is extremely difficult to make a general study involving all the terms appearing in the Horndeski model. Likewise, present model (2) may be treated as a special case of the same, which may be obtained under the choice, $\mathcal{L}_2 = K(X, \phi) = K(\phi)X - V(\phi)$, where, $K(\phi) = 2\frac{\omega(\phi)}{\phi}$, $G_3 = 0$, $G_4 = f(\phi)$, $G_{4X} = 0$, and $G_5 = 0$, and may be dubbed as the potential-driven G-inflation.

The field equations corresponding to action (2) are,

$$\left(R_{\mu\nu} - \frac{1}{2}g_{\mu\nu}R\right)f(\phi) + g_{\mu\nu}\Box f(\phi) - f_{;\mu;\nu} - \frac{\omega(\phi)}{\phi}\phi_{,\mu}\phi_{,\nu} + \frac{1}{2}g_{\mu\nu}\left(\phi_{,\alpha}\phi^{,\alpha} + V(\phi)\right) = T_{\mu\nu}, \tag{5}$$

$$Rf' + 2\frac{\omega(\phi)}{\phi}\Box\phi + \left(\frac{\omega'(\phi)}{\phi} - \frac{\omega(\phi)}{\phi^2}\right)\phi_{,\mu}\phi^{,\mu} - V'(\phi) = 0, \tag{6}$$

where prime denotes derivative with respect to ϕ, and $\Box$ denotes D'Alembertian, such that, $\Box f(\phi) = f''\phi_{,\mu}\phi^{,\mu} - f'\Box\phi$. The model involves three parameters viz. the coupling parameter $f(\phi)$, the Brans–Dicke parameter $\omega(\phi)$ and the potential $V(\phi)$. It is customary to choose these parameters by hand in order to study the evolution of the universe. However, we have proposed a unique technique to relate the parameters in such a manner, that choosing one of these may fix the rest [33–37]. This follows in view of a general conserved current which is admissible by the above pair of field equations, briefly enunciated below.

The trace of the field equation (5) reads as,

$$Rf - 3\Box f - \frac{\omega(\phi)}{\phi}\phi_{,\mu}\phi^{,\mu} - 2V = T_\mu^\mu = T. \tag{7}$$

Now eliminating the scalar curvature between Equations (6) and (7), one obtains,

$$\left(3f'^2 + \frac{2\omega f}{\phi}\right)'\phi_{,\mu}\phi^{,\mu} + \left(3f'^2 + \frac{2\omega f}{\phi}\right)\Box\phi + 2f'V - fV = f'T, \tag{8}$$

which may then be expressed as,

$$\left[\left(3f'^2 + \frac{2\omega f}{\phi}\right)^{\frac{1}{2}} \phi^{;\mu}\right]_{,\mu} - \frac{f^3}{2\left(3f'^2 + \frac{2\omega f}{\phi}\right)^{\frac{1}{2}}} \left(\frac{V}{f^2}\right)' = \frac{f'}{2\left(3f'^2 + \frac{2\omega f}{\phi}\right)^{\frac{1}{2}}} T, \tag{9}$$

and finally as,

$$\left(3f'^2 + \frac{2\omega f}{\phi}\right)^{1/2} \left[\left(3f'^2 + \frac{2\omega f}{\phi}\right)^{1/2} \phi^{;\mu}\right]_{;\mu} - f^3 \left(\frac{V}{f^2}\right)' = \frac{f'}{2} T^\mu_\mu. \tag{10}$$

Thus there exists a conserved current J^μ, where,

$$J^\mu_{;\mu} = \left[\left(3f'^2 + \frac{2\omega f}{\phi}\right)^{1/2} \phi^{;\mu}\right]_{;\mu} = 0. \tag{11}$$

for trace-less matter field ($T^\mu_\mu = T = 0$), provided

$$V(\phi) \propto f(\phi)^2. \tag{12}$$

To study cosmological consequence of such a conserved current, let us turn our attention to the minisuperspace Model (1), in which the conserved current (11), reads as

$$\sqrt{\left(3f'^2 + \frac{2\omega f}{\phi}\right)} a^3 \dot{\phi} = C_1, \tag{13}$$

in traceless vacuum dominated and also in radiation dominated eras. In the above, C_1 is the integration constant. Note that, fixing the form of the coupling parameter $f(\phi)$, the potential $V(\phi)$ is fixed in view of (12), once and forever. Further, we use a relation [37]

$$3f'^2 + \frac{2\omega f}{\phi} = \omega_0^2, \tag{14}$$

where, ω_0 is a constant, to fix the Brans–Dicke parameter as well. As a result, we obtain the relation

$$a^3 \dot{\phi} = \frac{C_1}{\omega_0} = C, \tag{15}$$

C being yet another non-vanishing constant. In the process, all the coupling parameters $f(\phi)$, $\omega(\phi)$ and the potential $V(\phi)$ are fixed a-priori. Note that one could have made other choice to fix the Brans–Dicke parameter, for example, an arbitrary functional form of the left hand side of (14). But then, additional functional parameter appears that would again require additional assumption. On the contrary, the above choice (14) finally leads to the conserved current associated with the canonical momenta conjugate to the scalar field ϕ, in the absence of a variable Brans–Dicke parameter, i.e., for $\left(\frac{\omega(\phi)}{\phi} = \frac{1}{2}\right)$ with minimal coupling $\left(f(\phi) = (16\pi G)^{-1} = \frac{M_p^2}{2}\right)$, which is physically meaningful. We shall work, in the typical unit, $\frac{M_p^2}{2} = c = 1$, and consider different forms of $f(\phi)$, that fixes the Brans–Dicke parameter $\omega(\phi)$ as well as the potential $V(\phi)$. In view of these known functional forms of the parameters of the theory, we focus our attention to study inflation, which must have occurred in the very early vacuum dominated universe. We relax the symmetry by adding an useful term in the potential $V(\phi)$, so that one of the terms act just as a constant in the effective potential. This ensures a constant value of the potential as the scalar field dies out, and this constant acts as an effective cosmological constant (Λ_e). In the process, the number of parameters increases to three (ω_0, V_0, V_1), which is essential to administer good fit with observation.

Scalar-Tensor Equivalence and Inflationary Parameters

As mentioned, it is convenient and hence customary to study inflationary evolution in the Einstein frame under suitable transformation of variables, where possible. Therefore, in order to study inflation, we consider very early vacuum dominated ($p = 0 = \rho$, for which trace of the matter field identically vanishes and symmetry holds) era, and express the action (2) in the form,

$$A = \int \left[f(\phi)R - \frac{K(\phi)}{2}\phi_{,\mu}\phi^{,\mu} - V(\phi) \right] \sqrt{-g}\, d^4x, \tag{16}$$

where, $K(\phi) = 2\frac{\omega(\phi)}{\phi}$. The above action (16) may be translated to the Einstein frame under the conformal transformation ($g_{E\mu\nu} = f(\phi)g_{\mu\nu}$) to take the form [51],

$$A = \int \left[R_E - \frac{1}{2}\sigma_{E,\mu}\sigma_E{}^{,\mu} - V_E(\sigma(\phi)) \right] \sqrt{-g_E}\, d^4x, \tag{17}$$

where, the subscript 'E' stands for Einstein's frame. The effective potential (V_E) and the field (σ) in the Einstein frame may be found from the following expressions,

$$V_E = \frac{V(\phi)}{f^2(\phi)}; \qquad \text{and,} \qquad \left(\frac{d\sigma}{d\phi}\right)^2 = \frac{K(\phi)}{f(\phi)} + 3\frac{f'^2(\phi)}{f^2(\phi)} = \frac{2\omega(\phi)}{\phi f(\phi)} + 3\frac{f'^2(\phi)}{f^2(\phi)}. \tag{18}$$

In view of the action (17), it is also possible to cast the field equations, viz. the Klein–Gordon and the $\binom{0}{0}$ equations of Einstein as,

$$\ddot{\sigma} + 3H\dot{\sigma} + V'_E = 0; \qquad 3H^2 = \frac{1}{2}\dot{\sigma}^2 + V_E, \tag{19}$$

where, $H = \frac{\dot{a}_E}{a_E}$ denotes the expansion rate, commonly known as the Hubble parameter. The slow-roll parameters and the number of e-foldings, then admit the following forms,

$$\epsilon = \left(\frac{V'_E}{V_E}\right)^2\left(\frac{d\sigma}{d\phi}\right)^{-2}; \quad \eta = 2\left[\left(\frac{V''_E}{V_E}\right)\left(\frac{d\sigma}{d\phi}\right)^{-2} - \left(\frac{V'_E}{V_E}\right)\left(\frac{d\sigma}{d\phi}\right)^{-3}\frac{d^2\sigma}{d\phi^2}\right]; \quad N = \int_{t_i}^{t_f} H dt = \frac{1}{2\sqrt{2}}\int_{\phi_e}^{\phi_b} \frac{d\phi}{\sqrt{\epsilon}}\frac{d\sigma}{d\phi}, \tag{20}$$

where, t_i, t_f stand for the initiation time and the end time, while ϕ_b, ϕ_e stand for the values of the scalar field at the beginning and at the end of inflation respectively. Comparing expression for the primordial curvature perturbation on super-Hubble scales produced by single-field inflation ($P_\zeta(k)$) with the primordial gravitational wave power spectrum ($P_t(k)$), one obtains the tensor-to-scalar ratio for single-field slow-roll inflation $r = \frac{P_t(k)}{P_\zeta(k)} = 16\epsilon$, while, the scalar tilt, conventionally defined as $n_s - 1$ may be expressed as $n_s - 1 = -6\epsilon + 2\eta$, or equivalently $n_s = 1 - 6\epsilon + 2\eta$, dubbed as scalar spectral index. According to the latest released results, the scalar to tensor ratio $r \leq 0.16$ (TT,TE,EE+lowEB+lensing), while $r \leq 0.07$ (TT,TE,EE+lowE+lensing+BK14+BAO) [40,41]. Further, combination of all the data (TT+lowE, EE+lowE, TE+lowE, TT,TE,EE+lowE, TT,TE,EE+lowE+lensing) constrain the scalar spectral index to $0.9569 \leq n_s \leq 0.9815$ [40,41]. It is useful to emphasize that under the present choice of unit $\frac{M_P^2}{2} = c = 1$ (which although appears to be a bit unusual but does not cause any harm), ϕ controls the cosmological evolution in the manner $\phi > 1$ corresponds to the inflationary stage, $\phi \sim 1$ describes the end of inflation while $\phi < 1$ is the low energy regime which triggers matter dominated era. We also point out that the above scalar-tensor equivalent form appearing in (17) may be achieved starting from $F(R)$ theory of gravity. For example, in view of $F(R) \propto R + R^n + R^m$, theory of gravity, a recent study of inflationary regime has been performed under conformal transformation to Einstein's frame and a good fit is obtained with the Planck's data by Bhattacharyya et al [52].

3. Inflation with Power Law and Exponential Potentials

In the non-minimal theory, the flat section of the potential $V(\phi)$ responsible for slow-roll is usually distorted. However, flat potential is still obtainable if the Einstein frame potential V_E is asymptotically constant [53,54]. Note that, the symmetry explored in Section 2, makes the potential (V_E) in the Einstein frame to be constant, once and forever. Thus, we need to relax the symmetry as required by the condition (12), by taking into account additional term in the potential, viz. a constant term V_0, or even a functional form, to ensure that the effective potential in the Einstein frame (V_E) is asymptotically constant. At the end of inflation, the universe becomes cool due to sudden large (exponential, in the present case) expansion. Therefore, in order that the structure we live in are formed, the universe must be reheated and take the state of a hot thick soup of plasma (the so called hot Big-Bang). This phenomena is possible if at the end of inflation, the scalar field starts oscillating rapidly on the Hubble time scale, about the minimum of the potential. In the process, particles are created under standard quantum field theoretic (in curved space-time) approach, which results in the re-heating of the universe. The universe then eventually transits to the radiation dominated era. At that epoch, the additional term may be absorbed in the potential if it is a constant term (V_0), or may even be neglected in case it is a function (since ϕ goes below the Planck's mass), without any loss of generality, to reassure symmetry. The symmetry leads to the first integral of certain combination of the field equations, which helps in solving the field equations leading to a Friedmann-like radiation dominated era, as shown earlier [37]. However, in the present manuscript, we only concentrate upon inflationary regime and of-course study the possibility of graceful exit from inflation. In the following subsection, we shall study different power law potentials, while in the next we shall deal with exponential potential. We consider de-Sitter solution in the form $a \propto e^{Ht}$, where the Hubble parameter H is slowly varying during inflation. We repeat that according to our current choice of units $(\frac{M_p^2}{2} = 1)$, which although is uncommon, but does not create any problem whatsoever, the value of $f(\phi)$ at the end of inflation must be a little greater than 1.

3.1. Power Law Potential $f(\phi) = \phi^n$:

Under the choice, $f(\phi) = \phi^n$, the potential is $V(\phi) \propto \phi^{2n}$. We shall take into account three different values of n, viz. $n = 1, \frac{3}{2}$ and 2, in the following three sub-subsections. For each value of n we shall study different cases taking into account different additive terms. In the first place however, we shall consider an additive constant V_0 in all the three cases, viz.

$$\text{Case-1: } V(\phi) = V_1\phi^{2n} + V_0, \tag{21}$$

corresponding to which, one can now find the expression for the Brans–Dicke parameter $\omega(\phi)$, the potential $V_E(\sigma)$ in the Einstein frame, the expression for $\frac{d\sigma}{d\phi}$, and the slow-roll parameters ϵ, η along with the number of e-foldings N, in view of the Equations (14), (18) and (20) respectively as,

$$\text{Case-1:} \begin{cases} \omega(\phi) = \dfrac{\omega_0^2 - 3n^2\phi^{2(n-1)}}{2\phi^{(n-1)}}, & V_E = V_1 + V_0\phi^{-2n}, & \left(\dfrac{d\sigma}{d\phi}\right)^2 = \dfrac{\omega_0^2}{\phi^{2n}}, \\[3mm] \epsilon = \dfrac{4n^2V_0^2\phi^{2(n-1)}}{\omega_0^2(V_0 + V_1\phi^{2n})^2}, & \eta = \dfrac{4n(n+1)V_0\phi^{2(n-1)}}{\omega_0^2(V_0 + V_1\phi^{2n})}, & N = \dfrac{\omega_0^2}{4\sqrt{2n}V_0}\displaystyle\int_{\phi_e}^{\phi_b}\dfrac{V_0 + V_1\phi^{2n}}{\phi^{2n-1}}d\phi. \end{cases} \tag{22}$$

The effect of the constant term V_0 is now clearly noticeable, since when ϕ is large, second term in the Einstein frame potential (V_E) becomes insignificantly small, for $n \geq 1$, and it almost becomes (non-zero) constant, assuring slow-roll. On the contrary, if V_0 is set to vanish from the very beginning, the Einstein frame potential $V_E = V_1$ would remain flat always, and the universe would have been ever-inflating.

We shall also consider a functional additive term in the potential for all the cases under consideration, such the the potential reads as

$$\text{Case-2: } V(\phi) = V_1 \phi^{2n} + V_0 \phi^m. \tag{23}$$

In view of the above potential (23) it is possible to find the expression for the Brans–Dicke parameter $\omega(\phi)$, the potential $V_E(\sigma)$ in the Einstein frame, the expression for $\frac{d\sigma}{d\phi}$, and the slow-roll parameters ϵ, η along with the number of e-foldings N, in view of the Equations (14), (18) and (20) respectively as,

$$\text{Case-2:} \begin{cases} \omega(\phi) = \dfrac{\omega_0^2 - 3n^2 \phi^{2(n-1)}}{2\phi^{(n-1)}}, & V_E = V_1 + V_0 \phi^{m-2n}, \quad \left(\dfrac{d\sigma}{d\phi}\right)^2 = \dfrac{\omega_0^2}{\phi^{2n}}, \quad \epsilon = \dfrac{(m-2n)^2 V_0^2 \phi^{2(m-n-1)}}{\omega_0^2 [V_1 + V_0 \phi^{(m-2n)}]^2}, \\[2ex] \eta = \dfrac{2V_0(m-2n)(m-n-1)\phi^{(m-2)}}{\omega_0^2 [V_1 + V_0 \phi^{(m-2n)}]}, & N = \dfrac{\omega_0^2}{2\sqrt{2}(m-2n)V_0} \displaystyle\int_{\phi_e}^{\phi_b} \dfrac{V_1 + V_0 \phi^{m-2n}}{\phi^{(m-1)}} d\phi. \end{cases} \tag{24}$$

In the following sub-subsections, we shall take three different values of n, as already mentioned, and present the data set in tabular form along with appropriate plots, to demonstrate the behaviour of the slow-roll parameters in comparison with the latest data set released by Planck [40,41]. Different additive terms, as indicated, will be considered in each subcase separately. In the subsection (3.1.3), we shall consider an additional case with a pair of additive terms in the form of a whole square.

3.1.1. $n = 1$, $f(\phi) = \phi$

Case-1: Under the choice $n = 1$, the potential (21) takes the form $V(\phi) = V_0 + V_1 \phi^2$, and thus the parameters of the theory under consideration (22) read as,

$$\begin{gathered} \omega(\phi) = \dfrac{\omega_0^2 - 3}{2}, \quad \dfrac{d\sigma}{d\phi} = \dfrac{\omega_0}{\phi}, \quad V_E = V_1 + V_0 \phi^{-2}, \quad \epsilon = \dfrac{4V_0^2}{\omega_0^2 (V_0 + V_1 \phi^2)^2}, \\[2ex] \eta = \dfrac{8V_0}{\omega_0^2 (V_0 + V_1 \phi^2)}, \quad N = \dfrac{\omega_0^2}{4\sqrt{2}V_0} \left[V_1 \left(\dfrac{\phi_b^2}{2} - \dfrac{\phi_e^2}{2} \right) + V_0 (\ln \phi_b - \ln \phi_e) \right]. \end{gathered} \tag{25}$$

In view of the above forms of the slow roll parameters (25), we present Tables 1 and 2, underneath, corresponding to two different values of the parameter $V_1 > 0$. The wonderful fit with the latest data sets released by Planck [40,41] is appreciable particularly because $0.968 \approx n_s < 0.982$, while $r < 0.0278$. Further, the number of e-fold ($36 \leq N \leq 62$) is sufficient to alleviate the horizon and flatness problems. Figures 1 and 2 are the two plots r versus n_s and r versus ω_0 respectively, presented for visualization. For example, the figures clearly depict that the plot which represents data sets corresponding to Table 2 appears to be even better.

One very interesting feature is that the above data sets remain unaltered even if the sign of V_0 and V_1 are interchanged. Note that, second derivative of the potential has to be positive, since it represents effective mass of the scalar field. In view of the forms of the potentials $V(\phi)$ and $V_E(\sigma)$ presented in (21) and (22) the effective mass of the scalar fields ϕ and σ respectively are,

$$\dfrac{d^2 V}{d\phi^2} = 2V_1; \quad \dfrac{d^2 V_E}{d\sigma^2} = 6\dfrac{V_0}{\phi^4}. \tag{26}$$

In our data set, we keep $V_1 > 0$, since ϕ is the scalar field under consideration, while translation to σ only amounts to handling the situation with considerable ease. However, as a matter of taste if one favours Einstein's frame over Jordan's frame, it is possible to revert the sign and keep $V_0 > 0$, without changing the data set.

Table 1. $f(\phi) = \phi$, (case-1): $\phi_b = 2.0$. $V_0 = -0.9 \times 10^{-13}\mathrm{T}^{-2}$, $V_1 = 0.9 \times 10^{-13}\mathrm{T}^{-2}$.

ω_0	$|\eta|$	$r = 16\epsilon$	ϕ_e	n_s	N
16.0	0.010418	0.0278	1.060	0.9688	36
16.5	0.009795	0.0261	1.059	0.9706	38
17.0	0.009227	0.0246	1.057	0.9723	41
17.5	0.008707	0.0232	1.056	0.9739	43
18.0	0.008230	0.0219	1.054	0.9753	46
18.5	0.007792	0.0208	1.053	0.9766	48
19.0	0.007387	0.0197	1.051	0.9778	51
19.5	0.007013	0.0187	1.050	0.9790	54
20.0	0.006667	0.0178	1.049	0.9800	56
20.5	0.006345	0.0169	1.048	0.9810	59
21.0	0.006047	0.0161	1.047	0.9819	62

Table 2. $f(\phi) = \phi$, (case-1): $\phi_b = 2.0$. $V_0 = -0.9 \times 10^{-13}\mathrm{T}^{-2}$, $V_1 = 1.0 \times 10^{-13}\mathrm{T}^{-2}$.

ω_0	$|\eta|$	$r = 16\epsilon$	ϕ_e	n_s	N
14.5	0.011047	0.0257	1.012	0.9682	36
15.0	0.01032	0.0239	1.009	0.9703	38
15.5	0.009667	0.0225	1.008	0.9722	41
16.0	0.009072	0.0210	1.006	0.9739	43
16.5	0.008531	0.0198	1.005	0.9755	46
17.0	0.008037	0.0187	1.003	0.9769	49
17.5	0.007584	0.0176	1.001	0.9782	52
18.0	0.007168	0.0166	1.000	0.9794	55
18.5	0.006786	0.0158	0.9986	0.9805	59
19.0	0.006433	0.0149	0.9973	0.9815	62

As mentioned, at the end of inflation, the scalar field must oscillate rapidly so that particles are produced and the universe turns to the phase of: a hot thick soup of plasma, commonly called the 'hot big-bang'. This phenomena is dubbed as graceful exit, which is required for the structure formation together with the formation of CMB. We therefore proceed to check if the present model admits graceful exit from inflation. Here, $V_E = V_1 + \frac{V_0}{\phi^2}$, and so one can express (19) as,

$$\frac{3H^2}{V_1} = \frac{\dot{\sigma}^2}{2V_1} + \left(1 + \frac{V_0}{V_1\phi^2}\right). \tag{27}$$

At large value of the scalar field, which in the present unit $\phi > 1$, we obtained slow-roll. However, as the scalar field falls below the Planck's mass M_p, then the Hubble rate H also decreases, and once it falls below the effective mass V_1, i.e., $H \ll V_1$, then the above equation may be approximated to,

$$\dot{\sigma}^2 = 2i^2\left(V_1 + \frac{V_0}{\phi^2}\right) \qquad \longrightarrow \qquad \dot{\phi} = i\frac{\phi}{\omega_0}\sqrt{2\left(V_1 + \frac{V_0}{\phi^2}\right)}, \tag{28}$$

where, $\dot{\sigma} = \frac{\omega_0}{\phi}\dot{\phi}$, in view of (25). Thus, finally we get,

$$\phi = \frac{1}{2V_1}\left[(1 - V_0 V_1)\cos\left(\frac{\sqrt{2V_1}}{\omega_0}t\right) + i(1 + V_0 V_1)\sin\left(\frac{\sqrt{2V_1}}{\omega_0}t\right)\right], \qquad (29)$$

which is an oscillatory solution, and the field then oscillates many times over a Hubble time. This coherent oscillating field corresponds to a condensate of non-relativistic massive (inflaton) particles, which ensures graceful exit from the inflationary regime, driving a matter-dominated era at the end of inflation. There is a long standing debate regarding the physical frame. It appears that most of the people favour Einstein's frame over Jordan's frame (we have briefly discussed the issue in conclusion). In this regard, it is important to mention that since in view of (25) $\sigma = \omega_0 \ln\phi$, therefore σ executes oscillatory behaviour as well.

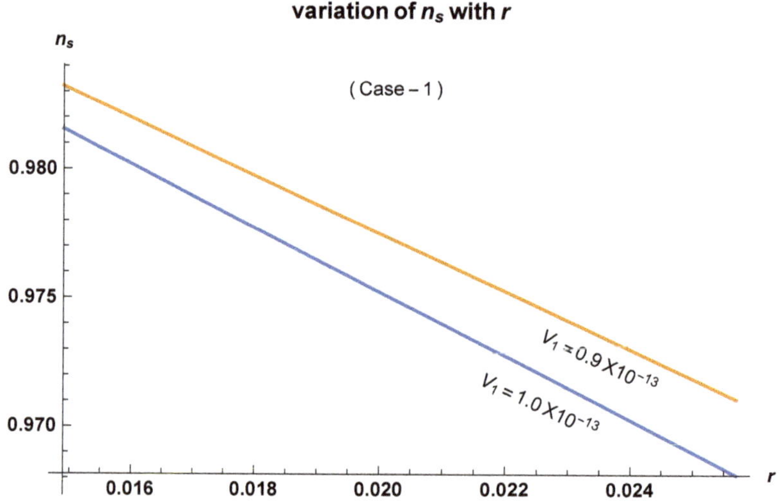

Figure 1. $f = \phi$, (case-1): Yellow ochre and blue colours represent Table 1 and 2 data, respectively.

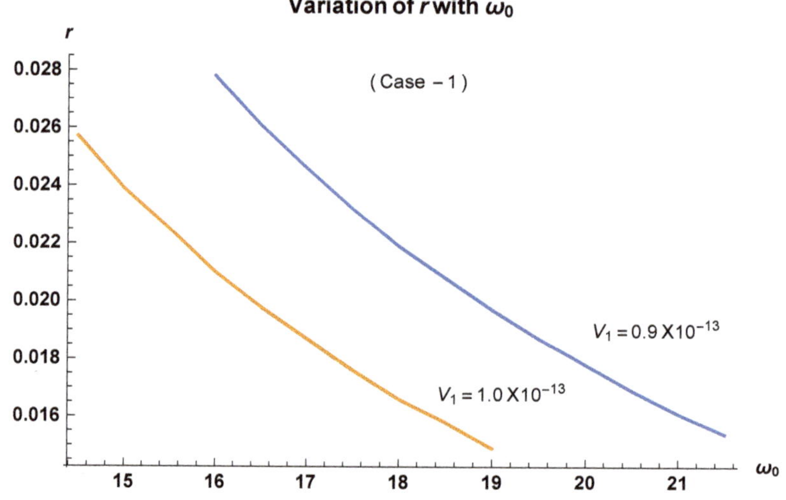

Figure 2. $f = \phi$, (case-1): Blue and yellow ochre colours represent Tables 1 and Table 2 data, respectively.

Case-2: Under the same situation $f(\phi) = \phi$, let us now consider, $V(\phi) = V_1\phi^2 + V_0\phi^4$, where instead of a constant term, we have added a quartic term in the potential. The expression for the Brans–Dicke parameter ($\omega(\phi)$), the potential (V_E) in the Einstein frame, $\frac{d\sigma}{d\phi}$, the slow-roll parameters ϵ, η and the number of e-foldings N, may then be found in view of the Equation (24), respectively as,

$$\omega(\phi) = \frac{\omega_0^2 - 3}{2\phi}, \quad \frac{d\sigma}{d\phi} = \frac{\omega_0}{\phi}, \quad V_E = V_1 + V_0\phi^2, \quad \epsilon = \frac{4V_0^2\phi^4}{\omega_0^2(V_1 + V_0\phi^2)^2},$$

$$\eta = \frac{8V_0\phi^2}{\omega_0^2(V_1 + V_0\phi^2)}, \quad N = \int_{\phi_e}^{\phi_b} \frac{\omega_0^2(V_1 + V_0\phi^2)}{4\sqrt{2}V_0\phi^3}d\phi. \tag{30}$$

Although, the potential V_E does not appear to attend a flat section, the smallness of the value of η confirms that there indeed exists a flat section, admitting slow-roll. In fact, in the Einstein frame (17), this is just the case of a standard inflation field theory with quadratic potential. Following Tables 3 and 4, for $V_1 > 0$, together with the associated plots n_s versus r in Figure 3 and r versus ω_0 in Figure 4 here again depict appreciably good fit with the recent released Planck's data set, particularly because $0.97 \leq n_s \leq 0.98$ while $r \leq 0.098$. Figure 3 depicts that the data of Table 3 are somewhat better.

As before here again we test if the model associated with a different potential admits graceful exit. Here $V_E = (V_1 + V_0\phi^2)$, and so from (19) one obtains,

$$\frac{3H^2}{V_1} = \frac{\dot{\sigma}^2}{2V_1} + \left(1 + \frac{V_0}{V_1}\phi^2\right) \tag{31}$$

As the Hubble rate $H \ll V_1$, the above equation can be approximated to, $\dot{\sigma}^2 = 2i^2(V_1 + V_0\phi^2)$, yielding $\dot{\phi} = i\frac{\phi}{\omega_0}\sqrt{2(V_1 + V_0\phi^2)}$, where, $\dot{\sigma} = \dot{\phi}\frac{\omega_0}{\phi}$. Finally we get,

$$\phi = \frac{2V_1\left[(1 - V_0V_1)\cos\left(\frac{\sqrt{2V_1}}{\omega_0}t\right) + i(1 + V_0V_1)\sin\left(\frac{\sqrt{2V_1}}{\omega_0}t\right)\right]}{1 + V_0^2V_1^2 - 2V_0V_1\cos\left(\frac{2\sqrt{2V_1}}{\omega_0}t\right)}. \tag{32}$$

The oscillatory behaviour of the scalar field clearly ensures graceful exit from inflationary regime, as already discussed, and in view of (30) σ also executes oscillatory behaviour.

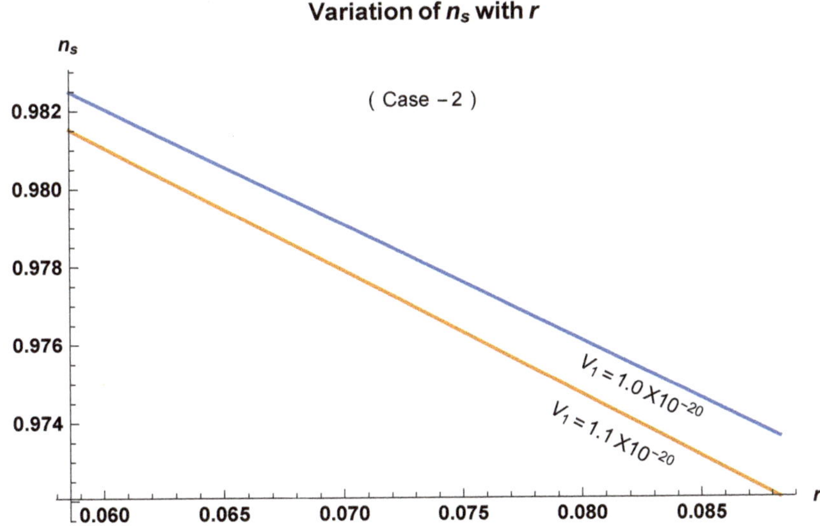

Figure 3. $f = \phi$. Yellow ochre and blue colours represent Tables 3 and 4 respectively.

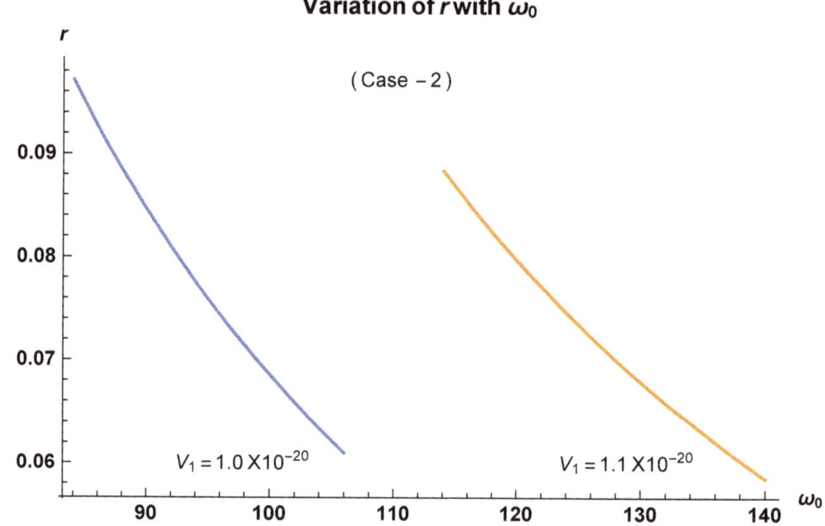

Figure 4. $f = \phi$. Yellow ochre and blue colours represent Tables 3 and 4 respectively, unlike previous cases.

Table 3. $f(\phi) = \phi$, (case-2): $\phi_b = 1.2$, $V_0 = -1.0 \times 10^{-20}\mathrm{T}^{-2}$; $V_1 = 1.1 \times 10^{-20}\mathrm{T}^{-2}$.

ω_0	$|\eta|$	$r = 16\epsilon$	ϕ_e	n_s	N
114	0.002607	0.08834	1.0581	0.9721	38
116	0.002518	0.08532	1.0580	0.9730	39
118	0.002433	0.08245	1.0578	0.9739	41
120	0.002353	0.07972	1.0577	0.9748	42
122	0.002276	0.07713	1.0575	0.9756	44
124	0.002204	0.07466	1.0574	0.9764	45
126	0.002134	0.07231	1.0572	0.9772	46
128	0.002068	0.07007	1.0571	0.9779	48
130	0.002005	0.06793	1.0570	0.9785	49
132	0.001945	0.06589	1.0569	0.9792	51
134	0.001887	0.06393	1.0567	0.9798	53
136	0.001832	0.06207	1.0566	0.9804	54
138	0.001780	0.06028	1.0565	0.9810	56
140	0.001729	0.05857	1.0564	0.9815	57

Table 4. $f(\phi) = \phi$, (case-2): $\phi_b = 1.2$, $V_0 = -1.0 \times 10^{-20}\text{T}^{-2}$; $V_1 = 1.0 \times 10^{-20}\text{T}^{-2}$.

ω_0	$|\eta|$	$r = 16\epsilon$	ϕ_e	n_s	N
84	0.003711	0.09715	1.0121	0.9710	36
86	0.003540	0.09268	1.0118	0.9723	38
88	0.003381	0.08852	1.0115	0.9736	40
90	0.003232	0.08463	1.0113	0.9747	42
92	0.003093	0.08099	1.0110	0.9758	44
94	0.002963	0.07758	1.0108	0.9768	46
96	0.002841	0.07438	1.0105	0.9778	47
98	0.002726	0.07138	1.0103	0.9787	49
100	0.002618	0.06855	1.0101	0.9795	51
102	0.002517	0.06589	1.0099	0.9803	54
104	0.002421	0.0634	1.0097	0.9810	56
106	0.002330	0.06101	1.0096	0.9818	58

3.1.2. $n = \frac{3}{2}$, $f(\phi) = \phi^{\frac{3}{2}}$

Case-1: Under the choice $n = \frac{3}{2}$, $f(\phi) = \phi^{\frac{3}{2}}$, and the potential (21) takes the cubic form, $V(\phi) = V_0 + V_1\phi^3$. Thus the expressions for the parameters of the theory under consideration along with the slow-roll parameters (22) are,

$$
\omega(\phi) = \frac{4\omega_0^2 - 27\phi}{8\sqrt{\phi}}, \quad \frac{d\sigma}{d\phi} = \frac{\omega_0}{\phi^{\frac{3}{2}}}; \quad V_E = V_1 + V_0\phi^{-3}, \quad \epsilon = \frac{9V_0^2\phi}{\omega_0^2(V_0 + V_1\phi^3)^2},
$$

$$
\eta = \frac{15V_0\phi}{\omega_0^2(V_0 + V_1\phi^3)}, \quad N = \frac{\omega_0^2}{6\sqrt{2}V_0}\left[V_0\left(\frac{1}{\phi_e} - \frac{1}{\phi_b}\right) + \frac{V_1}{2}(\phi_b^2 - \phi_e^2)\right].
\tag{33}
$$

As before, we present two sets of data in Tables 5 and 6, for two different values of $V_1 > 0$. Figures 5 and 6 depict the variations of the spectral index n_s with the scalar-tensor ratio r and the scalar-tensor ratio r with the Brans–Dicke parameter ω_0 respectively. Here again we observe that $r \leq 0.0278$, and $0.96 < n_s < 0.9815$, which are very much within the stipulated observational range [40,41], while number of e-folding N is sufficient to remove the flatness and the horizon problems.

Table 5. $f(\phi) = \phi^{\frac{3}{2}}$, (case-1): $\phi_b = 1.7$, $V_0 = -0.9 \times 10^{-13}\text{T}^{-2}$, $V_1 = 0.9 \times 10^{-13}\text{T}^{-2}$.

ω_0	$|\eta|$	$r = 16\epsilon$	ϕ_e	n_s	N
24	0.011314	0.0278	1.0408	0.9670	36
25	0.010427	0.0256	1.0392	0.9696	39
26	0.009640	0.0237	1.0377	0.9719	42
27	0.008939	0.0219	1.0364	0.9739	46
28	0.008312	0.0204	1.0351	0.9757	49
29	0.007749	0.0190	1.0339	0.9773	52
30	0.007241	0.0178	1.0328	0.9788	56
31	0.006781	0.0166	1.0318	0.9802	60
32	0.006364	0.0156	1.0308	0.9814	64

Table 6. $f(\phi) = \phi^{\frac{3}{2}}$, (case-1): $\phi_b = 1.7$, $V_0 = -0.9 \times 10^{-13}\text{T}^{-2}$, $V_1 = 1.0 \times 10^{-13}\text{T}^{-2}$.

ω_0	$|\eta|$	$r = 16\epsilon$	ϕ_e	n_s	N
22.0	0.011816	0.0254	1.0076	0.9668	36
22.5	0.011217	0.0243	1.0067	0.9683	38
23.0	0.010811	0.0233	1.0059	0.9697	39
23.5	0.010356	0.0223	1.0050	0.9709	41
24.0	0.009928	0.0214	1.0042	0.9721	43
24.5	0.009528	0.0205	1.0034	0.9733	45
25.0	0.009150	0.0197	1.0027	0.9743	47
25.5	0.008795	0.0189	1.0019	0.9753	48
26.0	0.008460	0.0182	1.0013	0.9762	50
26.5	0.008143	0.0175	1.0006	0.9771	52
27.0	0.007845	0.0169	1.000	0.9780	54

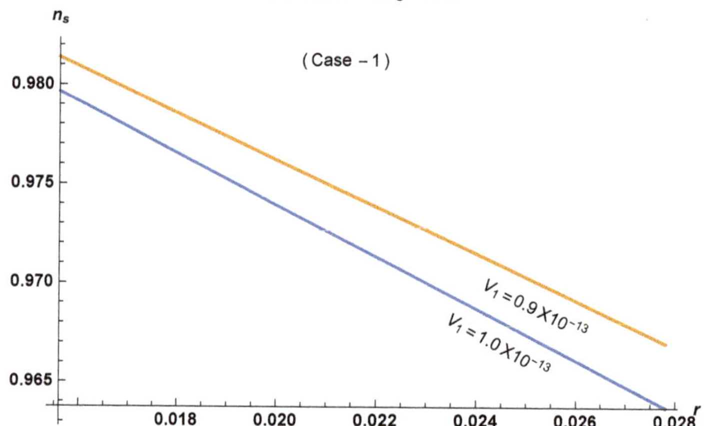

Figure 5. $f(\phi) = \phi^{\frac{3}{2}}$. Yellow ochre and blue colours represent Tables 5 and 6 respectively.

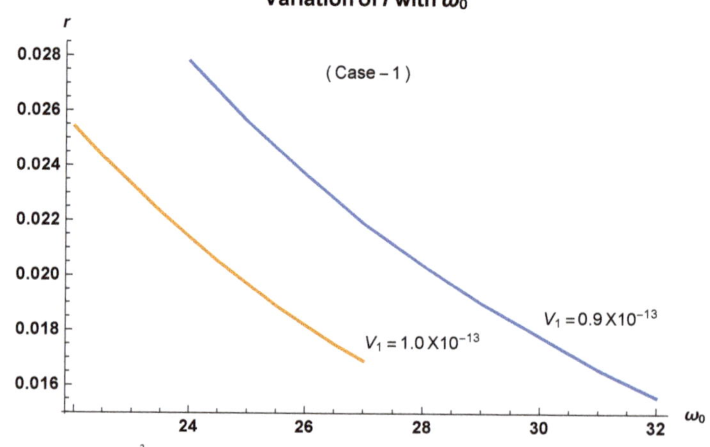

Figure 6. $f(\phi) = \phi^{\frac{3}{2}}$. Blue and yellow ochre colours represent Tables 5 and 6 respectively.

To check the behaviour of the scalar field, we proceed as before, and find,

$$\frac{3H^2}{V_1} = \frac{\omega_0^2 \dot{\phi}^2}{2V_1\phi^3} + \frac{V_0}{V_1\phi^3} + 1 \implies \dot{\phi}^2 = -2\frac{(V_0 + V_1\phi^3)}{\omega_0^2}, \tag{34}$$

under suitable approximation, as the Hubble rate $H \ll V_1$, and using the relation $\dot{\sigma} = \omega_0 \frac{\dot{\phi}}{\phi^{\frac{3}{2}}}$, in view of (33). The solution reads as,

$$\phi(t) = \text{InverseFunction}\left[\frac{2i}{\sqrt[4]{3}\sqrt[3]{V_1}\sqrt{V_0 - \#1^3 V_1}}\text{EllipticF}\left[\text{Sin}^{-1}\left(\frac{1}{\sqrt[4]{3}}\sqrt{-(-1)^{\frac{5}{6}} - i\#1\sqrt[3]{\frac{V_1}{V_0}}}\right), \sqrt[3]{-1}\right]\right.$$

$$\left.\sqrt[3]{V_0}\sqrt{(-1)^{5/6}\left(\#1\sqrt[3]{\frac{V_1}{V_0}} - 1\right)}\sqrt{\#1^2\sqrt[3]{\frac{V_1^2}{V_0^2}} + \#1\sqrt[3]{\frac{V_1}{V_0}} + 1\&}\right][c_1 - t]. \tag{35}$$

In the above the hash tag ($\#n$) denotes nth argument of a pure function, and c_1 is a constant. Although, the solution is not obtainable in closed form, rather is a complicated inverse elliptic function, nevertheless its oscillatory behaviour is quite apparent, and $\sigma = -2\frac{\omega_0}{\sqrt{\phi}}$ also oscillates as well.

Case-2: Under the choice $n = \frac{3}{2}$, $f(\phi) = \phi^{\frac{3}{2}}$, and taking the potential in the form, $V(\phi) = V_1\phi^3 + V_0\phi$, the expressions for the parameters of the theory under consideration along with the slow-roll parameters (24) are,

$$\omega(\phi) = \frac{4\omega_0^2 - 27\phi}{8\sqrt{\phi}}, \quad \frac{d\sigma}{d\phi} = \frac{\omega_0}{\phi^{\frac{3}{2}}}; \quad V_E = V_1 + \frac{V_0}{\phi^2}, \quad \epsilon = \frac{4V_0^2\phi}{\omega_0^2(V_0 + V_1\phi^2)^2},$$

$$\eta = \frac{6V_0\phi}{\omega_0^2(V_0 + V_1\phi^2)}, \quad N = \frac{\omega_0^2}{4\sqrt{2}V_0}\left[V_0\left(\frac{1}{\phi_e} - \frac{1}{\phi_b}\right) + V_1(\phi_b - \phi_e)\right]. \tag{36}$$

We present two sets of data in Tables 7 and 8 underneath, for two different values of $V_1 > 0$. Figures 7 and 8 depict the variations of the spectral index n_s with the scalar-tensor ratio r, and the scalar-tensor ratio r with the Brans–Dicke parameter ω_0, respectively. Here also we observe that $r \le 0.039$, and $0.971 < n_s < 0.983$, which are again in excellent agreement of Planck's data [40,41], while the number of e-folding N is also sufficient to remove the flatness and the horizon problems.

Table 7. $f(\phi) = \phi^{\frac{3}{2}}$, (case-2): $\phi_b = 1.7$, $V_0 = -0.9 \times 10^{-13}\text{T}^{-2}$, $V_1 = 0.90 \times 10^{-13}\text{T}^{-2}$.

ω_0	$|\eta|$	$r = 16\epsilon$	ϕ_e	n_s	N
28	0.006884	0.03885	1.0357	0.9717	40
29	0.006417	0.03623	1.0345	0.9736	43
30	0.005996	0.03384	1.0333	0.9753	46
31	0.005616	0.03169	1.0323	0.9769	49
32	0.005270	0.02974	1.0313	0.9783	52
33	0.004956	0.02797	1.0303	0.9796	55
34	0.004669	0.02635	1.0294	0.9808	59
35	0.004406	0.02486	1.0286	0.9819	62
36	0.004164	0.02350	1.0278	0.9829	66

Table 8. $f(\phi) = \phi^{\frac{3}{2}}$, (case-2): $\phi_b = 1.7$, $V_0 = -0.9 \times 10^{-13}\mathrm{T}^{-2}$, $V_1 = 0.95 \times 10^{-13}\mathrm{T}^{-2}$.

ω_0	$|\eta|$	$r = 16\epsilon$	ϕ_e	n_s	N
26	0.007358	0.03828	1.0103	0.9709	39
27	0.006823	0.03549	1.0089	0.9730	42
28	0.006345	0.03300	1.0076	0.9749	45
29	0.005915	0.03077	1.0064	0.9766	48
30	0.005527	0.02875	1.0053	0.9782	52
31	0.005176	0.02693	1.0043	0.9796	55
32	0.004858	0.02527	1.0033	0.9808	59
33	0.004568	0.02376	1.0024	0.9820	63
34	0.004303	0.02238	1.0016	0.9830	67

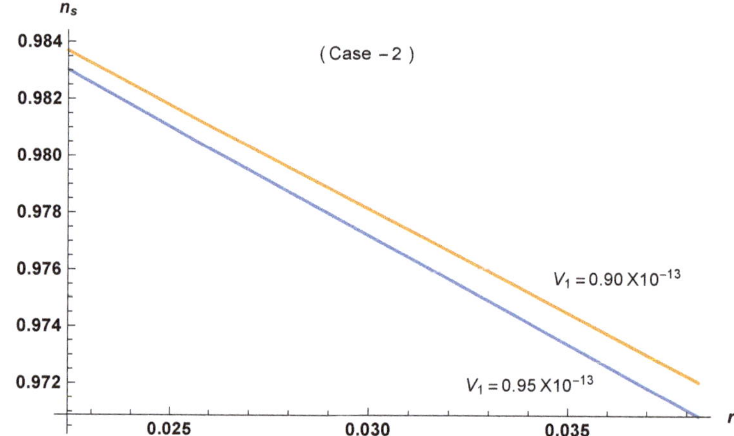

Figure 7. $f(\phi) = \phi^{\frac{3}{2}}$. Yellow ochre and blue colours represent Tables 7 and 8 respectively.

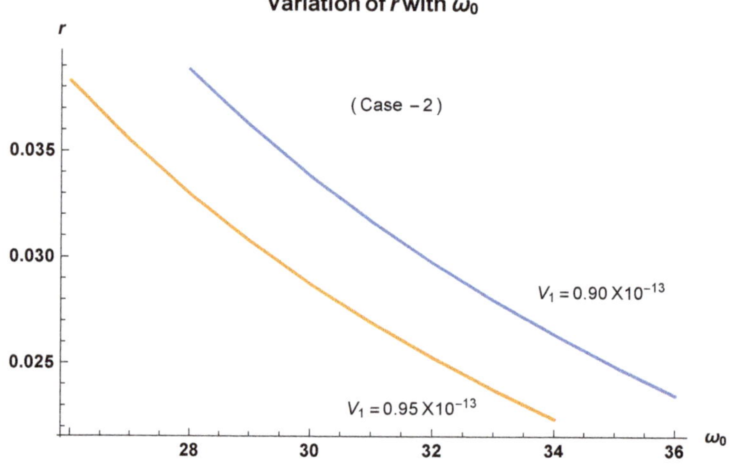

Figure 8. $f(\phi) = \phi^{\frac{3}{2}}$. Blue and yellow ochre colours represent Tables 7 and 8 respectively.

In order to study the behaviour of the scalar field at the end of inflation, we start with the Einstein frame potential as before, $V_E = (V_1 + \frac{V_0}{\phi^2})$, and express the field Equation (19) as,

$$\frac{3H^2}{V_1} = \frac{\dot{\sigma}^2}{2V_1} + \left(1 + \frac{V_0}{V_1\phi^2}\right) \tag{37}$$

As the Hubble rate falls, and $H \ll V_1$, the above equation may be approximated to, $\dot{\sigma}^2 = 2i^2(V_1 + \frac{V_0}{\phi^2})$, which in terms of the scalar field ϕ reads as,

$$\dot{\phi}^2 + \frac{2}{\omega_0^2}(V_0 + V_1\phi^2)\phi = 0. \tag{38}$$

The above equation when solved, is found to execute oscillatory behaviour as before, along with $\sigma = -2\frac{\omega_0}{\sqrt{\phi}}$, as well.

$$\phi(t) = \text{InverseFunction}\left[\frac{2i\#1^{3/2}\sqrt{\frac{V_0}{\#1^2V_1}} + 1\text{EllipticF}\left(i\sinh^{-1}\left(\frac{\sqrt{-\frac{i\sqrt{V_0}}{\sqrt{V_1}}}}{\sqrt{\#1}}\right)\Big| -1\right)}{\sqrt{-\frac{i\sqrt{V_0}}{\sqrt{V_1}}}\sqrt{-\#1\left(\#1^2V_1 + V_0\right)}} \,\&\right]\left[c_1 + \frac{\sqrt{2t}}{\omega_0}\right], \tag{39}$$

Case-3: Cubic potentials with additive term have important consequence. For example, a potential in the form $V = \frac{1}{2}m\omega^2 x^2 - \frac{1}{3}bx^3$ can be used to model decay of metastable states [55], and it also describes the global flow [56]. Further, the tunnelling rate in real time in the semiclassical limit may be found for arbitrary energy levels, while its ground state agrees well with the result found by the instanton method [57]. It is therefore worth to continue the present study in view of such an additive form in the cubic potential.

Under the choice $n = \frac{3}{2}$, $f(\phi) = \phi^{\frac{3}{2}}$, and taking the potential as cubic form added with a quadratic term, i.e., $V(\phi) = V_1\phi^3 + V_0\phi^2$, the expressions for the parameters of the theory under consideration along with the slow-roll parameters (24) are,

$$\omega(\phi) = \frac{4\omega_0^2 - 27\phi}{8\sqrt{\phi}}, \quad \frac{d\sigma}{d\phi} = \frac{\omega_0}{\phi^{\frac{3}{2}}}; \quad V_E = V_1 + \frac{V_0}{\phi}, \quad \epsilon = \frac{V_0^2\phi}{\omega_0^2(V_0 + V_1\phi)^2},$$

$$\eta = \frac{V_0\phi}{\omega_0^2(V_0 + V_1\phi)}, \quad N = \frac{\omega_0^2}{2\sqrt{2}V_0}\left[V_0\left(\frac{1}{\phi_e} - \frac{1}{\phi_b}\right) + V_1\ln(\phi_b - \phi_e)\right]. \tag{40}$$

We present two sets of data in Tables 9 and 10 underneath, for two different values of $V_1 > 0$. Figures 9 and 10 depict the variations of the spectral index n_s with the scalar-tensor ratio r and the scalar-tensor ratio r with the Brans–Dicke parameter ω_0 respectively. Here also we observe that $r \le 0.062$, and $0.973 < n_s < 0.983$, which are again in excellent agreement of Planck's data [40,41], while the number of e-folding N is also sufficient to remove the flatness and the horizon problems. It is interesting to note that the variation n_s with r for the two sets of data almost overlap in Figure 9.

Table 9. $f(\phi) = \phi^{\frac{3}{2}}$, (case-3): $\phi_b = 1.7$, $V_0 = -0.9 \times 10^{-13}\mathrm{T}^{-2}$, $V_1 = 0.9 \times 10^{-13}\mathrm{T}^{-2}$.

ω_0	$|\eta|$	$r = 16\epsilon$	ϕ_e	n_s	N
30	0.00270	0.0617	1.0339	0.9715	38
31	0.00253	0.0578	1.0328	0.9733	40
32	0.00237	0.0542	1.0317	0.9749	43
33	0.00223	0.0509	1.0308	0.9764	46
34	0.00210	0.0480	1.0299	0.9778	48
35	0.00198	0.0453	1.0290	0.9790	51
36	0.00187	0.0428	1.0282	0.9802	54
37	0.00177	0.0405	1.0273	0.9812	57
38	0.00168	0.0384	1.0267	0.9822	61

Table 10. $f(\phi) = \phi^{\frac{3}{2}}$, (case-3): $\phi_b = 1.7$, $V_0 = -0.9 \times 10^{-13}\mathrm{T}^{-2}$, $V_1 = 0.92 \times 10^{-13}\mathrm{T}^{-2}$.

ω_0	$|\eta|$	$r = 16\epsilon$	ϕ_e	n_s	N
29	0.00274	0.0594	1.0122	0.9723	39
30	0.00256	0.0555	1.0110	0.9741	41
31	0.00240	0.0520	1.0097	0.9757	44
32	0.00225	0.0488	1.0090	0.9772	47
33	0.00212	0.0459	1.0080	0.9786	50
34	0.00199	0.0432	1.0071	0.9798	53
35	0.00188	0.0408	1.0063	0.9809	56
36	0.00178	0.0386	1.0055	0.9820	60
37	0.00168	0.0365	1.0048	0.9829	63

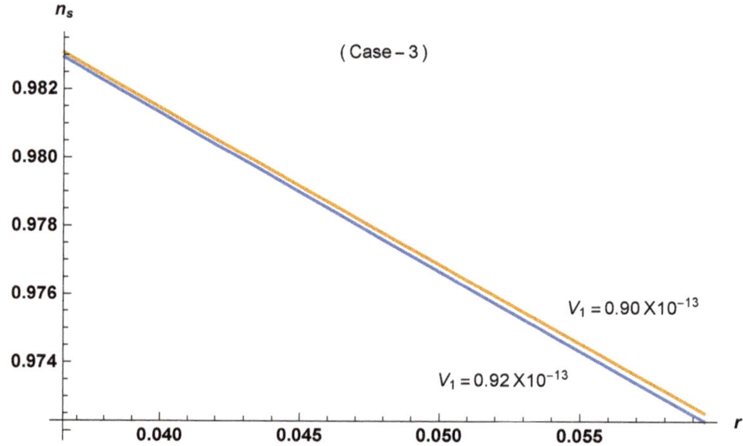

Figure 9. $f(\phi) = \phi^{\frac{3}{2}}$. Yellow ochre and blue colours represent Tables 9 and 10 respectively.

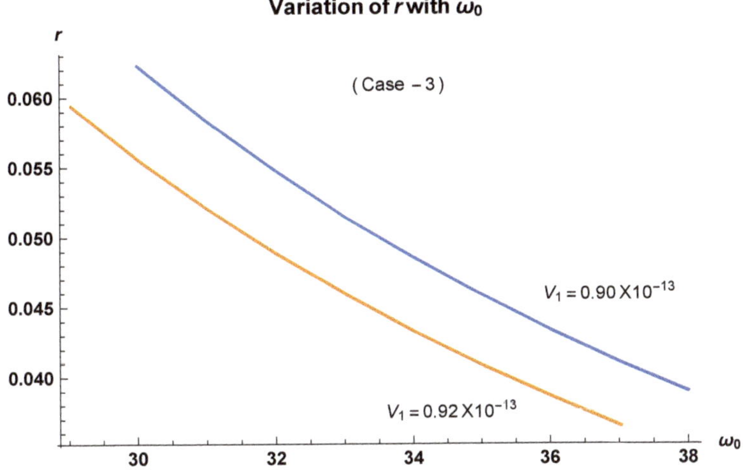

Figure 10. $f(\phi) = \phi^{\frac{3}{2}}$. Blue and yellow ochre colours represent Tables 9 and 10 respectively.

In order to study the behaviour of the scalar field at the end of inflation, we start with the Einstein frame potential as before, $V_E = (V_1 + \frac{V_0}{\phi})$, and express the field Equation (19) as,

$$\frac{3H^2}{V_1} = \frac{\dot{\sigma}^2}{2V_1} + \left(1 + \frac{V_0}{V_1\phi}\right) \tag{41}$$

As the Hubble rate falls, and $H \ll V_1$, the above equation may be approximated to, $\dot{\sigma}^2 = 2i^2(V_1 + \frac{V_0}{\phi})$, which in terms of the scalar field ϕ reads as,

$$\dot{\phi}^2 + \frac{2}{\omega_0^2}(V_0 + V_1\phi)\phi^2 = 0. \tag{42}$$

The above equation may be solved to find

$$\phi = \frac{V_0}{V_1}\left[-1 - \tan\left\{\frac{\sqrt{V_0}}{2}\left(c_1 + \frac{\sqrt{2}}{\omega_0}t\right)\right\}^2\right], \tag{43}$$

which unfortunately is not oscillatory. Perhaps, due to the asymmetry of the potential, the oscillatory behaviour of the scalar field with an additive quadratic term is not exhibited.

3.1.3. $n = 2$, $f(\phi) = \phi^2$

Case-1: Under the choice $n = 2$, $f(\phi) = \phi^2$, and the potential (21) is now chosen as $V(\phi) = V_0 + V_1\phi^4$. Therefore the Brans–Dicke parameter and the Einstein frame potential, together with the slow-roll parameters (22), take the following forms,

$$\omega(\phi) = \frac{\omega_0^2 - 12\phi^2}{2\phi}, \quad \frac{d\sigma}{d\phi} = \frac{\omega_0}{\phi^2}; \quad V_E = V_1 + V_0\phi^{-4}, \quad \epsilon = \frac{16V_0^2\phi^2}{\omega_0^2(V_0 + V_1\phi^4)^2},$$

$$\eta = \frac{24V_0\phi^2}{\omega_0^2(V_0 + V_1\phi^4)}, \quad N = \frac{\omega_0^2}{8\sqrt{2}V_0}\left[\frac{V_0}{2}\left(\frac{1}{\phi_e^2} - \frac{1}{\phi_b^2}\right) + \frac{V_1}{2}(\phi_b^2 - \phi_e^2)\right]. \tag{44}$$

It is quite transparent that for large value of the scalar field ϕ, a flat Einstein's frame potential is realizable here too. As before, we take two sets of data corresponding to two different values of $V_1 > 0$, and tabulate the parametric values in Tables 11 and 12. One can see that the scalar tensor ratio $r \leq 0.0202$, and the spectral index lies between $0.9645 \leq n_s \leq 0.9809$, which are in excellent agreement with Planck's data [40,41]. Further number of e-folding N is also sufficient to alleviate the flatness and the horizon problems. The n_s versus r and r versus ω_0 plots are presented in Figures 11 and 12, as well.

Equation (19) now reads as,

$$3H^2 = \frac{1}{2}\dot{\sigma}^2 + V_1 + \frac{V_0}{\phi^4},\tag{45}$$

which, as H falls below V_1, i.e., $H \ll V_1$, may be approximated to, $\frac{1}{2}\dot{\sigma}^2 + \frac{V_0}{\phi^4} + V_1 = 0$. In terms of the scalar field ϕ it is expressed as $\dot{\phi}^2 + \frac{2}{\omega_0^2}\left(V_1\phi^4 + V_0\right) = 0$, since, $\dot{\sigma} = \omega_0 \frac{\phi}{\phi^2}$, in view of (44). The solution is exhibited either as,

$$\phi(t) = -\frac{(-1)^{3/4}\sqrt[4]{V_0}\,\text{JacobiSN}\left(c_1\sqrt[4]{-V_0 V_1}\left(1 + i\frac{\sqrt{2}}{\omega_0}t\right)\Big| - 1\right)}{\sqrt[4]{V_1}},\tag{46}$$

(where JacobiSN is a meromorphic function in both arguments, which for certain special arguments may automatically be evaluated to exact values. In any case, under numerical simulation the above solution is found to exhibit oscillatory behaviour of the scalar field ϕ), or as inverse elliptic function as before,

$$\phi(t) = \text{InverseFunction}\left[-\frac{i\sqrt{\frac{\#1^4 V_1}{V_0} + 1}\,\text{EllipticF}\left(i\sinh^{-1}\left(\#1\sqrt{-\frac{i\sqrt{V_1}}{\sqrt{V_0}}}\right)\Big| - 1\right)}{\sqrt{-\frac{i\sqrt{V_1}}{\sqrt{V_0}}}\sqrt{\#1^4\left(-V_1\right) - V_0}}\ \&\right]\left[c_1 + \frac{\sqrt{2}t}{\omega_0}\right],\tag{47}$$

It is also clear that $\sigma = -\frac{\omega_0}{\phi}$ oscillates as well, and the universe transits from inflationary regime to the matter dominated era.

Table 11. $f(\phi) = \phi^2$, (case-1): $\phi_b = 1.7$, $V_0 = -0.9 \times 10^{-13}\text{T}^{-2}$, $V_1 = 0.9 \times 10^{-13}\text{T}^{-2}$.

ω_0	$|\eta|$	$r = 16\epsilon$	ϕ_e	n_s	N
26	0.013956	0.0202	1.0377	0.9645	37
27	0.012941	0.0188	1.0364	0.9671	39
28	0.012033	0.0175	1.0351	0.9694	42
29	0.01122	0.0163	1.0339	0.9715	45
30	0.010482	0.0152	1.0328	0.9733	49
31	0.009817	0.0142	1.0317	0.9750	52
32	0.009213	0.0134	1.0308	0.9766	56
33	0.008663	0.0126	1.0298	0.9780	59
34	0.008161	0.0118	1.0289	0.9792	63
35	0.007701	0.0112	1.0282	0.9804	67

Table 12. $f(\phi) = \phi^2$, (case-1): $\phi_b = 1.7$. $V_0 = -0.9 \times 10^{-13}\mathrm{T}^{-2}$, $V_1 = 1.0 \times 10^{-13}\mathrm{T}^{-2}$.

ω_0	$|\eta|$	$r = 16\epsilon$	ϕ_e	n_s	**N**
24	0.014543	0.0187	1.0127	0.9639	37
25	0.013403	0.0173	1.0112	0.9667	40
26	0.012392	0.0160	1.0098	0.9692	43
27	0.011491	0.0148	1.0085	0.9714	46
28	0.010685	0.0138	1.0073	0.9735	50
29	0.00996	0.0128	1.0062	0.9753	54
30	0.00931	0.0120	1.0051	0.9769	57
31	0.008717	0.0112	1.0041	0.9784	61
32	0.008180	0.0105	1.0032	0.9797	65
33	0.007692	0.0099	1.0023	0.9809	69

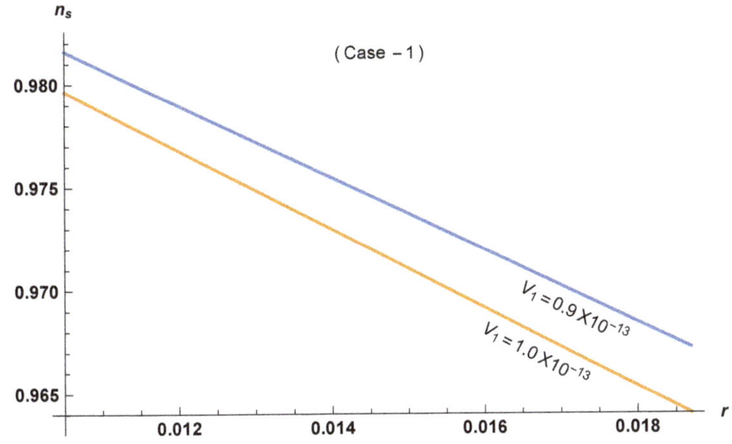

Figure 11. $f(\phi) = \phi^2$. Blue and yellow ochre colours represent Tables 11 and 12 respectively.

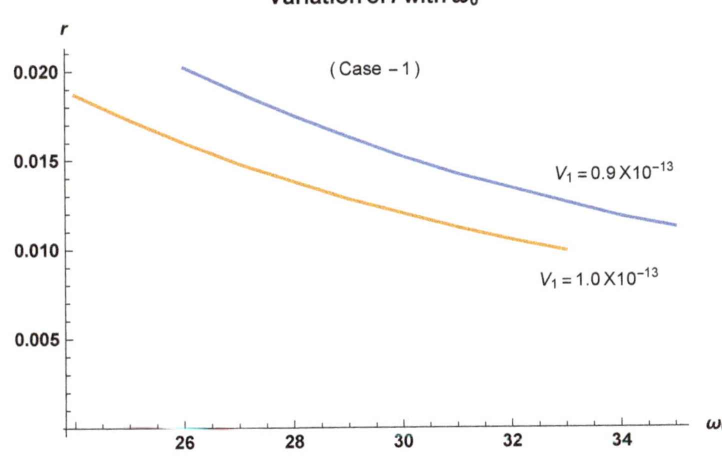

Figure 12. $f(\phi) = \phi^2$. Blue and yellow ochre colours represent Tables 11 and 12 respectively.

Case-2: Here, for $f(\phi) = \phi^2$, we consider the potential in the form, $V = V_1\phi^4 + V_0\phi^2$, i.e., instead of a constant additive term, we consider $V_0\phi^2$ in addition. This case was earlier studied in [37]. However, as already mentioned, over the years, Planck's data put up tighter constraints on inflationary parameters, and so it is quite reasonable to check if this form of potential passes the said constraints [40,41]. One can now find the expression for the Brans–Dicke parameter $\omega(\phi)$, the potential V_E in the Einstein frame, $\frac{d\sigma}{d\phi}$, the slow-roll parameters ϵ, η and the number of e-folding N, in view of the Equations (24), respectively as,

$$\omega(\phi) = \frac{\omega_0^2 - 12\phi^2}{2\phi}, \quad \frac{d\sigma}{d\phi} = \frac{\omega_0}{\phi^2}; \qquad V_E = V_1 + V_0\phi^{-2}, \qquad \epsilon = \frac{4V_0^2\phi^2}{\omega_0^2(V_1\phi^2 + V_0)^2},$$

$$\eta = \frac{4V_0\phi^2}{\omega_0^2(V_1\phi^2 + V_0)}, \qquad N = \frac{\omega_0^2}{4\sqrt{2}V_0}\left[V_1\ln(\phi_b - \phi_e) - \frac{V_0}{2}\left(\frac{1}{\phi_b^2} - \frac{1}{\phi_e^2}\right)\right]. \tag{48}$$

Note that the Einstein frame potential now takes the same form as in case-1 for $n = 1$, and a flat section of the potential is still realizable at large value of the scalar field ϕ. We present two Tables 13 and 14, as before for different values of $V_1 > 0$. The scalar to tensor ratio $r \leq 0.0636$ and the spectral index $0.9715 \leq n_s \leq 0.9831$ lie very much within the Planck's data, while the number of e-folding N is again sufficient to alleviate the horizon and flatness problems. The Figures 13 and 14 represent n_s versus r and r versus ω_0 respectively. In view of the plots, the data for Table 14, here appears to be even better.

To check if the scalar field executes oscillatory behaviour at the end of inflation, we note that here $V_E = V_1 + \frac{V_0}{\phi^2}$. So in view of Equation (19) one obtains,

$$\frac{3H^2}{V_1} = \frac{\dot{\sigma}^2}{2V_1} + \left(1 + \frac{V_0}{V_1\phi^2}\right) \tag{49}$$

As, H falls below V_1, and $H \ll V_1$, the above equation can be approximated to, $\dot{\sigma}^2 = 2i^2\left(V_1 + \frac{V_0}{\phi^2}\right)$, yielding, $\dot{\phi} = i\frac{\phi}{\omega_0}\sqrt{2(V_0 + V_1\phi^2)}$, where, $\dot{\sigma} = \omega_0\frac{\phi}{\phi^2}$. Thus, we obtain,

$$\frac{\phi}{\left(\sqrt{V_0} + \sqrt{(V_0 + V_1\phi^2)}\right)} = \sqrt{V_0}e^{i\frac{\sqrt{2V_0}}{\omega_0}t}. \tag{50}$$

It is also possible to solve for ϕ and express it in the following form,

$$\phi(t) = \frac{\sqrt{-\beta(t) + \sqrt{\beta(t)^2 - 4\alpha(t)\gamma(t)}}}{\sqrt{2\alpha(t)}}, \tag{51}$$

where, $\alpha(t) = 1 - V_0V_1e^{i2\frac{\sqrt{2V_0}}{\omega_0}t}$, $\beta(t) = -4V_0^3V_1e^{i4\frac{\sqrt{2V_0}}{\omega_0}t}$, $\gamma(t) = -4V_0^4e^{i4\frac{\sqrt{2V_0}}{\omega_0}t}$.

It is now quite apparent that the scalar field $\phi(t)$ executes oscillatory behaviour and therefore graceful exit from inflation is realizable. Since in view of (48) $\sigma = -\frac{\omega_0}{\phi}$, therefore σ also executes oscillatory behaviour.

Table 13. $f(\phi) = \phi^2$, (case-2): $\phi_b = 1.26$. $V_0 = -1.0 \times 10^{-20}\mathrm{T}^{-2}$; $V_1 = 1.0 \times 10^{-20}\mathrm{T}^{-2}$.

ω_0	$|\eta|$	$r = 16\epsilon$	ϕ_e	n_s	N
68	0.002337	0.0636	1.0148	0.9715	37
70	0.002206	0.0601	1.0144	0.9731	40
72	0.002085	0.0568	1.0140	0.9745	42
74	0.001974	0.0537	1.0136	0.9759	44
76	0.001871	0.0509	1.0132	0.9772	47
78	0.001776	0.0484	1.0129	0.9783	49
80	0.001689	0.0460	1.0126	0.9793	52
82	0.001607	0.0438	1.0122	0.9804	55
84	0.001532	0.0417	1.0119	0.9813	57
86	0.001461	0.0398	1.0117	0.9821	60
88	0.001396	0.0380	1.0114	0.9830	63

Table 14. $f(\phi) = \phi^2$, (case-2.): $\phi_b = 1.26$. $V_0 = -1.0 \times 10^{-20}\mathrm{T}^{-2}$; $V_1 = 1.1 \times 10^{-20}\mathrm{T}^{-2}$.

ω_0	$|\eta|$	$r = 16\epsilon$	ϕ_e	n_s	N
60	0.002363	0.0507	1.0656	0.9762	38
61	0.002287	0.0490	1.0653	0.9770	39
62	0.002213	0.0475	1.0650	0.9778	40
63	0.002144	0.0460	1.0648	0.9785	41
64	0.002078	0.0445	1.0646	0.9792	43
65	0.002014	0.0432	1.0643	0.9798	44
66	0.001953	0.0419	1.0641	0.9804	46
67	0.001895	0.0406	1.0638	0.9810	47
68	0.001840	0.0394	1.0636	0.9815	49
69	0.001787	0.0383	1.0634	0.9821	50
70	0.001737	0.0372	1.0632	0.9826	51
71	0.001688	0.0362	1.0630	0.9831	53

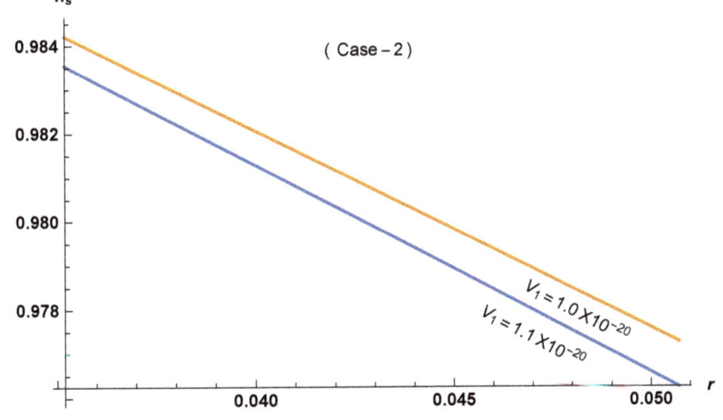

Figure 13. $f(\phi) = \phi^2$. Yellow ochre and blue colours represent Tables 13 and 14 respectively.

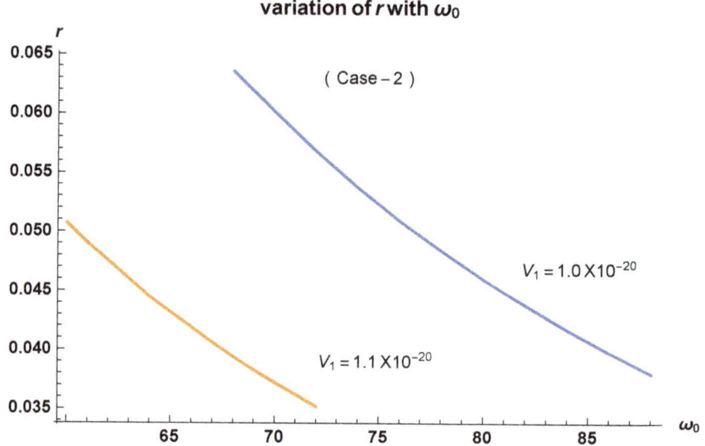

Figure 14. $f(\phi) = \phi^2$. Blue and yellow ochre colours represent Tables 13 and 14 respectively.

Case-3: We consider yet another case for $n = 2$, i.e., taking $f(\phi) = \phi^2$, with the potential $V(\phi)$ being represented by two additional terms apart from ϕ^4, as it should be, to make it a perfect square: $V(\phi) = (\sqrt{V_1}\phi^2 - \sqrt{V_0}\phi)^2 = V_1\phi^4 + V_0\phi^2 - 2\sqrt{V_0 V_1}\phi^3$. As before, one can now find the expression for the Brans–Dicke parameter $\omega(\phi)$, the potential V_E in the Einstein frame, $\frac{d\sigma}{d\phi}$, the slow-roll parameters ϵ, η and the number of e-foldings N, in view of the Equations (14), (18) and (20) respectively as,

$$\omega(\phi) = \frac{\omega_0^2 - 12\phi^2}{2\phi}, \quad \frac{d\sigma}{d\phi} = \frac{\omega_0}{\phi^2}; \quad V_E = V_1 - 2\frac{\sqrt{V_0 V_1}}{\phi} + \frac{V_0}{\phi^2}, \quad \epsilon = \frac{4V_0^2\phi^2}{\omega_0^2(\sqrt{V_0 V_1}\phi - V_0)^2},$$

$$\eta = \frac{4V_0^2\phi^2}{\omega_0^2(\sqrt{V_0 V_1}\phi - V_0)^2}, \quad N = \int_{\phi_e}^{\phi_b} \frac{\omega_0^2}{4\sqrt{2}V_0}\frac{(\sqrt{V_0 V_1}\phi - V_0)}{\phi^3} d\phi.$$

$$(52)$$

One can clearly see that the flat section of the potential is still attainable for large value of the scalar field ϕ. Tables 15 and 16 depict that the scalar to tensor ratio $r < 0.1$), is quite reasonable, while the spectral index $0.9752 \leq n_s \leq 0.981$ fits perfectly with Planck's data [40,41]. Figures 15 and 16 represent n_s versus r and r versus ω_0 plots respectively. Interestingly, two r versus n_s plots (Figure 15) corresponding to the two sets of data (Tables 15 and 16) merge almost perfectly.

The scalar field executes oscillatory behaviour here too, as we demonstrate below. Here, $V_E = V_1 - 2\frac{\sqrt{V_0 V_1}}{\phi} + \frac{V_0}{\phi^2}$, and so from (19) we find,

$$\frac{3H^2}{V_1} = \frac{\dot{\sigma}^2}{2V_1} + \left(1 - \frac{2}{\phi}\sqrt{\frac{V_0}{V_1}}\right) + \frac{V_0}{V_1\phi^2}.$$

$$(53)$$

As H falls below V_1, and $H \ll V_1$, the above equation can be approximated as, $\dot{\sigma}^2 = 2i^2(\frac{\sqrt{V_0}}{\phi} - \sqrt{V_1})^2$, which yields $\dot{\phi} = i\frac{\phi}{\omega_0}\sqrt{2}(\sqrt{V_0} - \sqrt{V_1}\phi)$ where, $\dot{\sigma} = \omega_0\frac{\dot{\phi}}{\phi^2}$. Therefore finally we obtain,

$$\phi(t) = \frac{\sqrt{V_0}e^{i\frac{\sqrt{2V_0}}{\omega_0}t}}{1 + \sqrt{V_1}e^{i\frac{\sqrt{2V_0}}{\omega_0}t}}.$$

$$(54)$$

Clearly, ϕ executes oscillatory behaviour, and graceful exit from inflation may be realized here too. Here again since in view of (52) $\sigma = -\frac{\omega_0}{\phi}$, therefore σ executes oscillatory behaviour, as well.

Table 15. $f(\phi) = \phi^2$, (case-3): $\phi_b = 1.3$. $V_1 = 0.9 \times 10^{-20} \mathrm{T}^{-2}$; $V_0 = 0.9 \times 10^{-20} \mathrm{T}^{-2}$.

ω_0	η	$r = 16\epsilon$	ϕ_e	n_s	N
110	0.006208	0.0993	1.0185	0.9752	57
112	0.005988	0.0958	1.0182	0.9760	59
114	0.005780	0.0925	1.0178	0.9769	61
116	0.005582	0.0893	1.0175	0.9777	63
118	0.005394	0.0863	1.0172	0.9784	65
120	0.005216	0.0835	1.0170	0.9791	67
122	0.005046	0.0807	1.0167	0.9798	70
124	0.004885	0.0782	1.0164	0.9804	72
126	0.004731	0.0757	1.0161	0.9810	74

Table 16. $f(\phi) = \phi^2$, (case-3): $\phi_b = 1.3$. $V_1 = 1.0 \times 10^{-20} \mathrm{T}^{-2}$; $V_0 - 0.9 \times 10^{-20} \mathrm{T}^{-2}$.

ω_0	η	$r = 16\epsilon$	ϕ_e	n_s	N
142	0.006160	0.0986	1.0386	0.9754	57
144	0.005990	0.0958	1.0388	0.9760	59
146	0.005827	0.0932	1.0391	0.9767	60
148	0.005671	0.0907	1.0393	0.9773	62
150	0.005521	0.0883	1.0395	0.9779	64
152	0.005376	0.0860	1.0397	0.9785	65
154	0.005237	0.0838	1.0399	0.9791	67
156	0.005104	0.0817	1.0400	0.9796	69
158	0.004976	0.0796	1.0402	0.9801	71
160	0.004852	0.0776	1.0404	0.9806	72

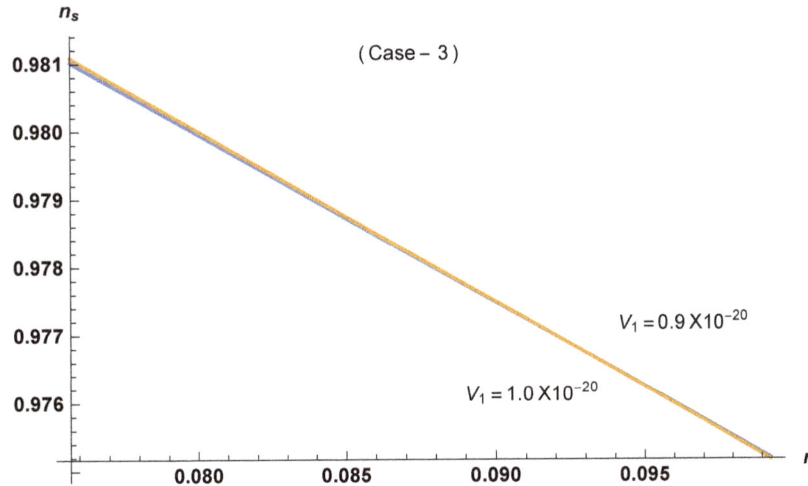

Figure 15. $f(\phi) = \phi^2$. Blue and yellow ochre colours represent Tables 15 and 16 respectively.

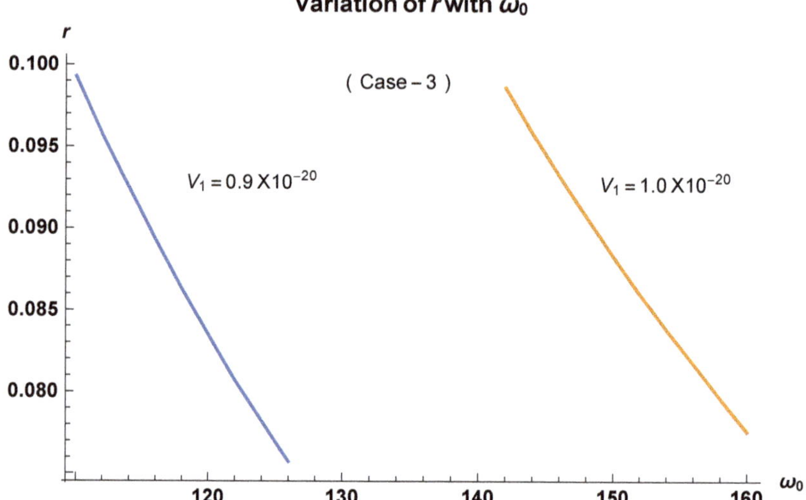

Figure 16. $f(\phi) = \phi^2$. Blue and yellow ochre colours represent Tables 15 and 16 respectively.

3.2. Exponential Potential

Finally, we consider an exponential form of the potential with: $f(\phi) = e^{\frac{\lambda\phi}{2}}$, with $V(\phi) = V_1 e^{\lambda\phi} + V_0$. It is possible to find the expression for the Brans–Dicke parameter $\omega(\phi)$, the potential V_E in the Einstein frame, $\frac{d\sigma}{d\phi}$, the slow-roll parameters ϵ, η and the number of e-folding N, in view of the Equations (14), (18) and (20) respectively as,

$$\omega(\phi) = \frac{(\omega_0^2 - \frac{3}{4}\lambda^2 e^{\lambda\phi})\phi}{2e^{\frac{\lambda\phi}{2}}}; \quad \frac{d\sigma}{d\phi} = \frac{\omega_0}{e^{\frac{\lambda\phi}{2}}}; \quad V_E = V_1 + V_0 e^{-\lambda\phi}, \quad \epsilon = \frac{\lambda^2 V_0^2 e^{\lambda\phi}}{\omega_0^2(V_1 e^{\lambda\phi} + V_0)^2},$$

$$\eta = \frac{\lambda^2 V_0 e^{\lambda\phi}}{\omega_0^2(V_1 e^{\lambda\phi} + V_0)}, \quad N = \int_{\phi_e}^{\phi_b} \frac{\omega_0^2(V_0 + V_1 e^{\lambda\phi})}{2\sqrt{2}V_0\lambda e^{\lambda\phi}} d\phi. \tag{55}$$

We present our results in the following Table 17, under the only choice of the parameter $\lambda = -1$. The data shows good agreement $r < 0.06$, and $0.9612 \leq n_s \leq 0.9836$ with Planck's data [40,41]. Figure 17, represents a plot for n_s versus r.

Table 17. $|\phi_b| = 1.2, V_0 = -1.0 \times 10^{-20}\text{T}^{-2}, V_1 = 1.0 \times 10^{-20}\text{T}^{-2}$.

ω_0	η	$r = 16\epsilon$	$|\phi_e|$	n_s	N
13	0.008468	0.0584	0.07690	0.9612	30
14	0.007301	0.0503	0.07141	0.9665	34
15	0.00636	0.0439	0.06665	0.9708	40
16	0.005589	0.0385	0.06249	0.9745	45
17	0.004952	0.0341	0.05882	0.9773	51
18	0.004417	0.0305	0.05555	0.9797	57
19	0.003964	0.02734	0.05263	0.9818	64
20	0.003578	0.02467	0.04999	0.9836	71

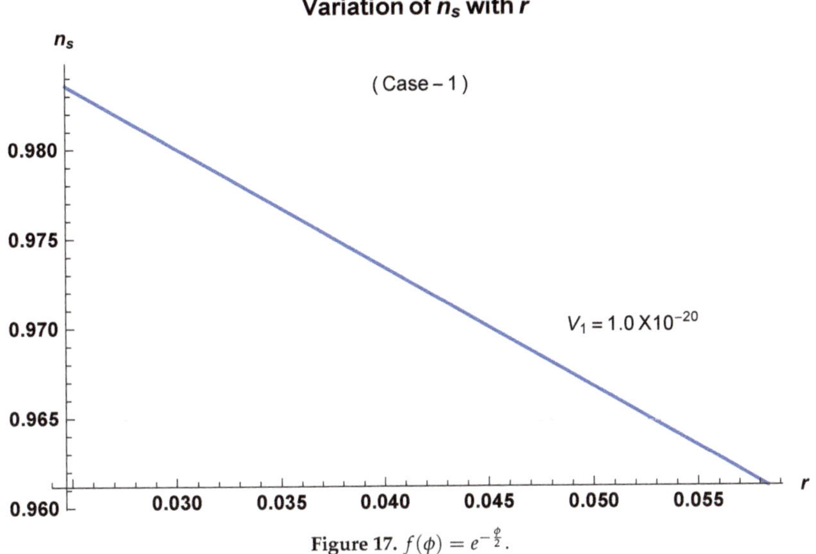

Figure 17. $f(\phi) = e^{-\frac{\phi}{2}}$.

The scalar field executes oscillatory behaviour here too, as we demonstrate below. Here $V_E = V_1 + V_0 e^{\phi}$, and so in view of Equation (19), one can calculate,

$$\frac{3H^2}{V_1} = \frac{\dot{\sigma}^2}{2V_1} + \left(1 + \frac{V_0}{V_1}e^{\phi}\right) \tag{56}$$

Again as H falls below V_1, and $H \ll V_1$, the above equation can be approximated to, $\dot{\sigma}^2 = 2i^2(V_1 + V_0 e^{\phi})$, which yields $\dot{\phi} = i\frac{e^{-\frac{\phi}{2}}}{\omega_0}\sqrt{2(V_1 + V_0 e^{\phi})}$, where, $\dot{\sigma} = \omega_0 \dot{\phi} e^{\frac{\phi}{2}}$. Finally, therefore

$$\phi = \ln\left[\frac{1}{4V_0^2}\left(e^{i\frac{\sqrt{2V_0}}{\omega_0}t} - 2V_0 V_1 + V_0^2 V_1^2 e^{-i\frac{\sqrt{2V_0}}{\omega_0}t}\right)\right]. \tag{57}$$

The oscillatory behaviour of the scalar field here again assures graceful exit from inflationary regime. In view of (55), $\sigma = 2\omega_0 e^{\frac{\phi}{2}}$, for $\lambda = -1$, which we have considered, hence σ also executes oscillatory behaviour.

4. Concluding Remarks

Scalar-tensor theories of gravity are generalizations of the Brans–Dicke theory, in which the coupling parameter is a function of the scalar field, i.e., $\omega_{BD} = \omega(\phi)$, and therefore is a variable. The requirement for such generalization of Brans–Dicke theory generated from the tight constraints on ω_{BD} established by the solar system experiments [58]. There exists various classification of scalar-tensor theory of gravity [59]. In the present manuscript we have considered standard non-minimal coupling, where the coupling parameter $f(\phi)$ is arbitrary. It has been noticed earlier that such a theory has an in-built symmetry being associated with a conserved current for trace-free fields, such as vacuum and radiation dominated eras for barotropic fluids. In view of such a symmetry, it is possible to fix all the variables of the theory, including the potential function, fixing the form of one of those. In this manuscript, we have chosen different forms of the coupling parameter $f(\phi)$, which fixed $\omega(\phi)$ and $V(\phi)$, to study the cosmological evolution of the very early universe in the context of inflation. Inflation is a quantum mechanical phenomenon, and has occurred around Planck's era. However, it has been argued that since the radiative corrections to the potential are negligible, hence the inflationary parameters can be computed using the classical Lagrangian [60]. This argument leads in general, to calculate inflationary parameters in view of the classical Lagrangian, which we have done in the present manuscript. The so called unification programmes, which essentially claim to unify early inflationary regime with late-time cosmic acceleration have no credentials, since none of the models passes through a well behaved radiation and early matter dominated era. However, a history of cosmic evolution starting from inflationary regime, followed by a Friedmann-like radiation ($a \propto \sqrt{t}$) and early matter dominated eras ($a \propto t^{\frac{2}{3}}$), that finally ends up to a late-time accelerated universe ($z = 0.75$), has already been explored in view of the present model [37]. In this connection, the present model makes a reasonably viable attempt to unify early inflation with late-time cosmic acceleration. Nevertheless earlier, only a single form of the coupling parameter $f(\phi)$ together with a particular form of $V(\phi)$ had been treated. Here, we have extended our work considering combinations of a host of power potentials ($V(\phi) = V_0 + V_1\phi^2$, $V_0 + V_1\phi^3$, $V_0 + V_1\phi^4$, $V_0\phi + V_1\phi^3$, $V_0\phi^2 + V_1\phi^3$, $V_0\phi^2 + V_1\phi^4$, $V_0\phi^2 + V_1\phi^4 - 2\sqrt{V_0 V_1}\phi^3$), together with an exponential potential ($V(\phi) = V_0 + V_1 e^{\lambda\phi}$). We find that all the quadratic, cubic, quartic and exponential potentials pass the tighter constraints on inflationary parameters released by latest Planck's data [40,41] quite comfortably. Further, all these potentials admit graceful exit from inflation, except one case of cubic potential associated with a square potential, i.e., ($V(\phi) = V_0\phi^2 + V_1\phi^3$), for which unfortunately the scalar field does not show up oscillatory behaviour at the end of inflation.

For the purpose of the present analysis, we have translated the non-minimally coupled Jordan's frame action to the Einstein frame, under conformal transformation. It is therefore worth to make certain comments in this regard. There is an age old debate regarding physical equivalence between the two: Jordan's and Einstein's frames, which are related under conformal transformation. Now, indeed if the two formulations are not equivalent, the problem arises in selecting the physically preferred frame. It emerges from the work of several authors, in different contexts on Kaluza–Klein and Brans–Dicke theories, that the formulations of a scalar-tensor theory in the two conformal frames are physically inequivalent [61–68]. Furthermore, the Jordan frame formulation of a scalar-tensor theory is not viable because the energy density of the gravitational scalar field present in the theory is not bounded from below, which amounts to the violation of the weak energy condition [69]). The system therefore is unstable and decays toward a lower and lower energy state ad infinitum [66–68]. Although, a quantum system may have states with negative energy density [69–71], such feature is not acceptable for a viable classical theory of gravity. In fact, a classical theory must have a ground state that is stable against small perturbations. The violation of the weak energy condition by scalar-tensor theories formulated in the Jordan conformal frame makes them unviable descriptions of classical gravity, while the Einstein frame formulation of scalar-tensor theories is free from such problem. However, in the Einstein frame also there is a violation of the equivalence principle due to the anomalous coupling of the scalar field to ordinary matter. Nevertheless, this violation is small and compatible with the available tests of the equivalence principle [72]. Further, Einstein's frame is indeed regarded as an important low energy manifestation of compactified theories [72–77]. However, in search of Noether symmetries of $F(R)$ theory of gravity, the two frames have been found to be physically equivalent [78]. So although the debate persists, but somehow it is quite relevant to consider Einstein's frame to be the physical frame. Therefore, in view of the above discussions, it is justified to study the physics associated with non-minimally coupled scalar-tensor theory of gravity, after translating it to the Einstein frame, as we have done in the present article.

Author Contributions: D.S., S.S. and A.K.S. have contributed to the extent they could. All authors have read and agreed to the published version of the manuscript.

Funding: This research received no external funding.

Conflicts of Interest: The authors declare no conflict of interest.

References

1. Partridge, R.B. *3K: The Cosmic Microwave Background Radiation*; Cambridge Univ. Press: Cambridge, UK, 1995.
2. Turner, M.S.; Kolb, E.W. *The Early Universe*; Addison-Wesley Publishing Company: Boston, MA, USA, 1990.
3. Peebles, P.J.E. *The Large Scale Structure of the Universe*; Princeton Univ. Press: Princeton, NJ, USA, 1980.
4. Peebles, P.J.E. *Principles of Physical Cosmology*; Princeton Univ. Press: Princeton, NJ, USA, 1993.
5. Baumann, D. The Physics of Inflation—A Course for Graduate Students in Particle Physics and Cosmology. *arXiv* **2015**, arXiv:0907.5424.
6. Sanyal, A.K. Enlightening the dark universe. *Indian J. Theor. Phys.* **2014**, *62*, 211–262.
7. Stelle, K.S. Renormalization of higher-derivative quantum gravity. *Phys. Rev.* **1977**, *953*, D16.
8. Sen, A. Recent Developments in Superstring Theory. *Nucl. Phys. Proc. Suppl.* **2001**, *94*, 35.
9. Mukhi, S. String theory: A perspective over the last 25 years. *Class. Quant. Grav.* **2011**, *28*, 153001.
10. Lahanas, A.B.; Nanopoulos, D.V. The road to no-scale supergravity. *Phys. Rep.* **1987**, *145*, 1.
11. Ferrara1, S.; Sagnotti, A. Supergravity at 40: Reflections and Perspectives. *J. Phys. Conf. Ser.* **2016**, 873.
12. Kenath, A.; Gudennavar, S.B.; Sivaram, C. Dark matter, dark energy, and alternate models: A review. *Adv. Space Res.* **2017**, *60*, 166.
13. Starobinsky, A.A. A new type of isotropic cosmological models without singularity. *Phys. Lett.* **1980**, *99*, B91.
14. Guth, A.H. Inflationary universe: A possible solution to the horizon and flatness problems. *Phys. Rev. D* **1981**, *23*, 347.
15. Olive, K.A. Inflation. *Phys. Rept.* **1990**, *190*, 307.
16. Lyth, D.H.; Riotto, A. Particle physics models of inflation and the cosmological density perturbation. *Phys. Rep.* **1999**, *314*, 1–146.
17. Liddle, A.R.; Lyth, D.H. *Cosmological Inflation and Large-Scale Structure*; Cambridge University Press: Cambridge, UK, 2000.
18. Baumann, D. TASI lectures on inflation. *arXiv* **2009**, arXiv:0907.5424.
19. Martin, J.; Ringeval, C.; Vennin, V. Encyclopaedia inflationaris. *Phys. Dark Univ.* **2014**, *5–6*, 75–235.
20. Martin, J.; Ringeval, C.; Trotta, R.; Vennin, V. The best inflationary models after Planck. *JCAP* **2014**, *1403*, 39.
21. Martin, J. The observational status of cosmic inflation after Planck. In *The Cosmic Microwave Background*; Fabris, J., Piattella, O., Rodrigues, D., Velten, H., Zimdahl, W., Eds.; Springer: Cham, Switzerland, 2015.
22. Linde, A.D. A new inflationary universe scenario: A possible solution of the horizon, flatness, homogeneity, isotropy and primordial monopole problems. *Phys. Lett. B* **1982**, *108*, 389.
23. Albrecht, A.; Steinhardt, P.J. Cosmology for grand unified theories with radiatively induced symmetry breaking. *Phys. Rev. Lett.* **1982**, *48*, 1220.
24. Press, W.H. Spontaneous production of the Zel'dovich spectrum of cosmological fluctuations. *Phys. Scr.* **1980**, *21*, 702.
25. Hawking, S.W. The development of irregularities in a single bubble inflationary universe. *Phys. Lett. B* **1982**, *115*, 295.
26. Starobinsky, A.A. Dynamics of phase transition in the new inflationary universe scenario and generation of perturbations. *Phys. Lett. B* **1982**, *117*, 175.
27. Guth, A.H.; Pi, S.Y. Fluctuations in the new inflationary universe. *Phys. Rev. Lett.* **1982**, *49*, 1110.
28. Bardeen, J.M.; Steinhardt, P.J.; Turner, M.S. Spontaneous creation of almost scale-free density perturbations in an inflationary universe. *Phys. Rev. D* **1983**, *28*, 679.
29. Mukhanov, V.F.; Chibisov, G.V. Quantum fluctuations and a nonsingular universe. *JETP Lett.* **1981**, *33*, 532.
30. Whitt, B. Fourth-order gravity as general relativity plus matter. *Phys. Lett. B* **1984**, *145*, 176.
31. Wands, D. Extended gravity theories and the Einstein–Hilbert action, *Class. Quant. Grav.* **1994**, *11*, 269.
32. Horndeski, G.W. Second-order Scalar-tensor field equations in a four-dimentional Space. *Int. J. Theor. Phys.* **1974**, *10*, 363.
33. Sanyal, A.K. Scalar–tensor theory of gravity carrying a conserved current. *Phys. Lett. B* **2005**, *624*, 81.
34. Sanyal, A.K. Study of symmetry in F(R) theory of gravity. *Mod. Phys. Lett. A* **2010**, *25*, 2667.
35. Sk, N.; Sanyal, A.K. Field independent cosmic evolution. *J. Astrophys.* **2013**, *2013*, 590171.
36. Sarkar, K.; Sk, N.; Debnath, S.; Sanyal, A.K. Viability of Noether symmetry of F(R) theory of gravity. *Int. J. Theor. Phys.* **2013**, *52*, 1194.
37. Tajahmad, B.; Sanyal, A.K. Unified cosmology with scalar–tensor theory of gravity. *Eur. Phys. J.* **2017**, *77*, 217.

38. Ade, P.A.; Aghanim, N.; Armitage-Caplan, C.; Arnaud, M.; Ashdown, M.; Atrio-Barandela, F.; Aumont, J.; Baccigalupi, C.; Banday, A.J.; Barreiro, R.B.; et al. Planck 2013 results. XVI. Cosmological parameters. *Astron. Astrophys.* **2014**, *571*, A16.
39. Ade, P.A.; Aghanim, N.; Arnaud, M.; Ashdown, M.; Aumont, J.; Baccigalupi, C.; Banday, A.J.; Barreiro, R.B.; Bartlett, J.G.; Bartolo, N.; et al. Planck 2015 resultss. XII. Cosmological parameters. *Astron. Astrophys.* **2016**, *594*, A13.
40. Aghanim, N.; Akrami, Y.; Ashdown, M.; Aumont, J.; Baccigalupi, C.; Ballardini, M.; Banday, A.J.; Barreiro, R.B.; Bartolo, N.; Basak, S.; et al. Planck 2018 Results. VI. Cosmological Parameters, (Planck Collaboration). *arXiv* **2018**, arXiv:1807.06209.
41. Akrami, Y.; Arroja, F.; Ashdown, M.; Aumont, J.; Baccigalupi, C.; Ballardini, M.; Banday, A.J.; Barreiro, R.B.; Bartolo, N.; Basak, S.; et al. Planck 2018 results. X. Constraints on inflation, (Planck Collaboration). *arXiv* **2018**, arXiv:1807.06211.
42. Kobayashi, T.; Yamaguchi, M.; Yokoyama, J. Generalised G-inflation: Inflation with most general second order field equation. *Prog. Theor. Phys.* **2011**, *126*, 511.
43. Kamada, K.; Kobayashi, T.; Yamaguchi, M.; Yokoyama, J. Higgs G-inflation. *Phys. Rev. D* **2011**, *83*, 083515.
44. Myrzakulov, R.; Sebastiani, L. Scalar tensor Horndeski Models: Simple cosmological applications. *Astrophys. Space Sci.* **2016**, *361*, 62.
45. Myrzakulov, R.; Sebastiani, L. K-Essence Non-Minimally Coupled with Gauss-Bonnet Invariant for Inflation. *Symmetry* **2016**, *8*, 57.
46. Gleyzes, J.; Langlois, D.; Piazza, F.; Vernizzi, F. Healthy theories beyond Horndeski. *Phys. Rev. Lett.* **2015**, *114*, 211101.
47. Gleyzes, J.; Langlois, D.; Piazza, F.; Vernizzi, F. Exploring gravitational theories beyond Horndeski. *JCAP* **2015**, *2*, 018.
48. Gao, X. Unifying framework for Scalar-tensor theories of gravity. *Phys. Rev. D* **2014**, *90*, 081501.
49. Zumalacárregui, M.; García-Bellido, J. Trasforming gravity: From derivative couplings to matter to second-order scalar-tensor theories beyond the Horndeski Lagrangian. *Phys. Rev. D* **2014**, *89*, 064046.
50. Crisostomi, M.; Hull, M.; Koyama, K.; Tasinato, G. Horndeski: Beyond, or not beyond? *JCAP* **2016**, *3*, 38.
51. Simone, A.D.; Hertzberg, M.P.; Wilczek, F. Running inflation in the standard model. *Phys. Lett. B* **2009**, *678*, 1.
52. Bhattcharyya, S.; Das, K.; Dutta, K. Attractor models in Scalar-Tensor Theories of Inflation. *Int. J. Mod. Phys.* **2018**, *27*, 1850079.
53. Park, S.C. Inflation in the nonminimal theory with 'K(phi)R' term. *AIP Conf. Proc.* **2009**, *1078*, 524.
54. Park, S.C.; Yamaguchi, S. Inflation by non-minimal coupling. *arXiv* **2008**, arXiv:0801.1722.
55. Kleinert, H.; Mustapic, I. Decay rates of metastable states in cubic potential by variational perturbation theory. *Int. J. Mod. Phys. A* **1996**, *11*, 4383.
56. Falconi, M.; Lacomba, E.A.; Vidal, C. The flow of classical mechanical cubic potential systems. *Discont. Cont. Dyn. Syst.* **2004**, *11*, 827.
57. Wartak, M.S.; Krzeminski, S. On tunnelling in the cubic potential. *J. Phys. A: Math. Gen.* **1989**, *22*, L1005.
58. Will, C.M. The confrontation between general relativity and experiment. *Living Rev. Rel.* **2014**, *17*, 4.
59. Quiros, I.; De Arcia, R.; García-Salcedo, R.; Gonzalez, T.; Horta-Rangel, F.A. An issue with the classification of the scalar-tensor theories of gravity. *Int. J. Mod. Phys. D* **2020**, *29*, 7.
60. Bezrukov, F.L.; Shaposhnikov, M. The standard model Higgs boson as the inflaton. *Phys. Lett. B* **2008**, *659*, 703.
61. Bombelli, L.; Koul, R.K.; Kunstatter, G.; Lee, J.; Sorkin, R.D. On energy in 5-dimensional gravity and the mass of the Kaluza-Klein monopole. *Nucl. Phys. B* **1987**, *289*, 735.
62. Sokolowski, L.M.; Golda, Z.A. Instability of Kaluza-Klein cosmology. *Phys. Lett. B* **1987**, *195*, 349.
63. Sokolowski, L.M. Uniqueness of the metric line element in dimensionally reduced theories. *Class. Quant. Grav.* **1989**, *6*, 59.
64. Cho, Y.M. Unified cosmology. *Phys. Rev. D* **1990**, *41*, 2462.
65. Cho, Y.M. Violation of equivalence principle in Brans–Dicke theory. *Class. Quant. Grav.* **1997**, *14*, 2963.
66. Magnano, G.; Sokolowski, L.M. On Physical Equivalence between Nonlinear Gravity Theories. *Phys. Rev. D* **1994**, *50*, 5039 .
67. Faraoni, V.; Gunzig, E.; Nardone, P. Conformal transformations in classical gravitational theories and in cosmology. *Fund. Cosmic Phys.* **1998**, *20*, 121.

68. Faraoni, V.; Gunzig, E.; Nardone, P. Einstein Frame or Jordan Frame? *Int. J. Theory Phys.* **1999**, *38*, 217.
69. Witten, E. Instability of the Kaluza-Klein vacuum. *Nucl. Phys. B* **1982**, *195*, 481.
70. Ford, L.H.; Roman, T.A. Cosmic flashing in four dimensions. *Phys. Rev. D* **1992**, *46*, 1328.
71. Ford, L.H.; Roman, T.A. Averaged energy conditions and quantum inequalities. *Phys. Rev. D* **1995**, *51*, 4277.
72. Cho, Y.M. Reinterpretation of Jordan-Brans–Dicke theory and Kaluza-Klein cosmology. *Phys. Rev. Lett.* **1992**, *68*, 3133.
73. Taylor, T.R.; Veneziano, G. Dilaton couplings at large distances. *Phys. Lett. B* **1988**, *213*, 450.
74. Cvetic, M. Low energy signals from moduli. *Phys. Lett. B* **1989**, *229*, 41.
75. Ellis, J.; Kalara, S.; Olive, K.A.; Wetterich, C. Density-dependent couplings and astrophysical bounds on light scalar particles. *Phys. Lett. B* **1989**, *228*, 264.
76. Damour, T.; Polyakov, A.M. The String Dilaton and a Least Coupling Principle. *Nucl. Phys. B* **1994**, *423*, 532.
77. Damour, T.; Polyakov, A.M. String theory and gravity. *Gen. Rel. Gravit.* **1994**, *26*, 1171.
78. Sk, N.; Sanyal, A.K. On the equivalence between different canonical forms of F(R) theory of gravity. *Int. J. Mod. Phys. D* **2018**, *27*, 1850085.

 *symmetry*

Review

Gravitational Dispersion Forces and Gravity Quantization

Fabrizio Pinto

Ekospace Center and Department of Aerospace Engineering, Faculty of Engineering,
Izmir University of Economics, Teleferik Mahallesi, Sakarya Cd. No:156, 35330 Balçova, İzmir, Turkey;
fabrizio.pinto@ieu.edu.tr

Abstract: The parallel development of the theories of electrodynamical and gravitational dispersion forces reveals important differences. The former arose earlier than the formulation of quantum electrodynamics so that expressions for the unretarded, van der Waals forces were obtained by treating the field as classical. Even after the derivation of quantum electrodynamics, semiclassical considerations continued to play a critical role in the interpretation of the full results, including in the retarded regime. On the other hand, recent predictions about the existence of gravitational dispersion forces were obtained without any consideration that the gravitational field might be fundamentally classical. This is an interesting contrast, as several semiclassical theories of electrodynamical dispersion forces exist although the electromagnetic field is well known to be quantized, whereas no semiclassical theory of gravitational dispersion forces was ever developed although a full quantum theory of gravity is lacking. In the first part of this paper, we explore this evolutionary process from a historical point of view, stressing that the existence of a Casimir effect is insufficient to demonstrate that a field is quantized. In the second part of the paper, we show that the recently published results about gravitational dispersion forces can be obtained without quantizing the gravitational field. This is done first in the unretarded regime by means of Margenau's treatment of multipole dispersion forces, also obtaining mixed potentials. These results are extended to the retarded regime by generalizing to the gravitational field the approach originally proposed by McLachlan. The paper closes with a discussion of experimental challenges and philosophical implications connected to gravitational dispersion forces.

Keywords: dispersion potentials; phenomenology of quantum gravity; novel experimental methods; epistemology; semiclassical methods

Citation: Pinto, F. Gravitational Dispersion Forces and Gravity Quantization. *Symmetry* **2021**, *13*, 40. https://doi.org/10.3390/sym13010040

Received: 30 November 2020
Accepted: 22 December 2020
Published: 29 December 2020

Publisher's Note: MDPI stays neutral with regard to jurisdictional claims in published maps and institutional affiliations.

1. Introduction

The Casimir effect is quite typically described as "an empirically verified quantum mechanical phenomenon involving an attractive force between two parallel uncharged mirrors in vacuum that exists even at zero temperature" [1]. The core of Casimir's original idea [2], which, with some variation of detail [3], he consistently traced to an early conversation with Bohr, is that a non-zero energy exists even in a vacuum. This is because the Uncertainty Principle of quantum mechanics prevents all components of the electric and magnetic fields to vanish identically at the same position in spacetime, so that contributions to the total energy density from electromagnetic fluctuations are expected [4]. Such energy is referred to as the zero-point energy (*Nullpunktsenergie*)—a concept that first emerged from Planck's "second theory" of the blackbody radiation in 1912 [5], then was famously considered in that context by Einstein and Stern [6] and by Bohr himself [7], and was finally extended to the electromagnetic field by Nernst [8] (for analyses of Nernst's work on zero-point energy see in [9–12]). The zero-point-energy exists also if two parallel plane reflectors separated by a gap of width s are introduced. However, the discontinuities at the boundaries cause the energy density to change as a function of the gap width. A relatively simple calculation quickly shows that this leads to an attractive force, more correctly a pressure, proportional to $-\hbar c/s^4$, between the two plates [13]), where $\hbar$ and c are Planck's

287

constant and the speed of light in vacuum, respectively. This is now referred to as the Casimir force.

As quantum mechanics plays a key role in these interpretations, scientists and philosophers alike are presented with experimental confirmation of the Casimir force as evidence of the quantum nature of the associated field, in this case, the electromagnetic field. Such a point of view extends, more in general, to all electrodynamical dispersion forces, that is, forces due to the frequency-dependent electromagnetic polarizability of the interacting systems, whether they be macroscopic or microscopic. Despite the above matter-of-fact description [1] quite representative of the literature in the field, a robust debate regarding the logical relationship between electrodynamical Casimir forces and the principles of quantum mechanics has long been part of the research landscape.

In recent years, the subject of dispersion forces between *gravitationally* polarizable objects driven by spacetime fluctuations has attracted steadily increasing attention. Despite close field theory methodological analogies, however, the developmental trajectory of this research subfield has notably differed from that dealing with electrodynamical dispersion forces. Such a different path has unfolded in the nearly total absence of a parallel debate, thus leading to unquestioned claims that the detection of gravitational dispersion forces would conclusively prove the quantum nature of gravitation, claims we wish to discuss in this paper.

The plan for our analysis is as follows. In the following section, we review from a historical perspective the vigorous debate about the relationship between quantum atomic theory, quantum electrodynamics, and the existence of dispersion forces in both the unretarded and retarded regimes. In particular, we recall heuristic approaches and semiclassical theories leading to results indistinguishable from those obtained from the full quantum electrodynamical treatment. As an illustration that these descriptions are far from fruitless speculation, we mention and reference applications demonstrated to engineer dispersion forces in classical fields for specific technological purposes. In the following section (Section 3), we consider, again historically, developments leading to proposals about the existence of dispersion forces due to the gravitational field. The emphasis is on highlighting the fact that consideration of possible gravitational dispersion forces has occurred in the complete absence of a debate as to whether gravitational field quantization is necessary as a prerequisite for the existence of such forces, unlike what occurred in the electrodynamical case.

Readers not interested in these arguments may want to skip directly to our computational illustrations that gravitational dispersion forces can exist in the presence of classical gravitation with quantized atoms. As a first step (Section 4), we recall London's treatment of dipole's electrodynamical forces (Section 4.1) and Margenau's extension of such a formulation to multipole dispersion forces in the unretarded regime (Section 4.2). Those expressions are then reformulated in the classical gravitational case and it is shown that the recently published results obtained in linearized quantum gravity are immediately recovered (Section 5). New relationships, first suggested by Spruch on dimensional grounds, are also explicitly derived for the mixed quadrupole–quadrupole and dipole–quadrupole electric gravitational potentials, which are tens of orders of magnitude larger than purely gravitational potentials (Section 8). The conclusions from this initial derivation are further strengthened by extending McLachlan's theory of dispersion forces to the gravitational case. For his purpose, we first recall McLachlan's theory (Section 6), in which the electromagnetic field is not quantized, and we show that it leads to the van der Waals and Casimir–Polder expressions in the unretarded and retarded regimes, respectively (Section 6.1). Then, restricting our treatment to the 1D case for succinctness, we consider the extension of McLachlan's approach to electric quadrupole dispersion forces (Section 6.2). The formulation of the same theory in the gravitational case is straightforward, as sketched out in Section 6.3.

We repeatedly stress throughout the paper that our efforts are *not* intended to support speculation that the electromagnetic field is not quantized. Instead, our starting point is that the conclusion that the electromagnetic field must be quantized could not be drawn based *solely* on the detection of dispersion forces. In fact, the existence of dispersion forces can be accommodated by quantizing the internal degrees of freedom of atoms interacting via classical electromagnetic fields. Therefore, the aim of this paper is to show that, analogously, any successful detection of gravitational dispersion forces could *not* be used to conclude that the gravitational field must be quantized. Quantized atoms interacting through a classical (general relativistic) gravitational field exhibit gravitational dispersion forces indistinguishable from those derived from linearized quantum gravity.

The challenging goal of detecting the gravitational equivalent of electrodynamical van der Waals, Casimir, and Casimir–Polder forces in the laboratory has been the subject of a few proposals, which are summarized in Section 7. The paper closes with a brief mention of connections between the main topic and two related subjects. On the one hand, the role played by dispersion forces within the philosophical idea of atomism is recalled, particularly as it regards the meaning of "void" in the presence of fluctuations of spacetime. On the other hand, some comments are provided about possible future technological applications of gravitational dispersion forces.

2. The Quantum Structure of the Atom, Quantum Electrodynamics, and Dispersion Forces

Historically, the first treatment of van der Waals forces between two hydrogen-like atoms was attempted by Wang [14] and later correctly completed by Eisenschitz and London [15]—indeed on the basis of the then newly developed principles of non-relativistic quantum mechanics. Such early studies were carried out contemporaneously with, but independently of, efforts to assemble the framework needed to describe the electromagnetic field in quantum mechanical language, the theory now referred to as quantum electrodynamics (QED) [16]. As lucidly clarified by London [17], who introduced the term "dispersion" in this context, the van der Waals force is explained in non-relativistic quantum theory as due to the existence of the zero-point-energy of perturbed harmonic oscillators, which schematically represent the interacting atoms. On the other hand, the electromagnetic field is assumed to be classical. Further developments, such as the calculation of atomic multipole effects, were built upon this same non-relativistic approach [18,19].

Within just a few years, experimental data on lyophobic colloids [20] clearly demonstrated that the decay of van der Waals potentials at large particle separations is more rapid than predicted by the London treatment. Overbeek provided a suggestive, albeit only intuitive, explanation based on the effect of the finiteness of the speed of light, which, at large interparticle distances, would cause interacting classical resonators to vibrate out-of-phase thus degrading their mutual attraction [21,22]. This concept brought to the fore the critical importance of retardation and led Casimir and Polder to attacking the problem of interatomic forces by means of the early tools of QED [23,24]. As recalled by Casimir, he followed up on this result by considering the attraction of two perfectly conducting neutral plates separated by a gap in terms of zero-point-energy of the electromagnetic field, so that "The problem in quantum electrodynamics is then reduced to a problem in classical electrodynamics" [25]. This approach, which Casimir humbly but entertainingly referred to as "Poor Man's QED" [26,27], led to his important discovery [2].

One important attribute of the "two parallel uncharged mirrors" to complete the above initial definition of the Casimir effect is that they are assumed to be *perfect* reflectors—clearly an extreme idealization. The strategy available to deal with intermolecular forces in real materials had been highlighted at least as early as in Einstein's very first published paper, in which he had been "...guided by the analogy with gravitational forces" ("Ich liess mich dabei von der Analogie der Gravitationskräfte leiten" [28]. Obviously, Einstein, writing even before the special relativity theory, is referring to Newtonian gravitation, which he would finally revise with his general relativity theory, in which gravitation is not trivially additive) [29]. Therefore, the van der Waals force between one atom and a slab or between two slabs was calculated by augmenting the theory by the ad hoc assumption of dispersion force additivity and by carrying out pairwise force summations in the continuum approximation [20,30–32]. However, it became quickly apparent that, unlike the case of Newtonian gravitation, additivity must actually be considered with suspicion. For instance Axilrod and Teller intriguingly proved that each of three isolated atoms interacting via the unretarded van der Waals–London force not only do not experience forces trivially given by the pairwise sum, but can even mutually *repel* simply depending on their specific geometric arrangement [33].

Both the problem of retardation and of additivity in real materials, regardless of scale or distance and including absorption, were eventually addressed by Lifshitz through the introduction of a random electromagnetic field into the Maxwell equations [34,35] Although the status of the Lifshitz theory in relationship to the fundamental principles of quantum mechanics has raised long-standing questions, as shown in the references above, the quantitative success of the resulting equations for the dispersion force between polarizable objects in explaining experimental data from a widespread variety of systems has been remarkable [36–39].

It is crucial to what follows that London's calculation of the near-range van der Waals force assumes a *true* vacuum, that is, "…a state with all physical properties equal to zero" [40], within which to consider the classical electrostatic interaction of two instantaneous dipoles described by non-relativistic quantum mechanics. On the other hand, the treatment of the long-range potential presented by Casimir and Polder is explicitly rooted in the concept of the zero-point-energy of the quantum vacuum [41]. This seems to imply, as is often stated, that the existence of retarded interactions cannot be explained outside the framework of quantum field theory. However, this is not the case. Indeed, somewhat in the path Overbeek had envisioned, theories leading to the correct retarded potentials have been developed "in spite of having used the nonretarded Schrödinger equation" [42] to treat the interacting atoms but retaining the electromagnetic field as classical. As the quantum nature of the electromagnetic field appears beyond doubt *on independent considerations*, such semiclassical endeavors are at times presented self-deprecatingly, in a manner mindful of, or perhaps in order to pre-empty, Osiander's opinion that, for the sole purpose of saving the phenomena, "…these hypotheses need not be true nor even probable …" [43] ("Neque enim necesse est, eas hypotheses esse veras, imo ne verisimiles quidem …" [44]). For instance, in a short pedagogical article aimed at illustrating the basic idea of the Casimir effect, Kleppner instead discusses, "delightfully" [45], a one-dimensional version [46] of London's unretarded model of interacting harmonic oscillators [17]. The obvious extension of that approach to include full retardation—the actual Casimir effect—is disposed of as to "flog the argument beyond the point of diminishing returns" [47].

One recurring apologetic argument for all such efforts is that pedagogy can take advantage of "…the considerable simplification in the mathematical aspects of the framework over the approach based on quantum electrodynamics" [48]. A slightly more ambitious perspective holds that "The various interpretations of the dispersion effect are aimed at minimizing the quantum theoretical effort" but "It is one of the objectives of the quantum electrodynamics procedure to check the errors and limits of these semiclassical approaches" [42].

Along these lines, additional motivation was provided by Casimir and Polder themselves in their early speculation that "…it might be possible to derive these expressions, perhaps apart from the numerical factors, by more elementary considerations. This would be desirable since it would also give a more physical background to our result, a result which in our opinion is rather remarkable" [24]. This goal was adopted by McLachlan, who explained: "In one sense the theory of molecular attractions is now complete, but there is still a need for an elementary discussion which gives better physical insight. The aim of this paper is to find the dispersion force between two molecules by a new method which uses elementary quantum mechanics and assumes almost no knowledge of quantum field theory" [49]. Indeed, Margenau, commenting in the aftermath of such work, stated that

"The approach just considered, for which we are indebted to McLachlan, is essentially classical, the only quantum mechanics occurring in the definition of the polarizability" [50].

At the far end of the spectrum, one finds efforts not just aimed at recovering known results with a less onerous mathematical apparatus for reasons of succinctness or insight but claiming to do so, regardless of complexity, in order to show that the nature of the electromagnetic field is not fundamentally quantum mechanical. In this view, referred to as "stochastic" or "random" electrodynamics (SED), there actually exists a fluctuating zero-point electromagnetic field but such an object is completely classical and it appears as a solution of the homogeneous wave equation for the potential vector [11]. As pointed out by Marshall [51,52] and Boyer [53], the theory is Lorentz-invariant, and it has been shown not only to recover the major expressions from "standard" dispersion force theory but also—stimulated by a question from none other than Casimir himself—to lead to the earliest prediction of a repulsive Casimir force between a perfectly electrically conducting and a perfectly magnetically permeable slab [54] (see also in [55]).

The fact that this semiclassical approch and "standard" QED calculations lead to the same expressions was stressed by Spruch, who observed: "Why vacuum fluctuation arguments worked in the past in the problems to which they were applied is, to our knowledge, not completely understood, but the simplicity of the approach gives it considerable appeal, as a means of providing physical insight into known results and as a means of suggesting new results" [56]. Indeed, Spruch adds that "it is not necessary to go to the trouble for almost all of the work has been done, and we simply note down an extension of a version given by Boyer". Interestingly, the result used therein had been obtained by Boyer by pointing out that "we may regard the fields equally well as classical fields subject to a random walk or as quantum fields" [53]. Several years later, Spruch presented these arguments in a remarkable pedagogical article written for *Physics Today* to illustrate that "…the essential idea is to proceed entirely classically…until that last step …in which we quantize the energy of the modes of vibration of the electromagnetic field" [57]. This approach yields a completely classical expression for the dispersion energy for any arbitrary specific energy density, and to the intriguing comment that such is "…a result Maxwell could have derived, and perhaps did". (The reasons that this "counterfactual history" statement cannot be considered realistic were explored in [22]) Finally, Spruch states that "[T]he quantum version follows immediately on making $\mathbf{E}_0$ the field of vacuum fluctuations …in that case the energy in a mode of the field is simply … $\frac{1}{2}\hbar\omega$ …". We must notice, however, that whereas that "last step" is attributed the actual meaning of field quantization by Spruch, within the SED framework the fields remain classical and Planck's constant "…$\hbar$ is regarded as nothing more than a number chosen to obtain consistency of the predictions of the theory with experiment" [11].

Predictably, a classical electromagnetic zero-point field is perceived as a highly controversial proposal [3]—indeed Boyer noted that "some readers …are distressed, even indignant at the idea …" [58]—and Milonni concluded that "In spite of the successes of SED, it cannot at this time be considered a serious alternative to QED" [11]. Such peremptory rejection, supported by the argument that, for instance, "no *classical* theory of the electromagnetic field can account for such experimentally observed phenomena as the photon polarization correlations in a cascade radiative decay of atomic states" [11], has, however, not resulted in the disappearance of the SED approach. From the theoretical standpoint, this is continues to be presented as being due, in a typical literature example, to "*computational* advantages" [59], couched in Osiander's language of profuse apology despite excellent agreement with experimental data, confessing that the purpose "…is not to propound some rival to QED" [59].

Even more importantly, continued pursuit of the SED standpoint eventually inspired novel experimental physics and engineering applications of the Casimir effect. One development was motivated by a description of dispersion forces in terms of the net radiation pressure of all modes of the zero-point-field first suggested by Casimir in his landmark publication [2], quoted verbatim much later by Debye [9] (see in [60], Note 5), and fully explored by Gonzáles [61] and, finally, by Milonni [11,62]. Despite some commentary critical of this famously lucid and oft-cited interpretation, delivered by Barton in a section tellingly entitled "Exorcism" [63], the prediction of the existence of forces between macroscopic boundaries does not appear to demand a quantum origin of the field, which can thus be classical. Therefore, both Boyer's SED papers and Milonni's more recent radiation pressure formulation are consistently cited as cornerstones in theses and articles leading to the successful detection of the *acoustic* Casimir effect [64–66], that is, "the force between two plates in a homogeneous, isotropic, acoustic field" [67]. The development of this research subfield led to a deeper understanding of the connection between the detailed features of the random vibration spectrum with the Casimir force as a function of the gap width between the two plates. Whereas, as we stated above, in the electromagnetic case such spectrum is uniquely determined by the requirements of Lorentz invariance, in the acoustic case it can be arbitrarily set by the experimenter.

The outcome of these studies was the mathematical proof that a narrow band random noise spectrum may result in large oscillations of the force with distance, later independently re-discovered by Ford in the electromagnetic case [68], which finally led to the first experimental detection of *repulsive* acoustic Casimir forces [66,67]. The remarkable possibility to tailor such interactions in intensity and *sign* suggested that engineering the noise spectrum might lead to novel applications in micromanipulation and particle levitation. More recently, this interesting result has led to the theoretical suggestion [69] that high-frequency noise fields might represent a tool to control *stiction*, the phenomenon responsible for non-linearities and a failure mode in microelectromechanical systems (MEMS) [70]. The Casimir effect with classical random fields—referred to by some as the "analog" Casimir effect— was further demonstrated by measuring the force between two vertically plates partially submerged in a dish containing a liquid (This interesting experiment has been presented as indirect support of the veracity of historical reports of a "maritime" Casimir effect, that is, the force attraction of two side-by-side ships in a rough sea [71]. Despite the correctness of the experiments reported in [72], the present author has shown [73] that there exists no basis to believe that the drawing shown in [71] suggests that "It was believed in the days of the clipper ships that . . . two vessels at close distance [in a strong swell] will attract each other". In fact, as also remarked by a reader in a letter to the Editor [74], the traditional belief was the *opposite*, that is, that ships in a completely flat sea attract. However, the captions of the two figures in the historical source, one showing rough seas and other one flat calm, were inexplicably swapped by the author of [71]) excited by a shake table [72].

As a final example of productivity of the SED point of view, again represented by citation of Boyer's papers as the logical starting point, we mention the very recent experiments aimed at engineering dispersion forces by means of arbitrary fluctuating *electromagnetic* fields in optical nanoparticle manipulation applications [75]. This remarkable work has generated renewed impetus in long standing efforts to shape dispersion forces by radiation [76] resulting in applications in bacterial screening [77] and in further progress in "gravitational-like" interactions [78–80] and "mock gravity" [81].

We conclude these introductory remarks by quoting leading research protagonists implicitly or explicitly exposing the need to accept the limitations presented by using forces between polarizable particles as a tool to ascertain the ultimate nature of the fields. For instance, the discoverers of "optical binding" offered the following commentary applicable to the issue at hand.

> "Finally we note that many readers of this journal view forces between elementary particles of nature as originating from the exchange of virtual quanta of fields to which they are coupled. The induced interaction discussed in this paper fits nicely into that scheme, but with real quanta being exchanged. We wonder whether other particles and fields may be substituted for our dipoles and light, yielding analogous effects in other domains of physics" [82].

An obvious illustration of this argument is the view of the Casimir effect as due to radiation pressure of a zero-point-field photons gas—an interpretation that cannot differentiate between real or virtual photons. Speaking more broadly, DeWitt has remarked that "There must be nearly two dozen ways of calculating the Casimir effect" [83]. Indeed, approaches as diverse as QED and SED lead to the same expressions for dispersion forces in various regimes, thus suggesting that the existence of dispersion forces is *not* a valid tool to determine whether a field is quantized. This was explicitly noted by the researcher who designed the first modern experiment to accurately test the predictions of the Lifshitz theory [84], S. K. Lamoreaux, who stated: "There are physical phenomena that truly require a quantization of the electromagnetic field for their explanation; the Casimir force is not among these phenomena, because the predictions based on different points of view are identical" [85].

As is tradition, the present author also proffers his own apology of quantum electro-dynamics, which has in fact never been falsified and is consistently hailed, with reason, as possibly the most successful physical theory ever devised by the human mind [86]. However, special care must be used as we attempt to employ dispersion forces to draw conclusions about the nature of "other particles and fields". In this sense, it is appropriate to extensively quote some words by Richard Feynman in his Nobel Prize acceptance speech [87]:

> "Physical reasoning does help some people to generate suggestions as to how the unknown may be related to the known. Theories of the known, which are described by different physical ideas may be equivalent in all their predictions and are hence scientifically indistinguishable. However, they are not psychologically identical when trying to move from that base into the unknown. For different views suggest different kinds of modifications which might be made and hence are not equivalent in the hypotheses one generates from them in ones attempt to understand what is not yet understood. I, therefore, think that a good theoretical physicist today might find it useful to have a wide range of physical viewpoints and mathematical expressions of the same theory (for example, of quantum electrodynamics) ..."

It may appear surprising to see an extraordinarily objective physicist as Feynman ponder about such concepts as "psychologically identical" ideas—a criticism he seems to pre-empty at the onset by stating that he just wants "to make the lecture more entertaining". However, considering that Julian Schwinger, seated in that same audience that day, would later refer to the Casimir effect as "One of the least intuitive consequences of quantum electrodynamics" [40], Feynman's warning must be taken very seriously as we move from the known, QED, to the unknown, the ultimate structure of spacetime.

3. Gravitational Dispersion Forces

It is a puzzling fact in the history of this subfield that no treatment of gravitational dispersion forces was, to the best knowledge of this author, ever published till very recently although, for instance, the non-relativistic regime does not require a treatment remarkably more complex than that developed by London. This silence was not due to a lack of sophisticated theoretical tools. The calculation of classical (i.e., non-quantum) corrections to the gravitational force upon a charge distribution in various gravitational fields was pursued intermittently, often without awareness of past results [88], as the first attempt by Enrico Fermi to study electrostatics within the then-novel theory of general relativity while at at the Scuola Normale Superiore at Pisa [89].

Even as QED was being developed, the problem of quantizing linearized gravity had also emerged [90–92], most prominently, in the pioneering works of Rosenfeld [93] and Bronstein [94]. The results by the latter are remarkable from our perspective even in the present day because, as pointed out by the editors [95] of his recently reprinted work [96], Bronstein even addressed the presence of a gravitational zero-point-energy and implemented the proper operator ordering [11] needed to avoid it (This now famous paper was Bronstein's PhD dissertation, presented in 1935. He was arrested in August 1937 during the Great Purge, then tried, sentenced, and shot on the same day in a Leningrad prison on 18 February 1938 [91,97]). His results, achieved at a remarkably early stage in the exploration of the challenges of a theory of quantum gravity and "Published in German in a journal inaccessible today" [95], remained relatively unknown. Indeed, as late as 1992, Gorelik still found the need to openly contradict Steven Weinberg's taking "liberties with history" [90] for rashly awarding to Planck [98] priority in identifying the "inconsistency between quantum mechanics and general relativity", an achievement which Gorelik attributed to Bronstein.

In the several decades that followed, corrections to the Coulomb and post-Newtonian potentials, as well as mixed potentials, were derived both within classical general relativity and quantum field theory frameworks [99–104] (the contemporary state-of-the-art is reviewed in [105–107]). The relatively late development of the gravitational potential corrections was explicitly acknowledged by Feinberg, Sucher, and Au as late as 1989: "Two-graviton exchange has also been studied in some approximations ... but, a general expression for the potential arising from this exchange, similar to that ... for two-photon exchange between charges has, to our knowledge, not been derived" [108]. Specifically regarding corrections to the Coulomb potential, the same authors observe "It is interesting to note that the r^{-2} term is independent of Planck's constant. Indeed, this term can be obtained in a purely classical treatment of electrodynamics ... This is not the case for the r^{-3} term". It is a further demonstration of the tortuous history of the subject that the *classical* r^{-2} term had already been exhibited by Berends and Gastmans in 1976 [100] but was recognized by them, "in proof", as having being published by Cécile and Bryce DeWitt as early as 1964 [99].

As far as the *quantum* corrections to the gravitational as well as the mixed potentials ($\propto r^{-3}$), remarkably early comments are found in the aforementioned paper by Spruch [57] and reiterated in a later review chapter on Casimir forces, also of a pedagogical nature [109]. For instance, after having obtained the order of magnitude of the correction to the Coulomb potential, $V_{ee} \sim \hbar e^4/c^3 m^2 r^3$, Spruch instructs the reader: "To obtain the gravitational analog of the electron–electron interaction, replace both factors of e^2 by Gm^2; to obtain the gravitational-electromagnetic interference term in the interaction, replace only one factor of e^2 with Gm^2". As noted later, as this mixed contribution is linear in G, it is "much larger and more interesting" (see in [57], Equation (6), box on p. 42, and p. 43; see also in [109], p. 28). However, having come so close to the subject of dispersion forces in gravitationally polarizable systems, Spruch walks away from it with the justification that "... there is no gravitationally neutral system and therefore no gravitational analog of the electrically neutral atom ...".

An element crucial to the focus of the present paper was added to the debate by J. P. Dowling in a letter written in response to Spruch's article in *Physics Today* [110]. Dowling reminded readers that, as we repeatedly mentioned above, there exist other points of view than explaining the existence of retarded dispersion forces by coupling of the interacting systems with the electromagnetic zero-point fluctuations, such as the Lifshitz theory [34] and the source theory by Schwinger, DeRaad, and Milton [40]. To these, he added yet another framework, that is, the "self-field approach" in which no second quantization is carried out, as proposed by Barut [111,112] and pursued by him and his collaborators, including J. F. Van Huele [113] and Dowling himself [114,115]. In the apologetic tone typical of authors straying from the standard theory, Barut and Dowling responded by citing the facts: "If a semiclassical theory is defined as a theory which is not second quantized, then self-field QED has been quite a successful semiclassical theory (at least to order α) in accounting for quite an array of phenomena thought to require at least the second quantization of the radiation field for their explanation" [115]. Among such phenomena, were also the "long-range Casimir–Polder van der Waals forces near boundaries" [114]. The obvious consequence of such observations is, once again, that if the existence of electrodynamical dispersion forces does not demand second quantization but can be correctly described within a semiclassical theory, neither does the existence of gravitational dispersion forces.

In the last forty years, the subject of dispersion forces involving the gravitationally field has been slowly, and rather haphazardly, rediscovered. Among earliest studies in this phase were calculations of the gravitational Casimir energy in very specific contexts of cosmology and field theory [116–118]. As late as 1993, Spruch had dismissed the problem stating: "We have been concerned almost exclusively with electromagnetic effects but it is amusing to consider gravitational effects, and it is trivial to do so" [109]. Apparently unaware of Spruch's assessment, Panella and Widom wrote, shortly thereafter, that "...to the best of our knowledge, the study of long range gravitational interactions between massive bodies (e.g., calculation of gravitational retarded static potentials) has not yet been undertaken". In their article, these authors presented possibly the earliest detailed calculation of gravitational dispersion forces—"a new finite effect of linearized quantum gravity namely, the retarded (Casimir) potential of a test mass interacting with a condensed matter system" [119]. Disappointingly, their result—the retarded gravitational potential between a point mass and an extended mass distribution found by an approach that parallels that used to analyze the analogous electrostatic problem [120]—remained virtually uncited till 2017.

Approximately two decades after Panella and Widom, the problem reemerged from two independent directions. On the macroscopic scale, the original motivation was provided by theoretical speculation—tempered by extreme skepticism in the absence of any experimental confirmation—that superconductors might act as nearly-ideal high frequency gravitational wave reflectors, thus opening the door to a new field of optics [121–124]. Following a suggestion by Bouwmeester, reported by Minter et al. [125], the possibility of gravitational wave reflection led to a treatment of the gravitational Casimir effect (Importantly, Casimir's approach was later explicitly mentioned by Sakharov (reprinted in [126]) in his theory of "induced gravity" [127]. Although effects connected to fluctuations in curved spacetime are sometimes referred to as a "gravitational Casimir effect" [128], in this paper, we consider gravitation as a fundamental, and not an emergent, interaction) analogous to that of the Lifshitz theory in QED [129,130]. Very shortly thereafter, and without any reference to the results by Quach [129,130] in the macroscopic regime, the computation of the gravitational van der Waals and Casimir–Polder potentials appeared, carried out by means of quantum field theoretical methods by Ford, Hertzberg, and Karouby [131], confirmed a few months later by Wu, Hu, and Yu [132]; Hu and Yu [133]; and Holstein [134].

On the one hand, the macroscopic scale results [129,130] have stimulated a justifiably lively debate as to whether a hypothetical detection of such a gravitational Casimir effect would in fact represent a conclusive confirmation of the quantum nature of spacetime [22,60,135–138]. On the other hand, the microscopic scale calculations have been accompanied by a unanimous consensus on the part of those who carried them out that any experimental confirmation of those predictions would reveal a signature of the quantum nature of the gravitational field. Consider the following examples of matter-of-fact commentary provided regarding the first black hole merger detection, announced within the same time frame as the above results [139], and gravitational dispersion forces. Holstein forcefully stated that "The recent observation at LIGO of gravitational radiation has verified the existence of gravitons and has emphasized the importance of studying processes involving their interactions" [134]. Ford and collaborators sweepingly concluded that their own work "...is a rare, precise, and definite prediction of quantum gravity, independent of the details of its UV completion ...(analogous to the electromagnetic Casimir–Polder and London–van der Waals forces) ..." [131]. Along these lines, Wu, Hu, and Yu introduced their calculations stating that "Although a full theory of quantum gravity is absent, one can still use linearized quantum gravity to find quantum gravitational corrections to classical physics which an ultimate quantum gravity theory must produce at low energies" [132]. Finally, Hu and Yu commented that "One naturally expects that, if gravity has a quantum nature, it should also generate Casimir-like forces" [133].

As must appear obvious in light of the facts we recalled so far, all such strong claims deserve careful analysis to avoid logical pitfalls. For instance, as we have repeatedly seen,

Casimir forces are also predicted by theories without second quantization and even with entirely classical fields [140,141]. Therefore, a non-quantum gravitational field should still very much be expected to generate Casimir forces. Furthermore, in electrodynamics one would find it obviously untenable to state that "the observation of light verifies the existence of photons". Indeed, Dev and Mazumdar voiced the opinion that "The LIGO detection as such does not confirm whether the observed gravitational wave is classical or quantum" [142].

From the methodological standpoint, as the present author has previously discussed [60], the adoption by Ross and Moreau of the SED mathematical tools developed by Boyer to carry out an analogous study of "stochastic gravity" [143] is neither less rigorous nor less promising than far more popular linear quantum gravitational methods. For instance, in addition to providing a useful framework to treat gravitation approaching the Planck scale, the possibility must be examined that contamination of gravitational dispersion force measurements [60] might occur from an as yet undetected classical gravitational stochastic field background of cosmological and astrophysical origins [144–146].

In summary, the historical evidence has shown that the developmental trajectory of the quantum gravity research subfield intriguingly differed from that of QED. Whereas at least half-a-century of extensive analysis in the electrodynamical domain established that "...the Casimir effect reveals nothing conclusive about the nature of the vacuum" [147], claims that "If gravity truly has a quantum nature, then gravitational waves should also generate Casimir-like forces" [135] were left largely unchallenged.

In the remainder of this paper, we discuss some examples of calculation of gravitational dispersion forces carried out assuming that the gravitational field is *classical*.

4. Unretarded Higher Multipole Electrodynamical van der Waals Forces

The computational strategy adopted below is part of the ongoing program by this author to bring computer algebra system (CAS) technology to bear in attacking prohibitively complex Casimir force problems [22,148]. This approach is particularly effective when generalizing computations to systems in which simplifying symmetries may be lost, as in the case of electrodynamical dispersion forces in the presence of arbitrary gravitational fields [88], to avoid errors due to very extensive algebraic calculations or to identify errors in the published literature, as well as for pedagogical reasons. In order to demonstrate the algorithm, first we verify several well known expressions for the generalized electric multipole van der Waals potential by means of the *Mathematica*™ system (v. 11.3.0.0). Second, we proceed by adapting this approach to compute the gravitational van der Waals potential expressions. Finally, we report on the explicit calculation of the mixed electric-gravitational potentials in the unretarded regime.

4.1. Margenau's Algorithm: London Potential

The interatomic London potential [15,17] will be calculated by the procedure developed by Margenau [18], who first computed the quadrupole contributions to the van der Waals interatomic forces on Frenkel's suggestion. This is summarized here for convenience. Let us consider the Coulomb potential (Gaussian units), $V(\mathbf{r}) = e/|\mathbf{r} - \mathbf{r}_0|$, at $\mathbf{r}$ (components x^i) due to a proton of charge $e = +|e|$ placed at $\mathbf{r}_0$ (components x_0^i). The total potential V_{pe} at $\mathbf{r}$ due to this proton and to its atomic electron of charge $-e$, placed at a position $\mathbf{r}_1$ away from the proton is, to second order in the components $x_1^i = x^i - x_0^i$ (index summation convention is used) and eventually setting $\mathbf{r}_0 = \mathbf{0}$:

$$V_{\text{pe}}(\mathbf{r}) = \frac{e}{r} - \frac{e}{r} + e\left(x_1^i \frac{\partial}{\partial x^i}\right)\left(\frac{1}{r}\right) + \dots , \tag{1}$$

where $\partial/\partial x = -\partial/\partial x_0$. Let us now place at second atom with proton at $\mathbf{r}$ and electron at $\mathbf{r}_2$ from the proton. The classical dipole–dipole potential energy of this atom in the potential $V_{\text{pe}}(\mathbf{r})$ is, also to first order in x_2^k:

$$W_{dd}(\mathbf{r_1}, \mathbf{r_2}, \mathbf{r}) = eV_{pe}(\mathbf{r}) - eV_{pe}(\mathbf{r}) - e\left(x_2^k \frac{\partial}{\partial x^k}\right)V_{pe}(\mathbf{r}) + \cdots = -e^2\left(x_1^i x_2^k \frac{\partial}{\partial x^i}\frac{\partial}{\partial x^k}\right)\left(\frac{1}{r}\right)\cdots . \tag{2}$$

By implementing the above steps in *Mathematica*, and by choosing $\mathbf{r} = (R, 0, 0)$, we find

$$W_{dd}(\mathbf{r_1}, \mathbf{r_2}, R) = -\frac{e^2}{R^3}(2x_1^1 x_2^1 - x_1^2 x_2^2 - x_1^3 x_2^3), \tag{3}$$

which is the classical perturbing field used by London. As is well known [18,149], the calculation next proceeds to obtain a good approximation of the expression for the van der Waals force by evaluating the following six-dimensional integral,

$$U_{vdW}(R) \simeq -\frac{1}{2E_I} \int_{V_1} \int_{V_2} \phi_{1,0,0}^*(\mathbf{r_1})\phi_{1,0,0}^*(\mathbf{r_2})\, W_{dd}^2(\mathbf{r_1}, \mathbf{r_2}, R)\, \phi_{1,0,0}(\mathbf{r_1})\phi_{1,0,0}(\mathbf{r_2})\, d\mathbf{r_1}\, d\mathbf{r_2}, \tag{4}$$

where E_I is the atomic ionization energy $E_I = e^2/2a_0$ and $a_0 = \hbar^2/m_e e^2$ is the Bohr radius and the normalized ground state wave function $\phi_{1,0,0}$ is:

$$\phi_{1,0,0}(\mathbf{r}) = \frac{e^{-r/a_0}}{\sqrt{a_0^3 \pi}}. \tag{5}$$

The elementary process consists of considering the symmetry properties of the integrals of the 6 terms produced by squaring the classical energy at Equation (3) and recognizing that some vanish while all others are identical to one another. Here, instead, we proceed by brute force as all such integrals are elementary and can be computed by *Mathematica* in ~ 1 s of CPU time on a typical laptop. With the above approximations, one finds the standard expressions:

$$U_{vdW}(R) = -\frac{3a_0^4 e^4}{E_I R^6} = -\frac{6a_0^5 e^2}{R^6}. \tag{6}$$

This quantity can be expressed in terms of $< 1, 0, 0 | r^2 | 1, 0, 0 > = r_{1,0,0}^2 = 3a_0^2$, so that we can finally write:

$$U_{vdW}(R) = -\frac{e^4 (r_{1,0,0}^2)^2}{3E_I R^6}, \tag{7}$$

which is the expression given by Margenau (see in [18], Equation (8) and in [50], p. 24). Further expressions of this result useful for comparison with the literature can be obtained by writing the square of the norm of the ground state dipole expectation value from the result above as $||\mu_{1,0,0}||^2 = 3e^2 a_0^2$, so that

$$U_{vdW}(R) = -\frac{(||\mu_{1,0,0}||^2)(||\mu_{1,0,0}||^2)}{3E_I R^6}, \tag{8}$$

which agrees with the complete expression at Equation (1.1) of the work in [150]:

$$U_{vdW}(R) = -\frac{2}{3R^6} \sum_{r,s} \frac{||\mu_{0,r}(A)||^2 ||\mu_{0,s}(B)||^2}{E_{r,0} + E_{s,0}}, \tag{9}$$

in the approximation in which one only retains the ground state contribution of two identical atoms, A and B. Finally, let us write the electric dipole polarizability from the Kramers–Heisenberg formula [11,150,151]:

$$\alpha_E^{(1)}(\omega) = \frac{2}{3} \sum_r \frac{E_{r,0}||\mu_{0,r}||^2}{E_{r,0}^2 - (\hbar\omega)^2}. \tag{10}$$

In the static limit for $\omega \to 0$, again only considering the contribution of the ground state and by using the results above, we find the classical static dipolar polarizability, $\alpha_{0,E}^{(1)} = 4a_0^3$ [11,152], and we obtain another standard expression:

$$U_{\text{vdW}}(R) = -\frac{3E_I(\alpha_{0,E}^{(1)})^2}{4R^6}. \tag{11}$$

4.2. Margenau's Algorithm: Quadrupole van der Waals Forces

The next objective to validate our approach is to generalize the above procedure to higher multipoles by first expanding $V_{\text{pe}}(\mathbf{r})$ to second order at Equation (1):

$$V_{\text{pe}}(\mathbf{r}) = e\left(x_1^i \frac{\partial}{\partial x^i} - \frac{1}{2}x_1^i x_1^j \frac{\partial}{\partial x^i}\frac{\partial}{\partial x^j}\right)\left(\frac{1}{r}\right) + \dots. \tag{12}$$

Therefore, the electrostatic potential energy of the two atoms is

$$
\begin{aligned}
W_{\text{dd}}(\mathbf{r_1}, \mathbf{r_2}, \mathbf{r}) &= -e\left(x_2^k \frac{\partial}{\partial x^k} + \frac{1}{2}x_2^k x_2^l \frac{\partial}{\partial x^k}\frac{\partial}{\partial x^l}\right) V_{\text{pe}}(\mathbf{r}) + \dots \\
&= -e^2\left(x_2^k \frac{\partial}{\partial x^k} + \frac{1}{2}x_2^k x_2^l \frac{\partial}{\partial x^k}\frac{\partial}{\partial x^l}\right)\left(x_1^i \frac{\partial}{\partial x^i} - \frac{1}{2}x_1^i x_1^j \frac{\partial}{\partial x^i}\frac{\partial}{\partial x^j}\right)\left(\frac{1}{r}\right).
\end{aligned}
\tag{13}
$$

By implementing these expansions in *Mathematica* and again choosing $\mathbf{r} = (R, 0, 0)$, we obtain the sum of three polynomials proportional to $1/R^3$, $1/R^4$, and $1/R^5$, where the first of course recovers the dipole–dipole potential energy seen above. Squaring this quantity to obtain $W_{\text{dd}}(\mathbf{r_1}, \mathbf{r_2}, R)$ leads to a polynomial that *Mathematica* measures as having 229 terms. The 6-dimensional integration shown at Equation (4), carried out in ≈ 180 s of CPU time, yields the following van der Waals potential,

$$U_{\text{vdW}}(R) = -\frac{3a_0^4 e^4}{E_I R^6} - \frac{135a_0^6 e^4}{2E_I R^8} - \frac{2835a_0^8 e^4}{4E_I R^{10}} = -\frac{6a_0^5 e^2}{R^6} - \frac{135a_0^7 e^2}{R^8} - \frac{2835a_0^9 e^2}{2R^{10}}, \tag{14}$$

coinciding with the results by Margenau [18,50] and Pauling [153] (In Equation (47.7) [154], the coefficient of the term in $1/R^{10}$ is given as 1416 instead of 1417.5). Margenau also expresses this result [18] in terms of the quantity $< 1, 0, 0|r^4|1, 0, 0 >= r_{1,0,0}^4 = \frac{45}{2}a_0^4$ and of $r_{1,0,0}^2$ already found above:

$$U_{\text{vdW}}(R) = -\frac{e^4(r_{1,0,0}^2)^2}{3E_I R^6} - \frac{e^4(r_{1,0,0}^2 r_{1,0,0}^4)}{E_I R^8} - \frac{7e^4(r_{1,0,0}^4)^2}{5E_I R^{10}}. \tag{15}$$

Finally, let us express this result in terms of the dipole moment found above and of the traceless quadruple moment tensor defined as [150,155–157]

$$Q_{ij} = -\frac{e}{2!}\left[x^i x^j - \tfrac{1}{3}r^2 \delta_{ij}\right] \tag{16}$$

Computing the expectation values $< 1, 0, 0||Q_{ij}|^2|1, 0, 0 >$ by means of *Mathematica*, the square of the Frobenius norm [158] of the ground state traceless quadruple moment tensor can be obtained:

$$||Q||_F^2 \equiv \sum_i \sum_j < 1, 0, 0||Q_{ij}|^2|1, 0, 0 >= \frac{15}{4}a_0^4 e^2. \tag{17}$$

Let us now use this result in the quadrupole-quadrupole, $1/R^{10}$ term at Equation (14), indicated as $V_{22}(R)$ in [150]:

$$V_{22}(R) = -\frac{2835a_0^8 e^4}{4E_I R^{10}} = -\frac{504}{5R^{10}}\frac{(||Q||_F^2)(||Q||_F^2)}{2E_I}, \tag{18}$$

in agreement with Equation (1.3) of the work in [150]. By now using Equation (17) and the norm squared of the ground state dipole expectation, $||\mu_{1,0,0}||^2 = 3e^2a_0^2$, we can rewrite the dipole-quadrupole, $1/R^8$ term at Equation (14), indicated as $V_{12}(R)$ in [150], as

$$V_{12}(R) = -\frac{135a_0^6e^4}{2E_IR^8} = -\frac{135}{R^8}\frac{1}{2E_I}\frac{4}{15}(||Q||_F^2)\frac{1}{3}||(\mu_{1,0,0}||^2) = -\frac{12}{R^8}\frac{(||Q||_F^2)(\mu_{1,0,0}||^2)}{2E_I}, \quad (19)$$

which indicates the factor of 3 in front of the summation at Equation (1.2) of the work in [150] should be 12.

Finally, let us express these results in terms of the static dipole ($\alpha_{0,E}^{(1)} = \frac{2}{3}||\mu_{0,r}||^2/E_{r,0}$) and static electric quadrupole polarizabilities, by using the standard expression for the latter (i.e., see Equation (6.5) in [159]), where $\mu_{0,r}$ and $E_{r,0}$ are the transition moments and the transition energies, and again only considering the ground state contribution [150]:

$$\alpha_{0,E}^{(2)} = \frac{1}{5}\sum_r \frac{E_{r,0}||Q||_F^2}{E_{r,0}^2 - (\hbar\omega)^2} \longrightarrow 2\frac{(||Q||_F^2)}{E_I}. \quad (20)$$

In summary, we find, including also Equation (21) rewritten in the notation of the work in [150]:

$$U_{\text{vdW}}(R) = V_{11}(R) = -\frac{3E_I(\alpha_0^{(1)})^2}{4R^6}; \quad (21)$$

$$V_{12}(R) = -\frac{9E_I\alpha_0^{(1)}\alpha_{0,E}^{(2)}}{R^8}; \quad (22)$$

$$V_{22}(R) = -\frac{504E_I(\alpha_{0,E}^{(2)})^2}{40R^{10}}. \quad (23)$$

Although in the literature all such results are obtained from rigorous calculations within the framework of QED, it is also pointed out, without proof, that "The near-zone result ... may also be obtained with second-order perturbation theory and the electrostatic potentials coupling two electric dipoles ..., an electric dipole and an electric quadrupole ..., and two electric quadrupoles ..." (see in [159], see also in [157], p. 64). Such an approach, which extends London's calculations to higher multipoles as shown by Margenau, does *not* rely on the quantization of the electromagnetic field.

5. Unretarded Gravitational van der Waals Forces

The classical electrostatic dipolar polarizability, $\alpha_{0,E}^{(1)}$, is typically introduced by means of a classical model in which an electron of mass m_e, bound to a proton by a spring of constant $K = m_e\omega_0^2$, is displaced by an external electric field, **E**, so that the position of static equilibrium is given by $-e\mathbf{E} = -K\mathbf{r}$. The corresponding induced dipole moment $\mathbf{p} = e\mathbf{r}$ is, therefore, $\mathbf{p} = e^2/m_e\omega_0^2\mathbf{E}$ and the polarizability is, by definition, $\alpha_{0,,E}^{(1)} = e^2/m_e\omega_0^2$.

Let us now develop a similar oscillator model to exhibit the static *gravitational* polarizability, $\alpha_G^{(2)}$. It is important to notice at the onset that this calculation differs from the electrostatic case in subtle ways that require careful consideration. For instance, as first pointed out by Szekeres almost half-a-century ago in what was possibly the first attempt to calculate the "gravitational dielectric constant" [160,161], and as later rediscovered by this author in related contexts [162,163], a model of a harmonic oscillator interacting with gravitational waves naturally leads to a Mathieu-type equation of motion. In what follows, as also done by Szekeres, we shall assume that the circumstances possibly triggering parametric resonance are not verified.

We shall again consider an atom in which the electron is connected to the proton by a spring aligned, for instance, along the direction pointing to another identical atom at distance R. Again, although the electrodynamical model and this gravitational model appear equivalent, it is not widely appreciated that the evolution of the latter in response to external gravitational fields depends *critically* on the assumed dynamics of the atoms. For instance, the gravitational effects by a mass on "a hydrogen atom whose point proton is immobile" [164] ("Le modèle envisagé est celui d'un atome d'hydrogène dont le proton ponctuel est immobile à l'extérieur de la matière créant le champ" [164]) are drastically different than those on an atom "…in free fall along a geodesic of the spacetime during the time required for an atomic transition" (see in [165], see also in [166]). For the purpose of defining the gravitational polarizability, here we shall assume that *both* atoms are in free fall although it is clear that, rigorously speaking, this assumption is contradiction with the existence of dispersion forces.

In this 1D model (see in [11], p. 100, Note 11), we shall assume the unperturbed electron to be oscillating harmonically with amplitude equal to $r_{1,0,0} = \frac{3}{2}a_0$ so that the position time average is equal to $\bar{z} = r_{1,0,0}/\sqrt{2}$. The displacement from this average due to a gravitational tidal force produced by a mass M is $\overline{\Delta z} = m_e R_{0z0z}\bar{z}/K$, where R_{0z0z} are the appropriate components of the Riemann tensor [167,168]. This yields the following *induced* trace-free quadrupole moment tensor [167,168]:

$$Q_{zz} = \frac{2}{3}m_e[(\bar{z}+\overline{\Delta z})^2 - \bar{z}^2] \simeq \frac{2}{3}m_e 2\,\bar{z}\,\overline{\Delta z} = \frac{4}{3}m_e^2 R_{0z0z}\frac{\bar{z}^2}{K}. \tag{24}$$

Therefore, the gravitational polarizability, in the static limit, can now be defined as (Equation (11) in [131]):

$$\alpha_{0,G}^{(2)} \equiv \frac{Q_{zz}}{R_{0z0z}} = \frac{4}{3}m_e^2\frac{\bar{z}^2}{K} = \frac{4}{3}m_e^2\frac{r_{1,0,0}^2}{2K} = \frac{4}{3}m_e^2(\frac{3}{2})^2\frac{a_0^2}{2K} = \frac{3}{2}\frac{m_e^2 a_0^2}{K} = \frac{3}{2}\frac{m_e a_0^2}{\omega_0^2}, \tag{25}$$

where the units also correspond to those given therein (see comment above Equation (32) in [131]).

For the purpose of succinctness, supported by explicit verification with *Mathematica*, we shall now return to our electrostatic results to extract the expression of the gravitational dispersion force. Obviously, as pointed out by Spruch [57,109], in this case we cannot have a neutral atom. However, the assumption of free fall, by the Principle of Equivalence, implies that all non-tidal gravitational interactions must not affect the dynamics of the system (Notice that introducing a negative mass is a standard procedure employed formally "to eliminate the mass monopole" [169] in analogy with an electric quadrupole. See also Section 6.2). Therefore, the leading contribution left in this unretarded regime is provided by the quadrupole–quadrupole term at Equation (18), upon carrying out Spruch's $(e^2)^2 \to (Gm_e^2)^2$ substitution, and with appropriate identification of the gravitational polarizability given above:

$$V_{22,near}^{GG}(R) = -\frac{2835\,a_0^8\,G^2 m_e^4}{4E_I R^{10}} = -\frac{315}{4}\frac{E_I G^2 (\alpha_{0,G}^{(2)})^2}{R^{10}}. \tag{26}$$

This equation, obtained by quantizing the interacting oscillators but with *classical* gravitation, is identical to that found by Ford, Hertzberg, and Karouby (see Equation (31) in [131]) from quantum field theory provided that the two atoms are identical and with $E_I = \hbar\omega_0$.

Finally, as also hinted to by Spruch, we provide new results, that is, the expressions for the mixed quadrupole–quadrupole and dipole–quadrupole electric-gravitational potentials in the unretarded regime. These can be obtained analogously to what done above by carrying out only one $e^2 \to Gm_e^2$ substitution while leaving one e^2 factor unchanged in Equations (18) and (19), respectively:

$$V_{22,near}^{EG}(R) = -\frac{2835\,a_0^8\,e^2 Gm_e^2}{2E_I R^{10}} = -\frac{63 E_I G \alpha_{0,E}^{(2)}\alpha_{0,G}^{(2)}}{R^{10}}. \tag{27}$$

$$V_{12,\text{near}}^{\text{EG}}(R) = -\frac{135\, a_0^6\, e^2 G m_e^2}{E_I R^8} = -\frac{45 E_I G \alpha_{0,\text{E}}^{(1)} \alpha_{0,\text{G}}^{(2)}}{2R^8}. \tag{28}$$

Notice, however, that the explicit *Mathematica* calculation shows a factor of 2 difference in addition to the $e^2 \to G m_e^2$ substitution.

6. McLachlan's Semiclassical Calculation of Electrodynamical Dispersion Forces

As we have mentioned above (Section 1), McLachlan developed an approach to the calculation of dispersion forces between polarizable bodies requiring only the quantization of the atomic degrees of freedom while leaving the interacting electromagnetic field classical [49,170]. Within this semiclassical framework, all results from the Lifshitz theory, including the potential of two or more molecules or that of infinite parallel dielectric surfaces, both in the unretarded and retarded regimes, are correctly recovered. As typical, a method that "...gives the main results of Lifshitz's treatment and does not use quantum field theory" is justified merely as having "the advantage of using simple physical concepts" [171]. In the case of this paper, however, we intend to employ this strategy to show that a completely classical gravitational field is expected to produce dispersion forces equal to those predicted in the low-energy limit of a hypothetical theory of quantum gravity. Therefore, a hypothetical experimental verification of the existence of such forces would *not* represent proof that the gravitational field is quantized. For the purpose of considering such a generalization of McLachlan's theory to the gravitational case, we first focus on the electromagnetic case, sketching his calculation of the interatomic dipole–dipole dispersion potential, which leads back to the expressions by London and by Casimir and Polder. We then broaden this treatment to dipole–quadrupole and quadrupole–quadrupole electrodynamical dispersion forces in simplified 1D models to highlight the physical principles; thus, reaching expressions obtained approximately four decades after Casimir and Polder by Thirunamachandran [172] from non-relativistic QED [79], later in collaboration with Salam, Power, and Jenkins [150,156,157,159,173].

6.1. Electric Dipole–Electric Dipole Dispersion Forces

Here, we closely follow the analyses of the relevant physical arguments in McLachlan's approach provided by Margenau (see in [50], especially Section 6.3), by Renne [174,175], and by Langbein [42], adapting all notation and definitions therein to compare our results to those above and elsewhere in the published literature, and to extend this treatment to the case of gravitational dispersion forces.

Let us consider two coupled harmonic oscillators of displacement components $u_{A,i}$ and $u_{B,j}$ of masses $m_{A,B}$ and natural frequencies $\omega_{A,B}$. By writing the equation of motion of each oscillator in the absence of friction and under the action of an external force $F_{\text{ext}} = F(\omega)\exp(-i\omega t)$, with $i = \sqrt{-1}$, the usual amplitude of the forced oscillation is found, $u_{A,B} = \chi_{A,B}(\omega) F_{\text{ext}}$, where $\chi_{A,B}(\omega)$ is the generalized scalar susceptibility (As discussed below, the dimensionality of this definition differs from that at Equation (10) by [Charge]^2. For this reason, we add a factor e^2 to the force equation, $F_{BA,j} = u_{A,i} e^2 T_{ij}(\omega)$, as explicitly done in all treatments following that by London [17] (for instance, see Equations (3.67–68) in [11]). The definitions in McLachlan are consistent with those by Langbein (compare Equation (5.3) in [170] with our Equation (10)). On the other hand, Margenau adopts our definition at Equation (10) (see Ch. 2, Equation (83) in [50])):

$$\chi_j(\omega) = \frac{1}{m_j(\omega_j^2 - \omega^2)}. \tag{29}$$

In the hypothesis of linear oscillator coupling, the force $F_{BA,j}$ exerted by oscillator A on B is given by the displacement $u_{A,i}$ as $F_{BA,j} = u_{A,i} e^2 T_{ij}(\omega)$, where $T_{ij}(\omega)$ will represent the *classical* interaction tensor and $T_{ij}(\omega) = T_{ji}(\omega)$ due to Newton's third law. By writing the equations of motion for the two oscillators and assuming solutions of the form $u_{A,B} = u_{A,B}(\omega)\exp(-i\omega t)$, with $i = \sqrt{-1}$, we find the standard homogeneous system of two equations whose determinant, set equal to zero, yields the secular equation for the normal modes [176]:

$$(\omega^2 - \omega_A^2)(\omega^2 - \omega_B^2) = e^4 \frac{1}{m_A} T_{ij} \frac{1}{m_B} T_{ji} \, . \tag{30}$$

At this point, as clearly articulated by London (see in [17], § 4; see also in [22,46,47,177]), the quantization of the oscillation modes of the harmonic oscillator, consistently with the Uncertainty Principle, leads to a non-vanishing ground state energy, which is shifted by the interatomic coupling according to the following expression,

$$\Delta E = \tfrac{1}{2}\hbar(\Omega_A + \Omega_B) - \tfrac{1}{2}\hbar(\omega_A + \omega_B) \, , \tag{31}$$

where $\Omega_{A,B}$ are the solutions of the secular equation. By solving Equation (30) and expanding the result in the parameter $T_{ij} T_{ji}/m_A m_B$, we find

$$\Delta E = -\frac{e^4}{4}\hbar \frac{1}{m_A} T_{ij} \frac{1}{m_B} T_{ji} \frac{1}{\omega_A \omega_B (\omega_A + \omega_B)} \, . \tag{32}$$

In order to verify this result, let us first discuss the structure of the dipole interaction tensor $\mathbf{T_{ij}}$ in the unretarded case. In the present system, the electrostatic potential at $\mathbf{r_B}$ of a dipole A located at $\mathbf{r_A}$ is given by the elementary expression [178,179]:

$$U_{\text{dip}} = -e\mathbf{u}_A \cdot \nabla_A \left(\frac{1}{|\mathbf{r_A} - \mathbf{r_B}|} \right) , \tag{33}$$

so that the force acting on the elastically bound electron of dipole B is

$$F_{BA,j} = -e^2 u_{A,i} \nabla_{A,i} \nabla_{B,j} \left(\frac{1}{|\mathbf{r_A} - \mathbf{r_B}|} \right) \tag{34}$$

and the the dipole interaction tensor can be read out as

$$\mathbf{T_{ij}} = -\nabla_{A,i} \nabla_{B,j} \frac{1}{|\mathbf{r_A} - \mathbf{r_B}|} \, . \tag{35}$$

A generalization of the quantity $T_{ij} T_{ji}$ to the full 3D case requires the explicit calculation of the quantity $\text{tr}\left(\mathbf{T_{ij} T_{ji}}\right) = 6/R_{AB}^6$, where $R_{AB} = |\mathbf{r_A} - \mathbf{r_B}|$. Therefore, for $\omega_A = \omega_B = \omega_0$ and $m_A = m_B = m_e$, we finally have

$$\Delta E(R) = U_{\text{vdW}}(R) = -\frac{3\hbar e^4}{4m_e^2 \omega_0^3 R_{AB}^6} = -\frac{3\hbar^4 e^4}{4m_e^2 E_I^2}\frac{1}{E_I R_{AB}^6} = -\frac{3a_0^4 e^4}{E_I R_{AB}^6} \, , \tag{36}$$

as already found at Equation (6) and where, again, $E_I = \hbar\omega_0 = e^2/2a_0$ and $a_0 = \hbar^2/m_e e^2$.

Importantly, a different route is to express the natural frequencies at Equation (32) in integral form by means of the identity:

$$\frac{1}{\omega_A \omega_B (\omega_A + \omega_B)} = \frac{1}{\pi} \int_{-\infty}^{\infty} d\omega \frac{1}{(\omega^2 + \omega_A^2)(\omega^2 + \omega_B^2)} \, , \tag{37}$$

so that, equivalently, by integrating along the imaginary frequency axis, where $\omega_C = \omega_R + i\omega_I$:

$$\Delta E = -\frac{\hbar}{2\pi} \int_0^{\infty} d\omega_I \, e^2 \chi_A(i\omega_I) T_{ij} \, e^2 \chi_B(i\omega_I) T_{ji} = -\frac{\hbar}{2\pi} \int_0^{\infty} d\omega_I \, \alpha_{A,E}^{(1)}(i\omega_I) T_{ij} \, \alpha_{B,E}^{(1)}(i\omega_I) T_{ji} \, . \tag{38}$$

Again, a generalization of this result to fully independent three-dimensional (3D) oscillators leads to the well-known result by McLachlan in terms of polarizability tensors:

$$\Delta E = -\frac{\hbar}{2\pi} \int_0^{\infty} d\omega_I \, \text{tr} \left(\alpha_{A,E}^{(1)}(i\omega_I) \cdot \mathbf{T}_{ij} \cdot \alpha_{B,E}^{(1)}(i\omega_I) \cdot \mathbf{T}_{ji} \right) . \tag{39}$$

In the case of isotropic molecules, the polarizability tensors again reduce to scalars and we can write, in the unretarded case:

$$\Delta E = -\frac{3\hbar}{\pi} \frac{1}{R_{ij}^6} \int_0^\infty d\omega_I \, \alpha_{A,E}^{(1)})(i\omega_I) \alpha_{B,E}^{(1)})(i\omega_I) \,. \tag{40}$$

In order to consider the retarded case, we must again compute the force on dipole B by means of the dipole *classical* radiation fields. This can be done starting from the dipole Hertz vector, $\mathbf{Z}(\mathbf{r_A}, \mathbf{r_B}, \omega)$, given by (see in [178], Sections 14.5–7):

$$\mathbf{Z}(\mathbf{r_A}, \mathbf{r_B}, \omega) = \mathbf{p_A} \frac{e^{i(\omega/c)|\mathbf{r_A}-\mathbf{r_B}|}}{|\mathbf{r_A} - \mathbf{r_B}|} \,, \tag{41}$$

The electric field is then found by means of the auxiliary vector, $\mathbf{C} = \nabla \times \mathbf{Z}$, as $\mathbf{E} = \nabla \times \nabla \times \mathbf{Z}$. In Cartesian coordinates, this leads to the following expression for the tensor to be multiplied by $\mathbf{p_A} - eu_{A,i}$ in order to obtain the force on dipole B, $F_{BA,j} = u_{A,i}e^2 T_{ij}(\omega)$,

$$\mathbf{T_{ij}} = \left[\nabla_i\nabla_j - \delta_{ij}\nabla^2\right]\frac{e^{i(\omega/c)|\mathbf{r_A}-\mathbf{r_B}|}}{|\mathbf{r_A} - \mathbf{r_B}|} \,, \tag{42}$$

where all derivatives are respect to the $\mathbf{r_A} = (x_A, y_A, z_A)$ coordinates. Let us now assume, without loss of generality, dipole A to be at the origin ($\mathbf{r_A} = 0$) and dipole B at $\mathbf{r_B} = (r, 0, 0)$. Direct calculation in this geometry shows that the only non-zero components of the interaction tensor are, after the customary rotation to the imaginary frequency axis (see Equation (101) in [50]):

$$\mathbf{T_{11}} = \left(\frac{2}{R^3}\right)\left(1 + R\frac{\omega_I}{c}\right)e^{-R\omega_I/c} \tag{43}$$

$$\mathbf{T_{22}} = \mathbf{T_{33}} = -\left(\frac{1}{R^3}\right)\left(1 + R\frac{\omega_I}{c} + R\frac{\omega_I^2}{c^2}\right)e^{-R\omega_I/c} \,. \tag{44}$$

Therefore, for isotropic atoms, the dipole polarizabilities are again scalars and the trace at Equation (39) becomes

$$\text{tr}\,(\mathbf{T_{ij}T_{ji}}) = e^{-2R\omega_I/c}\left(\frac{2\omega_I^4}{c^4R^2}\right)\left(1 + \frac{2c}{\omega_I R} + \frac{5c^2}{\omega_I^2 R^2} + \frac{6c^3}{\omega_I^3 R^3} + \frac{3c^4}{\omega_I^4 R^4}\right) \tag{45}$$

By using this result and Equation (10) in Equation (39), we find

$$\Delta E(R) = -\frac{\hbar}{2\pi}\left(\frac{2}{3\hbar}\right)^2\left(\frac{2}{c^4R^2}\right)\sum_{r,s}\omega_{r,0}\,\omega_{s,0}||\mu_{0,r}||^2||\mu_{0,s}||^2 \int_0^\infty d\omega_I \frac{\omega_I^4 e^{-2R\omega_I/c}}{(\omega_{r,0}^2 + \omega_I^2)(\omega_{s,0}^2 + \omega_I^2)} \times$$
$$\left(1 + \frac{2c}{\omega_I R} + \frac{5c^2}{\omega_I^2 R^2} + \frac{6c^3}{\omega_I^3 R^3} + \frac{3c^4}{\omega_I^4 R^4}\right) \,. \tag{46}$$

This expression for the dispersion force in terms of the dynamical polarizability as a function of the imaginary frequency was first obtained by Casimir and Polder [24] (see also Equation (42) in [50] and Equation (3.87) in [11,180]).

In the simplified case of two identical atoms in which only one transition is dominant and in the unretarded limit ($e^{-2R\omega_I/c} \to 1$), only the integral corresponding to the last term $\propto R^{-4}$ in Equation (45) must be considered so that we recover the well-known expression at Equation (21):

$$\Delta E(R) = U_{\text{vdW}}(R) = -\frac{3\hbar}{\pi}\left(\frac{2}{3\hbar}\right)^2\left(\frac{1}{R^6}\right)\omega_0^2\,||\mu_{0,r}||^4 \int_0^\infty d\omega_I \frac{1}{(\omega_0^2 + \omega_I^2)^2} = -\frac{3\hbar\omega_0(\alpha_{0,E}^{(1)})^2}{4R^6} \,, \tag{47}$$

where we used the expression for the static polarizability in terms of the dipole matrix element equivalent to that given above immediately preceding Equation (20), $\alpha_{0,E}^{(1)} = \frac{2}{3}||\mu_{0,r}||^2/\hbar\omega_0$.

The famous limit of the result at Equation (46) referred to as the Casimir–Polder potential can be obtained by realizing that, in the fully retarded regime, the static polarizability provides the leading contribution (See footnote 14, p. 104 of the work in [11]) so that by using Equation (45), Equation (39) becomes

$$\Delta E(R) = U_{CP}(R) = -\frac{\hbar}{2\pi}(\alpha_{0,E}^{(1)})^2 \int_0^\infty d\omega_I \, \mathrm{tr}\left(T_{ij} \cdot T_{ji}\right) = -\frac{23\,\hbar c}{4\pi R^7}(\alpha_{0,E}^{(1)})^2 . \tag{48}$$

Importantly for our further generalizations, in the isotropic case, Equation (46) can be rewritten by means of Equation (42) as [156,181]:

$$\Delta E = -\frac{\hbar}{2\pi}\int_0^\infty \alpha_{A,E}^{(1)})(i\omega_I)\alpha_{B,E}^{(1)})(i\omega_I)\left[\left(\nabla_i\nabla_j - \delta_{ij}\nabla^2\right)\frac{e^{-R\omega_I/c}}{R}\right]\left[\left(\nabla_i\nabla_j - \delta_{ij}\nabla^2\right)\frac{e^{-R\omega_I/c}}{R}\right] d\omega_I . \tag{49}$$

6.2. Application to Dispersion Forces with Electric Quadrupoles: One-Dimensional Case

As anticipated by McLachlan, the above strategy can be generalized to the case of atoms interacting through higher order electric and magnetic multipole fields (see in [49] Section 7 and Appendix). To the best of this author's knowledge, however, such a semiclassical program has never been appeared in the published literature. In what follows in support of our analysis of the gravitational case, we consider electric dipole–electric quadrupole and electric quadrupole–electric quadrupole interactions. In order to expose the physical processes involved behind this approach while avoiding lengthy calculations, we shall not analyze the full 3D geometry but, as often done in survey presentations about the unretarded and retarded regimes [46,47,177], we shall restrict ourselves to the 1D case. Extension to the full 3D case is promptly achieved through straightforward, though technically more intricate, calculations as just demonstrated in the previous section.

In analogy with the process leading to Equation (36), let us now consider one simple 1D classical quadrupole, A, represented by two pairs of identical, opposing, elastically bound point dipoles, $\mathbf{p}_A/\delta$ at a mutual vector distance $\delta\mathbf{u_A}$, where δ is a limit parameter (Section 4.4 in [182]) (Another useful simple quadrupole model is given by three charges $(+q, -2q, +q)$ arranged along a straight line at charge-to-charge distances equal to d [169]) Let us first assume the two opposing dipoles to be aligned along the x-axis and interacting with another single dipole, $\mathbf{p}_B$, like those used in the previous section, also parallel to the the x-axis.

The only non-vanishing component of the point quadrupole tensor, in the limit $\delta \to 0$, is $\mathcal{Q}_{xx} = u_A p_A$ and the *traceless* quadrupole tensor becomes $Q_{xx} = 3\mathcal{Q}_{xx} - \mathcal{Q}_{xx} = 2\mathcal{Q}_{xx} = 2u_A p_A$. The quadrupole potential can then be written as [183]:

$$U_{\text{quad}} = \frac{1}{6}Q_{xx}\frac{\partial^2}{\partial x^2}\left(\frac{1}{|\mathbf{r_A} - \mathbf{r_B}|}\right), \tag{50}$$

and, analogously to Equation (34), the force on the oscillating charge of dipole $\mathbf{p}_B$ is:

$$F_{BA,x} = \frac{1}{3}eu_A p_A\frac{\partial^3}{\partial x^3}\left(\frac{1}{|\mathbf{r_A} - \mathbf{r_B}|}\right). \tag{51}$$

Finally, since the 1D 'trace' $T_{ij}T_{ji} = (\partial^3/\partial x_A^3)(1/|x_A - x_B|) = 36/x^8$, we obtain, indicating $r = |x_A - x_B|$ and by identifying $p_A^2 = \frac{1}{3}||\mu_{1,0,0}||^2 = a_0^2 e^2$:

$$\Delta E_{12}^{1D} = -\frac{e^4}{4}\hbar\frac{1}{9}e^2||p_A||^2\frac{1}{2\omega_0^3}\frac{1}{m_e^2}\frac{36}{x^8} = -6\frac{a_0^6 e^4}{E_I}\frac{1}{r^8}, \tag{52}$$

to be compared to Equation (14). By again exploiting the identity at Equation (37) and by comparing the middle term of the above equation to that at Equation (36), it can be recognized that $||\mu_{1,0,0}||^2 \chi_A(i\omega_I)T_{ij} = \alpha_{B,E}^{(2)})$ and $e^2\chi_B(i\omega_I)T_{ji} = \alpha_{B,E}^{(1)})$ (see Equations (1.7)–(1.8) in [150]), so that this result can be put into the important integral form analogous to that as at Equation (38)

$$\Delta E_{12} = V_{12}(r) = -\frac{\hbar}{2\pi}\int_0^\infty d\omega_I\, \alpha_{A,E}^{(1)})(i\omega_I)T_{ij}\,\alpha_{B,E}^{(2)})(i\omega_I)T_{ji} = -\frac{18\hbar}{\pi}\frac{1}{r^8}\int_0^\infty d\omega_I\, \alpha_{A,E}^{(1)})(i\omega_I)\,\alpha_{B,E}^{(2)})(i\omega_I). \tag{53}$$

As mentioned above (Section 4.2), this result was previously obtained by a full QED approach by Jenkins (Equation (6.7), Ref. [159]), and then re-derived by appropriate generalizations by Salam (Equation (3.5) in [156]) and by Power (Equation (1.5) in [150]) all working with Thirunamachandran (Notice that units employed in these works differ from one another. Power and Thirunamachandran showed that the correct numerical factor in the full 3D case is 90. Also, the speed of light c factors in Equations (1.4)–(1.6) in [150], which refer to the unretarded limit, are erroneous).

In the case of electric unretarded electric quadrupole–quadrupole forces, the system is represented by two quadrupoles as A above, that is, a total of four dipoles elastically bound in opposing pairs and, in the 1D case, all parallel to the x-axis. The force due to the electric field produced by quadrupole A acting on the oscillating dipole of atom B is given by the well-known equation $\mathbf{F} = (\mathbf{p}\cdot\nabla)\mathbf{E}$ [179,182] so that the force becomes

$$F_{BA,x} = -\frac{1}{3}eu_A p_A p_B \frac{\partial^4}{\partial x^4}\left(\frac{1}{|\mathbf{r_A}-\mathbf{r_B}|}\right). \tag{54}$$

Therefore, $T_{ij}T_{ji} = (\partial^4/\partial x_A^4)(1/|x_A - x_B|) = 576/x^{10}$, leading to the unretarded potential

$$\Delta E_{22}^{1D} = -32\frac{a_0^8 e^4}{E_I}\frac{1}{r^{10}}, \tag{55}$$

again to be compared to Equation (14). The integral form immediately descends from this formulation as

$$\Delta E_{22} = V_{22,\text{near}}(r) = -\frac{288\hbar}{\pi}\frac{1}{r^{10}}\int_0^\infty d\omega_I\, \alpha_{A,E}^{(2)})(i\omega_I)\,\alpha_{B,E}^{(2)})(i\omega_I), \tag{56}$$

again in agreement, to within the constant numerical factor, with results previously found [150,156,159]. Importantly, a comparison of the quantity $T_{ij}T_{ji}$ at Equation (35) and in the cases above shows that the required order of the derivative of the potential $1/|\mathbf{r_A}-\mathbf{r_B}|$ increases by one unit for each integer order of the multipoles considered. This observation is reflected in the generalization of the results above to all multipole orders, including possibly anisotropic polarizabilities. For instance, to treat the general case of any interatomic distance, the factors $[(\nabla_i\nabla_j - \delta_{ij}\nabla^2)(e^{-R\omega_I/c}/R)]$ at Equation (49) must likewise be modified into $[(\nabla_i\nabla_j - \delta_{ij}\nabla^2)\nabla_k(e^{-R\omega_I/c}/R)]$ and $[(\nabla_i\nabla_j - \delta_{ij}\nabla^2)\nabla_k\nabla_l(e^{-R\omega_I/c}/R)]$ in the two systems analyzed, respectively, and "tumbling averages" must be taken over all orientations in the case of isotropic atoms (see Equation (2.9) in [156]).

To briefly illustrate the important fully retarded regime, let us consider the interactions ΔE_{12}^{1D} (Equation (52) and ΔE_{22}^{1D} (Equation (55) by starting from the quadrupole field Hertz vector j-component (see in [178], Section 14-8), as done above at Equation (41):

$$Z_{\text{quad},j} = \frac{\omega}{2c}\frac{e^{i(\omega/c)|\mathbf{r_A}-\mathbf{r_B}|}}{|\mathbf{r_A}-\mathbf{r_B}|^2}x_i\, Q_{ij}. \tag{57}$$

By proceeding again as above to calculate the electric field as $\mathbf{E} = \nabla\times\nabla\times\mathbf{Z}_{\text{quad}}$, analogously to Equation (45), by forming the trace of the interaction tensor in this 1D model, isolating the static polarizabilities $\alpha_{0,\mathrm{E}}^{(1)})$ and $\alpha_{0,\mathrm{E}}^{(2)})$, and by calculating an integral analogous to that at Equation (48), we find, for the dipole–quadrupole energy,

$$\Delta E_{12} = V_{12,\mathrm{far}}(r) = -C_{12}\frac{\hbar c}{\pi r^9}\alpha_{0,\mathrm{E}}^{(1)})\,\alpha_{0,\mathrm{E}}^{(2)}) \tag{58}$$

where C is a numerical constant. A similar procedure for the quadrupole–quadrupole interaction, seeking the quantity $(\mathbf{p}\cdot\nabla)\mathbf{E}$, leads to:

$$\Delta E_{22} = V_{22,\mathrm{far}}(r) = -C_{22}\frac{\hbar c}{\pi r^{11}}\alpha_{0,\mathrm{E}}^{(2)})\,\alpha_{0,\mathrm{E}}^{(2)}). \tag{59}$$

These results are in full agreement, to within the constants C_{12} and C_{22}, with those by Power and Thirumanamachandran (Equations (4.13)–(4.14) in [150]). Furthermore, the corresponding expressions in terms of integrals over the proper multipole polarizabilities as functions of the complex frequency, generalizing Equation (49), are promptly written and lead to the same expressions provided by Salam and collaborators (a full review is provided in [157] and references therein).

6.3. Application of McLachlan's Approach to Gravitational Dispersion Forces

The semiclassical computation of gravitational dispersion forces is a straightforward extension of the process just described. As we have already commented (Sections 5 and 6.2), strong analogies exist between the calculations with multipoles in the electrodynamic and in the gravitational case [169]. The calculation of the multipole fields in Cartesian coordinates in both electromagnetism and gravitation is mathematically relatively burdensome, as shown even in the much simplified 1D case, and it benefits from the computer algebra approach applied extensively in this paper. However, the physical principle of McLachlan's approach as applied to gravitation within the context of general relativity is quite clear. In analogy to electrodynamics, atoms interact through possibly retarded gravitational fields due to the oscillating sources and dispersion forces appear as the atomic energy levels are quantized, that is, one again envisions two systems of elastically bound dipole pairs aligned to the x-axis in near-free fall (considering accelerations due to dispersion forces as much smaller than that due to the monopoles). As shown in Section 5 in our simplified 1D model, the force acting on each oscillator is given by the appropriate Riemann tensor component, which can be obtained from the classical field metric solution given in general, for instance, by Ford et al. (Equations (4)–(9) in [131]) or, for a simplified model, by Price et al. (Equations (24)–(26) in [169]). Importantly, in the gravitational case, the Riemann tensor, and thus the interaction tensor, are determined through second derivatives with respect to spacetime coordinates of the metric tensor solution (Equations (7.27)–(7.29) in [184]), which does not directly determine the force. As in the electrodynamical case, this computation reproduces our results for the unretarded dispersion force in Section 5 and Ford's results in all distance regimes, to within numerical constants reflecting our dimensional simplification.

7. Experimental Detection of Gravitational Dispersion Forces

There is no doubt that identifying any effects of gravitation on dispersion forces is one of the most extreme contemporary experimental challenges. Generally speaking, these include not only the gravitational dispersion forces discussed in this paper but also the effects of time-independent spacetime curvature on electrodynamical dispersion forces, which can be viewed as a manifestation of the inertial equivalent of the Casimir potential energy [88,185]. Although these two avenues for gravitation to affect theoretical predictions are sometimes, puzzlingly, conflated [186], they are connected to drastically different phenomena. Whereas treatments of the latter hinge on relatively well-understood, uncontroversial concepts and measurements are deemed within reach [187–189], the former, which we briefly consider herein, depend on novel, exotic physical mechanisms.

As mentioned above (Section 3), the possibility to detect a macroscopic gravitational Casimir force has been recently brought to the fore [129] by the proposal that superconductors could be caused to act as nearly-ideal gravitational wave reflectors through a so-called "Heisenberg–Coulomb effect" [125]. This is a remarkably bold statement considering that even neutron stars are predicted to reflect gravitational waves with an index of reflection of only 4–25% [190]. The idea that superconductors might appreciably reflect and refract gravitational waves, thus enabling table-top gravitational wave optics [122,124], has been proposed by several authors [191] but it remains experimentally unverified. As already clarified by this author, in the theory of the Heisenberg–Coulomb effect, all radiation fields are treated classically [125] and the gravitational Casimir effect, even if detected, would not represent proof of the quantization of the gravitational field as extensively discussed in this paper [60]. Furthermore, initial predictions of the gravitational Casimir force as being even larger than its QED counterpart by almost one order of magnitude were shown to be due to a computational error [130]. Finally, recent measurements of the Casimir force between superconductors [186], although criticized [192], were reported to rule out the predicted gravitational Casimir effect in superconductors.

As regards systems in which detection of a gravitational Casimir–Polder effect might be feasible, Ford and collaborators [131] speculatively suggest "microscopic clumps built out of heavy sterile neutrinos" as possible, but as yet unconfirmed, interacting dark matter systems. An additional suggestion has been made by the present author to study Efimov states in dark matter once and if its components are definitely identified [193].

A more optimistic consideration is that identifying experimentally accessible phenomena that involve mixed potentials would represent an immense step towards fulfilling a detection goal since such quantities are larger by approximately 43 orders of magnitude ($\sim e^2 / G m_e^2$) than those due purely to gravitational fluctuations. This exploration is only now starting with indications, for instance, that low energy scattering experiments with neutrons in ground-based laboratories may prove useful [194].

As we have argued throughout this paper, any determination of the existence of gravitational dispersion forces will probably not represent conclusive proof of the quantum nature of the gravitational field. However, this makes the prospect of such an experiment in no way less exciting. What shall we find? Surprising scenarios are indeed possible, especially in the even more challenging retarded regime in which the existence of a minimal length may introduce modifications in the expressions found herein. Any theoretical estimates might be modified if fundamental assumptions are found to be inaccurate, for instance, because of a significant stochastic gravitational wave background [60] or the existence of Planck scale granularity [194]. Regardless of whether gravitational dispersion force behavior departs from or conforms to theoretical predictions, we shall enter a new era in our understanding of crucial parameters that characterize the structure of spacetime as well as the universe on astrophysical and cosmological scales.

8. Discussion and Conclusions

In closing, it is appropriate to discuss the connection between the main thesis of this paper and two apparently widely separate issues. Philosophically speaking, the history of forces between the basic constituents of matter is far older than that of the last few decades. In fact, interactions between "atoms" were deemed indispensable to the apparatus of philosophical atomism introduced by Leucippus and Democritus of Abdera. As this author has previously discussed [195], the existence of dispersion forces, whether within a quantum or a semiclassical description, ties the logical self-consistency of atomistic philosophy to the question of the nature of the vacuum. The challenge, as identified by Post, appears to be "…the difficulty of reconciling a world of interacting parts with atomism, which ideally requires independence for its atoms" [196]. As we have seen, physics, even at a semiclassical level, explains the existence of dispersion forces by introducing the concept of zero-point field, thus making a statement about another difficulty "…traditionally associated with atomism, …the problem of the void". J. A. Wheeler stated: "No point is more central than this, that empty space is not empty..." [167] and adopting his "foam-like structure" [197] must necessarily introduce new interactions among the basic constituents of matter. Atomism, identified by some with determinism [198], functions if we accept stochasticity or uncertainty so as to introduce needed forces between polarizable particles. Therefore the subject of gravitational dispersion forces, by introducing uncertainty in spacetime, adds one further layer of complexity to these reflections, as will be discussed elsewhere.

Technologically speaking, it would be natural to doubt that such fantastically small interactions as gravitational dispersion forces might lead to novel applications. Although we shall not even attempt to speculate about possible future developments here, we shall recall that, when Jordan Maclay and his collaborators suggested in 1995 that electrodynamical Casimir forces may have some important applications in microelectromechanical system (MEMS) engineering [70], it was certainly less than obvious that such apparently exotic interactions would, two decades later, enable a human to climb vertically on glass [199]. Such has been, however, the history of dispersion force research technology transfer [3].

Funding: This research received no external funding.

Institutional Review Board Statement: "Not applicable" for studies not involving humans or animals.

Informed Consent Statement: "Not applicable" for studies not involving humans.

Acknowledgments: An early sketch of the ideas fully developed in this paper was presented as an oral paper at the 4th Symposium on the Casimir Effect held in St. Petersburg [200]. It is my pleasure to acknowledge enlightening email correspondence with Professor Akbar Salam (Wake Forest University) on his work on long-range intermolecular forces. I am grateful to the issue editor Professor Maclay, for his gracious invitation to contribute this article.

Conflicts of Interest: The authors declare no conflicts of interest.

References

1. Simpson, W.M. Ontological aspects of the Casimir Effect. *Stud. Hist. Philos. Mod. Phys.* **2014**, *48*, 84–88. [CrossRef]
2. Casimir, H.B.G. On the attraction between two perfectly conducting plates. *Proc. Kon. Ned. Akad. Wetenshap* **1948**, *51*, 793–795.
3. Pinto, F. The future of van der Waals force enabled technology-transfer into the aerospace marketplace. In *Nanotube Superfiber Materials, Science to Commercialization*; Schulz, M.J., Shanov, V.N., Yin, J., Cahay, M., Eds.; Elsevier: Amsterdam, The Netherlands 2019; Chapter 29, pp. 729–794.
4. Sakurai, J.J. *Advanced Quantum Mechanics*; Addison-Wesley Publ. Co.: Redwood City, CA, USA, 1987.
5. Planck, M. Über die Begründung des Gesetzes der schwarzen Strahlung. *Ann. Phys.* **1912**, *342*, 642–656. [CrossRef]
6. Einstein, A.; Stern, O. Einige Argumente für die Annahme einer molekularen Agitation beim absoluten Nullpunkt. *Ann. Phys.* **1913**, *345*, 551–560. [CrossRef]
7. Bohr, N. *Niels Bohr Collected Works*; North-Holland Publishing Company: Amsterdam, The Netherlands, 1981; Volume 2.
8. Nernst, W. Uber einen versuch, von quantentheoretischen Betrachtungen zur Annahme stetiger Energieanderungen zuruckzukehren. *Verhandlungen Dtsch. Phys. Ges.* **1916**, *18*, 83–117.
9. Chu, B. *Molecular Forces, Based on the Baker Lectures of Peter J. W. Debye*; John Wiley & Sons: New York, NY, USA, 1967.
10. Kuhn, T.S. *Black-Body Theory and the Quantum Discontinuity, 1894–1912*; Clarendon Press: Oxford, UK, 1978.
11. Milonni, P.W. *The Quantum Vacuum*; Academic Press: San Diego, CA, USA, 1994; p. 522.
12. Kragh, H. Preludes to dark energy: Zero-point energy and vacuum speculations. *Arch. Hist. Exact Sci.* **2012**, *66*, 199–240. [CrossRef]
13. Power, E.A. *Introductory Quantum Electrodynamics*; American Elsevier Publishing Company: New York, NY, USA, 1965.
14. Wang, S.C. Die gegenseitige Einwirkung zweier Wasserstoffatome. *Phys. Z.* **1927**, *28*, 663–666.
15. Von Eisenschitz, R.; London, F. Uber das Verhaltnis der van der Waalsschen Krafte zu den homoopolaren Bindungskraften. *Z. Phys.* **1930**, *60*, 491–527. [CrossRef]
16. Pais, A. *Inward Bound*; Oxford University Press: Oxford, UK, 1986.
17. London, F. The general theory of molecular forces. *Trans. Faraday Soc.* **1937**, *33*, 8–26, [CrossRef]
18. Margenau, H. The role of quadrupole forces in van der Waals attraction. *Phys. Rev.* **1931**, *38*, 747–756. [CrossRef]
19. Margenau, H. Quadrupole Contributions to London's dispersion forces. *J. Chem. Phys.* **1938**, *6*, 896–899. [CrossRef]
20. Verwey, E.J.W.; Overbeek, J.T.G. *Theory of the Stability of Lyophobic Colloids*; Elsevier Publishing Company, Inc.: New York, NY, USA, 1948; pp. 1–108.
21. Casimir, H.B.G. Some remarks on the history of the so-called Casimir effect. In *The Casimir Effect 50 Years Later, Proceedings of the Fourth Workshop on Quantum Field Theory Under the Influence of External Conditions, Leipzig, Germany, 14–18 September 1998*; Bordag, M., Ed.; World Scientific Publishing Co., Pte. Ltd.: Singapore, 1999; pp. 3–9.

22. Pinto, F. Casimir forces: Fundamental theory, computation, and nanodevices applications. In *Quantum Nano-Photonics, NATO Science for Peace and Security Series B: Physics and Biophysics (Erice, Sicily, Italy)*; Di Bartolo, B., Ed.; Springer Nature B.V.: Berlin/Heidelberg, Germany, 2018; pp. 149–180.
23. Casimir, H.B.G.; Polder, D. Influence of retardation on the London-van-der-Waals forces. *Nature* **1946**, *158*, 787–788. [CrossRef]
24. Casimir, H.B.G.; Polder, D. The influence of retardation on the London-van der Waals forces. *Phys. Rev.* **1948**, *73*, 360–372. [CrossRef]
25. Casimir, H.B.G. On the history of the so-called Casimir effect. In *Comments on Atomic and Molecular Physics, Comments on Modern Physics. Part D (Special Issue: Casimir Forces)*; Babb, J.F., Milonni, P.W., Spruch, L., Eds.; Gordon and Breach Science Publishers: Kuala Lumpur, Malaysia, 2000; Volume 1, pp. 175–177.
26. Casimir, H.B.G. Van der Waals forces and zero point energy. In *Physics of Strong Fields*; Greiner, W., Ed.; Springer US: New York, NY, USA, 1987; pp. 957–964.
27. Casimir, H.B.G. Van der Waals forces and zero point energy. In *Essays in Honour of Victor Frederick Weisskopf (Physics and Society)*; Springer: New York, NY, USA, 1998; pp. 53–66.
28. Einstein, A. Folgerungen aus den Capillaritätserscheinungen. *Ann. Der Phys.* **1901**, *4*, 513–523, [CrossRef]
29. Einstein, A. Conclusions drawn from the phenomena of capillarity. In *The Collected Papers of Albert Einstein, Vol. 2 (The Swiss Years, Scientific Writings, 1900–1909)*; Princeton University Press: Princeton, NJ, USA, 1901; pp.1–11.
30. de Boer, J.H. The influence of van der Waals forces and primary bonds on binding energy, strength and special reference to some artificial resins. *Trans. Faraday Soc.* **1936**, *32*, 10–37. [CrossRef]
31. Hamaker, H. The London-van der Waals attraction between spherical particles. *Physica* **1937**, *4*, 1058–1072, [CrossRef]
32. Rowlinson, J.S. *Cohesion—A Scientific History of Intermolecular Forces*; Cambridge University Press: Cambridge, UK, 2002; pp. 1–333.
33. Axilrod, B.M.; Teller, E. Interaction of the van der Waals Type Between Three Atoms. *J. Chem. Phys.* **1943**, *11*, 299–300. [CrossRef]
34. Lifshitz, E.M. The theory of molecular attractive forces between solids. *Sov. Phys. JETP* **1956**, *2*, 73–83.
35. Dzyaloshinskii, I.; Lifshitz, E.M.; Pitaevskii, L.P. The general theory of van der Waals forces. *Adv. Phys.* **1961**, *10*, 165–209. [CrossRef]
36. Bordag, M.; Klimchitskaya, G.L.; Mohideen, U.; Mostepanenko, V.M. *Advances in the Casimir Effect*; Oxford University Press: Oxford, UK, 2009.
37. Klimchitskaya, G.L.; Mohideen, U.; Mostepanenko, V.M. The Casimir force between real materials: Experiment and theory. *Rev. Mod. Phys.* **2009**, *81*, 1827–1885, [CrossRef]
38. Dalvit, D.A.R.; Milonni, P.W.; Roberts, D.; da Rosa, F. *Casimir Physics*; Springer Lecture Notes in Physics 834; Springer: Berlin/Heidelberg, Germany, 2011.
39. Lamoreaux, S.K. The Casimir Force and Related Effects: The Status of the Finite Temperature Correction and Limits on New Long-Range Forces. *Annu. Rev. Nucl. Part. Sci.* **2012**, *62*, 37–56, [CrossRef]
40. Schwinger, J.; DeRaad, L.L.; Milton, K.A. Casimir effect in dielectrics. *Ann. Phys.* **1978**, *23*, 1–23. [CrossRef]
41. Saunders, S. Is the Zero-Point Energy Real? In *Ontological Aspects of Quantum Field Theory*; Kuhlmann, M., Lyre, H., Wayne, A., Eds.; World Scientific: Hackensack, NJ, USA, 2002; Chapter 16, pp. 313–343.
42. Langbein, D. *Theory of van der Waals Attraction*; Springer: Berlin/Heidelberg, Germany, 1974; pp. 1–144.
43. Copernicus, N. *On the Revolutions of the Heavenly Spheres*; Great Books of the Western World; William Benton: Chicago, IL, USA, 1952; Volume 16,.
44. Copernico, N. *De Revolutionibus Orbium Caelestium—La Costituzione Generale Dell'universo—A Cura di Alexandre Koyré*; Giulio Einaudi Editore s.p.a.: Torino, Italy, 1975.
45. Lamoreaux, S.K. Casimir forces: Still surprising after 60 years. *Phys. Today* **2007**, *60*, 40–45. [CrossRef]
46. Karplus, M.; Porter, R.N. *Atoms & Molecules*; The Benjamin/Cummings Publishing Company: Menlo Park, CA, USA, 1970.
47. Kleppner, D. With apologies to Casimir. *Phys. Today* **1990**, *43*, 9–11. [CrossRef]
48. Mahanty, J.; Ninham, B.W. *Dispersion Forces*; Academic Press: London, UK, 1976.
49. McLachlan, A.D. Retarded Dispersion Forces Between Molecules. *Proc. Roy. Soc. (Lond.) Ser. A* **1963**, *271*, 387–401.
50. Margenau, H.; Kestner, N.R. *Theory of Intermolecular Forces*, 2nd ed.; Pergamon Press: Oxford, UK, 1971; pp. 1–400.
51. Marshall, T.W. Random Electrodynamics. *Proc. R. Soc. Math. Phys. Eng. Sci.* **1963**, *276*, 475–491. [CrossRef]
52. Marshall, T.W. Statistical electrodynamics. *Proc. Camb. Phil. Soc.* **1965**, *61*, 537–546. [CrossRef]
53. Boyer, T.H. Recalculations of long-range van der Waals potentials. *Phys. Rev.* **1969**, *180*, 19–24. [CrossRef]
54. Boyer, T.H. Van der Waals forces and zero-point energy for dielectric and permeable materials. *Phys. Rev. A* **1974**, *9*, 2078–2084. [CrossRef]
55. Hushwater, V. Repulsive Casimir force as a result of vacuum radiation pressure. *Am. J. Phys.* **1997**, *65*, 381–384, [CrossRef]
56. Spruch, L.; Kelsey, J. Vacuum fluctuation and retardation effects on long-range potentials. *Phys. Rev. A* **1978**, *18*, 845–852. [CrossRef]
57. Spruch, L. Retarded, or Casimir, Long-Range Potentials. *Phys. Today* **1986**, *39*, 37–45. [CrossRef]
58. Boyer, T.H. Blackbody Radiation and the Scaling Symmetry of Relativistic Classical Electron Theory with Classical Electromagnetic Zero-Point Radiation. *Found. Phys.* **2010**, *40*, 1096–1098. [CrossRef]
59. Camparo, J. Semiclassical description of radiative decay in a colored vacuum. *Phys. Rev. A* **2001**, *65*, 13815. [CrossRef]

60. Pinto, F. Gravitational Casimir effect, the Lifshitz theory, and the existence of gravitons. *Class. Quantum Grav.* **2016**, *33*, 237001 [CrossRef]
61. González, A.E. On Casimir pressure, the Lorentz force and black body radiation. *Physica* **1985**, *131A*, 228–236. [CrossRef]
62. Milonni, P.W.; Cook, R.J.; Goggin, M.E. Radiation pressure from the vacuum: Physical interpretation of the Casimir force. *Phys. Rev. A* **1988**, *38*, 1621–1623. [CrossRef]
63. Barton, G. On the fluctuations of the Casimir force. *J. Phys. A Math. Gen.* **1991**, *24*, 991–1005. [CrossRef]
64. Susbilla, R.T. Casimir Acoustics. Ph.D. Thesis, Naval Postgraduate School, Monterey, CA, USA, 1996.
65. Holmes, C.D. Acoustic Casimir Effect. Ph.D. Thesis, Naval Postgraduate School, Monterey, CA, USA, 1997.
66. Larraza, A.; Denardo, B. An acoustic Casimir effect. *Phys. Lett. A* **1998**, *248*, 151–155. [CrossRef]
67. Larraza, A.; Holmes, C.D.; Susbilla, R.T.; Denardo, B.C. The force between two parallel rigid plates due to the radiation pressure of broadband noise: An acoustic Casimir effect. *J. Acoust. Soc. Am.* **1998**, *103*, 2267–2272. [CrossRef]
68. Ford, L.H. Spectrum of the Casimir effect and the Lifshitz theory. *Phys. Rev. A* **1993**, *48*, 2962–2967. [CrossRef] [PubMed]
69. Esquivel-Sirvent, R.; Reyes, L.I. Pull-in control in microswitches using acoustic Casimir forces. *EPL* **2008**, *84*, 48002. [CrossRef]
70. Serry, F.M.; Walliser, D.; Maclay, G. The anharmonic Casimir oscillator (ACO)-the Casimir effect in a model microelectromechanical system. *J. Microelectromech. Syst.* **1995**, *4*, 193–205. [CrossRef]
71. Boersma, S.L. A maritime analogy of the Casimir effect. *Am. J. Phys.* **1996**, *64*, 539–541. [CrossRef]
72. Denardo, B.C.; Puda, J.J.; Larraza, A. A water wave analog of the Casimir effect. *Am. J. Phys.* **2009**, *77*, 1095. [CrossRef]
73. Ball, P. Popular physics myth is all at sea. *Nature* **2006**, 2006–2008. [CrossRef]
74. Fisher, D.J. Maritime Casimir effect. *Am. J. Phys.* **1996**, *64*, 1228. [CrossRef]
75. Brügger, G.; Froufe-Pérez, L.S.; Scheffold, F.; Sáenz, J.J. Controlling dispersion forces between small particles with artificially created random light fields. *Nat. Commun.* **2015**, *6*, 7460. [CrossRef]
76. Holzmann, D.; Ritsch, H. Tailored long range forces on polarizable particles by collective scattering of broadband radiation. *New J. Phys.* **2016**, *18*, 103041. [CrossRef]
77. Shi, Y.Z.; Xiong, S.; Zhang, Y.; Chin, L.K.; Chen, Y.Y.; Zhang, J.B.; Zhang, T.H.; Ser, W.; Larrson, A.; Lim, S.H.; et al. Sculpting nanoparticle dynamics for single- bacteria-level screening and direct binding-efficiency measurement. *Nat. Commun.* **2018**, *9*, 815. [CrossRef]
78. Thirunamachandran, T. Intermolecular interactions in the presence of an intense radiation field. *Mol. Phys.* **1980**, *40*, 393–399. [CrossRef]
79. Craig, D.P.; Thirunamachandran, T. *Molecular Quantum Electrodynamics*; Dover Publications, Inc.: Mineola, NY, USA, 1998.
80. O'Dell, D.; Giovanazzi, S.; Kurizki, G.; Akulin, V.M. Bose-Einstein condensates with 1/r interatomic attraction: Electromagnetically induced "gravity". *Phys. Rev. Lett.* **2000**, *84*, 5687–5690. [CrossRef] [PubMed]
81. Luis-Hita, J.; Marqués, M.I.; Delgado-Buscalioni, R.; de Sousa, N.; Froufe-Pérez, L.S.; Scheffold, F.; Sáenz, J.J. Light Induced Inverse-Square Law Interactions between Nanoparticles: "Mock Gravity" at the Nanoscale. *Phys. Rev. Lett.* **2019**, *123*, 143201. [CrossRef] [PubMed]
82. Burns, M.M.; Fournier, J.M.; Golovchenko, J.A. Optical binding. *Phys. Rev. Lett.* **1989**, *63*, 1233–1236. [CrossRef] [PubMed]
83. DeWitt, B.S. The Casimir effect in field theory. In *Physics in the Making*; Sarlemijn, A., Sparnaay, M.J., Eds.; Elsevier Science Publishers: Amsterdam, The Netherlands, 1989; Chapter 9B, pp. 247–272.
84. Lamoreaux, S.K. Demonstration of the Casimir Force in the 0.6 to 6 μm Range. *Phys. Rev. Lett.* **1997**, *78*, 5–8. [CrossRef]
85. Lamoreaux, S.K. The Casimir force: background, experiments, and applications. *Rep. Prog. Phys.* **2004**, *68*, 201–236. [CrossRef]
86. Wright, A. Milestone 6. QED. *Nat. Mater.* **2010**, *9*, pS9. [CrossRef]
87. Feynman, R.P. The development of the space-time view of quantum-electrodynamics. *Science* **1966**, *153*, 699–708. [CrossRef]
88. Pinto, F. Resolution of a paradox in classical electrodynamics. *Phys. Rev. D* **2006**, *73*, 104020, [CrossRef]
89. Fermi, E. Sull'elettrostatica di un campo gravitazionale uniforme e sul peso delle masse elettromagnetiche. *Il Nuovo Cimento* **1921**, *22*, 176–188. [CrossRef]
90. Gorelik, G.E. The First Steps of Quantum Gravity and the Planck Values. In *Studies in the History of General Relativity*; Eisenstaedt, J., Kox, A., Eds.; Eistenin Studies; Birkhäuser: Basel, Switzerland, 1992; Volume 3, p. 367.
91. Stachel, J. The Early History of Quantum Gravity (1916–1940). In *Black Holes, Gravitational Radiation and the Universe*; Iyer, B.R., Bhawal, B., Eds.; Fundamental Theories of Physics; Springer Science: Dordrecht, the Netherlands, 1999; Volume 100, Chapter 31, pp. 525–534.
92. Kiefer, C. *Quantum Gravity*; Oxford University Press: Oxford, UK, 2007.
93. Rosenfeld, L. Zur Quantelung der Wellenfelder. *Ann. Phys.* **1930**, *397*, 113–152. [CrossRef]
94. Bronstein, M. Quantentheorie schwacher Gravitationsfelder. *Phys. Z. Sowjetunion* **1936**, *9*, 140–157.
95. Deser, S.; Starobinsky, A. Editorial note to: Matvei P. Bronstein, Quantum theory of weak gravitational field. *Gen. Relativ. Gravit.* **2012**, *44*, 263–265. [CrossRef]
96. Bronstein, M. Republication of: Quantum theory of weak gravitational fields. *Gen. Relativ. Gravit.* **2012**, *44*, 267–283. [CrossRef]
97. Gorelik, G.E.; Frenkel, V.Y. *Matvei Petrovich Bronstein and Soviet Theoretical Physics in the Thirties*, 1st ed.; Springer: Basel, Switzerland, 1994.
98. Planck, M. Über irreversible Strahlungsvorgänge. *Sitzungsberichte Königlich-Preußischen Akad. Wiss.* **1899**, *5*, 440–480.
99. DeWitt, C.M.; DeWitt, B.S. Falling charges. *Physics* **1964**, *1*, 3–20. [CrossRef]

100. Berends, F.A.; Gastmans, R. Quantum electrodynamical corrections to graviton-matter vertices. *Ann. Phys.* **1976**, *98*, 225–236. [CrossRef]
101. Milton, K.A. Quantum-electrodynamic corrections to the gravitational interaction of the electron. *Phys. Rev. D* **1977**, *15*, 538–540. [CrossRef]
102. Barker, B.M.; O'Connell, R.F. Post-Newtonian two-body and n-body problems with electric charge in general relativity. *J. Math. Phys.* **1977**, *18*, 1818–1824. [CrossRef]
103. Gupta, S.N.; Radford, S.F. Quantum field-theoretical electromagnetic and gravitational two-particle potentials. *Phys. Rev. D* **1980**, *21*, 2213–2225. [CrossRef]
104. Butt, M.S. Leading quantum gravitational corrections to QED. *Phys. Rev. D* **2006**, *74*, 125007. [CrossRef]
105. Holstein, B.R. Graviton physics. *Am. J. Phys.* **2006**, *74*, 1002–1011. [CrossRef]
106. Holstein, B.R.; Ross, A. Long Distance Effects in Mixed Electromagnetic-Gravitational Scattering. *arXiv* **2008**, arXiv:0802.0717.
107. Holstein, B.R. Analytical On-shell Calculation of Low Energy Higher Order Scattering. *J. Phys. G Nucl. Part. Phys.* **2017**, *44*, 1LT01. [CrossRef]
108. Feinberg, G.; Sucher, J.; Au, C.K. The dispersion theory of dispersion forces. *Phys. Rep.* **1989**, *180*, 83–157. [CrossRef]
109. Spruch, L. An overview of long-range Casimir interactions. In *Long-Range Casimir Forces*; Levin, F.S., Micha, D.A., Eds.; Springer Science + Business Media: New York, NY, USA, 1993; Chapter 1, pp. 1–71.
110. Dowling, J.P. Retarded potentials (Letter). *Phys. Today* **1991**, *39*, 13–15. [CrossRef]
111. Barut, A.O. Electromagnetic Interactions beyond Quantum Electrodynamics. In *Foundations of Radiation Theory and Quantum Electrodynamics*; Barut, A.O., Ed.; Springer Science + Business Media: Berlin/Heidelberg, Germany, 1980; Chapter 14, pp. 165–172.
112. Barut, A.O. Quantum-electrodynamics based on self-energy. *Phys. Scr.* **1988**, *T21*, 18–21. [CrossRef]
113. Barut, A.O.; Van Huele, J.F. Quantum electrodynamics based on self-energy: Lamb shift and spontaneous emission without field quantization. *Phys. Rev. A* **1985**, *32*, 3187–3195. [CrossRef]
114. Barut, A.O.; Dowling, J.P. Quantum electrodynamics based on self-energy, without second quantization: The Lamb shift and long-range Casimir-Polder van der Waals forces near boundaries. *Phys. Rev. A* **1987**, *36*, 2550–2556. [CrossRef]
115. Barut, A.O.; Dowling, J.P. Self-field quantum electrodynamics: The two-level atom. *Phys. Rev. A* **1990**, *41*, 2284–2294. [CrossRef]
116. Appelquist, T.; Chodos, A. Quantum effects in Kaluza-Klein theories. *Phys. Rev. Lett.* **1983**, *50*, 141–145. [CrossRef]
117. Appelquist, T.; Chodos, A. Quantum dynamics of Kaluza-Klein theories. *Phys. Rev. D* **1983**, *28*, 772–784. [CrossRef]
118. Abe, O. Casimir Energy in Quantum Gravity in R1 times T3 Space-Time. *Prog. Theor. Phys.* **1984**, *72*, 1225–1232. [CrossRef]
119. Panella, O.; Widom, A. Casimir effects in gravitational interactions. *Phys. Rev. D* **1994**, *49*, 917–922. [CrossRef]
120. Panella, O.; Widom, A.; Srivastava, Y.N. Casimir effects for charged particles. *Phys. Rev. B* **1990**, *42*, 9790–9793. [CrossRef]
121. Torr, D.G.; Li, N. Gravitoelectric-electric coupling via superconductivity. *Found. Phys. Lett.* **1993**, *6*, 371–383. [CrossRef]
122. Woods, R.C. Manipulation of gravitational waves for communications applications using superconductors. *Phys. C* **2005**, *433*, 101–107. [CrossRef]
123. Garcia-Cuadrado, G. Towards a New Era in Gravitational Wave Detection: High Frequency Gravitational Wave Research. In Proceedings of the Space, Propulsion & Energy Sciences International Forum SPESIF-2009, Huntsville, AL, USA, 24–27 February 2009; Volume 1103, pp. 553–563, [CrossRef]
124. Woods, R.C. Interactions between supeconductors and high-frequency gravitational waves. In *Gravity-Superconductors Interactions: Theory and Experiment*; Modanese, G., Robertson, G.A., Eds.; Bentham: Sharjah, UAE, 2012; pp. 23–57.
125. Minter, S.J.; Wegter-McNelly, K.; Chiao, R.Y. Do mirrors for gravitational waves exist? *Phys. E Low-Dimens. Syst. Nanostruct.* **2010**, *42*, 234–255. [CrossRef]
126. Sakharov, A.D. Vacuum Quantum Fluctuations in Curved Space and the Theory of Gravitation. *Gen. Relativ. Gravit.* **2000**, *32*, 365–367. [CrossRef]
127. Visser, M. Sakharov's induced gravity: A modern perspective. *Mod. Phys. Lett. A* **2002**, *17*, 977–991. [CrossRef]
128. Belgiorno, F.; Liberati, S. Black hole thermodynamics, Casimir effect and Induced Gravity. *Gen. Relativ. Gravit.* **1997**, *29*, 1181–1194. [CrossRef]
129. Quach, J.Q. Gravitational Casimir Effect. *Phys. Rev. Lett.* **2015**, *114*, 81104. [CrossRef] [PubMed]
130. Quach, J.Q. Erratum: Gravitational Casimir Effect. *Phys. Rev. Lett.* **2017**, *118*, 139901. [CrossRef] [PubMed]
131. Ford, L.H.; Hertzberg, M.P.; Karouby, J. Quantum Gravitational Force Between Polarizable Objects. *Phys. Rev. Lett.* **2016**, *116*, 151301. [CrossRef]
132. Wu, P.; Hu, J.; Yu, H. Quantum correction to classical gravitational interaction between two polarizable objects. *Phys. Lett. B* **2016**, *763*, 40–44. [CrossRef]
133. Hu, J.; Yu, H. Gravitational Casimir—Polder effect. *Phys. Lett. B* **2017**, *767*, 16–19. [CrossRef]
134. Holstein, B.R. Connecting Compton and Gravitational Compton Scattering. *EPJ Web Conf.* **2017**, *134*, 1003. [CrossRef]
135. Rini, M. Synopsis: A Casimir Effect Caused by Gravity. *Physics* **2015**, *8*, s23.
136. Becker, A. Ultra-cold mirrors could reveal gravity's quantum side. *New Sci.* **2015**. [CrossRef]
137. Hossenfelder, S. Can we prove the quantization of gravity with the Casimir effect? Probably not. *arXiv* **2015**, arXiv:1502.07429.
138. Pinto, F. If detected, would hypothetical gravitational Casimir effects prove gravity quantization? In Proceedings of the Fourteenth Marcel Grossman Meeting on General Relativity, Rome, Italy, 12–18 July 2015; Bianchi, M., Jantzen, R.T., Ruffini, R., Eds.; World Scientific: Singapore, 2017; pp. 3966–3973. [CrossRef]

139. Abbott, B.P. Observation of Gravitational Waves from a Binary Black Hole Merger. *Phys. Rev. Lett.* **2016**, *116*, 61102. [CrossRef] [PubMed]
140. Griffiths, D.J.; Ho, E. Classical Casimir effect for beads on a string. *Am. J. Phys.* **2001**, *69*, 1173–1176. [CrossRef]
141. Boyer, T.H. Casimir forces and boundary conditions in one dimension: Attraction, repulsion, Planck spectrum, and entropy. *Am. J. Phys.* **2003**, *71*, 990–998. [CrossRef]
142. Dev, P.S.B.; Mazumdar, A. Probing the scale of new physics by Advanced LIGO/VIRGO. *Phys. Rev. D* **2016**, *93*, 104001. [CrossRef]
143. Ross, D.K.; Moreau, W. Stochastic Gravity. *Gen. Relativ. Gravit.* **1995**, *27*, 845–858. [CrossRef]
144. Allen, B. The stochastic gravity-wave background: Sources and detection. In Proceedings of the Les Houches School on Astrophysical Sources of Gravitational Waves, Les Houches, France, 26 September–6 October 1995; pp. 373–417.
145. Lasky, P.D.; Mingarelli, C.M.F.; Smith, T.L.; Giblin, J.T.; Thrane, E.; Reardon, D.J.; Caldwell, R.; Bailes, M.; Bhat, N.D.R.; Burke-spolaor, S.; et al. Gravitational-Wave Cosmology across 29 Decades in Frequency. *Phys. Rev. X* **2016**, *6*, 11035. [CrossRef]
146. Quiñones, D.A.; Oniga, T.; Varcoe, B.T.H.; Wang, C.H. Quantum principle of sensing gravitational waves: From the zero-point fluctuations to the cosmological stochastic background of spacetime. *Phys. Rev. D* **2017**, *96*, 44018. [CrossRef]
147. Rugh, S.E.; Zinkernagel, H.; Cao, T.Y. The Casimir Effect and the Interpretation of the Vacuum. *Stud. Hist. Philos. Sci. Part Stud. Hist. Philos. Mod. Phys.* **1999**, *30*, 111–139. [CrossRef]
148. Pinto, F. Improved finite-difference computation of the van der Waals force: One-dimensional case. *Phys. Rev. A* **2009**, *80*, 42113. [CrossRef]
149. Cohen-Tannoudji, C.; Diu, B.; Laloe, F. *Quantum Mechanics (Two Volumes)*; John Wiley & Sons: New York, NY, USA, 1977.
150. Power, E.A.; Thirunamachandran, T. Dispersion interactions between atoms involving electric quadrupole polarizabilities. *Phys. Rev. A* **1996**, *53*, 1567–1575. [CrossRef]
151. Davydov, A.S. *Quantum Mechanics*; Pergamon Press: Oxford, UK, 1965; Volume 1.
152. Jackson, J.D. *Classical Electrodynamics (Second Edition)*; John Wiley & Sons: New York, NY, USA, 1975.
153. Pauling, L.; Beach, J.Y. The van der Waals interaction of hydrogen atoms. *Phys. Rev.* **1935**, *47*, 686–692. [CrossRef]
154. Pauling, L.; Wilson, B.E.J. *Introduction to Quantum Mechanics*; Dover Publications, Inc.: Mineola, NY, USA, 1985; pp. 1–468.
155. Raab, R.E.; de Lange, O.L. *Multipole Theory in Electromagnetism*; Oxford University Press: Oxford, UK, 2005.
156. Salam, A.; Thirunamachandran, T. A new generalization of the Casimir—Polder potential to higher electric multipole polarizabilities. *J. Chem. Phys.* **1996**, *104*, 5094–5099. [CrossRef]
157. Salam, A. *Non-Relativistic QED Theory of the van der Waals Dispersion Interaction*; Springer: Cham, Switzerland, 2016.
158. Golub, G.H.; Van Loan, C.F. *Matrix Computations*, 3rd ed.; The Johns Hopkins University Press: Baltimore, MD, USA; London, UK, 1996.
159. Jenkins, J.K.; Salam, A.; Thirunamachandran, T. Retarded dispersion interaction energies between chiral molecules. *Phys. Rev. A* **1994**, *50*, 4767–4777. [CrossRef] [PubMed]
160. Szekeres, P. Linearized gravitation theory in macroscopic media. *Ann. Phys.* **1971**, *64*, 599–630. [CrossRef]
161. Szekeres, P. Gravitational fields in matter. In *Relativity and Gravitation*; Kuper, C.G., Peres, A., Eds.; Gordon and Breach Science Publishers: New York, NY, USA, 1971; pp. 305–308.
162. Pinto, F. Rydberg atoms as gravitational-wave antennas. *Gen. Relativ. Gravit.* **1995**, *27*, 9–14, [CrossRef]
163. Pinto, F. Gravitational-wave response of parametric amplifiers driven by radiation-induced dispersion force modulation. In Proceedings of the Fourteenth Marcel Grossmann Meeting on General Relativity, Rome, Italy, 12–18 July 2015; Bianchi, M., Jantzen, R.T., Ruffini, R., Eds.; World Scientific: Singapore, 2017; pp. 3175–3182, [CrossRef]
164. Tourrenc, P.; Grossiord, J.L. Modification du spectre de l'hydrogène atomique sous l'influence de champs de gravitation. *Il Nuovo Cimento* **1976**, *32B*, 163–176. [CrossRef]
165. Parker, L. One-electron atom in curved space-time. *Phys. Rev. Lett.* **1980**, *44*, 1559–1562. [CrossRef]
166. Pinto, F. Rydberg atoms in curved space-time. *Phys. Rev. Lett.* **1993**, *70*, 3839–3843. [CrossRef]
167. Misner, C.W.; Thorne, K.S.; Wheeler, J.A. *Gravitation*; W. H. Freeman and Company: San Francisco, CA, USA, 1970; pp. 1–1304.
168. Schutz, B.F. *A First Course in General Relativity*; Cambridge University Press: Cambridge, UK, 2009; pp. 1–393.
169. Price, R.H.; Belcher, J.W.; Nichols, D.A. Comparison of electromagnetic and gravitational radiation: What we can learn about each from the other. *Am. J. Phys.* **2013**, *81*, 575–584. [CrossRef]
170. McLachlan, A.D.; Gregory, R.D.; Ball, M.A. Molecular interactions by the time-dependent Hartree method. *Mol. Phys.* **1964**, *7*, 119–129. [CrossRef]
171. McLachlan, A.D. Retarded dispersion forces in dielectrics at finite temperatures. *Proc. R. Soc. Lond. Ser. Math. Phys. Sci.* **1963**, *274*, 80–90.
172. Thirunamachandran, T. Vacuum fluctuations and intermolecular interactions. *Phys. Scr.* **1988**, *T21*, 123–128. [CrossRef]
173. Power, E.A.; Thirunamachandran, T. A new insight into the mechanism of intermolecular forces. *Chem. Phys.* **1993**, *171*, 1–7. [CrossRef]
174. Renne, M.J. Retarded van der Waals interaction in a system of harmonic oscillators. *Physica* **1971**, *53*, 193–209. [CrossRef]
175. Renne, M.J. Microscopic theory of retarded van der Waals forces between macroscopic dielectric bodies. *Physica* **1971**, *56*, 125–137. [CrossRef]
176. Goldstein, H.; Poole, C.; Safko, J. *Classical Mechanics*, 3rd ed.; Addison-Wesley: San Francisco, CA, USA, 2002; pp. 1–647.
177. Hinshelwood, C.N. *The Structure of Physical Chemistry*; Oxford University Press: London, UK, 1951.

178. Panofski, W.K.H.; Phillips, M. *Classical Electricity and Magnetism*, 2nd ed.; Addison-Wesley Publ. Co., Inc.: Reading, MA, USA, 1962; pp. 1–503.
179. Shadowitz, A. *The Electromagnetic Field*; Dover Publications, Inc.: New York, NY, USA, 1975.
180. Passante, R. Dispersion Interactions between Neutral Atoms and the Quantum Electrodynamical Vacuum. *Symmetry* **2018**, *10*, 735. [CrossRef]
181. Power, E.A.; Thirunamachandran, T. Dispersion forces between molecules with one or both molecules excited. *Phys. Rev. A* **1995**, *51*, 3660–3666. [CrossRef]
182. Zangwill, A. *Modern Electrodynamics*; Cambridge University Press: Cambridge, UK, 2013; pp. 1–977.
183. Good, R.H.; Nelson, T.J. *Classical Theory of Electric and Magnetic Fields*; Academic Press: New York, NY, USA, 1971.
184. Weber, J. *General Relativity and Gravitational Waves*; Interscience Publishers, Inc.: New York, NY, USA, 1961.
185. Bimonte, G.; Calloni, E.; Esposito, G.; Rosa, L. Energy-momentum tensor for a Casimir apparatus in a weak gravitational field. *Phys. Rev. D* **2006**, *74*, 085011. [CrossRef]
186. Norte, R.A.; Forsch, M.; Wallucks, A.; Marinković, I.; Gröblacher, S. Platform for Measurements of the Casimir Force between Two Superconductors. *Phys. Rev. Lett.* **2018**, *121*, 30405. [CrossRef]
187. Pinto, F. A trapped dipolar BEC interferometry test of $E = mc^2$. *Int. J. Mod. Phys. D* **2006**, *15*, 2235–2240. [CrossRef]
188. Bimonte, G.; Born, D.; Calloni, E.; Esposito, G.; Il'ichev, E.; Rosa, L.; Scaldaferri, O.; Tafuri, F.; Vaglio, R.; Hübner, U. The Aladin2 experiment: status and perspectives. *J. Phys. Math. Gen.* **2006**, *39*, 6153–6159. [CrossRef]
189. Allocca, A.; Bimonte, G.; Born, D.; Calloni, E.; Esposito, G.; Huebner, U.; Il, E.; Francesco, R. Results of Measuring the Influence of Casimir Energy on Superconducting Phase Transitions. *J. Supercond. Nov. Magn.* **2012**, *25*, 2557–2565. [CrossRef]
190. Ignat'ev, Y.G.; Zakharov, A.V. The reflection of gravitational waves from compact stars. *Phys. Lett.* **1978**, *66A*, 3–4. [CrossRef]
191. Li, N.; Torr, G. Gravitational effects on the magnetic attenuation of superconductors. *Phys. Rev. B* **1992**, *46*, 5489–5495. [CrossRef]
192. Klimchitskaya, G.L.; Mostepanenko, V.M. Recent measurements of the Casimir force: Comparison between experiment and theory. *Mod. Phys. Lett. A* **2020**, *35*, 2040007. [CrossRef]
193. Pinto, F. Efimov physics in curved spacetime: Field fluctuations and exotic matter. *Int. J. Mod. Phys. D* **2018**, *27*, 1847001. [CrossRef]
194. Pinto, F. Signatures of minimal length from Casimir-Polder forces with neutrons. *Int. J. Mod. Phys. D* **2020**, 2043026. [CrossRef]
195. Pinto, F. The history of technology transfer of the Casimir effect and van der Waals forces: From exotic, weak, and undesirable to enabling, emerging, and irresistible. In Proceedings of the Atti del XXXIX Convegno Annuale Della SISFA, Pisa, Italy, 9–12 September 2019; La Rana, A., Possi, P., Eds.; Pisa University Press: Pisa, Italy, 2020; pp. 281–287. [CrossRef]
196. Post, H.R. The Problem of Atomism. *Br. J. Philos. Sci.* **1975**, *26*, 19–26. [CrossRef]
197. Wheeler, J.A. From relativity to mutability. In *The Physicist's Conception of Nature*; Mehra, J., Ed.; D. Reidel Publishing Company: Dordrecht, The Netherlands, 1973; Chapter 9, pp. 202–247.
198. Pinto, F. Dispersion force engineering and next-generation spacecraft: Case studies in a nanoscale emerging enabling general-purpose technology. Materials Today. In Proceedings of the 17th International Conference on Nanosciences & Nanotechnologies (NN20), Thessaloniki, Greece, 7–10 July 2020.
199. Hawkes, E.W.; Eason, E.V.; Christensen, D.L.; Cutkosky, M.R. Human climbing with efficiently scaled gecko-inspired dry adhesives. *J. R. Soc. Interface* **2015**, *12*, 20140675, [CrossRef]
200. Pinto, F. Mixed electric-gravitational Casimir-Polder potentials. In Proceedings of the 4th Symposium on the Casimir Effect, Saint Petersburg, Russia, 23–29 June 2019.